Discrete Mathematical Structures

Uma Shanker Gupta
MSc. (Alld.), PhD. (Roorkee)
Ex-Professor and Ex-Emeritus Fellow
Department of Mathematics
Indian Institute of Technology
Roorkee

Pearson

ISBN 9789332521391
First Impression, 2014
Sixth Impression, 2024
Seventh Impression, 2025

Published by Pearson India Education Services Pvt. Ltd, CIN: U72200TN2005PTC057128.

Head Office: 1st Floor, Berger Tower, Plot No. C-001A/2, Sector 16B, Noida - 201 301, Uttar Pradesh, India.
Registered Office: Featherlite, 'The Address' 5th Floor, Survey No 203/10B, 200 Ft MMRD Road, Zamin Pallavaram, Chennai – 600 044.
Website: in.pearson.com; Email: companysecretary.india@pearson.com

Printed in India at: Manipal Technologies Limited, Manipal

To
My loving parents
whose blessings and guidance
made my journey to the
pioneering temple of engineering education

Contents

6. Recurrence Relations and Generating Functions 6-1

7. Algebraic Structures 7-1

Preface

Discrete Mathematical Structures aims to introduce, in a reasonably rigorous and lucid way, a hybrid subject that has gained an important place in mathematics and theoretical computer science. Although the content majorly deals with mathematics but it also has numerous applications in computer science and engineering.

The word 'discrete' means individually separate and distinct, and so it is regarded as opposite to continuous. Thus, discrete mathematics is the study of distinct objects and their relationships. The development of modern digital computers has greatly enhanced the applications of discrete mathematics. It includes topics dealing with abstraction such as abstract algebra, linear algebra, boolean algebra, graph theory and mathematical logic. Discrete mathematics has applications in various courses such as data structures, theory of compilers, analysis of algorithms and formal languages. As a result, it has become a core subject for undergraduate and postgraduate courses in computer science.

In addition to various applications, the study of discrete mathematical structures will help students to develop the ability of logical reasoning and enhance capability for visualizing abstraction.

Uma Shanker Gupta

Acknowledgements

This book is a culmination of several years of study and discussion I had with faculty members at IIT-Roorkee and ABES Engineering College, Ghaziabad. I acknowledge with gratitude the positive criticism and high appreciation of anonymous reviewers, which encouraged me to complete the book.

I would like to express my sincere thanks to my grand daughter Ayushi Gupta who persisted me to write this book in the interest of computer professionals and mathematicians of the future.

It is a pleasure to acknowledge the help I received from Parag Singhal, a colleague at ABES, in making valuable suggestions for improvement.

I would like to express my love and appreciation to my wife Dharmavati for her patience and encouragement without which this book would not have been possible. My thanks are also due to my son R.K. Gupta and daughter-in-law Vineeta Gupta for extending all possible help and support during the preparation of the book.

All efforts have been made to keep the text pedagogically sound and error free. However, criticism and suggestions would be most welcome to further improve this book.

I extend my sincere thanks to the editorial and production staff of Pearson Education, particularly Neha Goomer, for their care and enthusiasm in the preparation of this book.

I do hope that the book will be useful to students, teachers as well as to those using discrete mathematics as a tool for the solution of their problems.

Uma Shankar Gupta

About the Author

Uma Shanker Gupta joined the department of mathematics, the University of Roorkee (presently IIT-Roorkee), in 1967, after teaching for five years at Ewing Christian Degree College, Allahabad (an associated college of Allahabad University and now an autonomous college). He was awarded PhD (Mathematics) by the University of Roorkee in 1971.

He joined the Department of Applied Science and Humanities, ABES Engineering College, Ghaziabad, as professor in 2006 and subsequently held positions of Professor and Head, Applied Science and Humanities, Professor and Head MCA Dept and Director, MCA Institute. He was a visiting faculty of Mathematics in the School of Vocational Studies and Applied Science, Gautam Buddha University, UP, in the autumn semester of 2010–2011.

He has been a reviewer of many International journals like *Journal of Applied Mechanics*, *Journal of Sound and Vibration* to name a few. He became EMERITUS FELLOW in 2004 and held this position till 2006.

1

Set Theory

LEARNING OBJECTIVES

After reading this chapter, you will be conversant with:

- Introduction
- Sets
- Cartesian Product of Sets
- Multiset

1.1 INTRODUCTION

Set theory is one of the important tools in modern mathematics. It has not only revolutionized mathematical thinking but also unified various branches of mathematics. Every branch of mathematics can be considered as a study of sets of objects of one kind or the other. For example, algebra and number theory are concerned with the set of numbers whereas analysis deals with the set of functions. In this chapter, the concept of sets, subsets, operations on sets, inclusion and exclusion principle, multisets, operations on multisets, and Cartesian product of sets has been discussed.

1.2 SETS

Definition 1.1 A set is a collection of objects called elements that are 'distinct' and 'distinguishable'. By distinct, we mean that no element is repeated and by distinguishable, we mean that given any element, we can say whether it belongs to the set or not.

Sets are denoted by capital letters A, B, C,… and its elements by lower case letters a, b, c,…

If a is an element of set A, we write $a \in A$ where the symbol $\in$ is a Greek alphabet. The symbol stands for 'is an element of'. We also say that a belongs to A. If a is not an element of A, it is written as $a \notin A$. There are following four ways of defining a set.

Listing or Roaster Method

A set can be described by listing all its elements within braces and separating by commas.

For example

1. {1, 2, 3, 4, 5}, the set of first five natural numbers.
2. {1, −1, *i*, −*i*}, the set of fourth roots of unity.
3. {a, e, i, o, u}, the set of English vowel.

Sometimes, it is not possible to list all the members, then the set is described by listing its few elements at the beginning followed by dots and enclosing within braces.

For example

1. The set of odd natural numbers {1, 3, 5,…}
2. The set of integers that are multiples of 3{0, ±3, ±6…}
3. The set of squares of natural number {1, 4, 9,…}

Set Builder Method

In this method, the set is defined by specifying the property possessed by each element. Its general form is $\{x: P(x)\}$, which is read as 'set of elements x such that it has property $P(x)$'. The symbol ':' stands for 'such that'.

For example

1. The set of English vowel can be written as {*x*: *x* is a vowel of the English alphabet}
2. The set of fourth roots of unity can be written as $\{x: x^4 = 1\}$
3. The set of integers that are multiples of 3 can be written as

$$\{x: x = 3k, \text{ where } k = 0, \pm 1, \pm 2,\ldots\}$$

4. The set of squares of natural numbers can be written as

$$\{x: x = n^2, \text{ where '}n\text{' is a natural number}\}$$

Recurrence Relation Method

A set can be defined by a recursive formula. One or more elements are given along with a rule by which the rest of the elements of the set can be generated.

Example: If A is the set of first 10 natural numbers, then it can be described as

$$A = \{n_{i+1} = n_i + 1,\ i = 1, 2,\ldots, 9, \text{ where } n_1 = 1\}$$

Method of Characteristic Function

A set can be described by its characteristic function

$$\begin{aligned} \mu_A(x) &= 1 \quad \text{if } x \in A \\ &= 0 \quad \text{if } x \notin A \end{aligned}$$

The above example of the set A of the first 10 natural numbers can be described as

$$\begin{aligned} \mu_A(x) &= 1 \quad \text{if } x = 1, 2,\ldots, 10 \\ &= 0 \quad \text{otherwise} \end{aligned}$$

1.2.1 Types of Sets

Finite and Infinite Sets
A set containing finite number of elements is called a finite set. A set that contains infinitely many elements is called an infinite set.

For example

1. The set of days in a week.
2. The set of banks in a city.
3. The set of hospitals in a district.
4. The set of teachers in a college.
5. $\{x : 6x^2 - 7x + 2 = 0\} = \left\{\frac{1}{2}, \frac{2}{3}\right\}$ all are finite sets.

Examples of Infinite Sets
$N = \{1, 2, 3, \ldots\}$, the set of natural numbers.
$Z = \{0, \pm 1, \pm 2, \ldots\}$, the set of integers.
$Q = \{x\colon x = a/b,\ a, b \in Z \text{ and } b \neq 0\}$, the set of rational numbers.
R = Set of real numbers, R^+ = Set of positive real numbers = $\{x\colon x \in R \text{ and } x > 0\}$.
C = Set of complex numbers $\{x + iy\colon x, y \in R\}$.

Null Set or Empty Set
A set that contains no element is called a null set or an empty set and is denoted by ϕ. For example

1. $\phi = \{x\colon x \neq x\}$
2. $\phi = \{n\colon n \in N \text{ and } n + 1 = 0\}$
3. $\phi = \{n\colon n \in N \text{ and } n^2 + 1 = 0\}$
4. ϕ = The set of prime numbers between 31 and 37
5. ϕ = The set of straight lines in a plane passing through three non-collinear points

Singleton Set
A set that contains only one element is called a singleton set.

For example

1. $A = \{0\}$
2. $B = \{x\colon x \in N \text{ and } 10 < x < 12\}$
3. $C = \{x\colon 2x = 10 \text{ where } x \text{ is a positive integer}\}$

Universal Set
It is a special set in which every set A under discussion is a subset of U, where U is a universal set. Thus, $A \subseteq U$ for every set A.

Equality of Sets
Two sets A and B are said to be equal if they have the same elements and are written as $A = B$.

Mathematically $A \subseteq B$ and $B \subseteq A \Rightarrow A = B$ or $A = B$ if and only if $x \in A \Leftrightarrow x \in B$.

1.2.2 Subset

Let A and B be any two non-empty sets. The set A is called a subset of B if every element of A is an element of B and we write $A \subseteq B$. Mathematically $A \subseteq B$ if $x \in A \Rightarrow x \in B$.

1.2.3 Proper Subset

The set A is called a proper subset of B and is denoted by $A \subset B$ if every element of A is an element of B and there is at least one element in B that is not in A. For example

1. If $A = \{1, 2, 3\}$ and $B = \{1, 2, 3, 4, 5\}$, then $A \subset B$.
2. If $A = \{2, 4, 6\}$ and $B = \{6, 2, 4\}$, then each element of A is also an element of B and there is no element of B that is not in A and so A is a subset of B. The sets A and B are same. This shows that every set is a subset of itself.
3. The set of a natural numbers N is a proper subset of set of integers Z, that is, $N \subseteq Z$.
4. The set of rational numbers Q is a proper subset of the set of real numbers R.
5. The set of real numbers is a proper subset of the set of complex numbers

$$C = \{z: z = x + iy, x, y \in R\}$$

6. If $A = \{1, 2, 3, 4\}$ and $B = \{1, 4, 5, 6\}$, then A is not a subset of B as all elements of A are not in B. This is denoted by $A \not\subseteq B$.

1.2.4 Power Set

The subsets of a set can be used to form a new set.

Definition 1.2 If A is any set, then the set of all its subsets is called the power set of A and is denoted by $P(A)$. For example

1. Let $A = \{1, 2, 3\}$, then $P(A) = \{\phi, \{1\}, \{2\}, \{3\}, \{1, 2\}, \{2, 3\}, \{3, 1\}, \{1, 2, 3\}\}$
2. Let the set A contain n elements, then the number of subsets is equal to

$$^nc_0 + {}^nc_1 + {}^nc_2 + \ldots + {}^nc_n = (1 + 1)n = 2n$$

THEOREM 1.1 The null set ϕ is the subset of every set.

Proof 1: There are no elements in ϕ and therefore every element in ϕ belongs to a given set A. Thus, $\phi \subseteq A$.

Proof 2: Suppose ϕ is not the subset of the set A. Then there exist some elements $x \in \phi$ and $x \in A$. But this is not possible as there is no element $x \in \phi$. Hence, our assumption is wrong, that is, ϕ is the subset of A.

THEOREM 1.2 The null set ϕ is unique.

Proof 1: Let ϕ_1 and ϕ_2 be two empty sets, then null set being the subset of every set, we have $\phi_1 \subseteq \phi_2$ and $\phi_2 \subseteq \phi_1$.

This leads to the conclusion that $\phi_1 = \phi_2$, that is, there is only one null set.

Note 1.1 It is to be noted that $\{\phi\} = \phi$ is not true since $\{\phi\}$ contains an element ϕ whereas $\phi = \{\ \}$.

1.2.5 Venn Diagram

A pictorial representation of a set is given by Venn diagram named after English logician John Venn. In this diagram, the set U is represented by the interior of a rectangle and its subsets are represented by circles. Relationships between sets and also the operations on them can be visualised by Venn diagrams.

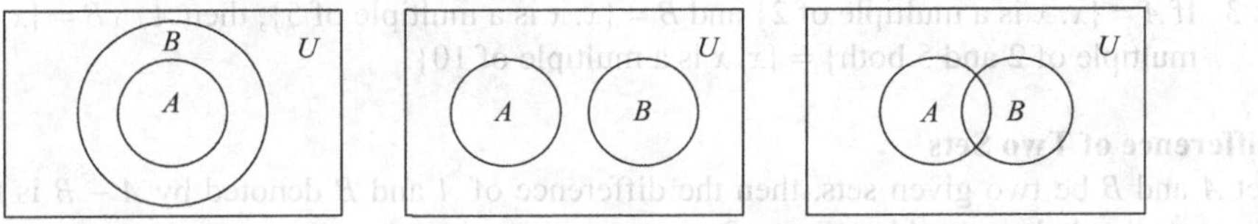

(a) $A \subseteq B$ (b) A and B are disjoint (c) A and B are not disjoint

1.2.6 Set Operations

In this section, we discuss ways in which two or more sets can be combined to form a new set.

Union of Two Sets

The union of sets A and B denoted by $A \cup B$ is the set of all elements that are in A or in B or in both.

Thus, $A \cup B = \{x: x \in A \text{ or } x \in B\}$. The word 'or' is used in the sense of $x \in B$ or to both A and B.

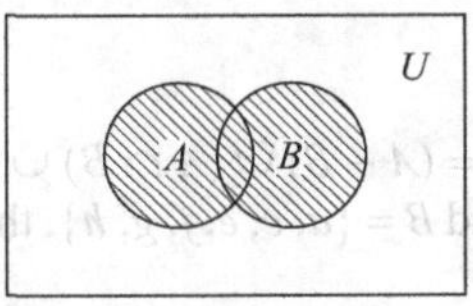

$A \cup B$ is shaded.

More generally, the union of n sets written as $A_1 \cup A_2 \cup \ldots \cup A_n = \bigcup_{i=1}^{n} A_i = \{x: x \in A_i$ for some $i\}$.

Intersection of Two Sets

The intersection of sets A and B denoted by $A \cap B$ is the set of elements that belong to both A and B. Thus, $A \cap B = \{x: x \in A \text{ and } x \in B\}$.

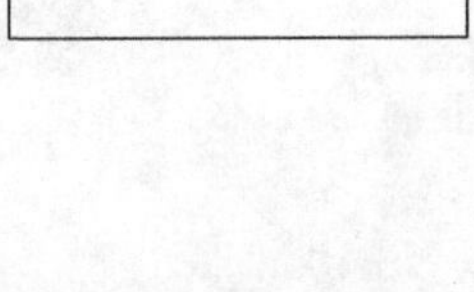

$A \cap B$ is shaded.
$(A \cap B)'$ is area other than shaded.
More generally, the intersection of sets $A_1, A_2, \ldots, A_n$ is denoted by,

$A_1 \cap A_2 \cap \ldots \cap A_n = \bigcap_{i=1}^{n} A_i = \{x: x \in A_i \text{ for each } i\}$, that is, the set of all elements that belong to each of them.

Examples

1. If $A = \{1, 2, 3\}$ and $B = \{2, 3, 4, 5, 6\}$, then $A \cup B = \{1, 2, 3, 4, 5, 6\}$ and $A \cap B = \{2, 3\}$.
2. If $A = \{a, b, c\}$ and $B = \{x, y, z\}$, then $A \cup B = \{a, b, c, x, y, z\}$ and $A \cap B = \Phi$.
3. If $A = \{x: x$ is a multiple of 2$\}$ and $B = \{x: x$ is a multiple of 5$\}$, then $A \cap B = \{x: x$ is a multiple of 2 and 5 both$\} = \{x: x$ is a multiple of 10$\}$.

Difference of Two Sets

Let A and B be two given sets, then the difference of A and B denoted by $A - B$ is the set of elements that belong to A but not to B.

Thus, $A - B = \{x: x \in A \text{ and } x \notin B\}$.

In Venn diagram, $A - B$ is represented by the shaded portion.

Clearly, $A = (A - B) \cup (A \cap B)$

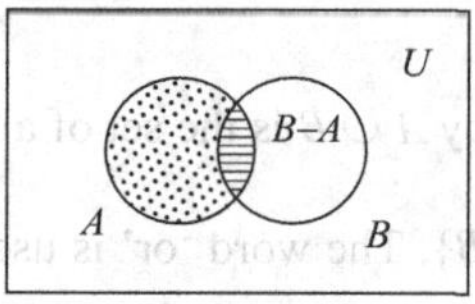

$A - B$ is the shaded portion.

Venn diagram shows that $A \cup B = (A - B) \cup (A \cap B) \cup (B - A)$

For example, if $A = \{a, b, c, d\}$ and $B = \{a, c, e, f, g, h\}$, then $A - B = \{b, d\}$, $B - A = \{e, f, g, h\}$

1.2.7 Disjoint Sets

Two sets that have no common elements are called disjoint sets. For example

1. If $A = \{1, 3, 5\}$ and $B = \{2, 4, 6\}$, then sets A and B are disjoint.
2. If $A = \{$BASIC, FORTRAN$\}$ and $B = \{$C++, JAVA$\}$, then sets A and B are disjoint.
3. The sets R_+ (i.e. set of positive real numbers) and R_- (i.e. set of negative real numbers) are disjoint sets.

1.2.8 Complement of a Set

Let U be the universal set and A be any subset of U. Then the complement of A is the set $U - A$ and is denoted by A' or A^C.

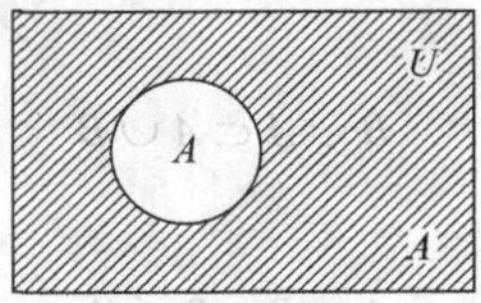

Thus,

$$A' = \{x: x \in U \text{ and } x \notin A\}$$

The shaded part in the Venn diagram represents A'. For example

1. If $U = R$, then the complement of the set of rational numbers Q say is the set of irrational numbers.
2. $U' = \phi$ and $\phi' = U$.

Note 1.2 The set $A - B$ is also called 'The complement of B with respect to A'. Further, $A - B$, $A \cap B$ and $B - A$ are disjoint sets.

1.2.9 Laws of Sets

Let A, B and C be any three sets and U the universal set, then

Idempotent Laws: $A \cup A = A$ and $A \cap A = A$
Commutative Laws: $A \cup B = B \cup A$ and $A \cap B = B \cap A$
Associative Laws: $A \cup (B \cup C) = (A \cup B) \cup C$ and $A \cap (B \cap C) = (A \cap B) \cap C$
Distributive Laws: (1) $A \cup (B \cap C) = (A \cup B) \cap (A \cup C)$
(2) $A \cap (B \cup C) = (A \cap B) \cup (A \cap C)$
De Morgan's Laws: $(A \cup B)' = A' \cap B'$ and $(A \cap B)' = A' \cup B'$

Proof: For Idempotent laws: (a) $A \cup A = A$ (b) $A \cap A = A$

(a) Let $x \in A \cup A \Leftrightarrow x \in A$ or $x \in A \Leftrightarrow x \in A$ and so $A \cup A = A$

(b) Let $x \in A \cap A \Leftrightarrow x \in A$ and $x \in A \Leftrightarrow x \in A$ and so $A \cap A = A$

To prove Commutative Laws: (a) $A \cup B = B \cup A$ (b) $A \cap B = B \cap A$

Proof: In order to prove $A \cup B = B \cup A$, we must prove that $A \cup B \subseteq B \cup A$ and $B \cup A \subseteq A \cup B$.

Let x be any element in $A \cup B$, then $x \in A \cup B$
$\Rightarrow x \in A$ or $x \in B$
$\Rightarrow x \in B$ or $x \in A$
$\Rightarrow x \in B \cup A$

Hence,

$$A \cup B \subseteq B \cup A \tag{1.1}$$

Again let y be any element in $B \cup A$, then $y \in B \cup A \Rightarrow y \in B$ or $y \in A$
$\Rightarrow y \in A$ or $y \in B$
$\Rightarrow y \in A \cup B$

Hence,

$$B \cup A \subseteq A \cup B \tag{1.2}$$

From (1.1) and (1.2),

$$A \cup B = B \cup A$$

To prove: $(A \cup B) \cup C = A \cup (B \cup C)$

Proof: Let x be any element in $(A \cup B) \cup C$, then $x \in (A \cup B) \cup C \Rightarrow x \in A \cup B$ or $x \in C$

$\Rightarrow (x \in A$ or $x \in B)$ or $x \in C$

$\Rightarrow x \in A$ or $(x \in B$ or $x \in C)$

$\Rightarrow x \in A$ or $x \in B \cup C$

$\Rightarrow x \in A \cup (B \cup C)$

Hence,

$$(A \cup B) \cup C \subseteq A \cup (B \cup C) \tag{1.3}$$

Let y be any element in $A \cup (B \cup C)$, then $y \in A \cup (B \cup C) \Rightarrow y \in A$ or $y \in B \cup C$

$\Rightarrow y \in A$ or $(y \in B$ or $y \in C)$

$\Rightarrow (y \in A$ or $y \in B)$ or $y \in C$

$\Rightarrow y \in (A \cup B)$ or $y \in C$

$\Rightarrow y \in (A \cup B) \cup C$

Hence,

$$A \cup (B \cup C) \subseteq A \cup (B \cup C) \tag{1.4}$$

From (1.3) and (1.4),

$$A \cup (B \cup C) = A \cup (B \cup C)$$

To prove De Morgan's Laws
(a) $(A \cup B)' = B' \cup A'$ and (b) $(A \cap B)' = A' \cup B'$

Proof: (a) Let x be any element in $(A \cup B)'$, then $x \in (A \cup B)' \Rightarrow x \notin A \cup B$

$\Rightarrow x \notin A$ and $x \notin B$

$\Rightarrow x \in A'$ and $x \in B'$

$\Rightarrow x \in A' \cap B'$

Hence,

$$(A \cup B)' \subseteq A' \cap B' \tag{1.5}$$

Again, let $y \in A' \cap B'$, then $y \in A' \cap B' \Rightarrow y \in A'$ and $y \in B'$

$\Rightarrow y \notin A$ and $y \notin B$

$\Rightarrow y \notin A \cup B$

$\Rightarrow y \in (A \cup B)'$

Hence,

$$(A' \cap B') \subseteq (A \cup B)' \quad (1.6)$$

From (1.5) and (1.6) $(A \cup B)' = A' \cap B'$ proved. Similarly, we can prove that $(A \cap B)' = A' \cup B'$.

1.2.10 Symmetric Difference of Two Sets

The symmetric difference of two sets A and B denoted by $A \oplus B$ is the set of elements that belong to either A or B but not to both. Thus,

$A \oplus B = \{x: x \in A \text{ or } x \in B \text{ but not to both}\}$

$$= (A - B) \cup (B - A)$$
$$= (A \cup B) - (A \cap B)$$

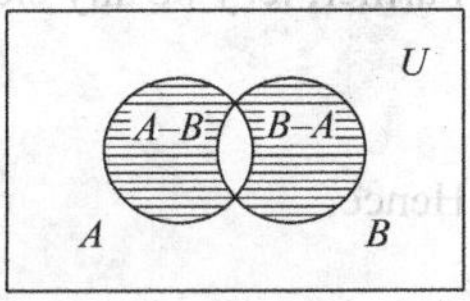

Clearly $A \oplus B = B \oplus A$

Example: If $A = \{a, b, c, d\}$ and $B = \{d, e, f, g\}$, then $A \oplus B = \{a, b, c, e, f, g\}$.

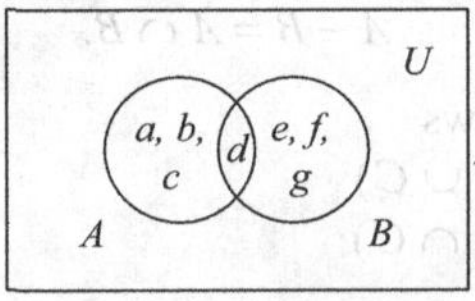

Some Illustrative Examples

Example 1 Prove that $(A - B) \cup (B - A) = (A \cup B) - (A \cap B)$

Solution: $x \in (A - B) \cup (B - A) \Rightarrow x \in (A - B) \text{ or } x \in (B - A)$

$\Rightarrow (x \in A \text{ and } x \notin B) \text{ or } (x \in B \text{ and } x \notin A)$

$\Rightarrow (x \in A \text{ or } x \in B) \text{ and } (x \notin A \text{ and } x \notin B)$

$\Rightarrow x \in (A \cup B) \text{ and } x \notin (A \cap B)$

$\Rightarrow x \in [(A \cup B) - (A \cap B)]$

Hence,

$$(A - B) \cup (B - A) \subseteq (A \cup B) - (A \cap B) \quad \text{(i)}$$

Similarly, we can prove that

$$(A \cup B) - (A \cap B) \subseteq (A - B) \cup (B - A) \quad \text{(ii)}$$

From (i) and (ii), $(A - B) \cup (B - A) = (A \cup B) - (A \cap B)$ proved.

Example 2 $A - (B \cup C) = (A - B) \cap (A - C)$

Solution: Let x be any element in $A - (B \cup C)$,

then $x \in A - (B \cup C) \Leftrightarrow x \in A \text{ and } x \notin (B \cup C)$

$\Leftrightarrow x \in A \text{ and } (x \notin B \text{ or } x \notin C)$

$\Leftrightarrow (x \in A, x \notin B) \text{ and } (x \in A \text{ or } x \notin C)$

$\Leftrightarrow (x \in A - B) \text{ and } (x \in A - C)$

$\Leftrightarrow x \in (A - B) \cap (A - C)$

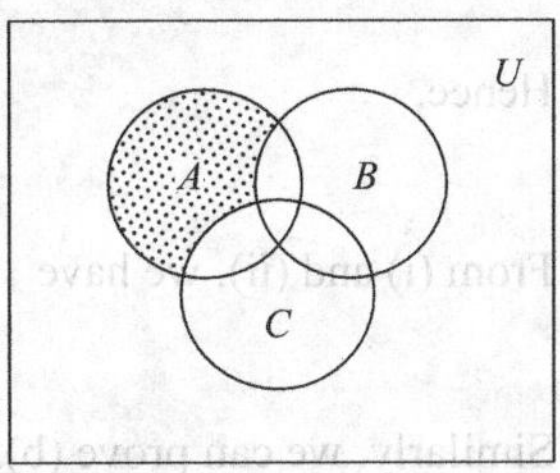

Hence, $A - (B \cup C) = (A - B) \cap (A - C)$. Shaded Area is $A - (B \cup C)$.

Example 3 If A and B are any two sets, then $A - B = A \cap B'$

Let x be any element in $A - B$, that is, $x \in (A - B) \Rightarrow x \in A$ and $x \notin B$,

$$\Rightarrow x \in A \text{ and } x \in B'$$
$$\Rightarrow x \in A \cap B'.$$

Hence,

$$(A - B) \subseteq A \cap B' \tag{i}$$

Further, let y be any element in $A \cap B'$, then $y \in A \cap B' \Rightarrow y \in A$ and $y \in B'$

$$\Rightarrow y \in A \text{ and } y \notin B$$
$$\Rightarrow y \in (A - B)$$

Hence,

$$A \cap B' \subseteq (A - B) \tag{ii}$$

From (i) and (ii), we have

$$A - B = A \cap B.$$

Example 4 Prove the distributive laws

(a) $A \cup (B \cap C) = (A \cup B) \cap (A \cup C)$
(b) $A \cap (B \cup C) = (A \cap B) \cup (A \cap C)$

Proof: (a) Let x be any element in $A \cup (B \cap C)$, then $x \in A \cup (B \cap C)$

$$\Rightarrow x \in A \text{ or } x \in (B \cap C)$$
$$\Rightarrow x \in A \text{ or } (x \in B \text{ and } x \in C)$$
$$\Rightarrow (x \in A \text{ or } x \in B) \text{ and } (x \in A \text{ or } x \in C)$$
$$\Rightarrow (x \in A \cup B) \text{ and } (x \in A \cup C)$$
$$\Rightarrow x \in (A \cup B) \cap (A \cup C)$$

Hence,

$$A \cup (B \cap C) \subseteq (A \cup B) \cap (A \cup C) \tag{i}$$

Further, let y be any element in $(A \cup B) \cap (A \cup C)$, then $y \in (A \cup B) \cap (A \cup C)$

$$\Rightarrow (y \in A \cup B) \text{ and } (y \in A \cup C)$$
$$\Rightarrow (y \in A \text{ or } y \in B) \text{ and } (y \in A \text{ or } y \in C)$$
$$\Rightarrow y \in A \text{ or } (y \in B \text{ and } y \in C)$$
$$\Rightarrow y \in A \text{ or } y \in B \cap C$$
$$\Rightarrow y \in A \cup (B \cap C)$$

Hence,

$$(A \cup B) \cap (A \cup C) \subseteq A \cup (B \cap C) \tag{ii}$$

From (i) and (ii), we have

$$A \cup (B \cap C) = (A \cup B) \cap (A \cap C).$$

Similarly, we can prove (b), that is,

$$A \cap (B \cup C) = (A \cap B) \cup (A \cap C).$$

1.2.11 The Inclusion and Exclusion Principle

Whenever two or more tasks are performed simultaneously, then the usual sum and product rules cannot be used. In the case of two tasks, if we add the number of ways in which each task is performed, then the number of ways to perform both the tasks is counted twice. In order to correct this double counting and find the number of ways of doing either of the tasks, the number of ways of performing each task is added, that is, *inclusion* and the number of ways of performing both the tasks is subtracted from this sum, that is, *exclusion*. This method of counting is called the *principle of inclusion and exclusion*.

Let A and B be two finite sets consisting of p and q elements. An element from the set A can be selected in p ways. Similarly, an element from the set B can be selected in q ways. The number of ways to select an element from A or B is $(p + q)$ minus the number of ways to select an element that is in both A and B, say r. Thus, the number of ways in which an element is selected from A or B is $p + q - r$.
That is

$$\begin{aligned} n(A \cup B) &= p + q - r \\ &= n(A) + n(B) - n(A \cap B) \end{aligned}$$

where $n(A)$ denotes the number of elements in the set A. It is also called its *cardinal number*.

If $n(A \cap B) = \phi$, then we have the *sum rule,* that is, $n(A \cup B) = n(A) + n(B)$. Thus, the principle of inclusion–exclusion is basically a counting method that comes into play in case of overlapping sets.
For example

(i) If $A = \{a, b, c, d\}$, then (i) $n(A) = 4$ (ii) $n(\phi) = 0$ and (iii) $n\{\{\phi\}\} = 1$.
(iv) If $A = \{a, b, c, d\}$ and $B = \{c, d, e, f, g\}$, then $n(A \cup B) = 7$ and $n(A \cap B) = 2$.

1.2.12 Some Simple Results on Cardinality of Sets

1. $n(A \cup B) = n(A) + n(B) - n(A \cap B)$
2. $n(A \cup B) = n(A) + n(B)$ if $A \cap B = \phi$
3. $n(A \cup B \cup C) = n(A) + n(B) + n(C) - n(A \cap B) - n(B \cap C) - n(C \cap A) + n(A \cap B \cap C)$ (inclusion–exclusion principle for three sets A, B, and C)
4. The general statement of inclusion–exclusion principle for n sets is as follows:
   If $A_i (i = 1, 2, \ldots, n)$ are finite subsets of a universal set U, then

$$\begin{aligned} n(A_1 \cup A_2 \cup \ldots \cup A_n) = \Sigma_{i=1}^{n}\, n(A_i) - \Sigma_{i,j}\, n(A_i \cap A_j) + \Sigma_{i,j,k}\, n(A_i \cap A_j \cap A_k) \\ + \ldots + (-1)^{n-1}\, n(A_1 \cap A_2 \cap \ldots \cap A_n) \end{aligned}$$

Proof: If A and B are any two sets, then $n(A \cup B) = n(A) + n(B) - n(A \cap B)$
Let $n(A \cap B) = k$
Since $A \cap B \subseteq A$ and $A \cap B \subseteq B$, we can assume $n(A) = a + k$

$n(B) = b + k$, for some positive integers a and b.

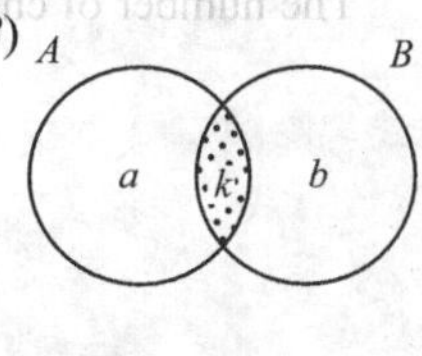

Then,
$$\begin{aligned} n(A \cup B) &= a + b + k \\ &= (a + k) + (b + k) - k \\ &= n(A) + n(B) - n(A \cap B) \end{aligned}$$

If A and B are disjoint, then $A \cap B = \phi$ and $n(A \cap B) = n(\phi) = 0$.
Then, $n(A \cup B) = n(A) + n(B)$. This is known as Addition Principle.

Note 1.3 The result (3) can be obtained by using (1).

Example 1 Find the number of positive integers ≤ 200 and divisible by 2 or 5.

Solution: Let $A = \{x: x \in N \text{ and } x \leq 200 \text{ and is divisible by } 2\}$
and $B = \{x: x \in N \text{ and } x \leq 200 \text{ and is divisible by } 5\}$
Thus, $A = \{2, 4, 6, 8, \ldots, 200\}$, $B = \{5, 10, 15, 20, \ldots, 200\}$ and

$A \cap B = \{10, 20, 30, \ldots, 200\}$. Now, $n(A) = 100$, $n(B) = 40$, and $n(A \cap B) = 20$.

Hence, by inclusion–exclusion principle

$$n(A \cup B) = n(A) + n(B) - n(A \cap B) = 100 + 40 - 20 = 120$$

Example 2 In a group of 100 persons, 70 can speak in English and 45 can speak in Hindi. Find the number of persons (a) who can speak in English only; (b) who can speak in Hindi only; and (c) who can speak in both English and Hindi.

Solution: Let E denote the set of persons who can speak in English and H the set of persons who can speak in Hindi. Then, $n(E \cup H) = 100$, $n(E) = 70$ and $n(H) = 45$.

(a) The number of persons who can speak in English only $= n(E - H)$
$$\begin{aligned} &= n(E) - n(E \cap H) \\ &= 70 - 15 = 55 \end{aligned}$$

Hence, the number of persons who can speak in English only = 55.

(b) The number of persons who can speak in Hindi only $= n(H - E) = n(H) - n(H \cap E)$
$$= 45 - 15 = 30$$

Hence, the number of persons who can speak in Hindi only = 30.

(c) $\therefore\ n(E \cap H) = n(E) + n(H) - n(E \cup H) = 70 + 45 - 100 = 15$
Hence, the number of persons who can speak in both English and Hindi is 15.

Example 3 In a survey of 500 school children, it was reported that 350 liked chocolate A and 275 liked chocolate B. Find the number of children who liked both the chocolates.

Solution: Let $n(A)$ and $n(B)$ denote the number of children who liked chocolates A and B, respectively, then $n(A) = 350$, $n(B) = 275$, and $n(A \cup B) = 500$.

The number of children who liked both the chocolates is

$$\begin{aligned} &= n(A \cap B) = n(A) + n(B) - n(A \cup B) \\ &= 350 + 275 - 500 \\ &= 625 - 500 = 125 \end{aligned}$$

Example 4 A software company received 50 applications for a job of a programmer. Of these, 30 knew JAVA and 25 knew C and 5 did not know any of the languages. Find the number of the applicants who knew both the languages.

Solution: Here, $n(J) = 30$, $n(C) = 25$, and $n(J \cup C) = 45$.

The number of applicants who knew both the languages is

$$n(J \cap C) = n(J) + n(C) - n(J \cup C)$$
$$= 30 + 25 - 45 = 10$$

1.3 CARTESIAN PRODUCT OF SETS

There is yet another way in which two sets can be combined to obtain another set. The Cartesian product of sets A and B is the set denoted by $A \times B$, which is defined by,

$$A \times B = \{(a, b): a \in A \text{ and } b \in B\}$$

The above definition can be generalised for more than two sets. If $A_1, A_2, A_3, \ldots, A_n$ are n given sets, then their Cartesian product is defined by,

$$A_1 \times A_2 \times A_3 \times \ldots \times A_n = \{a_1, a_2, \ldots, a_n): a_i \in A_i, i = 1, 2, \ldots, n\}$$

The expression $(a_1, a_2, \ldots, a_n)$ is called an ordered n-tuple. In general, $A \times B \neq B \times A$.

Example: Let $A = \{1, 2, 3\}$ and $B = \{3, 4\}$

Then $A \times B = \{(1, 3) (1, 4), (2, 3), (2, 4), (3, 3), (3, 4)\}$

$B \times A = \{(3, 1) (3, 2), (3, 3), (4, 1), (4, 2), (4, 3)\}$

Clearly $A \times B \neq B \times A$ and $n(A \times B) = n(A).n(B)$

Now $A \times A = \{(1, 1), (1, 2), (1, 3), (2, 1), (2, 2), (2, 3), (3, 1), (3, 2), (3, 3)\}$

and $B \times B = \{(3, 3), (3, 4), (4, 3), (4, 4)\}$

Note 1.4

- The Cartesian product of two sets is not commutative in general.
- If set A has m_1 elements and set B has m_2 elements, then the product $A \times B$ has m_1m_2 elements.
- If any one of A and B is infinite, then $A \times B$ is also an infinite set.
- If $A = \phi$ or $B = \phi$, then $A \times B = \phi$.
- If R is the set of real numbers, then $R \times R$, that is, R^2 represents the set of points on the Cartesian plane. $R \times R = \{(x, y): x, y \in R\}$.
- $R \times R \times R = R^3 = \{x, y, z): x, y, z \in R\}$ represents the set of points in three-dimensional space.

Example 1 If $A \times B = \{(1, 7), (1, 9), (3, 7), (3, 9), (5, 7), (5, 9)\}$, then find the sets A and B.

Solution: $A = \{1, 3, 5\}$ and $B = \{7, 9\}$

Example 2 If A, B, and C are three sets and $A \subseteq B$, then prove that $A \times C \subseteq B \times C$.

Solution: Let (a, c) be any element in $A \times C$, then

$$\begin{aligned}(a, c) \in A \times C &\Rightarrow a \in A \text{ and } c \in C \\ &\Rightarrow a \in B \text{ and } c \in C \qquad \text{(since } A \subseteq B\text{)} \\ &\Rightarrow (a, c) \in B \times C\end{aligned}$$

Hence,

$$A \times C \subseteq B \times C$$

Example 3 Prove that $A \times (B - C) = A \times B - A \times C$

Proof: Let (x, y) be any element in $A \times (B - C)$, then

$$\begin{aligned}(x, y) \in A \times (B - C) &\Leftrightarrow x \in A \text{ and } y \in (B - C) \\ &\Leftrightarrow x \in A \text{ and } (y \in B \text{ and } y \notin C) \\ &\Leftrightarrow (x \in A \text{ and } y \in B) \text{ and } (x \in A \text{ and } y \notin C) \\ &\Leftrightarrow (x, y) \in A \times B) \text{ and } (x, y) \notin (A \times C) \\ &\Leftrightarrow (x, y) \in [(A \times B) - (A \times C)]\end{aligned}$$

Hence,

$$A \times (B - C) = (A \times B) - (A \times C)$$

Example 4 Prove that $A \times (B \cup C) = (A \times B) \cup (A \times C)$

Proof: Let (x, y) be any element in $A \times (B \cup C)$, then

$$\begin{aligned}(x, y) \in A \times (B \cup C) &\Leftrightarrow x \in A \text{ and } y \in B \cup C \\ &\Leftrightarrow x \in A \text{ and } (y \in B \text{ or } y \in C) \\ &\Leftrightarrow (x \in A \text{ and } y \in B) \text{ or } (x \in A \text{ and } y \in C) \\ &\Leftrightarrow ((x, y) \in A \times B) \text{ or } ((x, y) \in A \times C) \\ &\Leftrightarrow (x, y) \in (A \times B) \cup (A \times C)\end{aligned}$$

Similarly we can prove that,

$$A \times (B \cap C) = (A \times B) \cap (A \times C)$$

1.4 MULTISET

A set is a collection of distinct objects. There are many situations where we come across a collection of non-distinct objects. For example, consider the names of students in a class where we may have two or more students have the same name. We may be required to talk about the collection of names of the students. A similar situation can arise in an industry where two or more employees have the same name. Thus, we may have to deal with a collection of objects that are not necessarily distinct. Such a set is called a multiset. *A multiset is a set in which an element can occur as a member any number of times.*

For example, if $A = \{a, a, b, c, c, d\}$ and $B = \{a, a, a, b, b, c, c, c\}$ are two non-empty sets, then sets A and B can also be written as $A = \{2a, b, 2c, d\}$ and $B = \{3a, 2b, 3c\}$.

The multiplicity of an element in a multiset is defined as the number of times an element appears in the set.

1.4.1 Operations on Multisets

Union of Multisets

The union of two multisets A and B denoted by $A \cup B$ is the multiset such that the multiplicity of an element is equal to the maximum of the multiplicity of an element in A and B.

Intersection of Multisets

The intersection of two multisets A and B denoted by $A \cap B$ is the multiset of elements such that the multiplicity of an element is equal to the minimum of multiplicity of the element in A and B.

Difference of Multisets

The difference of two multisets A and B denoted by $A - B$ is the multiset where multiplicity of an element in A minus the multiplicity of element in B is associated with each element if the difference is positive. The multiplicity is zero if the difference is negative.

Sum of Multisets

The sum of two multisets A and B denoted by $A + B$ is a multiset such that the multiplicity of an element is equal to the sum of multiplicity of the element in A and B.

Example: Let $A = \{4a, 3b, 2c\}$ and $B = \{2a, b, d, 3c\}$ be two multiplicity sets.
Then, $A \cup B = \{4a, 3b, 3c, d\}$, $A \cap B = \{2a, b, 2c\}$
$A - B = \{2a, 2b\}$ and $B - A = \{c, d\}$
$A + B = \{6a, 4b, 5c, d\}$

Cardinality of a Multiset

The cardinality of a multiset is the number of its distinct elements without considering the multiplicity of an element.

Let $A = \{a, a, b, b, c, c, c, c, d, d, e\}$ be set. The cardinality of multiset A is 5.

Example 1.5 A survey of 200 students shows that 100 students study Mathematics, 90 study Chemistry, 60 study Physics, 26 study Mathematics and Chemistry, 22 study Chemistry and Physics, 24 study Physics and Mathematics and 10 study all the three.

Find the number of students who study the following:

(a) Mathematics and Chemistry
(b) Chemistry and Physics
(c) Mathematics and Physics
(d) Physics only

Solution: Let M, C and P denote the set of students who study Mathematics, Chemistry, and Physics, respectively. We draw three intersecting circles to represent these sets.
We are given $n(M) = 100$, $n(C) = 90$ and $n(P) = 60$, $n(M \cap C) = 26$,

$$n(C \cap P) = 22,\ n(P \cap M) = 24, \text{ and } n(M \cap C \cap P) = 10.$$

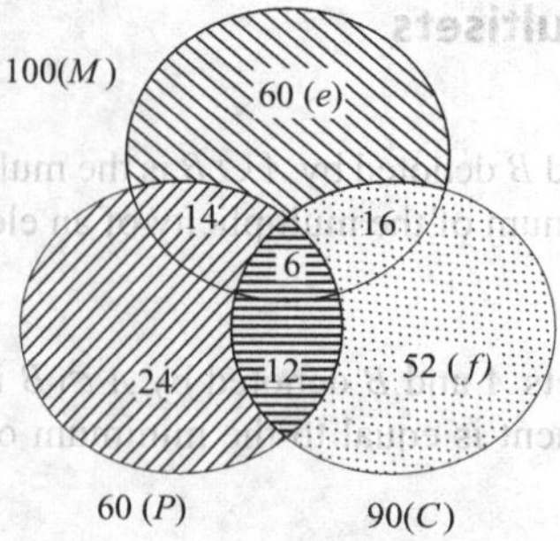

We first distribute 200 students into various regions. From the Venn diagram, we can find $e + a + b + d = 100$, $b + c + d + g = 60$, $a + b + c + f = 90$, $a + b = 26$, $b + d = 24$, $b + c = 22$. These relations give $b = 10$, $c = 12$, $d = 14$, $a = 16$,

$$e = 100 - (16 + 10 + 14) = 100 - 40 = 60$$
$$g = 60 - (10 + 12 + 14) = 60 - 36 = 24$$
$$f = 90 - (10 + 16 + 12) = 90 - 38 = 52$$

Hence, $n(M \cup C - P) = 16$, $n(C \cap P - M) = 12$, $n(M \cap P - C) = 14$, and $n(P) = 24$.

So far, we have accounted for 188 students out of 200. The remaining 12 students lie outside the region $M \cup C \cup P$.

EXERCISES

1. List the elements of the following sets:
   (a) $\{x|x \in N$ and is less than $5\}$ (b) $\{x|x \in Z$ and $x^2 = 1\}$
   (c) $\{n|n \in N$ and $2 < n < 7\}$ (d) $\{x|x \in R$ and $x^2 = 2\}$
   (e) $\{x|x \in Z$ and $|x| < 3\}$
2. List three elements of the set.
   $\{n \in N|n^2 + n$ is a multiple of $3\}$
3. Are the following sets, null sets?
   (a) $\{n \in N|2n = 3\}$ (b) $\{x \in R|x^2 + 1 = 0\}$
   (c) $\{n \in N|n + 5 = 1\}$ (d) $\{x|x$ is a multiple of 2 and is odd$\}$
4. If N, Z, Q, and R denote the sets of natural numbers, integers, rational numbers and real numbers, respectively, find whether $N \subseteq Z \subseteq Q \subseteq R$ is true.
5. Are the sets A and B given below equal?
   (a) $A = \{1, 3, 5\}$ and $B = \{5, 3, 1\}$ (b) $A = \{1, 7, 9\}$ and $B = \{9, 7, 1\}$
6. Let $S = \{1, 2, 3, 4\}$, find the set A of all subsets of S that has exactly three elements.
7. Are the following statements for power set $P(A)$ for the set A true?
   (a) If $A = \{1, 2\}$, then $P(A) = \{\Phi, \{1\}, \{2\}, \{1, 2\}\}$.
   (b) If $A = \{1, 2, 3\}$, then $P(A) = \{\Phi, \{1\}, \{2\}, \{3\}, \{1, 2\}, \{2, 3\}, \{1, 3\}, \{1, 2, 3\}\}$.

8. Is the set $A = \{4, 5, 6, 8\}$ a subset of set $B = \{x: x \in N \text{ and } x \text{ is even}\}$?
9. If the set $A = \{1, 2, 3, 5\}$, $B = \{6, 7, 8, 5\}$, and $C = \{2, 4, 6, 8\}$. Find (a) $A \cup B$ and $A \cap B$; (b) $A \cup C$ and $A \cap C$; (c) $B \cup C$ and $B \cap C$.
10. Let $A = \{1, 5\}$, $B = \{5, 7\}$, and $C = \{5, 9\}$. Find $A \cap B$ and $A \cap C$. Show that $A \cap B = A \cap C$ but $B \neq C$.
11. Show with the help of Venn diagrams, the distributive law
$A \cap (B \cup C) = (A \cap B) \cup (A \cap C)$

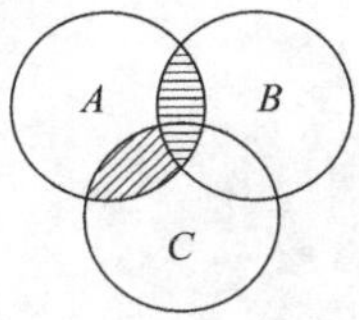

$A \cap (B \cup C)$ is shaded.

12. Let $U = \{1, 2, 3, 4, 5, 6\}$, $A = \{1, 2, 3, 4\}$, $B = \{4, 5, 6\}$, and $C = \{2, 4, 6\}$. Find the following sets:

(a) A' (b) B' (c) $(A \cup B)'$ (d) $A - B$ (e) $(A \cup B) - C$ (f) $A \oplus B$

13. If A and B be any two sets, then using Venn diagram,
(a) Show that A is the disjoint union of the sets $A - B$ and $A \cap B$.
(b) $A \cup B$ is the disjoint union of $A - B$, $A \cap B$ and $B - A$.
14. For the Venn diagram, showing sets A, B, and C, shade the following sets:

(a) $B - (A \cup C)$ (b) $B' - (A \cup C)$

15. Determine the power set $P(A)$ of the set $A = \{1, 2, 3, 4\}$. How many elements are in $P(A)$?
16. Which of the following statements are true?
(a) $A \cup A' = U$ (b) $A \cap A' = \phi$ (c) $A \cap A = \phi$ (d) $A \cap \phi = A$
(e) $A \cap \phi = \phi$ (f) $\phi' = U$ (h) $U' = \phi$
17. Let $A = (2, 3)$ and $B = (4, 5, 6)$. List the elements of
(a) $A \times B$ (b) $B \times A$ (c) Is $A \times B = B \times A$?
Find the number of elements in $A \times B$
18. Show that if A_1 has n_1 elements, A_2 has n_2 elements and A_3 has n_3 elements, then $A_1 \times A_2 \times A_3$ has n_1, n_2, n_3 elements.
19. If $P = \{1, 2\}$, $Q = \{a, b, c\}$ and $R = \{p, q\}$, list all the elements of $P \times Q \times R$.

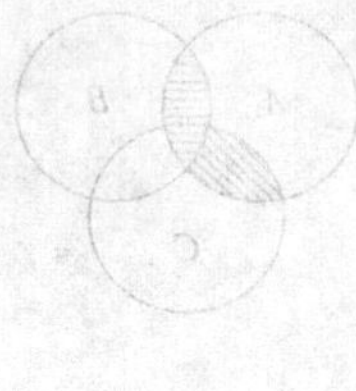

2

Relations and Digraphs

LEARNING OBJECTIVES

After reading this chapter, you will be conversant with:

- Introduction
- Binary Relation
- Equivalence Class
- Partition of a Set
- Congruence Modulo Relation
- Pictorial Representation of Relation
- Digraphs
- Power of Relation (R)
- Paths in Relations and Digraphs
- Matrix Representation of Composite Relations
- Connectivity Relation

2.1 INTRODUCTION

The concept of relation plays a very important role in mathematics and its applications. The often used relations are less than, greater than, parallel to, subset of, and so on. A relation between two sets of objects can be defined by listing the elements as ordered pairs which are in given relation. On the human level various relationships exist such as Doctor–Patient, Lawyer–Clint, Husband–Wife, Teacher–Student, Father–Son, and so on. Two ordered pairs (a, b) and (c, d) are said to be equal if and only if $a = c$ and $b = d$.

2.2 BINARY RELATION

Definition 2.1 Let A and B be two given non-empty sets. The binary relation R from the set A to B is defined to be the subset of $A \times B$.

Thus, if $R \subseteq A \times B$ and $(a, b) \in R$, then we say that a is related to b by relation R and write $a\,R\,b$.

Symbolically, $R = \{(a, b): a \in A, b \in B \text{ and } a\,R\,b\}$

If $A = B$, then R is said to be a binary relation on A and $R = \{(a, b): a, b \in A \text{ and } a\,R\,b\}$

If an element a is not related to b by relation R, we write $a \not R\, b$. If A is the set of all real numbers, then we can define different relations on A.

For example,

1. $x\,R\,y$ if $x < y$
2. $x\,R\,y$ if $x \leq y$
3. $x\,R\,y$ if $x > y$
4. $x\,R\,y$ if $x \geq y$

Example 1 Let $A = \{1, 2, 3, 4\}$. Define R by $x\,R\,y$ if $x < y$

Then $1R2$, $1R3$, $1R4$, $2R3$, $2R4$, $3R4$.

Thus, the relation R is completely known if we know all R – related pairs.

$$R = \{(1, 2), (1, 3), (1, 4), (2, 3), (2, 4), (3, 4)\}$$

Obviously, R is a subset of $A \times A$.

Example 2 Let $A = \{1, 2, 3\}$ and $B = \{r, s\}$.
Then $R = \{(1, r), (2, s), (3, r)\}$ is a relation from A to B as R is a subset of $A \times B$.

Example 3 Define relation R on the set of natural numbers N by $a\,R\,b$ if $b = 2a$, $a, b \in N$.

Then $1R2$, $2R4$, $3R6$, …, But $2 \not R\, 3$, $3 \not R\, 4$. Now, $R = \{(a, 2a)$ for every $a \in N\}$

Example 4 Let $A = R$, the set of real numbers. Define relation R^* on a by $x\,R^*y$ if $(x^2/64) + (y^2/25) = 1$, that is, R^* consists of all points on the ellipse.

Note 2.1

- If a set A has n elements, then $A \times A$ has $(n \times n = n^2)$ elements. So the total number of relations is $2^{(n \times n)}$ as the total number of subsets of $A \times A$ is $2^{(n \times n)}$.
- If A has m elements and B has n elements, then there are mn elements in the set $A \times B$ and the number of relations is $2^{m \times n}$.

2.2.1 Domain and Range of Relation

The domain of the relation R denoted by dom (R) consists of the first elements in R and the range denoted by Range (R) consists of the second elements.

Thus, Dom $(R) = \{a: (a, b) \in R, a \in A\}$ and Range $(R) = \{b: (a, b) \in R, b \in B\}$

Example 1 Let $A = \{3, 5, 7\}$ and $B = \{3, 6, 10, 14, 15\}$
Define a relation R from A to B by, $R = \{(a, b): a \text{ divides } b\}$
Then, $R = \{(3, 3), (3, 6), (3, 15), (5, 10), (5, 15), (7, 14)\}$
Dom $(R) = \{3, 5, 7\}$ and Range (R) $\{3, 6, 10, 14, 15\}$

Example 2 Let $A = \{1, 2, 3, 4\}$, $B = \{x, y, z, p, q\}$, and $R = \{(1, x), (2, y), (3, p)\}$
Then, Dom $(R) = \{1, 2, 3\}$ and Range $(R) = \{x, y, p\}$

2.2.2 Types of Relations

Inverse Relation: Let R be any relation from a set A to a set B. The inverse relation denoted by R^{-1} is the relation from B to A defined by,

$$R^{-1} = \{(b, a): (a, b) \in R\}$$

that is, the inverse relation is obtained by interchanging the order in each ordered pair of R.

$$(a, b) \in R \Leftrightarrow (b, a) \in R^{-1}$$

The Dom (R) = Range (R^{-1}) and vice versa.
For example: Let the relation R from $A = \{2, 3, 5\}$ to $B = \{4, 7, 9\}$
defined by $R = \{(2, 4), (3, 7), (5, 9)\}$, then $R^{-1} = \{(4, 2), (7, 3), (9, 5)\}$
Further, let R be the relation from $A = \{a, b, c\}$ to $B = \{x, y, z\}$
defined by $R = \{(a, x), (a, y), (c, y), (c, z)\}$
Then, R^{-1} is given by $R^{-1} = \{(x, a), (y, a), (y, c), (z, c)\}$.

Void Relation: A relation R is said to be void relation if $R = \varphi$.

Example: Let $A = \{3, 5, 7, 11\}$ and let relation R be defined on A by $a\,R\,b$, if a is a multiple of b. Here $R = \varphi$ and hence the relation R defined on A is a void relation.

Identity Relation: The relation R on a set A denoted by I_A is called identity relation if $I_A = \{(a, b): b = a\}$.

Example 1 Let $A = \{a, b, c\}$, then the identity relation on A is $I_A = \{(a, a), (b, b), (c, c)\}$
Example 2 Let $A = \{1, 2, 3, 4\}$ be the set on which the relation R is defined by $R = \{(1, 1), (2, 2), (3, 3), (4, 4)\}$, then $I_A = \{(1, 1), (2, 2), (3, 3), (4, 4)\}$.

Universal Relation: A relation on a set A is called a universal relation if $R = A \times A$.
Example: Let $A = \{a, b, c\}$, then the universal relation R is given by,

$$R = A \times A = \{(a, a), (a, b), (a, c), (b, c), (b, a), (b, b), (c, a), (c, b), (c, c)\}$$

Example: If $A = \{1, 2\}$, then the universal relation $R = A \times A = \{(1, 1), (1, 2), (2, 1), (2, 2)\}$

2.2.3 Properties of Relation on a Set

Reflexive Relation: A binary relation R on a set A is reflexive, if $a\,Ra$ for every $a \in A$.
The relation is not reflexive if there exists an $a \in A$ such that $a \not R\, a$, that is, $(a, a) \notin R$.

Example 1 Consider the following relations on the set $A = \{1, 2, 3, 4\}$

1. $R_1 = \{(1, 2), (1, 3), (2, 3), (3, 3), (4, 4)\}$
2. $R_2 = \{(1, 1), (2, 2), (3, 3), (4, 4), (1, 4), (2, 3)\}$

3. $R_3 = \{(1, 3), (2, 4)\}$
4. R_4 = Universal relation.

Determine which of the relations are reflexive.

Solution:

1. R_1 is not reflexive as (2, 2), (1, 1) are not in R_1.
2. R_2 is reflexive as it contains {(1, 1), (2, 2), (3, 3), (4, 4)}, that is, $(a, a) \in R_2$ for each $a \in A$.
3. R_3 is not reflexive as it does not contain elements that are related to itself.
4. R_4 = Universal set = $A \times A$ and thus elements (1, 1), (2, 2), (3, 3), (4, 4) $\in R$ and so R_4 is reflexive.

Example 2 Let L be the set of straight lines in a plane. Define relation R by

1. xRy if x is parallel to y, $x, y \in L$.
2. xRy if x is perpendicular to y, $x, y \in L$.

Are these relations reflexive?

Solution:

1. Since every line can be regarded parallel to itself, $x\ R\ x,\ \forall\ x \in L$. Hence the relation R of parallelism is reflexive.
2. Since no straight line is perpendicular to itself, $x\ \not R\ x, \forall\ x \in L$. Hence the relation R of perpendicularity is not reflexive.

Example 3 Let S be the collection of sets. Define relation R by $A\ R\ B$ if $A \subseteq B$, that is, set inclusion relation. Is this relation reflexive?

Solution: Since every set is a subset of itself, ARA for every set in S and hence the set inclusion relation is reflexive.

Example 4 Let N be the set of natural numbers. Define relation R by $x\ R\ y$ if x is a divisor of y where $x, y \in N$. Find whether the relation is reflexive.

Solution: The relation R is reflexive since every natural number is a divisor of itself.

Irreflexive Relation: A binary relation R on a set A is called irreflexive, if $a\ \not R\ a,\ \forall\ a \in A$ (i.e for every $a \in A$), that is, $(a, a) \notin R,\ \forall\ a \in A$.

Example 1 Let $A = \{1, 2, 3, 4\}$ and define relation R by $R = \{(1, 2), (1, 3), (2, 3), (3, 4)\}$.

Solution: The relation R is irreflexive because $(a, a) \notin R,\ \forall\ a \in A$.

Example 2 Let $A = \{1, 2, 3\}$ and $R = \{(1, 1), (1, 2), (3, 3), (2, 3)\}$. Is the relation reflexive or irreflexive?

Solution: The relation is not reflexive as for every $a \in A$, $(a, a) \notin R$, that is, $(2, 2) \notin R$. The relation R is not irreflexive as $(a, a) \notin R,\ \forall\, a \in R$ is not true because (1, 1) and $(3, 3) \in R$.

Symmetric Relation: A binary relation R on a set A is symmetric, if $a\ R\ b \Rightarrow b\ R\ a,\ \forall\ a, b \in R$ (i.e., for all $a, b \in R$), that is, $(a, b) \in R \Rightarrow (b, a) \in R$. Thus, a relation R is symmetric if $R = R^{-1}$.

Example 1 Let $A = \{a, b, c\}$. Define the relation $R = \{(a, b), (b, a), (a, c), (a, a), (c, a)\}$
The relation R is symmetric as $(a, a) \in R \Rightarrow (a, a) \in R$
$(a, b) \in R \Rightarrow (b, a) \in R$ and $(a, c) \in R \Rightarrow (c, a) \in R$.

Example 2 Let $A = \{1, 2, 3, 4\}$. Define relation $R = \{(1, 1), (2, 2), (1, 3), (2, 3), (3, 2)\}$

The relation R is not symmetric as $(1, 3) \in R \not\Rightarrow (3, 1) \in R$.

Asymmetric Relation: A relation R on a set A is asymmetric, if whenever $a\ R\ b$ then $b \not R\ a$. Thus, a relation is not asymmetric when for some $a, b \in A$ both $a\ R\ b$ and $b\ R\ a$.

Anti-symmetric Relation: A binary relation R on a set A is anti-symmetric, if $a\ R\ b$ and $b\ R\ a \Rightarrow a = b$ for all $a, b \in R$.

Example 1 Let N denote the set of natural numbers. Define R by $a\ R\ b$, if a divides b. The relation R is anti-asymmetric because if a divides b and b divides a, then $a = b$.

Example 2 Let R denote the set of real numbers. Define relation R by $xR\ y$ if $x \le y$. The relation R is anti-symmetric because $x\ R\ y$ and $y\ R\ x \Rightarrow x \le y$ and $y \le x \Rightarrow x = y$.

Example 3 Let $A = \{3, 4, 5, 6\}$. Define $R = \{(3, 3), (3, 5), (5, 3), (4, 4), (4, 6)\}$
The relation R is not anti-symmetric because $(3, 5) \in R$ and $(5, 3) \in R \not\Rightarrow 3 = 5$.

Example 4 Let $A = \{1, 2, 3,\}$. Define $R = \{(1, 2), (2, 1), (2, 3), (1, 1)\}$. The relation R is not anti-symmetric because $(1, 2) \in R$ and $(2, 1) \in R$ but $1 \ne 2$.

Note 2.2 Anti-symmetric is not the same as 'not symmetric'. The relation in Example 4 is not symmetric but neither it is anti-symmetric.

Transitive Relation: A binary relation R on a set A is transitive if $a\ R\ b$, $b\ R\ c \Rightarrow a\ R\ c\ \forall\ a, b, c \in R$.

Example 1 Let R denote the set of real numbers. Define relation R by $x\ R\ y$ if $x < y$. The relation less than is transitive because $x < y$ and $y < z \Rightarrow x < z$, that is, $xR\ y$ and $yR\ z \Rightarrow xR\ z$.

Example 2 Let S be the set of straight lines in a plane. Define relation R by $x\ R\ y$, if x and y are parallel. The relation 'is parallel to' is transitive because if line x is parallel to line y which is parallel to line z, then x is also parallel to z. That is, $x\ R\ y$, $yR\ z \Rightarrow x\ R\ z$ for all $x, y, z \in S$.

Example 3 Let $A = \{a, b, c, d\}$ and define $R = \{(a, b), (a, c), (b, d), (a, d)\}$. Find whether R is a transitive relation.

Solution: There is only one pair (a, b), (b, d) belonging to R and for this pair $(a, d) \in R$ is true and hence the relation R is transitive.

Example 4 Let $A = \{a, b\}$. Define $R = \{(a, b), (b, a), (b, b)\}$. Is R is a transitive relation?

Solution: The relation R is not transitive because $(a, b) \in R$ and $(b, a) \in R \not\Rightarrow (a, a) \in R$.

Example 5 Let S be the set of people. Define relation R by $x\ R\ y$, if x is an ancestor of y.

Solution: The relation R is transitive because if x is an ancestor of y and y is an ancestor of z, then x is an ancestor of z. Then, $x\ R\ y$ and $y\ R\ z \Rightarrow x\ R\ z$.

Example 6 Let S be the set of people. Define the following relations on S:

(a) $x \, R \, y$ if x is father of y. (b) $x \, R \, y$ if x is a friend of y. (c) $x \, R \, y$ if x is a brother of y.

Solution:

(a) The relation 'is father of' is not transitive because $x \, R \, y$ and $y \, R \, z \not\Rightarrow x \, R \, z$.

(b) The relation 'is friend of' is not transitive because $x \, R \, y$ and $y \, R \, z \not\Rightarrow x \, R \, z$.

(c) The relation 'is brother of' is transitive because $x \, R \, y$ and $y \, R \, z \Rightarrow x \, R \, z$.

2.2.4 Equivalence Relation

A relation R on a set A is called an equivalence relation (denoted by ~) if it satisfies the following conditions:

1. R is reflexive, that is, $a \, R \, a, \ \forall \ a \in A$.
2. R is symmetric, that is, if $a \, R \, b$ then $b \, R \, a$.
3. R is transitive, that is, $a \, R \, b$ and $b \, R \, c \Rightarrow a \, R \, c$.

Example 1 Let S be the set of all triangles in a plane. Define relation R by $a \, R \, b$, if the triangles a and b are similar. Prove that R is an equivalence relation.

Solution:

1. ***R* is reflexive** because every triangle is similar to itself, that is, $a \, R \, a$ for every $a \in S$.
2. ***R* is symmetric** because if triangles a and b are similar, then so also are b and a, that is, $a \, R \, b \Rightarrow b \, R \, a$.
3. ***R* is transitive** because if triangles a and b are similar and b and c are similar, then triangles a and c are also similar, that is, $a \, R \, b$ and $b \, R \, c \Rightarrow a \, R \, c$.

Hence, the relation R of similarity of triangles is an equivalence relation.

Example 2 Let S be the set of all triangle in a plane. Define relation R by $x \, R \, y$, if the triangle x is congruent to y. Prove that R is an equivalence relation.

Example 3 Let A be the set of all residents in a given state. Define relation R by $x \, R \, y$, if x and y belong to the same district. Prove that R is an equivalence relation.

Example 4 The relation R defined as set inclusion ($\subseteq$) on the set of all sets A is not an equivalence relation because it is reflexive and transitive but not symmetric.

Example 5 The relation of equality on a set of sets A is an equivalence relation as it is reflexive, symmetric and transitive.

Example 6 Let N be the set of natural numbers. Define relation R on $N \times N$ by $(a, b) \, R \, (c, d)$ if and only if $ad = bc$. Prove that R is an equivalence relation.

Solution:

1. ***R* is reflexive** since for each $(a, b) \in N \times N$, we have $ab = ba$, that is, $(a, b) \, R \, (a, b)$.
2. Let $(a, b) \, R \, (c, d) \Rightarrow ad = bc \Rightarrow (c, d) \, R \, (a, b)$ showing that R is symmetric.
3. Let $(a, b) \, R \, (c, d)$ and $(c, d) \, R \, (e, f)$.

$\Rightarrow ad = bc$ and $cf = ed$. Thus, $(ad)(cf) = (bc)(ed)$.
$\Rightarrow af = be \Rightarrow (a, b)\,R\,(e, f)$, showing that R is transitive. Hence, R is an equivalence relation.

Example 7 If R and S are two equivalence relations on a set A, prove that $R \cap S$ is also an equivalence relation on A.

Solution:

1. Since R and S are equivalence relations, R and S are reflexive, that is, $(a, a) \in R$ and $(a, a) \in S$ which implies $(a, a) \in R \cap S$. Thus, $R \cap S$ is reflexive.
2. Let $(a, b) \in R \cap S \Rightarrow (a, b) \in R$ and $(a, b) \in S$
   $\Rightarrow (b, a) \in R$ and $(b, a) \in S$ (since R and S are equivalence relations)
   $\Rightarrow \in (b, a) \in R \cap S$, which shows that $R \cap S$ is symmetric.
3. Let $(a, b) \in R \cap S$ and $(b, c) \in R \cap S$
   $\Rightarrow [(a, b) \in R$ and $(a, b) \in S]$ and $[(b, c) \in R$ and $(b, c) \in S]$
   $\Rightarrow [(a, b) \in R$ and $(b, c) \in R]$ and $[(a, b) \in S$ and $(b, c) \in S]$
   $\Rightarrow [(a, c) \in R]$ and $[(a, c) \in S]$
   $\Rightarrow (a, c) \in R \cap S$, which shows that $R \cap S$ is transitive.

Since $R \cap S$ is reflexive, symmetric and transitive, $R \cap S$ is an equivalence relation.

Example 8 The union of two equivalence relations R and S, that is, $R \cup S$ on a set A is always reflexive and symmetric.

Solution: Since R and S are equivalence relations on A, both are reflexive and symmetric.

1. Let $(a, b) \in R \cup S \Rightarrow (a, b) \in R$ or $(a, b) \in S$
   $\Rightarrow (b, a) \in R$ or $(b, a) \in S$
   $\Rightarrow (b, a) \in R \cup S$

   Hence, $R \cup S$ is symmetric.
2. Since R and S are equivalence relations, $(a, a) \in R$ and $(a, a) \in S \Rightarrow (a, a) \in R \cup S$ for every $a \in A$ hence $R \cup S$ is reflexive.

Note 2.3 The relation $R \cup S$ is not transitive.

Example: Let $A = \{a, b, c\}$. Define relation R and S by $R = \{(a, b)\}$ and $S = \{(b, c)\}$, then $R \cup S = \{(a, b), (b, c)\}$ which is not transitive.

Example 9 If R and S are relations from A to B, then prove that

(a) $(R^{-1})^{-1} = R$
(b) If $R \subseteq S$, then $R^{-1} \subseteq S^{-1}$
(c) $(R \cup S)^{-1} = R^{-1} \cup S^{-1}$
(d) $(R \cap S)^{-1} = R^{-1} \cap S^{-1}$

Proof:

(a) Let $(x, y) \in (R^{-1})^{-1}$, then $(y, x) \in R^{-1} \Rightarrow (x, y) \in R$ and $(R^{-1})^{-1} \subseteq R$ (i)
Conversely let $(x, y) \in R$, then $(y, x) \in R^{-1} \Rightarrow (x, y) \in (R^{-1})^{-1}$

Hence,

$$(x, y) \in R \Rightarrow (x, y) \in (R^{-1})^{-1} \text{ and so } R \subseteq (R^{-1})^{-1} \quad \text{(ii)}$$

From (i) and (ii), $(R^{-1})^{-1} = R$. Proved

(b) Since $R \subseteq S$, an element $(x, y) \in R^{-1} \Rightarrow (y, x) \in R$
$\Rightarrow (y, x) \in S$
$\Rightarrow (x, y) \in S^{-1}$

Hence,

$$R^{-1} \subseteq S^{-1} \text{ ---------------- Proved}$$

(c) To prove $(R \cup S)^{-1} = R^{-1} \cup S^{-1}$
Let $(x, y) \in (R \cup S)^{-1} \Leftrightarrow (y, x) \in R \cup S$
$\Leftrightarrow (y, x) \in R$ or $(y, x) \in S$
$\Leftrightarrow (x, y) \in R^{-1}$ or $(x, y) \in S^{-1}$
$\Leftrightarrow (x, y) \in R^{-1} \cup S^{-1}$

Hence,

$$(R \cup S)^{-1} = R^{-1} \cup S^{-1} \text{ ---------------- Proved}$$

(d) To prove $(R \cap S)^{-1} = R^{-1} \cap S^{-1}$

Let $(x, y) \in (R \cap S)^{-1} \Leftrightarrow (y, x) \in R \cap S$
$\Leftrightarrow (y, x) \in R$ and $(y, x) \in S$
$\Leftrightarrow (x, y) \in R^{-1}$ and $(x, y) \in S^{-1}$
$\Leftrightarrow (x, y) \in R^{-1} \cap S^{-1}$

Hence,

$$(R \cap S)^{-1} = R^{-1} \cap S^{-1} \text{ ---------------- Proved}$$

Example 10 If R is a relation on a set A, then prove that

(a) R is reflexive $\Rightarrow R^{-1}$ is reflexive.
(b) R is symmetric $\Leftrightarrow R = R^{-1}$.

Proof:

(a) Suppose that R is reflexive, and hence, $(x, x) \in R$ for every $x \in A$ which implies that $(x, x) \in R^{-1}$ for each $x \in A$. Hence, R^{-1} is reflexive.

(b) Suppose that R is symmetric, that is, $(x, y) \in R \Rightarrow (y, x) \in R \Rightarrow (x, y) \in R^{-1}$.
Hence,

$$R \subseteq R^{-1} \quad \text{(i)}$$

Further,

$$(x, y) \in R^{-1} \Rightarrow (y, x) \in R^{-1} \Rightarrow (x, y) \in R \text{ and so } R^{-1} \subseteq R \quad \text{(ii)}$$

From (i) and (ii), $R = R^{-1}$.

Conversely Let $R = R^{-1}$, then $(y, x) \in R \Rightarrow (x, y) \in R^{-1}$

$\Rightarrow (y, x) \in R$, which shows that R is symmetric.

Example 11 If R is an equivalence relation on a set A, then prove that R^{-1} is also an equivalence relation on A.

Proof: Since R is an equivalence relation on *a* set A, it must be reflexive, symmetric, and transitive.

1. Since R is reflexive, for every a, $(a, a) \in R \Rightarrow (a, a) \in R^{-1}$ and so R^{-1} is reflexive.
2. Since R is symmetric $(a, b) \in R \Rightarrow (b, a) \in R$
   Now $(a, b) \in R \Rightarrow (b, a) \in R^{-1}$ and $(b, a) \in R \Rightarrow (a, b) \in R^{-1}$,
   which shows that $(a, b) \in R^{-1}$ and $(b, a) \in R^{-1}$ and so R^{-1} is symmetric.
3. Since R is transitive $(a, b) \in R$ and $(b, c) \in R \Rightarrow (a, c) \in R$
   Now $(a, b) \in R \Rightarrow (b, a) \in R^{-1}$
   $(b, c) \in R \Rightarrow (c, b) \in R^{-1}$
   $(a, c) \in R \Rightarrow (c, a) \in R^{-1}$

From above, we have $(c, b) \in R^{-1}$ and $(b, a) \in R^{-1} \Rightarrow (c, a) \in R^{-1}$. Hence, R^{-1} is transitive. R^{-1} is an equivalence relation.

2.2.5 Complementary Relation $\bar{R}$

The complement of relation R from A to B denoted by $\bar{R}$ is defined by $a\ \bar{R}\ b$ if and only if $a \not R\ b$.

Example: Let $A = \{1, 2, 3\}$, $B = \{a, b, c\}$, $R = \{(1, a), (1, b), (2, b), (3, c)\}$ and $S = \{(1, b), (2, b), (3, c)\}$. Find (i) The complement $\bar{R}$ of R (ii) R^{-1} (iii) $R \cap S$ (iv) $R \cap S$

Solution:

1. $A \times B = \{(1, a), (1, b), (1, c), (2, a), (2, b), (2, c), (3, a), (3, b), (3, c)\}$
   The complement $\bar{R}$ = Set of element of $A \times B$ which are not in R.
   $= \{(1, c), (2, a), (2, c), (3, a), (3, b)\}$
2. $R^{-1} = \{(a, 1) (b, 1), (b, 2), (c, 3)\}$
3. $R \cap S = \{(1, b), (2, b), (3, c)\}$
4. $R \cup S = \{(1, a), (1, b), (2, b), (3, c)\}$

2.2.6 Closure of Relation

Let R be a given relation on a set A that may not be reflexive, symmetric or transitive. Naturally, if we add some pairs then we have the desired property. The smallest relation S on A that contains R and possesses the desired property is called the closure of R with respect to that property.

2.2.7 Reflexive Closure

Let R be a given relation on a set A. The reflexive closure S of R is obtained by adding to R all pairs of the form (a, a), which are not in R for every $a \in A$.

Example: Let $R = \{(1, 1), (2, 2), (1, 2), (2, 4)\}$ be defined on the set $A = \{1, 2, 3, 4\}$. Find the reflexive closure of R.

Solution: Reflexive closure $S = R \cup \{(3, 3), (4, 4)\} = \{(1, 1), (2, 2), (1, 2), (2, 4), (3, 3), (4, 4)\}$
S is the reflexive closure of R.

2.2.8 Symmetric Closure

Let R be a given relation on the set A. The symmetric closure S of R is obtained by adding to R all pairs of R^{-1}, that is, all pairs (b, a) if $(a, b) \in R$, that is, $S = R \cup R^{-1}$.

Example: Let $R = \{(2, 2), (1, 2), (1, 3), (2, 4), (4, 4)\}$ be a relation on $A = \{1, 2, 3, 4\}$. Find the symmetric closure of R.

Solution: Since R is not symmetric, we find R^{-1}
Required symmetric closure $S = R \cup R^{-1} = \{(2, 2), (1, 2), (2, 4), (4, 4), (1, 3), (2, 1), (4, 2), (3, 1)\}$

2.2.9 Composition of Relations

Let A, B, and C be given sets and let R be a relation from A to B and S be a relation from B to C. Thus R is a subset of $A \times B$ and S is a subset of $B \times C$. The relations R and S give rise to another relation from A to C denoted by ROS defined by $ROS = \{(a, c)$: there exists $b \in B$ for which $(a, b) \in R$ and $(b, c) \in S\}$ or $a\ (ROS)\ c$ if for some $b \in B$, $a\ R\ b$ and $b\ S\ c$. The relation ROS is called composition of R and S.

2.2.10 Transitive Closure

Let R be a given relation on the set A that is not transitive. The smallest transitive relation that contains R is called the transitive closure of R and is denoted by R^*.

To Obtain R^*, We Proceed as Follows: Since R is not transitive, it should have ordered pairs (a, b) and (b, c) such that $(a, c) \notin R$ and so add (a, c) to R. We can continue this with every such pair in the new relation. The resulting relation is the transitive closure of R.

Example: Find the transitive closure of the relations $R = \{(b, c), (c, b), (c, a)\}$

$$S = \{(a, a), (b, b), (c, c)\} \text{ and } T = \phi \text{ on the set } \{a, b, c\}.$$

Solution: $R = \{(b, c), (c, b), (c, a)\}$. Since $(b, c) \in R$ and $(c, a) \in R$, it requires the element (b, a) to be transitive. So, we add (b, a) to R. The new relation is $R_1 = \{(b, c), (c, b), (c, a), (b, a)\}$. It contains both (b, c) and (c, b) but not (b, b) or (c, c). Add these to R_1. The new relation is

$$R_2 = \{(b, c), (c, b), (c, a), (b, a), (b, b), (c, c)\}$$

It is transitive and contains R and so it is the transitive closure of R, that is, R^*. The relations S is transitive and $S^* = S$. The transitive closure of φ is itself.

THEOREM 2.1 The following theorem can also be used to find the transitive closure R^* of a relation R.

Proof: Let R be a relation on a set A with n elements, then

$$R^* = R \cup R^2 \cup R^3 \cup, \ldots \cup R^n$$

where $R^2 = ROR$, $R^3 = R^2OR$, … $R^n = R^{n-1}O\,R$.

For a set with infinite number of elements, $R^* = \sum_{i=1}^{\infty} R^i$

Example: Find the transitive closure R^* of the relation $R = \{(1, 2), (2, 3), (3, 3)\}$ defined on the set $A = \{1, 2, 3\}$.

Solution: The transitive closure R^* of the relation R is given by $R^* = R \cup R^2 \cup R^3$ as A contains three elements.

Now,

$$R^2 = ROR = \{(1, 3), (2, 3), (3, 3)\}$$
$$R^3 = ROR = \{(1, 3), (2, 3), (3, 3)\} \text{ o } \{(1, 3), (2, 3), (3, 3)\}$$
$$= \{(1, 3), (2, 3), (3, 3)\}$$

R^* = Transitive closure of $R = R \cup R^2 \cup R^3 = \{(1, 2), (2, 3), (3, 3), (1, 3)\}$

Alternatively: R contains (1, 2) and (2, 3) and so (1, 3) is to be added for R to be transitive. The new relation is $R_1 = \{(1, 2), (2, 3), (3, 3), (1, 3)\}$. The relation R_1 is transitive, and hence, transitive closure $R^* = \{(1, 2), (2, 3), (3, 3), (1, 3)\}$

2.3 EQUIVALENCE CLASS

Let R be a relation on a non-empty set A and let a be any element of A. Then, the set of all those elements of R that are related to a is called the equivalence class of a related to R and is denoted by $[a]$ or $\bar{a}$. Thus, $[\bar{a}] = \{x{:}x \in A \text{ and } x\,R\,a\}$.

Since R is an equivalence relation, it is symmetric, that is, $x\,R\,a = a\,R\,x$ and so $[a]$ can also be defined by $[a] = \{x{:}x \in A \text{ and } a\,R\,x\}$.

2.3.1 Properties of Equivalence Classes

Let R be an equivalence relation on a set A. Let a and b be two elements in A.

1. $a \in [a]$
2. $b \in [a] \Leftrightarrow [b] = [a]$
3. if a and b are distinct elements, then either $[a] \cap [b] = \phi$ or $[a] = [b]$

Proof:

1. Since R is an equivalence relation defined on the set, it is reflexive and hence $a\,R\,a$ for every $a \in A$. But $[a] = \{x{:}x \in A \text{ and } x\,R\,a\}$, so $a\,R\,a \Rightarrow a \in [a]$.
2. Let $b \in [a] \Rightarrow b\,R\,a$

   Now,

$$x \in [b] \Rightarrow x\,R\,b \text{ and } b\,R\,a$$
$$\Rightarrow x\,R\,a \Rightarrow x \in [a] \qquad \text{[Since } R \text{ is transitive]}$$

Hence,

$$[b] \subseteq [a] \qquad (2.1)$$

Again, $x \in [a] \Rightarrow x\, R\, a$

$\Rightarrow x\, R\, a$ and $a\, R\, b$ [Since R is symmetric]

$\Rightarrow x\, R\, b$

$\Rightarrow x \in [b]$

Hence,

$$[a] \subseteq [b] \qquad (2.2)$$

From (2.1) and (2.2), we have $[a] = [b]$

Conversely: Suppose $[b] = [a]$

Since R is reflexive $b\, R\, b \Rightarrow b \in [b]$

$\Rightarrow b \in [a]$ (since $[b] = [a]$)

3. Let $[a] \cap [b] \neq \varphi$, then we have to prove that $[a] = [b]$

   Let $x \in [a] \cap [b] \Rightarrow x \in [a]$ and $x \in [b]$

   $\Rightarrow (x\, R\, a)$ and $(x\, R\, b)$

   $\Rightarrow (a\, R\, x)$ and $(x\, R\, b)$ (since R is symmetric $x\, R\, a \Rightarrow a\, R\, x$)

   $\Rightarrow a\, R\, b$

   $\Rightarrow [a] = [b]$

   Thus,

   $$[a] \cap [b] \neq \phi \Rightarrow [a] = [b]$$

   But, if $[a] \neq [b] \Rightarrow [a] \cap [b] = \phi$

2.4 PARTITION OF A SET

Let S be a given non-empty set. A partition P of S is a collection of non-empty subsets $\{A_i\}$ of S such that

1. Each element $a \in S$ is one of the subsets A_i and
2. The sets A_i are disjoint, that is, $A_i \cap A_j = \phi$ if $A_i \neq A_j$, then P is called a partition of S. Thus, a partition of a set is the collection of disjoint subsets of S whose union is S.

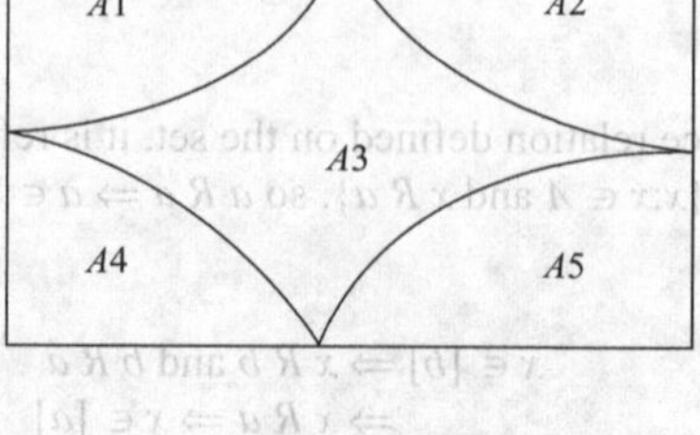

Example 1 Students in a class are partitioned according to their surnames.

Example 2 A country is partitioned into states.

Example 3 A deck of playing cards is partitioned into four suits.

2.5 CONGRUENCE MODULO RELATION

Two integers a and b are said to be congruent modulo a positive integer m, if $(a - b)$ is divisible by m, that is, $a - b = km$, where k is an integer. It is written as $a \equiv b \pmod{m}$.

THEOREM 2.2 The relation 'congruence modulo m' is an equivalence relation.

Proof:

1. The relation is reflexive as $a - a = 0$ is divisible by m, that is, $a \equiv a \pmod{m}$.
2. The relation is symmetric because $a \cong b \pmod{m} \Rightarrow a - b = km$
$$\Rightarrow b - a = (-k)\, m = k'\, m$$
$$\Rightarrow b \equiv a \pmod{m}.$$
3. The relation is transitive $a \equiv b \pmod{m}$ and $b \equiv c \pmod{m}$
$$\Rightarrow b - a = km \text{ and } b - c = k'm$$
$$\Rightarrow (b - c) - (b - a) = (k' - k)\, m$$
$$\Rightarrow a - c = q\, m, \text{ where } q = k' - k$$
$$\Rightarrow a \equiv c \pmod{m}.$$

Note 2.4 If a is any integer and m is a positive integer, then there exist integers q and r with $0 \le r < m$ such that $a = mq + r \Rightarrow a - r = mq \Rightarrow m$ divides $(a - r) \Rightarrow a \equiv r \pmod{m}$ showing that every integer a is congruent modulo to the remainder when a is divided by m. The remainders when divided by m are 0, 1, 2, ...$(m - 1)$.

Example: Let Z be the set of integers. Define $a \cong b \pmod 5$, then we have the set Z divided into five classes that are disjoint and elements belonging to the same class are in (congruent modulo 5) relation.

1. $\{0, 5, 10, \ldots\ldots\ldots\}$ $= \{5k\} = [0] = I_0$
2. $\{1, 6, 11, -4, -9, \ldots\ldots\ldots\}$ $= \{5k + 1\} = [1] = I_1$
3. $\{2, 7, 12, -3, -8, \ldots\ldots\ldots\}$ $= \{5k + 2\} = [2] = I_2$
4. $\{3, 8, 13, -2, -7, \ldots\ldots\ldots\}$ $= \{5k + 2\} = [3] = I_3$
5. $\{4, 9, 14, -1, -6, \ldots\ldots\ldots\}$ $= \{5k + 4\} = [4] = I_4$

where $k \in Z$. These classes are called equivalence classes.

The class I_0 is the set of integers that has remainder zero when divided by 5. Similarly, I_1 is the set of integers that has remainder 1 when divided by 5 and so on.

The set $\{I_{0,}\ I_{1,}\ I_{2,}\ I_{3,}\ I_4\}$ is the quotient set of Z for the equivalence relation 'congruence modulo 5', that is, Z/R.

2.6 PICTORIAL REPRESENTATION OF RELATION

The pictorial representation of a relation is called the graph of the relation. *A* relation can also be represented in a tabular form by a matrix or by an arrow diagram.

Example 1 Let $A = \{a, b, c\}$ and $B = \{1, 2, 3\}$ be two given sets. Define relation $R = \{(a, 2), (a, 3)\ (b, 1), (b, 3)\ (c, 1)\ (c, 2)\}$. The relation R can be represented in the following two ways:

	1	2	3
a	0	1	1
b	1	0	1
c	1	1	0

1. Relation matrix

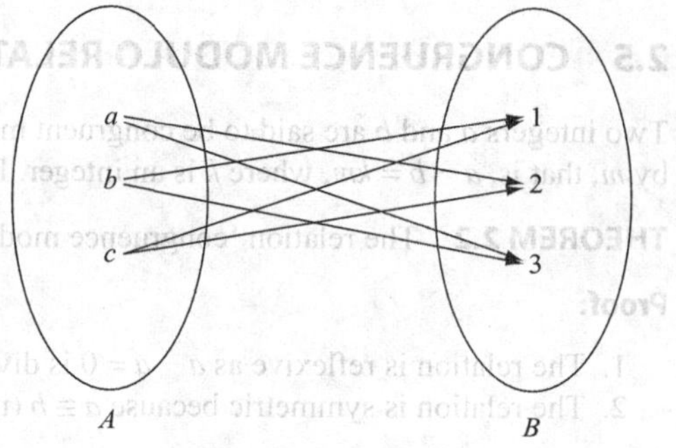

2. Arrow diagram.

The matrix of the relation R denoted by M_R from the set $A = \{a_1, a_2, \ldots\ldots\ldots\ldots\ldots\ldots, a_m\}$ to the set $B = \{b_1, b_2, \ldots, b_n\}$ is defined by $M_R = [m_{ij}]$ of order $m \times n$,

where
$$m_{ij} = 1 \text{ If } a_i \, R \, b_j$$
$$= 0 \text{ If } a_i \not R \, b_j$$

The matrix M_R is called adjacency matrix or relation matrix or Boolean matrix. For the arrow diagram, the elements of A and B are represented in two disjoint discs and then draw an arrow from $a \in A$ to $b \in B$ whenever $a \, R \, b$.

Example 2 Let $A = \{1, 2, 3\}$ and $B = \{x, y, z, u\}$. Define $R = \{(1, x), (1, z), (2, y), (3, z)\}$. Find relation matrix M_R and also draw its arrow diagram.

Solution: The relation matrix is

$$M_R = \begin{array}{c} \\ 1 \\ 2 \\ 3 \end{array}\!\!\begin{array}{c} \begin{array}{cccc} x & y & z & u \end{array} \\ \begin{bmatrix} 1 & 0 & 1 & 0 \\ 0 & 1 & 0 & 0 \\ 0 & 0 & 1 & 0 \end{bmatrix} \end{array}$$

Arrow diagram is shown as follows:

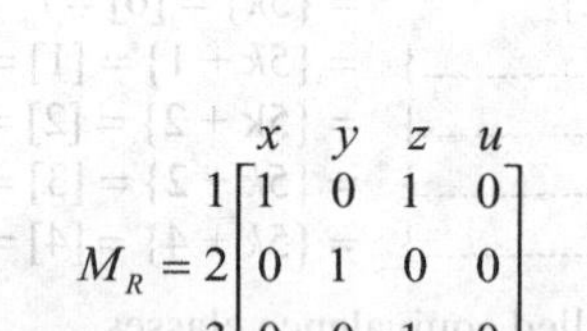

2.7 DIGRAPHS

(Directed Graph): Let R be a relation from a finite set A to itself. This relation can be represented pictorially as follows:

Draw a small circle for each element of A and label it with the corresponding element of A. Now draw an arrow from each element x to each element y whenever x is related to y. Such a pictorial representation is called a directed graph or digraph of R. Thus, digraph of a set is a geometrical representation of relation between its elements. The elements encircled are called vertices and the arrows are called edges.

Example 1 Let $A = \{1, 2, 3, 4\}$ and $R = \{(1, 2), (2, 1), (2, 4), (2, 3), (3, 2), (3, 3), (3, 4), (4, 4)\}$. Draw the digraph of R.

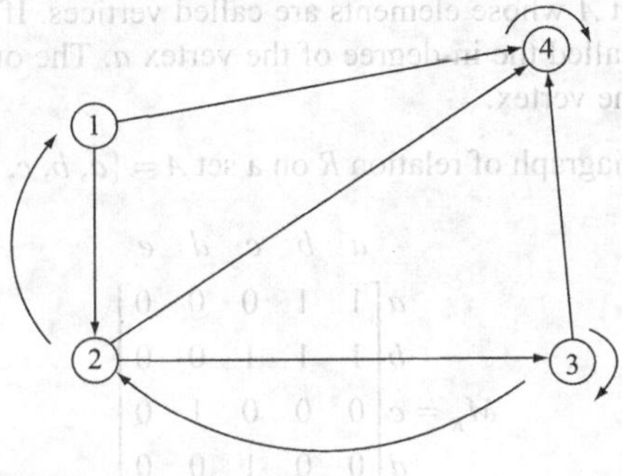

Example 2 Find the relation for the following diagraph.

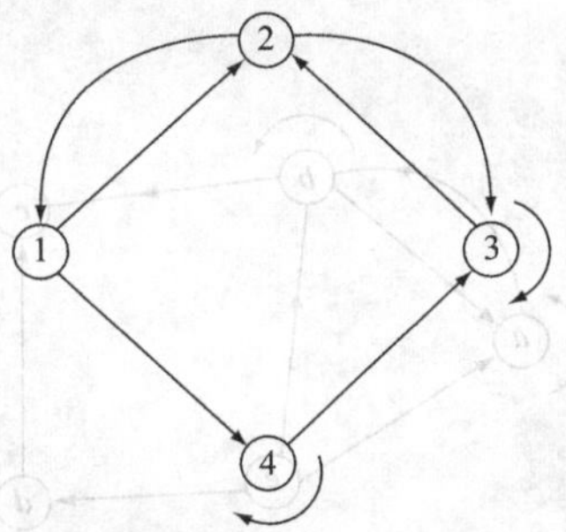

Solution: Since $x\,R\,y$ if and only if there is an edge from x to y.
We have $R = \{(1, 2), (1, 4), (2, 1), (2, 3), (3, 2), (3, 3), (4, 3), (4, 4)\}$

Example 3 Let $A = \{a, b, c\}$ and R be given by the following diagraph. Find R and M_R.

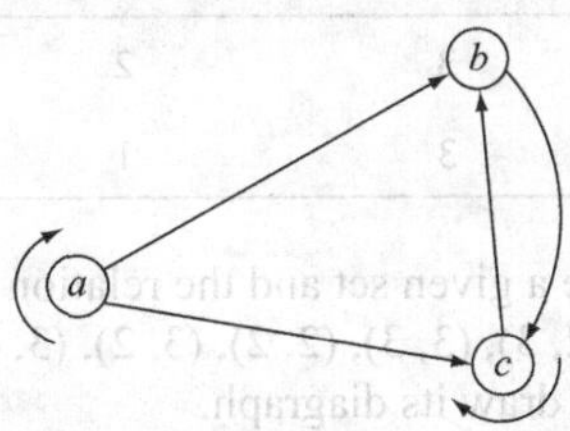

Solution: $R = \{(a, a), (a, b), (a, c), (b, c), (c, b), (c, c)\}$ and

$$M_R = \begin{array}{c} \\ a \\ b \\ c \end{array}\begin{array}{c} \begin{array}{ccc} a & b & c \end{array} \\ \begin{bmatrix} 1 & 1 & 1 \\ 0 & 0 & 1 \\ 0 & 1 & 1 \end{bmatrix} \end{array}$$

2.7.1 In-degree and Out-degree of Vertices

Let R be a relation on a set A whose elements are called vertices. If $a \in A$, then the number of edges terminating at a is called the in-degree of the vertex a. The out-degree of a vertex is the number of edges leaving the vertex.

Example 1 Consider the diagraph of relation R on a set $A = \{a, b, c, d, e\}$ that has the matrix.

$$M_R = \begin{array}{c} \\ a \\ b \\ c \\ d \\ e \end{array}\begin{array}{c} \begin{array}{ccccc} a & b & c & d & e \end{array} \\ \begin{bmatrix} 1 & 1 & 0 & 0 & 0 \\ 1 & 1 & 1 & 0 & 0 \\ 0 & 0 & 0 & 1 & 0 \\ 0 & 0 & 1 & 0 & 0 \\ 1 & 1 & 0 & 1 & 0 \end{bmatrix} \end{array}$$

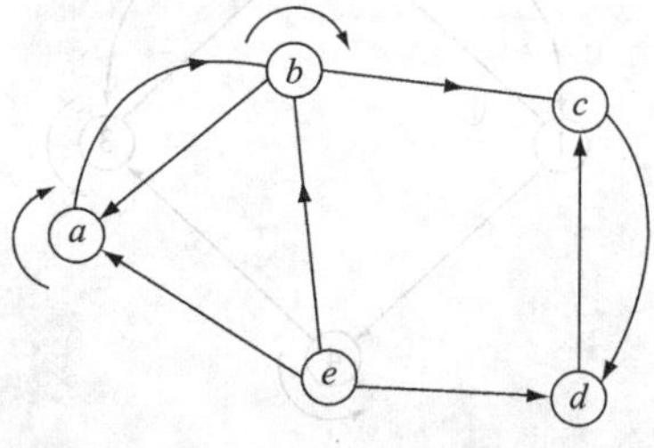

Find the in-degrees and out-degrees of all the vertices.

	a	b	c	d	e
In-degree	3	3	2	2	0
Out-degree	2	3	1	1	3

Example 2 Let $A = \{1, 2, 3, 4\}$ be a given set and the relation R on A be given by,
$R = \{(1, 1), (1, 2), (1, 3), (1, 4), (2, 3), (3, 3), (2, 2), (3, 2), (3, 4), (4, 4), (4, 1), (4, 3)\}$
Find the matrix for relation R and draw its diagraph.

Solution: The matrix for relation R is

$$M_R = \begin{array}{c} \\ 1 \\ 2 \\ 3 \\ 4 \end{array}\begin{array}{c} \begin{array}{cccc} 1 & 2 & 3 & 4 \end{array} \\ \begin{bmatrix} 1 & 1 & 1 & 1 \\ 0 & 1 & 1 & 0 \\ 0 & 1 & 1 & 1 \\ 1 & 0 & 1 & 1 \end{bmatrix} \end{array}$$

and the corresponding diagraph is as follows:

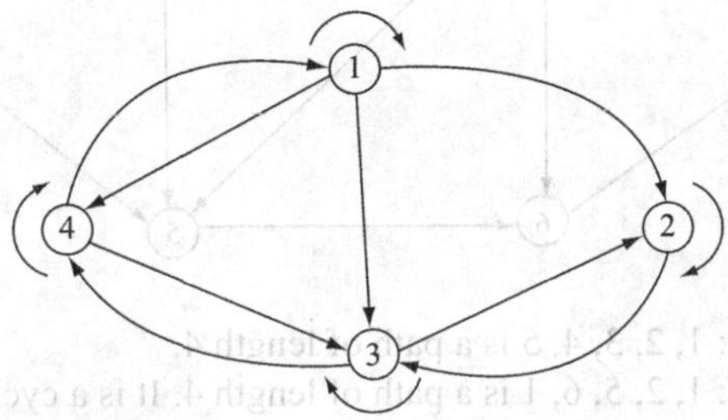

Note 2.5 The matrix representing a relation R can also be used to determine whether the relation is reflexive, symmetric, and transitive. The relation R is reflexive if there is a loop at every vertex of the directed graph so that pairs (a, a) occur in relation. All elements in the main diagonal of the adjacency or relation matrix are unity, that is, $m_{ij} = 1$ for all $i = j$.

Symmetric: The relation is symmetric of $m_{ij} = m_{ji}$ for all i and j in the relation matrix. That is the relation matrix is symmetric.

2.8 POWER OF RELATION

Let A be a given non-empty set and let R be a relation defined on A. Then

1. $R^1 = R$
2. $R^{n+1} = R \circ R^n$ for all positive integers n.

Example 1 Let $R = \{(1, 1), (1, 2), (1, 3), (2, 3), (3, 3), (2, 4)\}$ be a relation on the set $\{1, 2, 3, 4\}$. Find R^2, R^3, and R^4.

Solution:

$$R^2 = \{(1, 3), (2, 3), (1, 1), (1, 2)\}$$
$$R^3 = RoR^2 = \{(1, 1), (1, 2), (1, 3)\}$$
$$R^4 = RoR^3 = \{(1, 1), (1, 2), (1, 3)\}$$

2.9 PATHS IN RELATIONS AND DIGRAPHS

Let R be a relation on a set A. A path of length n in R from a to b is a finite sequence $a, x_1, x_2, \ldots, x_{n-1}, b$ beginning with a and ending with b such that $aRx_1, x_1Rx_2, \ldots, x_{n-1}R_b$.

Obviously, a path of length n involves $n + 1$ elements of A that may not be necessarily distinct. A diagraph of a relation can be used to identify a path. The succession of edges in a diagraph where the indicated direction is followed provides the path, whose length is the number of edges and where all the vertices need not be distinct.

Cycle: A path that begins and ends at the same vertex is called a cycle. Consider the following diagraph.

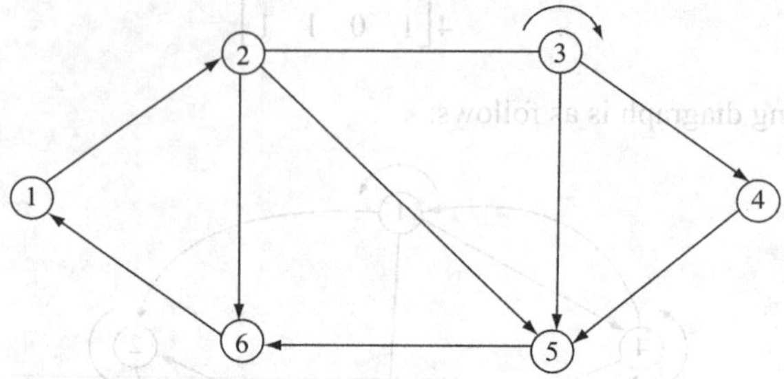

Then

π_1: 1, 2, 3, 4, 5 is a path of length 4.
π_2: 1, 2, 5, 6, 1 is a path of length 4. It is a cycle.
π_3: 3, 3 is a cycle of length 1.
π_4: 2, 5, 6, 1, 2 is a cycle of length 4.
π_5: 4, 5, 6, 1, 2, 3, is a path of length 5.
π_6: 2, 3, 4, 5, 6 is a path of length 4.

2.10 MATRIX REPRESENTATION OF COMPOSITE RELATIONS

Let A, B, and C be any three non-empty sets. Let R be a relation from A to B and S be a relation from B to C that can be represented by matrices M_R and M_S respectively.

Let $$M_R = [m_{ik}]_{pxq} \text{ and } M_S = [n_{kj}]_{q\times r}$$

Then matrix $$M_{RS} = M_R M_S = [m_{ij}]_{p\times r}$$

is obtained by multiplying the matrices in the usual manner. The non-zero entries in the product matrix tell us that elements are related by RoS, where $RoS = \{(a, c): aRb \text{ and } bSc \text{ for some } b \in B\}$.

If R is a relation from a set A to itself then the composition RoR is always defined and is denoted by R^2 The elements of R^2 and higher powers can be obtained by multiplication of matrix M_R by itself.

Example: Consider the relation $R = \{(1, 2), (2, 3), (3, 3)\}$ on the set $A = \{1, 2, 3\}$. Find R^2 using relation matrix.

Solution: If M_R denotes the relation matrix, then we have

$$M_R = \begin{array}{c} \\ 1 \\ 2 \\ 3 \end{array}\begin{array}{c} \begin{array}{ccc} 1 & 2 & 3 \end{array} \\ \begin{bmatrix} 0 & 1 & 0 \\ 0 & 0 & 1 \\ 0 & 0 & 1 \end{bmatrix} \end{array}$$

$$M_R^2 = \begin{bmatrix} 0 & 1 & 0 \\ 0 & 0 & 1 \\ 0 & 0 & 1 \end{bmatrix}\begin{bmatrix} 0 & 1 & 0 \\ 0 & 0 & 1 \\ 0 & 0 & 1 \end{bmatrix} = \begin{bmatrix} 0 & 0 & 1 \\ 0 & 0 & 1 \\ 0 & 0 & 1 \end{bmatrix}$$

$$\therefore R^2 = \{(1, 3), (2, 3), (3, 3)\}$$

$$\text{Similarly, } M_R^3 = RoR^2 = M_R^2 = \begin{bmatrix} 0 & 1 & 0 \\ 0 & 0 & 1 \\ 0 & 0 & 1 \end{bmatrix}\begin{bmatrix} 0 & 0 & 1 \\ 0 & 0 & 1 \\ 0 & 0 & 1 \end{bmatrix} = \begin{bmatrix} 0 & 0 & 1 \\ 0 & 0 & 1 \\ 0 & 0 & 1 \end{bmatrix}$$

$$\therefore R^3 = \{(1, 3), (2, 3), (3, 3)\}$$

2.11 CONNECTIVITY RELATION

If R is a relation defined on a set A then for a fixed positive integer n, the relation R^n on A is defined as follows:

$a\ R^n\ B$ means that there is a path of length n from a to b in R.

We also define a relation R on A by letting $a\ R^\infty\ b$ mean that there is some path in R from a to b. The relation R^∞ is called the connectivity relation for R.

Example 1 Let A be the set of cities and let $a\ R\ b$ if there is a direct flight from a to b on at least one airline.

Then $a\ R^n\ b$ means that one can book a flight from a to b having $(n - 1)$ intermediate steps. $a\ R^\infty\ b$ means that one can go from a to b by plane.

Example 2 Let $A = \{1, 2, 3, 4\}$ and $\{(1, 1), (1, 2), (2, 3), (2, 4), (3, 4)\}$. Find R^2 and R^3.

Solution:

$$R^2 = RoR = \{(1, 1), (1, 2), (1, 3), (1, 4), (2, 4)\}$$
$$R^3 = RoR^2 = \{(1, 1), (1, 4), (1, 2), (1, 3)\}.$$

EXERCISES

1. Find a or b so that the following statements are true.
   (a) $(a, 5) = (2, 5)$ (b) $(2, b) = (2, 4)$ (c) $(a, 5b) = (a, 10)$
   (d) $(2x + 3, 1) = (9, 1)$ (e) (JAVA, C++) = (x, y)
   Find x and y if $(x + 1, 6) = (3, x + 2y)$
2. Let $A = \{a, b, c\}$ and $B = \{1, 2, 3, 4\}$. Define relation R by $R = \{(a, 1), (a, 4), (b, 2), (c, 3), (c, 4)\}$. Find the domain and range of R.
3. Let $R = \{(2, 3), (2, 4), (1, 2), (3, 3)$ be the relation on the set $A = \{1, 2, 3, 4\}$. Find the domain and range of R.

4. Let R and S be the following relations on the set $A = \{1, 2, 3, 4\}$, $R = \{(1, 1), (1, 3), (3, 2), (3, 4), (4, 2)\}$ and $S = \{(2, 1), (3, 3), (3, 4), (4, 1)\}$. Find the following composition relations:
   (a) RoS (b) SoR (c) RoR (d) (SoS)

5. Let R be the relation defined on the set of natural numbers N by the equation $3x + y = 12$, that is, $R = \{(x, y): 3x + y = 12\}$.
   (a) Write R as a set of ordered pairs. Find its domain and range.
   (b) Find R^{-1}
   (c) Composition RoR.

6. If $A = \{1, 2, 3, 4\}$ and $B = \{1, 4, 5, 6, 16\}$ and aRb if and if $b = a^2$, find R. Find its
   (a) domain (b) range (c) matrix of relation.

7. If $A = (1, 2, 3, 4\}$ and $B = (1, 4, 9, 12\}$ and aRb if a/b (i.e., if a divides b). Find R, its domain, range and the matrix of relation.

8. Let $A = \{a, b, c, d\}$. Find whether the following relation R defined on A is reflexive, symmetric, transitive or irreflexive.
   (a) $R = \{(a, a), (b, b), (c, c)\}$
   (b) $R = \{(a, b), (a, c), (a, d), (b, c), (b, d), (c, d)\}$
   (c) $R = \{(a, b), (b, a), (a, c), (c, a), (d, d)\}$

9. Define the relation R over the set of integers Z by xRy, if $x + y$ is even. Show that the relation R is reflexive, symmetric, and transitive.

10. Find the relation determined by the following diagraphs.

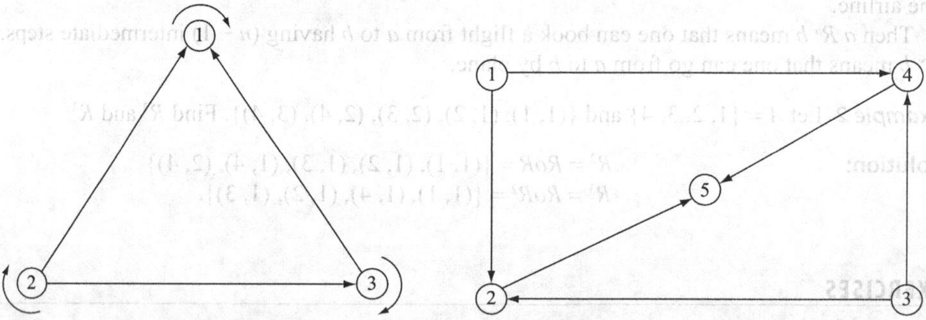

(a) (b)

11. Let $A = \{1, 2, 3\}$ be a set and the relation R on A be defined by,
    $R = \{(1, 1), (2, 2), (1, 2), (2, 3), (3, 3)\}$. Is R
    (a) reflexive (b) symmetric (c) transitive (d) an equivalence relation.
    Draw its diagraph and also find the relation matrix M_R. Compute $R^2 = RoR$.

12. Show that if a relation on a set A is transitive and irreflexive, then it is asymmetric.

13. If a relation R on a set A is symmetric, prove that the relation R^2 also symmetric.

14. Let $A = \{a, b, c\}$ be a set. Determine whether the relation R whose matrix $M_R = \begin{bmatrix} 1 & 1 & 0 \\ 1 & 1 & 0 \\ 0 & 0 & 1 \end{bmatrix}$ is given, is an equivalence relation.

15. Determine whether the relation R whose diagraph is given is an equivalence relation.

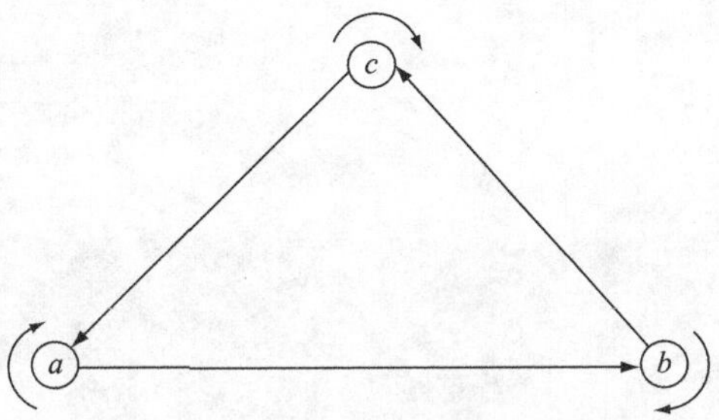

16. Let $P = \{1, 2, 3\}$ and let $A = P \times P$. Define relation R on A by $(a, b)\ R\ (a', b')$ if and only if $a + b = a' + b'$. Show that R is an equivalence relation. Find the quotient set A/R.

17. If $\{(a, b), (c, d)\}$ is a partition of the set $A = \{a, b, c, d\}$, determine the corresponding equivalence relation.

18. If $\{(1, 2, 4), (3, 5)\}$ is a partition of the set $A = \{1, 2, 3, 4, 5\}$, determine the corresponding equivalence relation.

19. Let R and S be the following relations on $A = \{1, 2, 3, 4\}$:

    $R = \{(1, 1), (1, 2), (1, 4), (2, 3), (2, 4), (4, 4)\}$, $S = \{(1, 2), (1, 3), (2, 1), (4, 4)\}$.

    Find (a) $R \cap S$ (b) $R \cap S$ (c) $\overline{R}$ (d) RoS (e) $S^2 = SoS$

14. Let $A = \{a, b, c\}$ be a set. Determine whether the relation R whose matrix $M_R = \begin{bmatrix} 1 & 1 & 0 \\ 1 & 1 & 0 \\ 0 & 0 & 1 \end{bmatrix}$ is given, is an equivalence relation.

15. Determine whether the relation R whose diagraph is given is an equivalence relation.

16. Let $P = \{1, 2, 3\}$ and let $A = P \times P$. Define relation R on A by $(a, b)\ R\ (a', b')$ if and only if $a + b = a' + b'$. Show that R is an equivalence relation. Find the quotient set A/R.

17. If $\{(a, b), (c, d)\}$ is a partition of the set $A = \{a, b, c, d\}$, determine the corresponding equivalence relation.

18. If $\{(1, 2, 4), (3, 5)\}$ is a partition of the set $A = \{1, 2, 3, 4, 5\}$, determine the corresponding equivalence relation.

19. Let R and S be the following relations on $A = \{1, 2, 3, 4\}$:

    $R = \{(1, 1), (1, 2), (1, 4), (2, 3), (2, 4), (4, 4)\}$, $S = \{(1, 2), (1, 3), (2, 1), (4, 4)\}$.

    Find (a) $R \cap S$ (b) $R \cap S$ (c) $\bar{R}$ (d) RoS (e) $S^2 = SoS$

3

Functions

LEARNING OBJECTIVES

After reading this chapter, you will be conversant with:

- Introduction
- Definition
- Domain and Range of a Function
- Difference Between Relation and Function
- Different Types of Functions (or Mappings) Constant Function
- Composition of Functions
- Functions for Computer Science
- Some Special Functions Used in Discrete Mathematics
- Some Important Theorems and Problems
- Ackermann's Function
- Fuzzy Sets
- Time Complexity of Algorithm
- Connectivity Relation

3.1 INTRODUCTION

A concept of fundamental importance next to the notation of 'set' is that of a function. The knowledge of function plays a key role in several branches of mathematics and computer science. The idea of function is also used in our everyday life. When we compute electricity or water bill or airfare or railway fare for various destinations, we use the concept of a function. When a number x is entered into a calculator and a function key f is pressed, we get another number $f(x)$.

Thus, a function can be defined as a rule say f that associates each element of a set A with a unique element of another set B. This rule is called a function or mapping or transformation from A to B and is written as $f: A \rightarrow B$.

3.2 DEFINITION

Let A and B be two given non-empty sets not necessarily distinct. A function from a set A to a set B is a binary relation from A to B such that for every $x \in A$, there corresponds a unique element $y \in B$ such that $y = f(x)$. The element $y \in B$ is called the f-image of $x \in A$ and x is called the pre-image of y under f.

Thus, the function f is a subset of $A \times B$ with the property that there is just one pair (x, y) in f having first coordinate x. Hence,

$$f = \{(x, y): y = f(x), x \in A, y \in B\}$$

An important point in the definition of the function is that every element of A must be the first coordinate of exactly one ordered pair, that is, *an element cannot have more than one image.*

3.3 DOMAIN AND RANGE OF A FUNCTION

Let $f: A \to B$ be a function defined by $y = f(x)$, $x \in A$ and $y \in B$. The set A is called the domain and B is called the co-domain of the function f. $f(x)$ is called the value of the function at x. The range of f is the set of all those elements of B that are image of some $x \in A$. It is denoted by $f(A)$. Thus, $f(A) = \{y: y = f(x), y \in B \text{ for all } x \in A\} = \{f(x): x \in A\}$

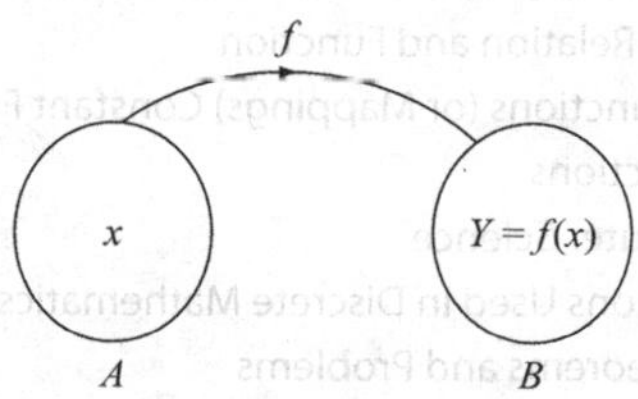

Example 1 Let $f: A \to B$ be a function defined by $f(1) = a, f(2) = b, f(3) = c, f(4) = d$ where $A = \{1, 2, 3, 4,\}$ and $B = (a, b, c, d\}$.

Solution: The function can also be denoted by $f = \{(1, a), (2, b), (3, c), (4, d)\}$. Since for each element in A, there is a corresponding unique element in B, f is a function.
The range of $f = \{a, b, c, d\}$, Domain = $\{1, 2, 3, 4\}$

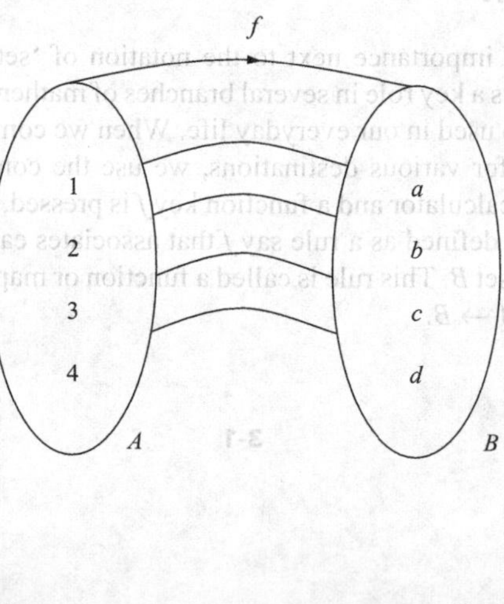

Example 2 Let f: $N \to N$ be a function defined *by* $f(x) = x + 2$, for all $x \in N$. Prove that f is a function. Find domain, co-domain and range of f.

Solution: f is a function since every natural number is assigned to a unique natural number which is obtained by adding 2 to the number.
For example, $f(1) = 1 + 2 = 3, f(2) = 2 + 2 = 4, f(3) = 3 + 2 = 5$, and so on.
Domain = N, Co-domain = N, Range = $N - \{1, 2\}$.

Example 3 Let f: $A \to A$ be defined by $f = \{(1, 3), (2, 3), (3, 1), (4, 4)\}$, where $A = \{1, 2, 3, 4\}$. Find domain, co-domain, and range of the function f.

Solution: Since each element of set A is assigned a unique element in co-domain set, f is a function. Domain = $\{1, 2, 3, 4\}$, Co-domain = $\{1, 2, 3, 4\}$, Range = $\{1, 3, 4\}$.

Example 4 Let f: $R \to R$ be defined by $f(x) = |x|$. Prove that f is a function and find the range of f.

Solution: Every real number is assigned to its absolute value which is a real number and so is a function. Range of f is the set of non-negative real numbers, which is a subset of co-domain set R of real numbers.

Example 5 Let A and B be two given sets where $A = \{1, 2, 3, 4\}$ and $B = \{x, y, z,\}$.
Define $f_1 = \{(1, x), (1, y), (3, z), (4, z)\}$and $f_2 = \{(1, x), (2, x), (3, z),(4, y)\}$. Find whether f_1 and f_2 are functions.

Solution:

(1) f_1 is not a function as the element 1 corresponds to two elements x and y, that is, element 1 has two images.
(2) f_2 is a function since each element in A has a unique image in B.

3.4 DIFFERENCE BETWEEN RELATION AND FUNCTION

1. A relation R from set A to B is a subset of $A \times B$ and so the domain of R may be a subset of A. In the case of function f from A to B, domain of f must be equal to A.
2. In a relation, one element of A may be related to more than one element of B but in a function, each element of A is related to one and only one element of B.
3. In a relation, some elements of A may not be related to any element of B and so every function is a relation but the converse is not true.

3.5 DIFFERENT TYPES OF FUNCTIONS (OR MAPPINGS) CONSTANT FUNCTION

A function f: $A \to B$ is called a constant function, if $f(x) = k$ for every $x \in A$ and k is a constant. In this case, range consists of only one element, that is, k.

Example: Let f: $R \to R$ be defined by $f(x) = 1$. The function f is a constant function since 1 is assigned to every element in R.

3.5.1 Identity Function

Identity function is a function on a set to itself that is denoted by I and is defined by,

$$I: A \to A, \text{ where } I(x) = x \text{ for every } x \in A$$

that is, every element is mapped to itself. The identity function on a set A is also denoted by I_A.

Example: Let $A = \{a, b, c, d\}$, then $I_A: A \to A$ is defined by $I(a) = a$, $I(b) = b$, $I(c) = c$, $I(d) = d$.

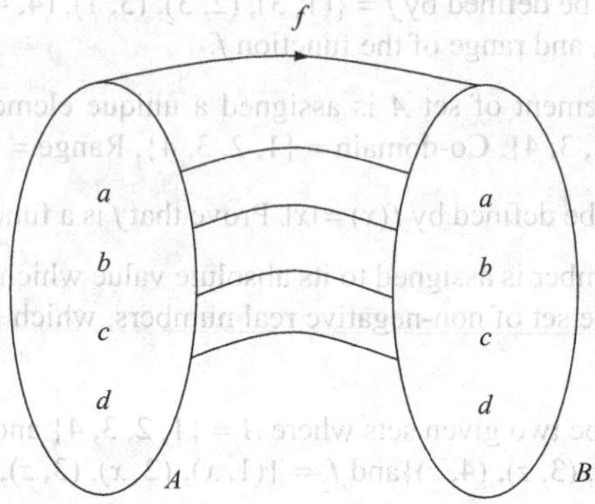

3.5.2 One-One and Many-One Function

A function $f: A \to B$ is said to be one-one (or injective), if different elements in A have different images in B. That is, if $f(x_1) = f(x_2) \Rightarrow x_1 = x_2, x_1, x_2 \in A$ or $f(x_1) \neq f(x_2) \Rightarrow x_1 \neq x_2, x_1, x_2 \in A$.

If two or more elements in A have the same f-image in B, then f is called many-one function. That is, if $f(x_1) = f(x_2) = f(x_3) = \ldots, x_1 \neq x_2 \neq x_3, \ldots$, for all $x_1, x_2, x_3, \ldots, \in A$, then f is many-one function.

In one-one mapping, an element in B has only one pre-image in A whereas in many-one mapping, some elements in B have more than one pre-image in A.

Example 1 Let $f: N \to N$ be defined by $f(n) = n + 1 \ \forall \ n \in N$
The function f is one-one as $f(n_1) = f(n_2) \Rightarrow n_1 + 1 = n_2 + 1 \Rightarrow n_1 = n_2$, n_1 and $n_2 \in N$.

Example 2 Let $f: R \to R+$ be defined by $f(x) = |x| \ \forall \ x \in R$. The function f is many-one as $f(x_1) = f(-x_1) = x_1, x_1 \neq -x_1$, that is, two distinct elements in R have the same image.

Example 3 Let $f: N \to B$, where N denotes the set of natural numbers and $B = \{0, 1\}$ be defined by,

$$\text{and} \quad \begin{aligned} f(2n) &= 0 \\ f(2n+1) &= 1 \end{aligned} \quad \text{for all } n \in N.$$

The function f is many-one as all even natural numbers are mapped to the same element zero and all odd natural numbers are mapped to one.

3.5.3 INTO and ONTO (Surjective) Functions (or Mappings)

A function $f: A \to B$ is said to be ONTO (or surjective), if each element $y \in B$ is the image of some element $x \in A$. In this case, range of $f = B$, that is, $f(A) = B$. In other words, the function is ONTO if its range is the same as its co-domain. The function is INTO if there is at least one element y in B which is not the f-image of any $x \in A$. In this case, $f(A)$ is a proper subset of B, that is, $f(A) \subset B$.

One-One INTO Function:

1. The function $f: A \to B$ is one-one INTO as different elements have different image. The function is into as the element $d \in B$ is not the image of any element of A, that is, $f(A) \subset B$.
2. The function $f: N \to Z$ is defined by $f(n) = 2n$, $n \in N$ is one-one INTO.

The function f is one-one as $f(n_1) = f(n_2) \Rightarrow 2n_1 = 2n_2 \Rightarrow n_1 = n_2$.

The function is INTO as all negative and all positive integers including zero do not correspond to any element $n \in N$, that is, $f(N) \subset Z$.

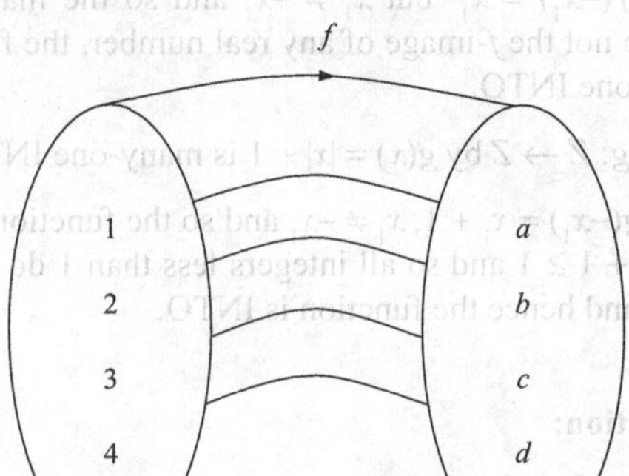

One-One ONTO (or Bijective) Function:

Example 1 The function $f: A \to B$ is one-one ONTO as different elements in A have different images in B and every element in B is the image of some element in A. Clearly $f(A) = B$.

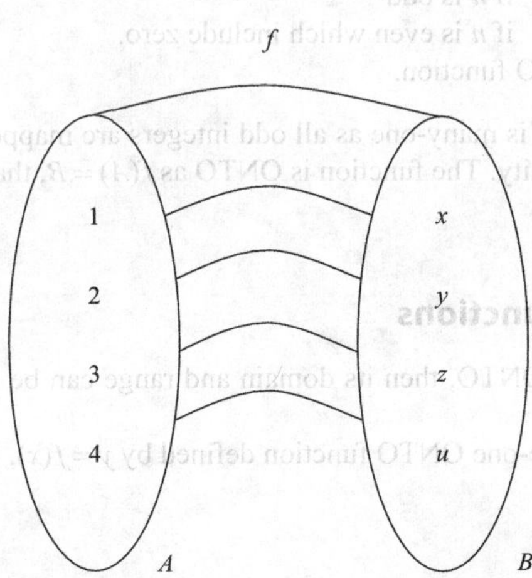

Example 2 The function $f: A \rightarrow B$ defined by $f(x) = \sin x$ is one-one ONTO, where $A = \{x: x \in R \text{ and } -\pi/2 \leq x \leq \pi/2\}$ and $B = \{y: y \in R \text{ and } -1 \leq y \leq 1\}$.

Solution: We know that two different real numbers lying in the interval $[-\pi/2, \pi/2]$ will have different value of sines and so the function is one-one, that is, $x_1 \neq x_2 \Rightarrow \sin x_1 \neq \sin x_2$.

Further for any element y lying in the interval $[-1, 1]$, there is a unique number x in $[-\pi/2, \pi/2]$ such that $y = \sin x$ and hence the function is ONTO.

Many-One INTO Function:

Example 1 The function $f: R \rightarrow R$ defined by $f(x) = x^2$ is many-one INTO.

Solution: Since $f(x_1) = f(-x_1) = x_1^2$ but $x_1 \neq -x_1$ and so the mapping is many-one. Further negative real numbers are not the f-image of any real number, the function is INTO. Hence, the given function f is many-one INTO.

Example 2 The function $g: Z \rightarrow Z$ by $g(x) = |x| + 1$ is many-one INTO function.

Solution: Since $g(x_1) = g(-x_1) = x_1 + 1$, $x_1 \neq -x_1$ and so the function g is many-one. Further, for any $x \in Z$, $|x| \geq 0$ and $|x| + 1 \geq 1$ and so all integers less than 1 do not have the pre-image. For example, o $\notin$ range of g and hence the function is INTO.

Many-One ONTO Function:

Example 1 Let $A = \{a, b, c\}$ and $B = \{u, v\}$ and $f = \{(a, u), (b, u), (c, v)\}$, than $f: A \rightarrow B$ is a many-one ONTO function.

Solution: Since every element of A has f-image in B, f is a function. Further $f(a) = f(b) = u$, $a \neq b$, that is, two different elements $a, b \in A$ have the same f-image $u \in B$ and hence the function is many-one. As range of $f = f(A) = B$, the function is ONTO.

Example 2 Let $f: Z \rightarrow B$, where Z denotes the set of integers and $B = \{0, 1\}$ be defined by,

$$f(n) = 0, \quad \text{if } n \text{ is odd}$$
$$= 1, \quad \text{if } n \text{ is even which include zero.}$$

Then f is many-one ONTO function.

Solution: The function f is many-one as all odd integers are mapped to element zero and even integers are mapped to unity. The function is ONTO as $f(A) = B$, that is, range of f is equal to its co-domain.

3.5.4 Invertible Functions

If a function is one-one ONTO, then its domain and range can be interchanged to form a new function.

Let $f: A \rightarrow B$ be an one-one ONTO function defined by $y = f(x)$, $\forall\, x \in A$ and $y \in B$.

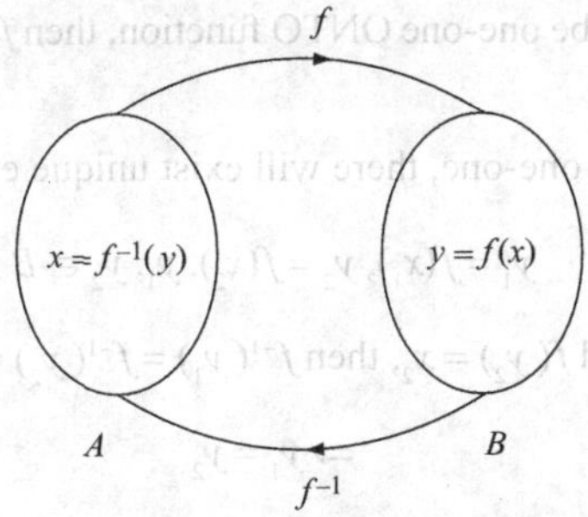

Since f is one-one ONTO, each element $y \in B$ is the f-image of exactly one element $x \in A$, such that $y = f(x)$. We can denote x by $f^{-1}(y)$. Thus, we can define a new correspondence f^{-1} that associates each element $y \in B$ with a unique element $x \in A$. Hence, $f^{-1}: B \to A$ defines a function called inverse of f.

Example 1 Let $A = \{1, 2, 3\}$ and $B = \{u, v, w\}$ be two sets and the function $f: A \to B$ defined by $f(1) = u, f(2) = v$, and $f(3) = w$. Represent f and f^{-1} diagrammatically.

Solution: $f^{-1}: B \to A$ will be defined by $f^{-1}(u) = 1, f^{-1}(v) = 2, f^{-1}(w) = 3$,

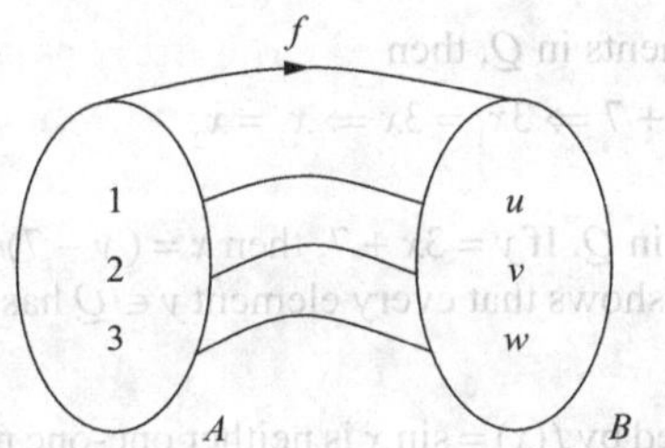

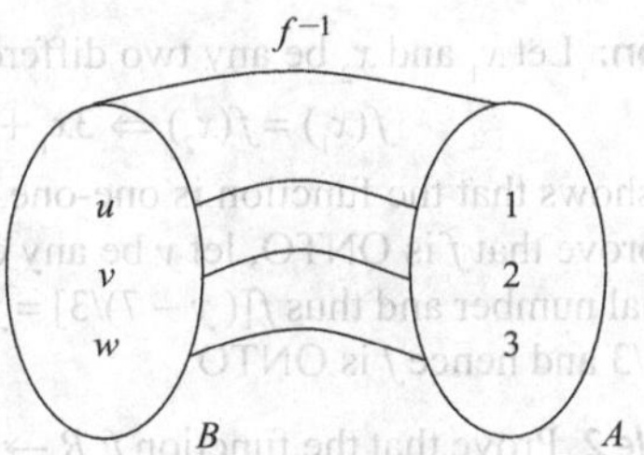

Example 2 Let $A = \{1, 2, 3, 4\}$ and $B = \{a, b, c, d\}$ be two sets and the function $f: A \to B$ defined by $f = \{(1, a), (2, b), (3, c), (4, d)\}$. Find f^{-1}. Represent f and f^{-1} diagrammatically.

Solution: The inverse function f^{-1} is given by $f^{-1} = \{(a, 1), (b, 2), (c, 3), (d, 4)\}$

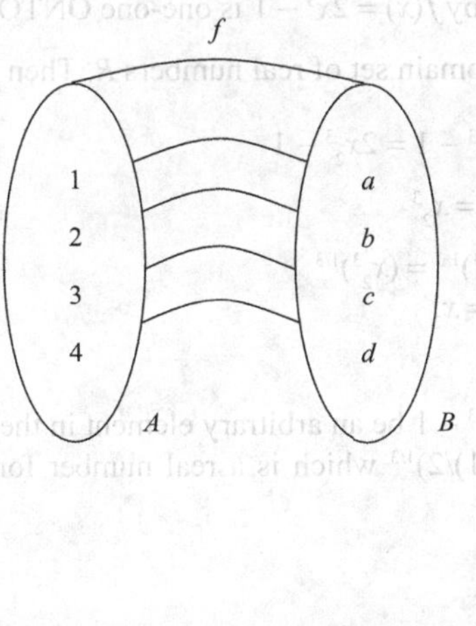

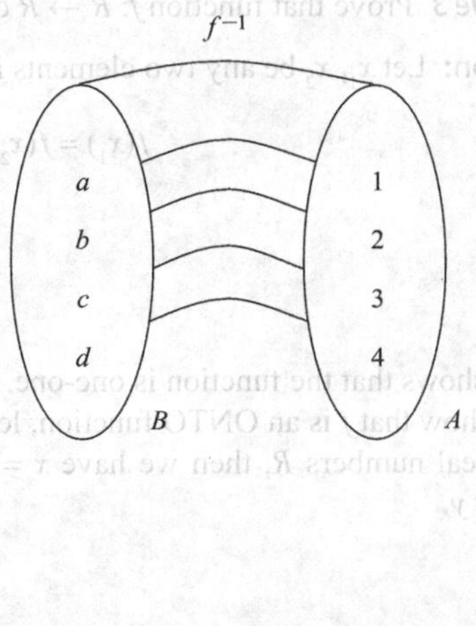

THEOREM 3.1 Let $f: A \to B$ be one-one ONTO function, then $f^{-1}: B \to A$ is also one-one ONTO function.

Proof: Since the function f is one-one, there will exist unique elements $x_1, x_2 \in A$ such that

$$y_1 = f(x_1), y_2 = f(x_2), y_1, y_2 \in B$$

Now, suppose $f^{-1}(y_1) = x_1$ and $f(y_2) = x_2$, then $f^{-1}(y_1) = f^{-1}(y_2) \Rightarrow x_1 = x_2 \Rightarrow f(x_1) = f(x_2)$

$$\Rightarrow y_1 = y_2$$

which shows that the function f^{-1} is one-one.

To prove that f^{-1} is ONTO, let x be an arbitrary element in A, then for every $x \in A$, there exists a unique element $y \in B$ such that $f(x) = y$ because f is one-one. Hence, for each $x \in A$, we have $y \in B$ such that $x = f^{-1}(y)$ which shows that f^{-1} is ONTO.

3.5.5 Some Illustrative Examples

Example 1 Prove that the function $f: Q \to Q$ defined by $f(x) = 3x + 7$ is one-one ONTO where Q is the set of rational numbers. Also find f^{-1}.

Solution: Let x_1 and x_2 be any two different elements in Q, then

$$f(x_1) = f(x_2) \Rightarrow 3x_1 + 7 = 3x_2 + 7 \Rightarrow 3x_1 = 3x \Rightarrow x_1 = x_2$$

which shows that the function is one-one.

To prove that f is ONTO, let y be any element in Q. If $y = 3x + 7$, then $x = (y - 7)/3$ which is a rational number and thus $f[(y - 7)/3] = y$ which shows that every element $y \in Q$ has pre-image $(y - 7)/3$ and hence f is ONTO.

Example 2 Prove that the function $f: R \to R$ defined by $f(x) = \sin x$ is neither one-one nor ONTO.

Solution: Since $\sin(\pi/6) = \sin(2\pi/3) = 1/2$, that is, $f(x_1) = f(x_2)$, $x_1 \neq x_2$ and so the function f is not one-one. The range of values of f is the set of real numbers in the interval $[-1, 1]$, which is a subset of co-domain R. Hence, f is not ONTO.

Example 3 Prove that function $f: R \to R$ defined by $f(x) = 2x^3 - 1$ is one-one ONTO.

Solution: Let x_1, x_2 be any two elements in the domain set of real numbers R. Then

$$\begin{aligned} f(x_1) = f(x_2) &\Rightarrow 2x_1^3 - 1 = 2x_2^3 - 1 \\ &\Rightarrow x_1^3 = x_2^3 \\ &\Rightarrow (x_1^3)^{1/3} = (x_2^3)^{1/3} \\ &\Rightarrow x_1 = x_2 \end{aligned}$$

which shows that the function is one-one.

To show that f is an ONTO function, let $y = 2x^3 - 1$ be an arbitrary element in the co-domain set of real numbers R, then we have $x = ((y + 1)/2)^{1/3}$ which is a real number for every real number y.

Now, $f\left(\left(\frac{y+1}{2}\right)^{\frac{1}{3}}\right) = 2\left(\frac{y+1}{2}\right) - 1 = y + 1 - 1 = y$, showing that each element $y \in R$ (range set) has the pre-image in the domain set R, and hence, the function is ONTO.

Example 4 Prove that the function $f: C \to R$ defined by $f(z) = |z|$ is neither one-one nor ONTO.

Solution: Let $z = x + iy$ be any complex number where $x, y \in R$. Then $|z| = |x + iy| = \sqrt{x^2 + y^2}$, which is a non-negative real number. Thus, $|z| \geq 0$.

The function is not one-one as

$$f(1 + i) = f(-1 + i) = f(1 - i) = f(-1 - i) = \sqrt{2}$$

which shows that many elements in C are mapped to a single element in R.

The function is not ONTO as all the negative real numbers of co-domain do not have any pre-image in the domain set C of complex numbers.

3.5.6 Equality of Functions

The functions f and g are said to be equal if and only if they have the same domain and co-domain and $f(x) = g(x)$ for every x in the domain.

3.6 COMPOSITION OF FUNCTIONS

Let $f: A \to B$ and $g: B \to C$ be two functions, then the product or composition of functions f and g done by gof is a function from A to C denoted by gof of gf.

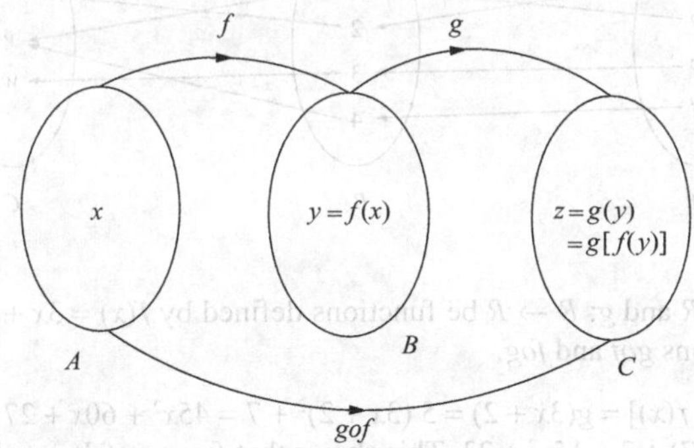

Thus, $gof: A \to C$, where $(gof)\,x = g[f(x)]$, for every $x \in A$.

Showing that each element $x \in A$ is mapped to a unique element $y = f(x) \in B$ which is mapped to a unique element $z = g(y) \in C$, that is, $x \in A$ is mapped to $g[f(x)] \in C$.

Example 1 Let $A = \{1, 2, 3\}$, $B = \{a, b, c\}$, and $C = \{u, v, w\}$ be three given sets. Consider $f: A \to B$ and $g: B \to C$ defined by $f = \{(1, a), (2, b), (3, c)\}$ and $g = \{(a, w)\ (b, v), (c, u)\}$. Then the composition function is given by $gof = \{(1, w), (2, v), (3, u)\}$

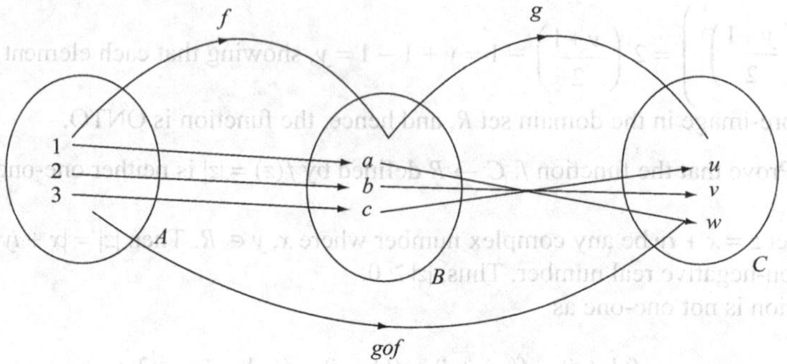

Example 2 Let $A = \{a, b, c, d\}$, $B = \{1, 2, 3, 4\}$, and $C = \{u, v, w\}$. Define functions $f: A \to B$ and $g: B \to C$ by $f = \{(a, 2), (b,1), (c, 3), (d, 4)\}$ and $g = \{(1, v), (2, u), (3, w), (4, v)\}$. Find the composite function *gof*.

Solution: The composition function *gof* is from the set A to the set C is defined as follows:

$$(gof)\,a = g[f(a)] = g(2) = u,\ (gof)\,b = g[f(b)] = g(1) = v$$
$$(gof)\,c = g[f(c)] = g(3) = w,\ (gof)\,d = g[f(d)] = g(4) = v$$

Hence, the function $gof = \{(a, u), (b, v), (c, w), (d, v)\}$

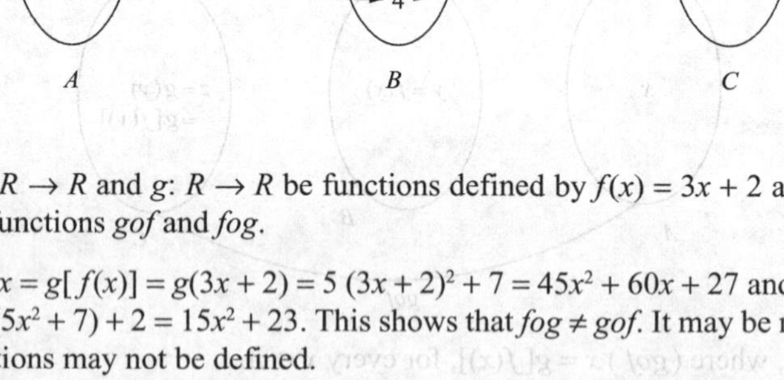

Example 3 If $f: R \to R$ and $g: R \to R$ be functions defined by $f(x) = 3x + 2$ and $g(x) = 5x^2 + 7$, find composite functions *gof* and *fog*.

Solution: $(gof)\,x = g[f(x)] = g(3x + 2) = 5\,(3x + 2)^2 + 7 = 45x^2 + 60x + 27$ and $(fog)\,x = f[g(x)] = f(5x^2 + 7) = 3\,(5x^2 + 7) + 2 = 15x^2 + 23$. This shows that $fog \neq gof$. It may be noted that in some cases, both functions may not be defined.

Example 4 Consider the function $f = R \to R$ and $g: R - \{1\} \to R$ defined by $f(x) = 3x - 5$ and $g(x) = \dfrac{x}{x-1}$. Now $(fog)\,x = f[g(x)] = f\left(\dfrac{x}{x-1}\right) = 3\left(\dfrac{x}{x-1}\right) - 5 = \dfrac{5-2x}{x-1}$

Here *gof* is not defined because range $(f) = R \not\subset$ Domain $(g) = R - \{1\}$.

THEOREM 3.2 Composition of functions is an associative operation.

Proof: Let $f: A \to B$, $g: B \to C$, and $h: C \to D$ be three functions, then $(fog)\, oh = fo\,(goh)$.

The domains of $(fog)\, oh$ and $fo\,(goh)$ are the same, that is, set A and the co-domain is the set D. Further,

$$[(fog)\, oh]\, x = (fog)\, h\,(x) = f\{g[h\,(x)]\} \qquad \text{(i)}$$

and

$$fo\,(goh)\, x = fo\,\{g[h\,(x)]\} = f\{g[h\,(x)]\} \qquad \text{(ii)}$$

From (i) and (ii), we have $(fog)\, h = fo\,(goh)$.

THEOREM 3.3 If $f: A \to B$ has an inverse $f^{-1}: B \to A$, then $f^{-1}of = I_A$ and $fof^{-1} = I_B$.

Proof: Since $f: A \to B$ is defined, we shall have $y = f(x)$ for some $x \in A$ and $y \in B$. As f^{-1} is defined, we have $x = f^{-1}(y)$ for some $y \in B$ and $x \in A$.

$$(f^{-1}of)\, x = f^{-1}[f(x)] = f^{-1}(y) = x$$

$$(f^{-1}of)\, x = I_A\, x. \text{ Hence, } f^{-1}of = I_A.$$

Similarly $(fof^{-1})\, y = fo\,[f^{-1}(y)] = f(x) = y = I_B\,(y)$. Hence, $fof^{-1} = I_B$

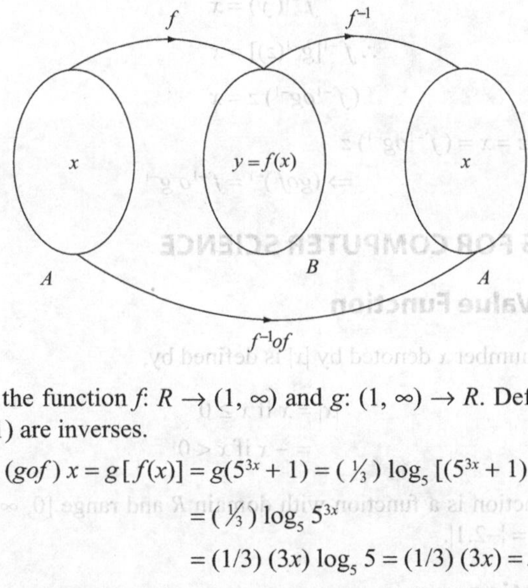

Example: Show that the function $f: R \to (1, \infty)$ and $g: (1, \infty) \to R$. Defined by $f(x) = 5^{3x} + 1$, $g(x) = (\tfrac{1}{3}) \log_5 (x - 1)$ are inverses.

Solution: For $x \in R$, $(gof)\, x = g[f(x)] = g(5^{3x} + 1) = (\tfrac{1}{3}) \log_5 [(5^{3x} + 1) - 1]$

$$= (\tfrac{1}{3}) \log_5 5^{3x}$$

$$= (1/3)\,(3x) \log_5 5 = (1/3)\,(3x) = x$$

Further for any $x \in (1, \infty)$, $(fog)\, x = f[g(x)] = f[\tfrac{1}{3} \log_5 (x - 1)]$

$$= 5^{3[\frac{1}{3}\log_5 (x-1)]} + 1$$

$$= 5^{\log_5 (x-1)} + 1 = (x - 1) + 1 = x$$

Hence, f and g are inverses.

THEOREM 3.4 Let f: $A \to B$ and g: $B \to C$ be two invertible functions, then the composite function gof: $A \to C$ is also invertible and $(gof)^{-1} = f^{-1} o\, g^{-1}$.

Proof: Since f and g are invertible functions, both are one-one and ONTO. Hence, we have $f(x_1) = f(x_2) \Rightarrow x_1 = x_2$ for all $x_1, x_2, \in A$ and $g(y_1) = g(y_2) \Rightarrow y_1 = y_2$ for all $y_1, y_2 \in B$.

$$\therefore (gof)(x_1) = (gof)\, x_2 \Rightarrow g[f(x_1)] = g[f(x_2)] \Rightarrow f(x_1) = f(x_2) \Rightarrow x_1 = x_2$$

which shows that gof is one-one function.

Further, since f and g are ONTO functions, for each $z \in C$, there will exist some $y \in B$ such that $g(y) = z$ and $f(x) = y$, $x \in A$. Hence every $z \in C$ has pre-image $x \in A$ under gof and therefore gof is an ONTO function.

We have proved that gof is one-one and ONTO and hence gof is invertible.

This implies that gof: $A \to C$ and so $(gof)^{-1}$: $C \to A$

Now

$$(gof)^{-1} z = x \Rightarrow (gof)\, x = z$$
$$\Rightarrow g[f(x)] = z$$
$$\Rightarrow g(y) = z \text{ or } g^{-1}(z) = y$$

But

$$f^{-1}(y) = x$$
$$\therefore f^{-1}[g^{-1}(z)] = x$$

or

$$(f^{-1} o g^{-1})\, z = x$$

Thus, we have $(gof)^{-1} z = x = (f^{-1} o g^{-1})\, z$

$$\Rightarrow (gof)^{-1} = f^{-1} o\, g^{-1}$$

3.7 FUNCTIONS FOR COMPUTER SCIENCE

3.7.1 Absolute Value Function

The absolute value of number x denoted by $|x|$ is defined by,

$$|x| = x \text{ if } x \geq 0$$
$$= -x \text{ if } x < 0$$

The absolute value function is a function with domain R and range $[0, \infty)$. It is not one to one because $|5| = |-5|$, $|2.1| = |-2.1|$.

3.7.2 Floor Function

The floor function $f(x)$ defined for a real number x and denoted by $\lfloor x \rfloor$ is the greatest integer less than or equal to x. Hence, if x is a real number and n is an integer then $\lfloor x \rfloor = n$ when $n \leq x < n + 1$

______ n ______ x ______ $n + 1$ ______

Floor of $x = \lfloor x \rfloor$

The function $\lfloor x \rfloor$ is read as the floor of x.

Examples: $\lfloor 5.1 \rfloor = 5, \lfloor 2.3 \rfloor = 2, \lfloor 8.6 \rfloor = 8, \lfloor -3.2 \rfloor = -4, \lfloor -7.1 \rfloor = -8, \lfloor \sqrt{5} \rfloor = 2, \lfloor -2 \rfloor = -2$.
The domain of floor function is R and its range is Z.

3.7.3 Ceiling Function

The ceiling function $g: R \to R$ is defined by $g(x) = \lceil x \rceil$, when $\lceil x \rceil$ is the least integer greater than or equal to x. Thus, $g(x) = \lceil x \rceil = n$, where $n - 1 < x \leq n$.

Examples: $\lceil 5.1 \rceil = 6, \lceil 2.3 \rceil = 3, \lceil 8.6 \rceil = 9, \lceil -3.2 \rceil = -3, \lceil -7.1 \rceil = -7, \lceil \sqrt{5} \rceil = 3, \lceil -2 \rceil = -2$.

Note 3.1 The floor and ceiling functions are often used in the study of algorithms. The floor of x rounds down x while ceiling of x rounds up. If x is not an integer, then the floor of x is the nearest integer to the left of x and ceiling of x is the nearest integer to the right of x. The programming languages C and Java provide the floor and ceiling function as built-in functions.

3.7.4 Mod and Div Functions

The mod function $f(x, y) = x \bmod y$ denotes the remainder when an integer x is divided by a positive integer y. In C++, the Mod operator is denoted by percent symbol %.

The div function $f(x, y) = x \operatorname{div} y$ denotes the quotient when x is divided by y. In C++, the div operator is denoted by the forward slash /.

Examples:

22 mod 4 = 2,	25 mod 5 = 0	37 mod 6 = 1
15/4 = 3,	27/5 = 5	37/9 = 4

3.7.5 Integer Value Function

Let x be any real number. The integer value of x denoted by INT (x) converts x into an integer truncating the fractional part of the number.

Thus, INT (5.32) = 5, INT ($\sqrt{2}$) = 1, INT (−7.5) = −7.

It can be observed that INT $(x) = \lfloor x \rfloor$ or INT $(x) = \lceil x \rceil$ according as x is positive or negative.

3.8 SOME SPECIAL FUNCTIONS USED IN DISCRETE MATHEMATICS

3.8.1 Polynomial Function

A function $f: R \to R$ defined by $f(x) = a_0 x^n + a_1 x^{n-1} + a_2 x^{n-2} + \dots + a_{n-1} x + a_n$ when $a_0, a_1, \dots, a_n \in R$ and $a_0 \neq 0$, $n \in W$ is a polynomial function of degree n. The set W denotes the set of whole numbers.

3.8.2 Exponential Functions

Let a be any positive real number and x any real number, then the function $f: R \to R^+$ defined by $f(x) = a^x$, $a \neq 1$ is called an exponential function.

The most commonly used function in computer science is the function $f(x) = 2^x$, whose graph is given as follows.

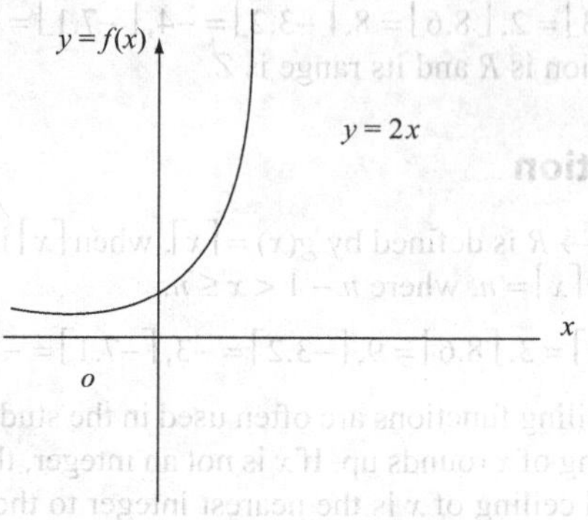

3.8.3 Logarithmic Function

If x and y are any two real numbers such that $x = a^y$, then y is called the logarithm of x to the base a and is denoted by $y = \log_a x$, a being a positive real number and not equal to one.

Thus, the function $f: R^+ \to R$ defined by $f(x) = \log_a x$ is the logarithmic function to the base a. The most commonly used logarithmic function in computer science is $\log_2 x$ which is denoted by $lg\, x$.

The graph of $y = lg\, x$ is given as follows:

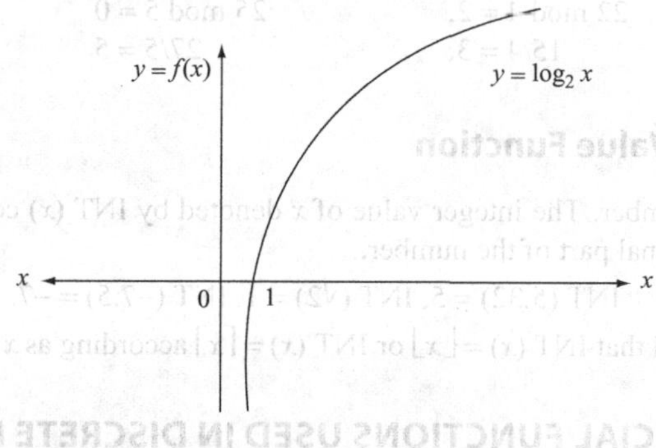

3.8.4 Recursively Defined Functions

A function is said to be recursively defined if it is defined employing the function itself.

The definition must have the following properties:

1. Let $f(n)$ be defined in terms of $f(n-1), f(n-2), \ldots$ where n is a positive integer.
2. The initial values $f(1), f(2), \ldots$ should be known.

Then $f: N \to R$ defines a recursive function. A popular example of a recursive function is the Ackermanns's function.

3.8.5 Sequence (or Discrete) Functions

A sequence is a function from the set of natural numbers $N = \{1, 2, 3, \ldots\}$ to the set A. A sequence is usually denoted by,

$$a_1, a_2, a_3, \ldots, a_n \ldots \text{ or by } \{a_n : n \in N\} \text{ or by } (a_n).$$

In some cases, the domain of the sequence is the set {0, 1, 2, …} of non-negative integers rather than N. In such a case, the sequence is denoted by $a_0, a_1, a_2, \ldots, a_n$.

If the elements listed in the sequence stop after m steps, $m \in N$, then the sequence is called a finite sequence and is denoted by $a_1, a_2, \ldots, a_m$.

A sequence that is not finite is called an infinite sequence.

Note 3.2 The elements in a sequence may be all different or some may be repeated.

Example 1 The sequence $1, \frac{1}{2}, \frac{1}{3}, \ldots$ may be represented by $a_n = \frac{1}{n}$, $n \in N$.

The sequence begins with $n = 1$.

Example 2 The sequence $1, \frac{1}{3}, \frac{1}{9}, \frac{1}{27}, \ldots$ may be represented by $a_n = 3^{-n}$. Here the first element begins with $n = 0$.

Example 3 The sequence 2, 5, 8, 11, … can be described by recursive relation.

$$a_n = a_{n-1} + 3,\ 2 \le n \le \infty \text{ and } a_1 = 2$$

Example 4 The important sequence, 1, –1, 1, –1, 1, … is described by,

$$a_n = (-1)^{n+1},\ n \in N$$

Example 5 The sequence 2, 4, 8, 16, … is described by,

$$a_n = 2a_{n-1},\ 2 \le n \le \infty \text{ and } a_1 = 2.$$

3.8.6 Strings

Let A be a finite set that can be taken as a character set or an alphabet. A finite sequence over the set A written as $a_1\, a_2 \ldots a_n$ without commas is called a string or word. The number of characters in the string is called its length.

Example 1 Define recursively the factorial function.

Solution: The function n factorial is defined by $n! = 1.2.3\ldots\ldots(n-2).(n-1).\, n$
The factorial function f is defined by $f(n) = n!$, where $f(0) = 1$. Since $n! = n(n-1)!$
We can define f as follows:

$$f(n) = nf(n-1), \qquad n \ge 1 \text{ (Recursive relation)}$$

and

$$f(0) = 1 \qquad \text{(Initial condition)}$$

Note 3.3 $f(0) = 0!$ is defined to be 1.

Suppose we want to calculate 5!

$$f(5) = 5! = 5.f(4) = 5[4.f(3)] = 5.4.3.f(2) = 5.4.3.2.f(1) = 5.4.3.2.1.f(0) = 5.4.3.2.1.1 = 120$$

Example 2 The recursive formula $a_1 = 2$, $a_n = 3a_{n-1}$, $2 \le n \le 5$, defines the finite sequence 2, 6, 18, 54, 162.

Example 3 The infinite sequence 3, 8, 13, 18, 23, … can be defined by the recursive formula $c_1 = 3, c_n = c_{n-1} + 5$.

Example 4 The explicit formula $a_n = (-2)^n, 1 \leq n < \infty$ describes the infinite sequence −2, 4, −8, 16, ...

3.8.7 Characteristic Function

If A is a subset of a universal set U, then the characteristic function f_A of A is defined by,

$$f_A(x) = 1 \text{ if } x \in A$$
$$= 0 \text{ if } x \notin A$$

We can add and multiply characteristic functions since their values are real numbers.

THEOREM 3.5 The characteristic function of subsets satisfies the following properties:

(1) $f_{A \cap B}(x) = f_A(x) f_B(x)$ for all x.

(2) $f_{A \cup B}(x) = f_A(x) + f_B(x) - f_A(x) f_B(x)$ for all x.

Proof:

(1) $f_A(x) f_B(x)$ is equal to 1 if both $f_A(x)$ and $f_B(x)$ are equal to 1, which is possible only when x is in A and also is in B, that is, x is in $A \cap B$. Further, since $f_A f_B$ is 1 if $x \in A \cap B$. Otherwise zero, it must be $f_{A \cap B}(x)$.

(2) If x is in A, $f_A(x) = 1$ and so $f_A(x) + f_B(x) - f_A(x) f_B(x) = 1 + f_B(x) - f_B(x) = 1$.

Similarly, if x is in B, $f_B(x) = 1$ and so $f_A(x) + f_B(x) - f_A(x) f_B(x) = 1 + f_A(x) - f_A(x) = 1$

If x is not in A or B, then $f_A(x)$ and $f_B(x)$ are 0 and so $f_A(x) + f_B(x) - f_A(x) f_B(x) = 0$ and so it must be $f_{A \cup B}(x)$.

3.8.8 Computer Representation of Sets

Characteristic function is used in representing sets in a computer. A set can be represented in a computer by arranging its elements in a sequence. The elements in the set are listed in particular order of no importance.

Let A be a subset of a universal set $U = \{x_1, x_2, \ldots, x_n\}$, then the characteristic function f_A assigns 1 to an element x_i that belongs to A and 0 to an element x_i that does not belong to A. Thus, f_A is represented by a sequence of 1's and 0's of length n.

Example: Let $U = \{1, 2, 3, 4, 5\}$, $A = \{1, 2, 3\}$, $B = \{4, 5\}$, and $C = \{1, 3, 5\}$.

Then f_A has value 1 when x is 1, 2, or 3 and otherwise is 0. Hence, f_A corresponds to the sequence 1 1 1 0 0. Similarly, f_B and f_C corresponds to the sequence 0 0 0 1 1 and 1 0 1 0 1, respectively.

Note 3.4 Any set containing n elements can be arranged in the form of a sequence of length n. To each of its subset, there corresponds a sequence of zeros and ones of length n representing the characteristic function of that subset. Thus, every universal set can be represented in a computer as an array of specific length.

3.9 SOME IMPORTANT THEOREMS AND PROBLEMS

THEOREM 3.6 If $f: A \to B$ is one-one and ONTO, then the inverse mapping of f is unique.

Proof: Let $f: A \to B$ be one-one and ONTO.
Let $g: B \to A$ and $h: B \to A$ be inverse mappings of f.
Let b be any element in B and let $g(b) = c$ and $h(b) = d$, where c and $d \in A$.
Since g is the inverse mapping of f, then $g(b) = c \Rightarrow f(c) = b$
Similarly, since h is inverse mapping of f, so, $h(b) = d \Rightarrow f(d) = b$

$$\therefore b = f(c) = f(d) \Rightarrow c = d \text{ since } f \text{ is one-one.}$$

Hence,

$$c = d \Rightarrow g(b) = h(b)$$
$$\Rightarrow g = h$$

Since g is the same function as h and so the inverse mapping of f is unique.

THEOREM 3.7 Let $f: A \to B$ and $g: B \to A$ be two functions such that $gof: A \to A$ is the identity function on I_A, then $g = f^{-1}$. If $fog: B \to B$ is the identity function I_B, then $f = g^{-1}$.

Proof: In order to prove that f^{-1} exists, we have to prove that f is one-one and ONTO.

(1) Let $a_1, a_2 \in A$, then

$$f(a_1) = f(a_2) \Rightarrow g[f(a_1)] = g[f(a_2)]$$
$$\Rightarrow (gof)\, a_1 = (gof)\, a_2$$
$$\Rightarrow a_1 = a_2 \text{ (since } gof \text{ is an identity function)}$$

f is one-one.

(2) Let b be any element in B. Since g is a function from B to A, $g(b) \in A$.
Let $g(b) = a$ for $a \in A$, then $f[g(b)] = f(a) \Rightarrow (fog)\, b = f(a)$
$\Rightarrow I_B(b) = f(a) \Rightarrow b = f(a)$
Thus, for every $b \in B$, there exists $a \in A$ such that $b = f(a)$ and so f is ONTO. Hence, f is invertible, that is, f^{-1} exists.

(3) To prove that $f^{-1} = g$
We have $gof = I_A$

$$(gof)\, of^{-1} = I_A\, of^{-1}$$
$$\Rightarrow go\,(fof^{-1}) = I_A\, of^{-1}$$
$$\Rightarrow goI = f^{-1}$$
$$\Rightarrow g = f^{-1}$$

In a similar manner, it can be proved that g is also invertible and $g^{-1} = f$.

Example 1 Let $A = \{a, b, c, d\}$ and $B = \{1, 2, 3, 4\}$ be two given sets. Find whether following are functions from A to B. In case, they are functions, find the range.

(1) $f = \{(a, 4), (b, 3), (c, 2), (d, 1)\}$
(2) $g = \{(a, 1), (b, 1), (c, 3), (d, 4)\}$
(3) $h = \{(a, 2), (b, 3), (c, 4)\}$

Solution:

(1) f is a function, since each element of A is associated with a unique element of B.
Range $(f) = \{1, 2, 3, 4\}$

(2) g is a function, since each element of A is associated with unique element of B.
Range $(g) = \{1, 3, 4\}$

(3) h is not a function, since $d \in A$ is not associated with any element in B.

Example 2 Find the range of the function $f: A \to R$ defined by $f(x) = x^2 + 5$, where $A = \{0, 1, 2, 3\}$.

Solution: The range of f consists of those elements of R which are f-image of different elements of A. Now, $f(0) = 5, f(1) = 1 + 5 = 6, f(2) = 4 + 5 = 9, f(3) = 14$. Thus, the range of $f = \{5, 6, 9, 14\}$.

3.10 ACKERMANN'S FUNCTION

A popular example of a recursive function is Ackermann's function.

Ackermann's Function: $A(x, y)$ is a function on $N_0 \times N_0$ defined as follows:

$$A(0, y) = y + 1$$
$$A(x + 1, 0) = A(x, 1)$$
$$A(x + 1, y + 1) = A[x, A(x + 1, y)]$$

$A(x, y)$ is a recursive function and can be computed for all finite integral values of x and y.

Example 1 Compute $A(1, 1), A(1, 2), A(2, 1)$, and $A(2, 2)$.

Solution: $A(1, 1) = A(0 + 1, 0 + 1) = A[0, A(1, 0)] = A[0, A(0, 1)] = A(0, 2) = 3$
$A(1, 2) = A(0 + 1, 1 + 1) = A[0, A(1, 1)] = A(0, 3) = 4.$
$A(2, 1) = A(1 + 1, 0 + 1) = A[1, A(2, 0)] = A[1, A(1, 1)] = A(1, 3) = A[0, A(1, 2)] = A(0, 4) = 5$
$A(2, 2) = A(1 + 1, 1 + 1) = A[1, A(2, 1)] = A(1, 5) = A(0 + 1, 4 + 1) = A[0, A(1, 4)] = 1 + A(1 + 4)$
$= 1 + A(0 + 1, 3 + 1) = 1 + A[0, A(1, 3)] = 1 + 1 + A(1, 3) = 1 + 1 + 1 + A(1, 2) = 1 + 1 + 1 + 4 = 7.$

Example 2 Using recurrence relation
$A(2, n) = A(2, n - 1) + 2$ for all integers $n \geq 1$, where $A(2, 0) = 3$, show that
$A(2, n) = 2n + 3$ for all $n \geq 0$.

Solution: $A(2, n) = A(2, n - 1) + 2 = A(2, n - 2) + 2 + 2 = A(2, n - 3) + 2 + 2 + 2 = \ldots$

$$= A(2, n - n) + (2 + 2 + 2 + \ldots + n \text{ times}) = A(2, 0) + 2n$$
$$\therefore A(2, n) = 3 + 2n, \qquad \text{for } n \geq 0$$

Example 3 Use the definition of Ackermann's function to show that

$$A(3, n) = 2^{n+3} - 3, \qquad \text{for } n \geq 0$$

Solution: Ackermann's function is defined by

$$A(x + 1, y + 1) = A[x, A(x + 1, y)]$$
$$\therefore A(3, n) = A(2 + 1, (n - 1) + 1)$$
$$= A[2, A(3, n - 1)]$$

$$= 3 + 2\,A(3, n-1) \text{ (using } \textit{Example 2}\text{)}$$
$$\therefore A(3, n) = 2A(3, n-1) + 3$$

Applying the relation successively

$$A(3, n) = 2\,(2A(3, n-2) + 3) + 3$$
$$= 2^2 A(3, n-2) + 2.3 + 3 \quad \text{(i)}$$
$$= 2^2\,(2A(3, n-3) + 3) + 2.3 + 3$$
$$= 2^3 A(3, n-3) + 2^2.3 + 2.3 + 3 \quad \text{(ii)}$$
$$\cdots\cdots\cdots\cdots\cdots\cdots\cdots\cdots$$
$$= 2^n A(3, n-n) + 2^{n-1}.3 + 2^{n-2}.3 + \ldots + 2.3 + 3$$
$$= 2^n A(3, 0) + (2^{n-1} + 2^{n-2} + \ldots + 2^2 + 2 + 1)3$$
$$= 2^n \cdot A(2, 1) + 3\cdot(2^n - 1) = 2^n\cdot 5 + 3(2^{n-1}) = (5 + 3)\,2^n - 3 = 8\cdot 2^n - 3 = 2^3\cdot 2^n - 3 = 2^{n+3} - 3.$$

3.11 FUZZY SETS

The concept of fuzzy set introduced by Lotfi A. Zadeh in 1965 is a generalization of the classical abstract or crisp set in which either the element belongs to the set or it does not belong. The realization that the above two-valued logic is inadequate for dealing with vagueness, imprecision, and uncertainty of the real life problems gave rise to the development of the concept of fuzzy sets. For example, the collections or classes are the set of tall people, beautiful women, wealthy persons, cold weather, warm days, sunny days, expensive televisions or mobiles, and so on, which do not constitute sets in the usual mathematical sense.

Zadeh introduced the concept of a fuzzy set that can be defined mathematically by assigning to each individual in the universe of discourse a value representing its grade of membership. This grade indicates the degree to which the individual is similar or compatible to the concept represented by the fuzzy set. The membership degree is generally represented by a real number ranging in the closed unit interval [0, 1]. The full membership and full non-membership is indicated by the values of 1 and 0, respectively. Thus, the concept of crisp set can be regarded as the restricted case of the concept of a fuzzy set in which only two grades of membership is allowed.

The capability of fuzzy sets to express transitions from membership to non-memberships and vice-versa has many applications as it is capable of providing meaningful representation of imprecise and vague concepts. For example, instead of describing the cloudy weather on a particular day in terms of exact percentage of cloud cover, we can say **that it is cloudy.** This description of weather is not specific but vague but it is useful in taking decisions. Thus, fuzzy sets allow us to represent vague concepts expressed in natural language. For a term such as **cloudy day**, we cannot use it to mean 100 per cent cloud cover nor 80 or 70 per cent precisely. One can accept a cloud cover of 60 per cent and above for a cloudy day. It is just not possible to draw a line that a particular percentage of cloud cover defines a cloudy weather. We can, therefore, add a qualification that a cloud cover greater than a particular percentage can be considered to be a **cloudy weather**.

Definition 3.1 Let A be a **crisp set** defined on the universe U. Then for any member of U, either $x \in A$ or $x \notin A$. However, for a **fuzzy set** F in U, it is not necessary that either $x \in F$ or $x \notin F$.

In fuzzy set theory, the characteristic function (membership function) associates each point $x \in U$ a real number in the interval [0, 1] instead of two elements set {0, 1}.

Definition 3.2 Let U be a universal set. *A* **fuzzy set** A in U characterized by membership function μ_A is

$$\mu_A: U \to [0, 1]$$

This implies that every element $x \in U$ has a membership degree μ_A in [0, 1].

Thus, a fuzzy set can be defined as a set of pairs consisting of each member in the universe and its degree of membership.

Illustrations:

1. A builder classifies the houses, he or she offers to his or her customers. An indicator of the comfort of these houses is the number of bedrooms in it.

   Let $U = \{1, 2, 3, 4, 5, 6\}$ be the set of types of houses described by x = number of bedrooms in the house. Then the fuzzy set of comfortable type of house for a 4 – person family can be described as, $A = \{(1, 0.1), (2, 0.5), (3, 0.7), (4, 1), (5, 0.3)\}$

2. Let a set P (of powerful people) be defined as:

   P = {(Ram, 0.2), (Shyam, 0.3), (Sonu, 0.5), (Seema, 0.95), (Rahul, 0.8)}

   The above set shows that Ram is least powerful and Seema is most powerful.

3. Let a set R (of rich people) be defined as:

   R = {(Ram, 0.5), (Shyam, 0.6), (Sonu, 0.55), (Seema, 0.9), (Rahu, 0.85)}

4. A = Set of real numbers close to 5

   $$= \left\{ x : \mu_A(x) = \frac{1}{1+(x-5)^2} \right\}$$

5. If $U = \{x_1, x_2, \ldots, x_n\}$, then a fuzzy set A is also written as

   $$A = \left\{ \frac{\mu_A(x_1)}{x_1}, \frac{\mu_A(x_2)}{x_2}, \ldots\ldots\ldots, \frac{\mu_A(x_n)}{x_n} \right\}$$, which alternatively is also written as

   $$A = \{[x_1, \mu_A(x_1)], [x_2, \mu_A(x_2)], \ldots, [x_n, \mu_A(x_n)]\}$$

6. Let $U = \{5, 6, 7, 8\}$ be a set with membership grades 0.1, 0.3, 0.6, and 1.

   The fuzzy set A can be written as $A = \{(5, 0.1), (7, 0.6), (8, 1)\}$

   $$= \left\{ \frac{0.1}{5}, \frac{0.6}{7}, \frac{1}{8} \right\}$$

3.11.1 Some Definitions

Cardinality of Fuzzy Set: The cardinality of a finite fuzzy set A is defined by,

$$|A| = \sum_{x \in U} \mu_A(x)$$

The relative cardinality of the fuzzy set A is defined by,

$$\|A\| = \frac{|A|}{|U|}$$

For example, if the fuzzy set A of $U = \{1, 2, 3, 4, 5, 6, 7, 8, 9, 10\}$ is given by,

$$A = \left\{\frac{0.1}{2}, \frac{0.4}{3}, \frac{0.7}{5}, \frac{0.8}{6}, \frac{0.2}{10}\right\}$$

The $|A| = 0.1 + 0.4 + 0.7 + 0.8 + 0.2 = 2.2$
The relative cardinality

$$\|A\| = \frac{|A|}{|U|} = \frac{2.2}{10} = 0.22$$

Normal Fuzzy Set: A fuzzy set A of U is said to be **normal** if and only if max $\mu_A(x) = 1$ for all $x \in U$ otherwise subnormal.

Subset: Let A and B be any two fuzzy sets in U with membership functions μ_A and μ_B, respectively. Then A is a subset of B or is contained in B written as $A \subseteq B$ if and only if $\mu_A(x) \le \mu_B(x)$ for every $x \in U$.

Example: If $U = \{10, 20, 30, 40, 50\}$ and A and B are two fuzzy sets defined by,

$$A = \left\{\frac{0.10}{10}, \frac{0.20}{20}, \frac{0.25}{30}, \frac{0.30}{40}, \frac{0.35}{50}\right\} \text{ and } B = \left\{\frac{0.4}{10}, \frac{0.5}{20}, \frac{0.8}{30}, \frac{0.9}{40}, \frac{1}{50}\right\}$$

Then, since $\mu_A(x) \le \mu_B(x)$ for every $x \in U$ and $A \subseteq B$. Further B is a normal fuzzy set of U whereas A is a subnormal fuzzy set of U.

Support of Fuzzy Set: Let A be the fuzzy set in a universal set U. The support of A is the crisp set that contains all elements of U for which membership function is positive. Thus, Support $(A) = \{x: \mu_A(x) > 0, x \in U\}$.

α-Cut: One of the most important concepts of fuzzy sets is the concept of an α-cut and its variant a strong α-cut.

The α-cut of a fuzzy set A is the crisp set of elements that belong to the fuzzy set A at least to degree α. It is denoted by A_α.
Thus,

$$A_\alpha = \{x \in U: \mu_A(x) \ge \alpha\}$$

The strong α-cut demoted by $A_{\alpha+}$ is defined by,

$$A_{\alpha+} = \{x \in U: \mu_A(x) > \alpha\}$$

Example 1 Let $U = \{1, 2, 3, 4, 5, 6, 7, 8\}$ and a fuzzy set A of U be given by,

$$A = \{(1, 0.1), (2, 0.3), (3, 0.4), (4, 0.5), (5, 0.7), (6, 0.75), (7, 0.8), (8, 0.2)\}$$

Then α-cuts are

$$A_{0.1} = \{1, 2, 3, 4, 5, 6, 7, 8\}$$
$$A_{0.4} = \{3, 4, 5, 6, 7\}$$
$$A_{0.7} = \{5, 6, 7\}$$

Clearly $A_{0.0} = U$ and $A_{\alpha} = \varphi$, for all $\alpha > 1$.
The strong α-level cuts are $A^{+}_{0.7} = \{6, 8\}$, $A^{+}_{0.5} = \{5, 6, 7\}$

Example 2 Let us consider three fuzzy sets that represent the concepts of a young, middle aged, and old person. A reasonable expression of membership functions A_1, A_2, and A_3 is given below. These functions are defined over the interval [0, 90].

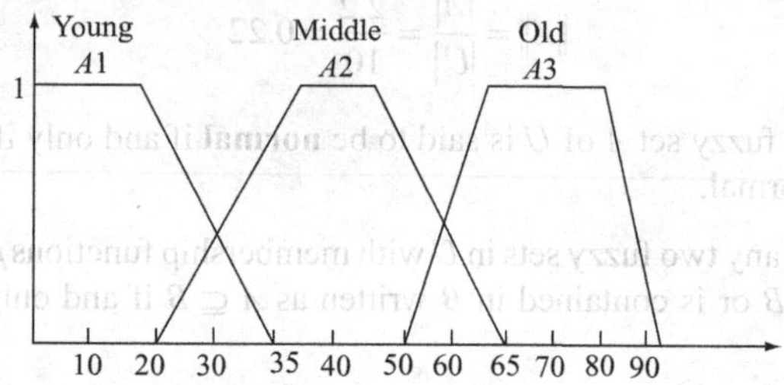

$$A_1(x) = 1, \qquad \text{when } x \le 20$$
$$= \frac{35 - x}{15}, \qquad 20 \le x \le 35$$
$$= 0 \qquad x > 35$$
$$A_2(x) = 0, \qquad \text{when } x \le 20 \text{ or } \ge 65$$
$$= \frac{x - 40}{20}, \qquad 20 < x < 40$$
$$= 1 \qquad 40 < x < 50$$
$$= \frac{65 - x}{15}, \qquad 50 < x < 65$$
$$A_3(x) = 0, \qquad \text{when } x \le 50$$
$$= \frac{65 - x}{15}, \qquad 50 \le x \le 65$$
$$= 1 \qquad x > 65$$

The above membership functions are taken to be trapezoidal as an approximation. The other membership functions can be triangular, bell-shaped, *S*-shaped, and so on.

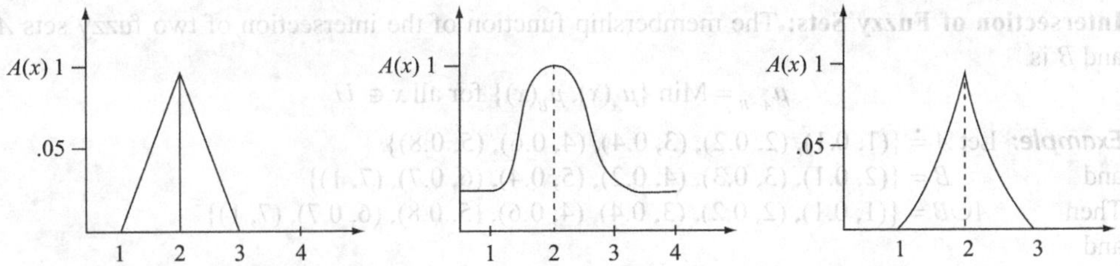

All the above membership functions may be used for characterizing fuzzy sets of real numbers close to 2 in different contexts.

In some physical problem such as warm temperature or cold temperature, the membership function is defined by *S*-function that is a piecewise quadratic function and can be defined as follows:

$$
\begin{aligned}
S(x) &= 0, & x \geq a \\
&= 1 - 2\left(\frac{x-c}{c-a}\right)^2, & b \ll x \ll c \\
&= 2\left(\frac{x-a}{c-a}\right)^2, & a \ll x \ll b \\
&= 1, & x > c
\end{aligned}
$$

where a and c are lower and upper fuzzy subset values and $b = (a + c)/2$ is crossover point, that is, the value of temperature x at which $S = 0.5$.

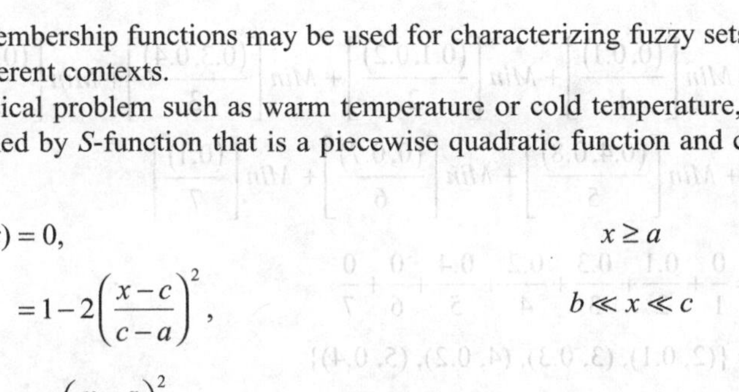

3.11.2 Operations on Fuzzy Sets

Complement: The complement of a fuzzy set A denoted by $\bar{A}$ is defined by,

$$\mu_{\bar{A}}(x) = 1 - \mu_A(x) \qquad \text{for all } x \in U$$

For example, if an element in the fuzzy set A has a membership grade of 0.7, then its membership grade in its complement $\bar{A}$ is 0.3.

Union of Fuzzy Sets: The membership function of the union of two fuzzy sets A and B with membership functions μ_A and μ_B is defined by,

$$\mu_{A \cup B} = \text{Max } \{\mu_A(x), \mu_B(x)\} \qquad \text{for all } x \in U$$

Intersection of Fuzzy Sets: The membership function of the intersection of two fuzzy sets A and B is

$$\mu_{A\cap B} = \text{Min}\ \{\mu_A(x), \mu_B(x)\} \text{ for all } x \in U$$

Example: Let $A \doteq \{(1, 0.1), (2, 0.2), (3, 0.4), (4, 0.6), (5, 0.8)\}$
and $B = \{(2, 0.1), (3, 0.3), (4, 0.2), (5, 0.4), (6, 0.7), (7, 1)\}$
Then $A\cup B = \{(1, 0.1), (2, 0.2), (3, 0.4), (4, 0.6), (5, 0.8), (6, 0.7), (7, 1)\}$
and

$$A \cap B = Min\left[\frac{(0,0.1)}{1}\right] + Min\left[\frac{(0.1,0.2)}{2}\right] + Min\left[\frac{(0.3,0.4)}{3}\right] + Min\left[\frac{(0.2,0.6)}{4}\right] + Min\left[\frac{(0.4,0.8)}{5}\right] + Min\left[\frac{(0,0.7)}{6}\right] + Min\left[\frac{(0,1)}{7}\right]$$

$$A \cap B = \frac{0}{1} + \frac{0.1}{2} + \frac{0.3}{3} + \frac{0.2}{4} + \frac{0.4}{5} + \frac{0}{6} + \frac{0}{7}.$$

$$= \{(2, 0.1), (3, 0.3), (4, 0.2), (5, 0.4)\}$$

Applications: The successful applications of fuzzy set have been growing steadily since 1990. The fuzzy set theory has been widely used in pattern recognition, cluster analysis, information and knowledge-based systems, various areas of engineering, and virtually problem areas of decision making. In addition, some scattered applications can also be found in medicine, biology, physics, chemistry, meteorology, and geology.

3.12 TIME COMPLEXITY OF ALGORITHM

Consider a problem given to two mathematicians known for their excellence in discrete mathematics to solve. The algorithms developed by them are different. One algorithm took much less time as compared to the other when implemented on a computer. This has led to the concept of efficiency of an algorithm. Another measure of efficiency of an algorithm is the memory space required to hold the numbers in memory for later use. The smaller the space required, better is the algorithm. These two requirements of an algorithm are termed as **computational complexity**. An analysis of memory space needed for an algorithm is called **space complexity of algorithm**. Similarly, the analysis for time of an algorithm is called **time complexity**. An algorithm is efficient if it requires less space and also the time is small.

Since the storage space required by an algorithm is simply the data size, the efficiency is measured by operation counts, namely the number of elementary operations of addition, subtraction, multiplication, division, assignment, and comparison. We shall, however, concentrate entirely upon time efficiency in our further discussion.

The running time of an algorithm can be estimated in the following ways:

Exactly (Empirically): The exact time of execution of an algorithm is obtained by setting a timer function. However, it is machine dependent.

Inexactly: The running time of an algorithm is estimated by the operation counts that are inherently inexact as all the operations addition, subtraction, multiplication, square root, and so on do not require equal amount of time. This is specified by a numerical function called **complexity function** expressed in terms of input size.

The **time complexity of an algorithm** can be explained by the following considerations:

Worst Case Complexity: For an algorithm, it is uncertain exactly how many operations will be needed and so we estimate the upper bound of the operations for all inputs of a given size. This estimate is called **worst case complexity**.

Best Case Complexity: When minimum number of operations is counted for the execution of the algorithm, it is called **best case complexity**.

Average Case Complexity: Average case complexity of an algorithm is the function defined by the average number of operations required for all inputs of size n. This method does not provide a good measure of the complexity as there is no specific way for the choice of the value of input size n and also the n inputs.

In view of the above considerations, the worst-case time complexity is regarded as the measure for any algorithm.

Definition 3.3 The time complexity function is defined by $f: N \rightarrow N$, where $f(n)$ is an upper bound for the number of operations to carry out an algorithm for some input of size n.

Example 1 Let us consider the time complexity function for adding two n-digit integers.

Solution: Let two n-digit integers be represented by $a = (a_{n-1}\, a_{n-2} \ldots a_2\, a_1\, a_0)$ and $b = (b_{n-1}\, b_{n-2} \ldots b_2\, b_1\, b_0)$ to the base 10. The unit digit of $a + b$ is obtained by adding unit digits a_0 and b_0 and thus a single operation. The tens digit of $a + b$ is obtained by adding a_1 and b_1 and then it may be required to add 1 to carry from the previous step. Hence at most two operations of addition are required. Thus, after unit digit, for each digit of $a + b$, at most two operations are required. Hence the maximum number of operation is $f(n) = 1 + 2\,(n - 1) = 2n - 1$.

Example 2 Consider the time complexity function for a linear search algorithm where our objective is to find a particular value from a list consisting of n values.

Solution: Let the input list consist of values $a_1, a_2, \ldots, a_n$ and a linear search is applied to find a. In the search, a is compared with the first element a_1 in the list, then with the second element a_2 and so on until it matches a_k. If it does not match any element in the list, then a is not in the list.

As the operation involved is the comparison of elements, at each step two comparisons are made, one to compare the element a and the other whether end element in the list is reached. Hence in the worst case, $2n$ comparisons are required. If the element a is not in the list when the loop has executed n times, an additional comparison is made to exit the loop. Hence, the maximum number of operations is $f(n) = 2n + 1$.

3.12.1 Rate of Growth of Functions

Let A be an algorithm and n be the size of the input data, then obviously the time complexity function $f(n)$ increases as n increases. We are interested to examine the rate of increase of $f(n)$. This is generally done by comparing $f(n)$ with some standard functions such as $\log_2 n$, n, $n \log_2 n$, n^2, n^3, 2^n.

The approximate rate of growth of above functions for certain values of n are given in the Table 3.1.

Table 3.1 Time comparison of different time complexity function

n \ g(n)	log2$_n$	N	n log 2^n	n^2	n^3	2^n
5	3	5	15	25	125	32
10	4	10	40	100	10^3	1024
100	7	100	700	10^4	10^6	10^{30}
1000	10	10^3	10^4	10^6	10^9	10^{300}

The functions listed in the table are in the increasing order of their rates of growth, the logarithmic function $\log_2 n$ grows most slowly, the functions n^k grow according to the exponent k and the exponential function 2^n grows most rapidly.

3.12.2 Determination of Growth Function

Since the running time of an algorithm cannot be measured exactly for all inputs of size n, we are interested in finding an upper bound for the complexity function $f(n)$ when n is very large. Thus, theoretically, we compute a function $g(n)$ to which $f(n)$ tends as $n \to \infty$.

We define a symbolic notation for the size of the function $f(n)$. Three notations: the big Oh, the big Ω (Omega), and the big Θ (Theta) are used.

The Big-*Oh* Notation: Let f and g be functions $N \to R$, then $f(n)$ is of order at most $g(n)$ if there exists an integer n_0 and a positive real number c such that

$$|f(n)| \leq c\,|g(n)|, \qquad \text{for all } n \geq n_0$$

In symbols, we write $f(n) = O(g(n))$ and read it as $f(n)$ is big *Oh* of $g(n)$.

When we say that the time of execution of an algorithm is $O(g(n))$, it means that the time required is not more than some constant times $|g(n)|$. In fact, $g(n)$ is an upper bound for $f(n)$. Graphically, it can be represented as follows.

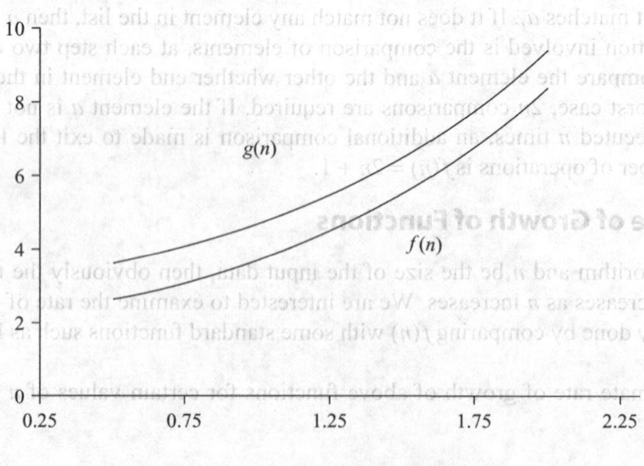

The big-*Oh* (*O*) notation has been widely used in the study of growth of functions but it has a drawback that it does not give exact order of growth. It only gives that *f* does not grow faster than *g*.

Example 1 Let $f(n) = 2n + 1$ show that $f(n) = O(n)$

Solution:

$$f(n) = 2n + 1 \le 2n + n, \quad \text{for all } n \ge n_o.$$
$$\le 3n$$

Hence,

$$f(n) \le cg(n), \text{ where } c = 3, g(n) = n \text{ for all } n \ge n_o = 1.$$

Hence,

$$f(n) = O(n).$$

Example 2 Let $f(n) = 5n^2 - n + 6$, show that $f(n) = O(n^2)$

Solution: $|f(n)| = |5n^2 - n + 6| \le |5n^2| + |-n| + 6$

$$\le 5n^2 + n + 6$$
$$\le 5n^2 + n^2 + 6n^2$$
$$\le 12n^2 \quad \text{for all } n \ge n_o = 1$$
$$\therefore |f(n)| \le c\,|g(n)|$$

where $\quad g(n) = n^2, c = 12 \quad$ for all $n \ge n_o = 1$

Hence,

$$f(n) = O(n^2).$$

Example 3 Use the definition of big *O* to prove that $2n^4 + n^3 + 3n^2 - 5 = O(n^4)$

Solution: Let $f(n) = 2n^4 + n^3 + 3n^2 - 5$

$$\therefore |f(n)| = |2n^4 + n^3 + 3n^2 - 5|$$
$$\le |2n^4| + |n^3| + |3n^2| + 5$$
$$\le 2n^4 + n^3 + 3n^2 + 5$$
$$\le 2n^4 + n^4 + 3n^4 + 5n^4 \quad \text{for all } n = n_o \ge 1$$
$$\le 11n^4$$
$$\therefore |f(n)| \le c\,|g(n)|$$

where $\quad c = 11, g(n) = n^4$

Hence,

$$f(n) = 0(n^4).$$

Example 4 Find big-*Oh* order for the function $f(n) = 1 + 2 + \ldots + n + 2n^2 + 3n^3$

Solution:

$$1 + 2 + 3 + \ldots + n \le n + n + n + \ldots + n$$
$$\le n\,(n)$$
$$\le n^2 \quad \text{for all } n \ge 1$$

Hence,

$$|f(n)| \le n^2 + 2n^2 + 3n^3$$
$$\le n^3 + 2n^3 + 3n^3$$

$\leq 6n^3$

$\therefore |f(n)| \leq c\ |g(n)|$

where $\quad c = 6, g(n) = n^3$

$\therefore f(n) = O(n^3)$

Example 5 Show that

(1) $n! = O(n^n)$

(2) $\lg n! = O(n \lg n)$

(3) $2^{n+1} = O(2^n)$

Solution:

(1) $n! = n\ (n-1)\ (n-2)\ldots\ldots\ldots\ldots\ldots 2.1$ $\quad$ (i)

$\leq n.\ n.\ n\ldots\ldots\ldots\ldots n.\ n$ $\quad$ for all $n \geq 1$

$\leq n^n$

$= O(n^n)$

(2) From (1) above, $n! \leq n^n$

$\therefore \lg n! \leq n \lg n > 1$

$= O(n \lg n)$

(3) $2^{n+1} = 2.2^n$

$\leq c.\ 2^n \quad c > 2, \quad n \geq 1$

$= O(2^n)$

3.12.3 Complexity of Well-known Algorithms

The complexity of well-known and widely used algorithms for searching and sorting in computer science are the following:

1. Linear search: $O(n)$
2. Binary search: $O(\log_2 n)$
3. Bubble sort: $O(n^2)$
4. Merge sort: $O(n \log_2 n)$

3.12.4 Some Properties of Time Complexity Functions

THEOREM 3.8 Let an algorithm consist of two sub-algorithms whose execution times are

$$f_1(n) = O[g_1(n)] \text{ and } f_2(n) = O[g_2(n)]$$

Then

(1) $f_1(n) + f_2(n) = O(\max\ \{|g_1(n)|, |g_2(n)|\})$

(2) $(f_1 \cdot f_2)\ (n) = O[g_1(n).\ g_2(n)]$

The plus (+) sign implies that the two sub-algorithms are distinctly placed (i.e., not cascaded) and so are executed separately. The sign $(\cdot)$ implies that they there are cascaded and executed dependently.

Proof:

(1) Since $f_1(n) = O[g_1(n)]$ and $f_2(n) = O[g_2(n)]$
There will exist positive constants c_1, n_1 and c_2, n_2 such that

$$|f_1(n)| \le c_1|g_1(n)| \qquad n \ge n_1$$

and

$$|f_2(n)| \le c_2|g_2(n)| \qquad n \ge n_2$$

Now choose $c = \max(c_1, c_2)$, $n_o = \max(n_1, n_2)$
and $g(n) = \max\{g_1(n), g_2(n)\}$
Then

$$\begin{aligned} |f_1(n) + f_2(n)| &\le |f_1(n)| + |f_2(n)| \\ &\le c_1|g_1(n)| + c_2|g_2(n)| \\ &\le c|g(n)| + c|g(n)|, \qquad n \ge n_o \\ &\le 2c\,|g(n)| \end{aligned}$$

$$\begin{aligned} \therefore f_1(n) + f_2(n) &= O[g(n)] \\ &= O(\max\{|g_1(n)|, |g_2(n)|\}) \end{aligned}$$

(2) $$\begin{aligned} |(f_1 \cdot f_2)\, n| &= |f_1(n)\, f_2(n)| \\ &= |f_1(n)| \cdot |f_2(n)| \\ &\le c_1\,|g_1(n)|\; c_2\,|g_2(n)| \\ &\le c\,|g_1(n)\, g_2(n)| \end{aligned}$$

Hence,

$$(f_1 f_2)(n) = O\,[g_1(n) \cdot g_2(n)]$$

Note 3.5 Equal Order: The functions f and g are of same order if $f = O(g)$ and $g = O(f)$.

Example 1 Let $f(n) = 5n^2 + 2n + 3\lg n!$ Estimate the growth of $f(n)$.

Solution: Since $5n^2 + 2n = O(n^2)$ and $3 = O(1)$ and $\lg n! = O(n \log n)$

$$\begin{aligned} \therefore 3 \lg n! &= O(1).\ O(n\lg n) \\ &= O(1.n \lg n) = O(n \lg n) \end{aligned}$$

Also, $\lg n \le n \therefore n \lg n \le n^2$ for $n \ge 1$

Hence,

$$\begin{aligned} f(n) &= O(n^2) + O(n \lg n) \\ &= O(n^2) \ \therefore \text{(maximum is } n^2) \end{aligned}$$

Example 2 Let $f(n) = (2n^2 + 3n + 4) \lg n$. Estimate the growth of $f(n)$

Solution: $$\begin{aligned} 2n^2 + 3n + 4 &\le 2n^2 + 3n^2 + 4n^2 \\ &\le 9n^2 \qquad \text{for all } n \ge 1 \\ &= O(n^2) \qquad \text{for all } n \ge 1 \end{aligned}$$

Clearly $\lg n = O(\lg n)$

Hence,

$$\begin{aligned} f(n) &= (2n^2 + 3n + 4)\lg n \\ &= O(n^2).\,O(\lg n) \\ &= O(n^2 \lg n) \end{aligned}$$

Example 3 Find the order of the polynomial function $f(n) = 2n^3 + 3n^2 - 5n + 7$

Solution: $|f(n)| = |2n^3 + 3n^2 - 5n + 7| \le |2n^3| + |3n^2| + |-5n| + 7$

$$\begin{aligned} &\le 2n^3 + 3n^2 + 5n + 7 \\ &\le 2n^3 + 3n^3 + 5n^3 + 7n^3 \\ &\le 17n^3 \qquad\qquad n = n_0 \ge 1. \end{aligned}$$

$\therefore f(n) = O(n^3)$ as there exists a constant $c = 17$ and $g(n) = n^3$ and $n = n_0 \ge 1$

Example 4 Let $f(n) = a_0 n^m + a_1 n^{m-1} + a_2 n^{m-2} + \ldots + a_{m-1} n + a_m$ be a polynomial of degree m. Show that $f(n) = O(n^m)$.

Solution: By triangle inequality $|f(n)| = |a_0 n^m + a_1 n^{m-1} + \ldots + a_{m-1} n + a_m|$

$$\begin{aligned} &\le |a_0|\, n^m + |a_1|\, n^{m-1} + \ldots + |a_{m-1}|\, n + |a_m| \\ &\le |a_0|\, n^m + |a_1|\, n^m + \ldots + |a_m|\, n^m \\ &\le \sum_{i=0}^{m} \left(|a_i|\right) n^m \le c\,n^m = O(n^m) \end{aligned}$$

Example 5 Show that the function $f(n) = 2n^3 + n$ and $g(n) = n^3$ are of the same order.

Solution: $f(n) = 2n^3 + n \le 2n^3 + n^3$

$$\le 3n^3, \qquad \text{for } n > 1$$

Further,

$$\begin{aligned} & n^3 = 2n^3 - n^3 \le 2n^3 + n, && n > 1 \\ & \therefore g(n) = cf(n) && \text{for } c = 1, \quad n \ge 1 \\ & \therefore g(n) = O[f(n)] \\ & \therefore f(n) = O(n^3) = O[g(n)] \\ & \therefore f(n) = g(n) \end{aligned}$$

3.12.5 The Big-Omega Ω (Lower Bound) and Big-Theta Θ (Tight Bound) Notations

The big-omega notation is employed to estimate the lower bound of the growth of f.

Let f and g be two functions $N \rightarrow R$ such that $|f(n)| \ge c\,|g(n)|$, for $n \ge n_0$, where c is a positive real constant and n_0 is some integer.

In symbols, we write as $f(n) = \Omega\,[g(n)]$ and read it as $f(n)$ is big-omega of $g(n)$.

Example 1 Find big-omega function for the function $f(n) = 5n + 6$

Solution: We can write $f(n) = 5n + 6 \geq 5n^0$ for $n = n_0 = 1$
The constant c can be taken as 5.
Hence,

$$f(n) \geq cn^0, \text{ that is, } f(n) \geq cg(n) \text{ where } g(n) = n^0 = 1.$$
$$\therefore f(n) = \Omega\,(g(n) = \Omega\,(1)$$

Example 2 Find big-omega function for the function $f(n) = 3n^2 + n + 6$

Solution: We can write $f(n) = 3n^2 + n + 6 \geq 3n^2$, for $n \geq 1$
The constant c can be taken as 3.
Hence

$$f(n) \geq cn^2$$
$$\geq c{\cdot}g(n) \quad \text{where } g(n) = n^2$$
$$\therefore f(n) = \Omega\,(n^2)$$

Big-Theta Θ Notation: The function $f(n) = \Theta\,[g(n)]$ if $g(n)$ is both an upper bound and lower bound for $f(n)$.

In other words, $f(n) = \Theta\,[g(n)]$ if there exist positive real constants c and c' and an integer n_0 such that

$$|f(n)| \leq c\,|g(n)| \text{and } |f(n)| \geq c'\,|g(n)|, \text{ for all } n \geq n_0$$

that is, $c'\,g(n) \leq f(n) \leq cg(n)$, for all $n \geq n_0$

Example 1 Let $f(n) = \dfrac{n(n+1)}{2}$, show that $f(n) = \theta(n^2)$

Solution: $f(n) = \dfrac{1}{2}\,n\,(n+1) = \dfrac{1}{2}\cdot n^2 + \dfrac{1}{2}\cdot n \leq \dfrac{1}{2}\cdot n^2 + \dfrac{1}{2}\cdot n^2, \ldots .\ n \geq 1$

$$\leq n^2$$

The constant c can be taken as 1, and $g(n) = n^2$

then

$$f(n) \leq cg\,(n) = O[g(n)] = O(n^2)$$

Further,

$$f(n) = \frac{n^2}{2} + \frac{n}{2} \geq \frac{n^2}{2}, \qquad n \geq 1$$

The constant c' can be taken as $\dfrac{1}{2}$ and $g(n) = n^2$

$$f(n) \geq c'\,g(n),\ c' = \frac{1}{2},\ g(n) = n^2$$

Thus,

$$c'\,g(n) \leq f(n) \leq cg(n)$$

Hence,

$$f(n) = \Theta\,[g(n)] = \Theta\,(n^2)$$

Example 2 Let $f(n) = n^2 (2n - 1)$, show that $f(n) = \Theta(n^3)$

Solution: $f(n) = 2n^3 - n^2$

$$\therefore |f(n)| = |2n^3 - n^2| \le |2n^3| + |n^2|$$
$$\le 2n^3 + n^3$$
$$\le 3n^3, \quad n \ge 1$$

The constant c can be taken as 3 and $g(n) = n^3$.
Further,

$$f(n) = 2n^3 - n^2 = n^3 + (n^3 - n^2) \ge n^3, \quad n \ge 1$$

The constant c' can be taken as 1 and $g(n) = n^3$.
Thus, we have $f(n) \le cg(n)$ and $f(n) \ge c'\, g(n)$, where $g(n) = n^3$.
Hence,

$$f(n) = \Theta\,[g(n)] = \Theta\,(n^3).$$

3.12.6 Little Oh (o) and Little Omega (ω)

Little Oh: Let f and g be functions $N \to R$. We say that $f(n) = o[g(n)]$ if $\lim_{n\to\infty} \frac{f(n)}{g(n)} = 0$

that is,

$$f(n) < cg(n)$$

Example 1 Let $f(n) = 5 + 3n$ and $g(n) = n^2$ then $f(n) = o[g(n)]$, since $\lim_{n\to\infty} \frac{f(n)}{g(n)} = \frac{5+3n}{n^2} = 0$

Little Omega (ω): Let f and g be functions $N \to R$. We say that $f(n) = \omega(g(n))$ if $\lim_{n\to\infty} \frac{g(n)}{f(n)} = 0$

that is,

$$f(n) > c\, g(n)$$

Example 2 Let $f(n) = 2n^2 + 1$ and $g(n) = n$
then

$$f(n) = \omega\,[g(n)], \text{ since } \lim_{n\to\infty} \frac{g(n)}{f(n)} = \lim_{n\to\infty} \frac{n}{2n^2+1} = 0$$

Note 3.6 Let $f(n)$ and $g(n)$ be two growth functions and let $\lim_{n\to\infty} \frac{f(n)}{g(n)} = l$
Then the following cases arise

1. If $0 < l < \infty$, that is, l has a finite value, then f and g are of the same order.
2. If $l = 0$, then f is of smaller order than that of g.
3. If $l \to \infty$, then f is of larger order than that of g.

Example 3 Since $\lim\limits_{n\to\infty} \dfrac{n+10}{n^2} = 0, \therefore n + 10 < n^2$ that is, $o(n + 10) < o(n^2)$

Since $\lim\limits_{n\to\infty} \dfrac{2n^2+5}{3n} \to\infty \therefore 2n^2 + 5 > 3n$, that is, $o(2n^2 + 5) > o(3n)$

Since $\lim\limits_{n\to\infty} \dfrac{5n+2}{2n+1} = \lim\limits_{n\to\infty} \dfrac{5+\frac{2}{n}}{2+\frac{1}{2}} = \dfrac{5}{2}$, which is finite, and hence $o(5n + 2) = o(2n + 1)$ that is, both functions are of the same order.

3.13 CONNECTIVITY RELATION

The various powers R^n of a relation can be used to construct a new relation called connectivity relation.

Let R be a relation on a set A. Then aR^nb means that there is a path of length n from a to b. We define a relation $aR^{\infty}b$ to mean that there is some path from a to b. The relation R^{∞} is called the connectivity relation for R and is defined as the union of all powers of R:

$$R^{\infty} = R \cup R^2 \cup R^3 \cup \ldots \cup R^n \cup \ldots$$
$$= \bigcup_{n=1}^{\infty} R^n$$

Hence,

$$M_{R^{\infty}} = M_R \vee M_{R^2} \vee M_{R^3} \vee \ldots$$

Connectivity relation consists of end points of all possible paths in R.

THEOREM 3.9 Let R be a relation on a set A, then R^{∞} is the transitive closure of R.

Proof: If a and b are in the set A, then $a\,R^{\infty}\,b$ if there exists a path from a to b. Now R^{∞} is certainly transitive since, if $a\,R^{\infty}b$ and $b\,R^{\infty}c$, the composition of paths from a to b and b to c leads to the path from a to c. Hence, R^{∞} is transitive.

To show that R^{∞} is the smallest transitive relation containing R, we prove that if there is a transitive relation S such that $R \subseteq S$, then $R^{\infty} \subseteq S$.

Now if S is transitive, then $S^n \subseteq S$ for every $n \geq 1$, that is, if a and b are connected by a path of length n, then $a\,S\,b$. It follows that $S^{\infty} = \bigcup_{n=1}^{\infty} S^n \subseteq S$. It is also true that if $R \subseteq$ S, then $R^n \subseteq S^n$ as any path in R is also a path in S. Combining these facts together, we find that if $R \subseteq S$ and S is transitive on A, then

$$R^{\infty} \subseteq S^n \subseteq S$$

This shows that R^{∞} is the smallest transitive relation on A that contains R.

Note 3.7 The connectivity relation R^∞ has several interpretations. From the geometric point of view, it specifies the vertices which are connected by path. From the algebraic point of view, R^∞ is the transitive closure of R, which plays an important role in the theory of equivalence relations and in the theory of certain languages such as BASIC, JAVA and C++.

Example 1 Let $A = \{1, 2, 3\}$ and $R\{(1, 1), (1, 2), (1, 3), (2, 3)\}$. Find the connectivity relation R^∞.

Solution: $R^\infty = R \cup R^2 \cup R^3$

Now

$$R^2 = RoR = \{(1, 1), (1, 2)\ (1, 3), (2, 3)\} \text{ o } \{(1, 1), (1, 2), (1, 3), (2, 3)\}$$
$$=\{(1, 1), (1, 2), (1, 3)\}$$
$$R^3 = R^2oR = \{(1, 1), (1, 2), (1, 3)\}$$
$$\therefore\ R^\infty = \{(1, 1), (1, 2), (1, 3), (2, 3)\}$$

Example 2 Find the connectivity relation R^∞ of the relation $R = \{(1, 2), (2, 3), (2, 2)\}$ on $A = \{1, 2, 3\}$.

Method 1

Solution: $R^\infty = R \cup R^2 \cup R^3$

Now

$$R^2 = RoR = \{(1, 2), (2, 3), (2, 2)\} \text{ o } \{(1, 2), (2, 3), (2, 2)\}$$
$$= \{(1, 2), (1, 3), (2, 2), (2, 3)\}$$
$$R^3 = R^2oR = \{(1, 2), (1, 3), (2, 2), (2, 3)\}$$
$$R^\infty = R \cup R^2 \cup R^3 = \{(1, 2), (2, 3), (2, 2), (1, 3)\}$$

Method 2 Solution by method of matrix representation. The matrix of relation R is

$$M_R = \begin{bmatrix} 0 & 1 & 0 \\ 0 & 1 & 1 \\ 0 & 0 & 0 \end{bmatrix}$$

$$M_R^{\ 2} = (M_R)^2 = \begin{bmatrix} 0 & 1 & 1 \\ 0 & 1 & 1 \\ 0 & 0 & 0 \end{bmatrix}, (M_R)^3 = \begin{bmatrix} 0 & 1 & 1 \\ 0 & 1 & 1 \\ 0 & 0 & 0 \end{bmatrix} = M_R^{\ 3}$$

$$\therefore R^\infty = M_R \vee M_R^{\ 2} \vee M_R^{\ 3}$$

$$= \begin{bmatrix} 0 & 1 & 1 \\ 0 & 1 & 1 \\ 0 & 0 & 0 \end{bmatrix}$$

$$\therefore R^\infty = \{(1, 2), (1, 3), (2, 2), (2, 3)\}$$

which gives the same relation as in method 1.

Method 3 The diagraph of the relation R is the following: since connectivity relation R^∞ is the transitive closure of R, we can proceed graphically by computing all paths. It can be observed that from vertex 1, the paths are formed by joining vertices 2 and 3. The paths from 1 proceeds to 2 and from 2 to 3. Thus ordered pair (1, 3) exists in R^∞. Again, beginning with vertex 2, there is only path to 3. Hence, we have

$$R^\infty = \{(1, 2),\ (1, 3),\ (2, 2),\ (2, 3)\}$$

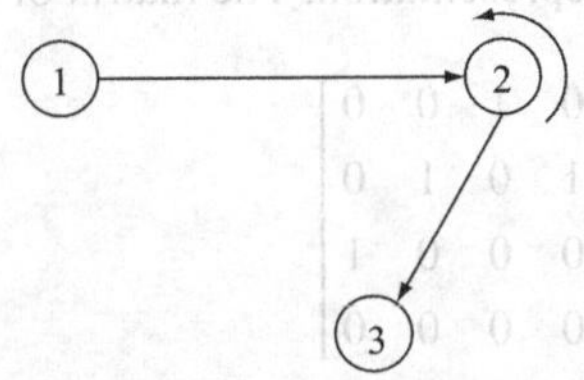

THEOREM 3.10 Let A be a finite set with $|A| = n$ and let R be a relation on A. Then, powers of R greater than n are not needed to compute R^∞.

The proof of theorem is given using simple examples of transitive closure for a set with few elements, that is, three or four elements.

Example 1 Find the connectivity relation R^∞ for relation $R = \{(1, 1), (1, 2) (2, 3)\}$ on $A = \{1, 2, 3\}$

Solution:

$$R^\infty = R \cup R^2 \cup R^3$$

Now

$$R^2 = RoR = \{(1, 1), (1, 2), (2, 3)\}o\{(1, 1), (1, 2), (2, 3)\}$$

or $$R^2 = \{(1, 1), (1, 2) (1, 3)\}$$

and $$R^3 = R^2oR = \{(1, 1), (1, 2) (1, 3)\}o\{(1, 1), (1, 2), (2, 3)\}$$

$$\therefore\quad R^3 = \{(1, 1), (1, 2), (1, 3)\}$$

Hence,

$$R^\infty = R \cup R^2 \cup R^3 = \{(1, 1),\ (1, 2)\ (1, 3),\ (2, 3)\}$$

Example 2 Find the connectivity relation R^∞ for the relation

$$R = \{(1, 2), (2, 3), (3, 4)\} \text{ on } A = \{1, 2, 3, 4\}$$

Solution:

$$R^2 = RoR = \{(1, 2), (2, 3) (3, 4)\}o\{(1, 2), (2, 3), (3, 4)\}$$

or $$R^2 = \{(1, 3), (2, 4)\}$$

$$R^3 = R^2oR = \{(1, 3), (2, 4)\}o\{(1, 2), (2, 3), (3, 4)\}$$

or $$R^3 = \{(1, 4)\}$$

and $$R^4 = R^3oR = \{(1, 4)\}o\{(1, 2), (2, 3), (3, 4)\} = \phi$$

Furthermore,

$$R^5 = R^6 = \ldots = \phi$$

Hence,

$$R^{\infty} = R \cup R^2 \cup R^3 = \{(1, 2), (1, 3)\ (1, 4), (2, 3), (2,4), (3,4)\}$$

Example 3 Find the connectivity relation R^{∞} of the relation $R = \{(1,2), (2,1), (2,3), (3,4)\}$ on $A = \{1,2,3,4\}$.

Solution: By method of matrix representation. The matrix of R is

$$M_R = \begin{bmatrix} 0 & 1 & 0 & 0 \\ 1 & 0 & 1 & 0 \\ 0 & 0 & 0 & 1 \\ 0 & 0 & 0 & 0 \end{bmatrix}$$

$$M_{R^2} = (M_R)^2 = \begin{bmatrix} 1 & 0 & 1 & 0 \\ 0 & 1 & 0 & 1 \\ 0 & 0 & 0 & 0 \\ 0 & 0 & 0 & 0 \end{bmatrix}, \quad M_{R^3} = (M_R)^3 = \begin{bmatrix} 0 & 1 & 0 & 1 \\ 1 & 0 & 1 & 0 \\ 0 & 0 & 0 & 0 \\ 0 & 0 & 0 & 0 \end{bmatrix}$$

$$M_{R^4} = (M_R)^4 = \begin{bmatrix} 1 & 0 & 1 & 0 \\ 0 & 1 & 0 & 1 \\ 0 & 0 & 0 & 0 \\ 0 & 0 & 0 & 0 \end{bmatrix}, \quad M_{R^5} = \begin{bmatrix} 0 & 1 & 0 & 1 \\ 1 & 0 & 1 & 0 \\ 0 & 0 & 0 & 0 \\ 0 & 0 & 0 & 0 \end{bmatrix}$$

Continuing in this way, it is seen that $(M_R)^n$ equals $(M_R)^2$ if n is even and equals $(M_R)^3$ if n is odd and greater than 1. Thus,

$$M_{R^{\infty}} = M_R \vee M_{R^2} \vee M_{R^3} = \begin{bmatrix} 1 & 1 & 1 & 1 \\ 1 & 1 & 1 & 1 \\ 0 & 0 & 0 & 1 \\ 0 & 0 & 0 & 0 \end{bmatrix}$$

∴ connectivity relation $R^{\infty}=\{(1, 1), (1, 2), (1, 3), (1, 4), (2, 1), (2, 2), (2, 3), (2, 4), (3, 4)\}$

Note 3.8 In this example, we did not need to consider all power of R^n to obtain R^{∞}. This is true whenever the set A is finite.

Three different methods, that is, graphical method (using diagraph), various powers of relation R and matrix method have been used to compute connectivity relation R^{∞} (transitive closure). Each of these methods have certain difficulties. The graphical method is impractical for large sets and relations. The matrix method can be used, which is systematic to be programmed for a computer. However, for large matrices, it is inefficient. A significant and more efficient algorithm for computing the transitive closure of the relation R, that is, R^{∞} was given by Stephen Warshall in 1962 and is known as Warshall's algorithm.

Warshall's Algorithm: The aim of the algorithm is to generate a sequence of matrices $W_0, W_1,\ldots, W_n$ for the relation R defined on the set $\{a_1, a_2,\ldots, a_k,\ldots, a_n\}$ beginning with $W_0 = M_R = [m_{i,j}]$.

The first iteration consists of searching the existence of paths from any vertex to any other vertex directly through an edge or indirectly through an intermediate vertex (or pivot vertex), say v_1. If W_1 denotes the resulting matrix then

$$W_1 = \left(m_{ij}\right)^1 = \begin{cases} 1 \text{ if there exists an edge from } v_i \text{ to } v_j \text{ or} \\ \quad \text{there is a path (of length 2) from } v_i \text{ to } v_1 \text{ and } v_1 \text{ to } v_j \\ 0 \text{ otherwise} \end{cases}$$

The second iteration is to search paths from any vertex to any other with v_1 or v_2 or both as pivots. We compute the resulting matrix W_2 defined as follows:

$$W_2 = \left(m_{ij}{}^2\right) = \begin{cases} 1 \text{ if there exists a path from } v_i \text{ to } v_j \text{ or a path from } v_i \text{ to } v_j \\ \quad \text{using only pivot (intermediate) values from } \{v_1, v_2\} \\ 0 \text{ otherwise} \end{cases}$$

In general, in the kth iteration, the pivots are used from the set $\{v_1, v_2,\ldots, v_k\}$ and the resulting matrix W_k is defined as follows:

$$W_k = \left(m_{ij}\right)^k = \begin{cases} 1 \text{ if there exists a path from } v_i \text{ to } v_j \text{ or a path from } v_i \text{ to } v_j \\ \quad \text{using only pivots from } \{v_1, v_2,\ldots, v_k\} \\ 0 \text{ otherwise} \end{cases}$$

This is the basis of Warshall's algorithm.

We can also compute $W_k = (m_{ij})^k$ from previous iteration $W_{k-1} = (m_{ij})^{k-1}$ by the following procedure:

Step 1: Transfer to W_k all 1s in W_{k-1}.

Step 2: List the locations $r_1, r_2,\ldots$ in columns k of W_{k-1} where entry is 1 and the location $s_1, s_2\ldots$ in row k of $W,_{k-1}$, where the entry is 1.

Step 3: Put 1s in all the positions $r_i s_j$ of W_k (if they are not already there).

In other words, if (i,j)th position in W_{k-1} has 1, then W_k has also 1 in the (i,j)th position. We add 1 in the (i,j)th position of W_k only if kth column of W_{k-1} has 1 in position i and kth row of W_{k-1} has 1 in position j.

Thus, we have

$$W_k = (m_{ij})^k = m_{ij}^{k-1} \vee (m_{ik}^{k-1} \wedge m_{kj}^{k-1})$$

Warshall's algorithm

```
Warshall              W_0 = M_R: n × n zero matrix
                      W = M_R(W = [w_ij])

For k = 1 to n
begin
        for i = 1 to n
        begin
        for j = 1 to n
W(i,j) = W(i,j) ∨ (W(i,k) ∧ W(k,j)
end
end {W=[W_ij] is R}
```

Example 1 Using Warshall's algorithm, find the transitive closure R^∞ of the relation $R = \{(1, 2), (2, 2), (2, 3), (3, 3)\}$ on the set $A = \{1, 2, 3\}$

Solution: The matrix of R is

$$W_0 = M_R = \begin{bmatrix} 0 & 1 & 0 \\ 0 & 1 & 1 \\ 0 & 0 & 1 \end{bmatrix}$$

(1) To compute W_1 so that $K = 1$ (i.e., column 1)
   (i) Transfer all 1's from W_0 to W_1.
   (ii) Non-zero entries in column 1 are none and non-zero entries in row 1 are at location 2.

   Hence, no new entry. Hence, $W_1 = W_0 = \begin{bmatrix} 0 & 1 & 0 \\ 0 & 1 & 1 \\ 0 & 0 & 1 \end{bmatrix}$

(2) To compute W_2 so that $K = 2$ (i.e., column 2)
   (i) Transfer all 1's from W_1 to W_2.
   (ii) Non-zero entries in column 2 are in locations 1 and 2 and non-zero entries in row 2 are at locations 2 and 3.

   Hence,

$$W_2 \text{ has 1 in positions } (1, 2), (1, 3), (2, 2), (2, 3).$$

$$\therefore\ W_2 = \begin{bmatrix} 0 & 1 & 1 \\ 0 & 1 & 1 \\ 0 & 0 & 1 \end{bmatrix}$$

(3) To compute W_3 so that $K = 3$ (i.e., column 3)
   (i) Transfer all 1's from W_2 to W_3.
   (ii) Non-zero entries in column 3 are in locations 1, 2 and 3 and non-zero entries in row 3 is in location 3.

   Hence,

$$W_3 \text{ has 1s in positions } (1, 3), (2, 3) \text{ and } (3, 3).$$

$$\therefore\ W_3 = \begin{bmatrix} 0 & 1 & 1 \\ 0 & 1 & 1 \\ 0 & 0 & 1 \end{bmatrix}$$

   Hence,

$$R^\infty = \{(1, 2), (1, 3), (2, 2), (2, 3), (3, 3)\}$$

Example 2 Using Warshall's algorithm, find the transitive closure for the relation $R = \{(1, 1), (1, 2), (2, 1), (2, 2), (3, 4), (1, 4), (2, 3)\}$ on the set $A = \{1, 2, 3, 4\}$.

Solution: The matrix of relation R is

$$W_0 = M_R = \begin{bmatrix} 1 & 1 & 0 & 1 \\ 1 & 1 & 1 & 0 \\ 0 & 0 & 0 & 1 \\ 0 & 0 & 0 & 0 \end{bmatrix}$$

(1) To compute W_1 so that $K = 1$ (column 1)

(i) Transfer all 1's from W_0 to W_1.

(ii) Non-zero entries in column 1 are in locations 1 and 2 and non-zero entries in row 1 are in locations 1, 2 and 4.

Hence,

W_1 has 1 in positions (1, 1), (1, 2), (2, 1), (2, 2), (1, 4), (2, 4)

$$\therefore\ W_1 = W_0 = \begin{bmatrix} 1 & 1 & 0 & 1 \\ 1 & 1 & 1 & 1 \\ 0 & 0 & 0 & 1 \\ 0 & 0 & 0 & 0 \end{bmatrix}$$

(2) To compute W_2 so that $k = 2$ (column 2)

(i) Transfer all 1's from W_1 to W_2.

(ii) Non-zero entries in column 2 are in locations 1 and 2 and non-zero entries in row 2 are in locations 1, 2, 3.

Hence,

W_2 has 1 in positions (1, 1), (1, 2), (1, 3), (2, 1), (2, 2), (2, 3).

$$\therefore\ W_2 = \begin{bmatrix} 1 & 1 & 1 & 1 \\ 1 & 1 & 1 & 1 \\ 0 & 0 & 0 & 1 \\ 0 & 0 & 0 & 0 \end{bmatrix}$$

(3) To compute W_3 so that $K = 3$ (column 3).

(i) Transfer all the entries from W_2 to W_3.

(ii) Non-zero entries in column 3 are in locations 1 and 2 and non-zero entries in row 3 are at location 4.

Hence, W_3 has 1 in positions (1, 4) and (2, 4).

As these values are already there, $W_3 = W_2$.

$$\therefore\ W_3 = \begin{bmatrix} 1 & 1 & 1 & 1 \\ 1 & 1 & 1 & 1 \\ 0 & 0 & 0 & 1 \\ 0 & 0 & 0 & 0 \end{bmatrix}$$

(4) To compute W_4 so that $K = 4$ (column 4)
   (i) Transfer all 1's from W_3 to W_4.
   (ii) Non-zero entries in column 4 are in locations 1, 2 and 3 and the non-zero entries in row 4 are nil.
   Hence, no entry is added.

$$\therefore W_4 = W_3$$

Hence

$$R^{\infty} = \{(1, 1), (1, 2), (1, 3), (1, 4), (2, 1), (2, 2), (2, 3), (2, 4), (3, 4)\}$$

Example 3 Using Warshall's algorithm, compute transitive closure of the relation $R = \{(1, 1), (1, 4), (2, 2), (2, 3), (3, 2), (3, 3), (4, 1), (4, 4)\}$ on the set $A = \{1, 2, 3, 4\}$

Solution: The matrix of relation R is

$$W_0 = M_R = \begin{bmatrix} 1 & 0 & 0 & 1 \\ 0 & 1 & 1 & 0 \\ 0 & 1 & 1 & 0 \\ 1 & 0 & 0 & 1 \end{bmatrix}$$

(1) To compute W_1 so that $K = 1$ (column 1)
   (i) Transfer all 1's from W_0 to W_1.
   (ii) Non-zero entries in column 1 are in locations 1 and 4 and non-zero entries in row 1 are in locations 1 and 4.
   Hence, W_1 has 1 in positions (1, 1), (1, 4), (4, 1), (4, 4).

$$\therefore W_1 = \begin{bmatrix} 1 & 0 & 0 & 1 \\ 0 & 1 & 1 & 0 \\ 0 & 1 & 1 & 0 \\ 1 & 0 & 0 & 1 \end{bmatrix}$$

(2) Compute W_2 so that $K = 2$ (column 2)
   (i) Transfer all 1's from W_1 to W_2.
   (ii) Non-zero entries in column 2 are in locations 2 and 3 and non-zero entries in row 2 are in locations 2 and 3.
   Hence, W_2 has 1 in positions (2, 2), (2, 3), (3, 2), (3, 3) and thus there is no new entry.

$$\therefore W_2 = W_1 = \begin{bmatrix} 1 & 0 & 0 & 1 \\ 0 & 1 & 1 & 0 \\ 0 & 1 & 1 & 0 \\ 1 & 0 & 0 & 1 \end{bmatrix}$$

(3) To compute W_3 so that $K = 3$ (column 3)

(i) Transfer all 1's from W_2 to W_3.

(ii) Non-zero entries in column 3 are in locations 2 and 3 and non-zero entries in row 3 are in locations 2 and 3.

Hence, w_3 has 1 in positions (2, 2), (2, 3), (3, 2), (3, 3).

As these values are there, no new entry is needed. Hence,

$$W_3 = W_2 = \begin{bmatrix} 1 & 0 & 0 & 1 \\ 0 & 1 & 1 & 0 \\ 0 & 1 & 1 & 0 \\ 1 & 0 & 0 & 1 \end{bmatrix}$$

(4) To compute W_4 so that $K = 4$ (column 4)

(i) Transfer all 1's from W_3 to W_4.

(ii) Non-zero entries in column 4 are in locations 1 and 4 and non-zero entries in row 4 are in locations 1 and 4.

Hence, W_4 has 1 in positions (1, 1), (1, 4), (4, 1), (4, 4).

As these values are already there, no new entry is needed. Hence,

$$W_4 = W_3 = \begin{bmatrix} 1 & 0 & 0 & 1 \\ 0 & 1 & 1 & 0 \\ 0 & 1 & 1 & 0 \\ 1 & 0 & 0 & 1 \end{bmatrix}$$

Hence, the required transitive closure.

$$R^{\infty} = \{(1, 1), (1, 4), (2, 2), (2, 3), (3, 2), (3, 3), (4, 1), (4, 4)\}$$

EXERCISE – A

1. Define a 'function' giving three examples.
2. Let A be set of workers in a factory. Determine which of the following assignments define a function
   (i) To each worker, assign his or her age.
   (ii) To each worker, assign his or her supervisor.
   (iii) To each worker, assign a machine to work.
3. Define the following giving one example in each case
   (i) INTO and ONTO functions.
   (ii) One-one into and one-one onto functions.
   (iii) Many-one into and many-one onto functions.
4. State whether the following diagrams define a function from $A = \{a, b, c, d\}$ to the set $B = \{x, y, z, w\}$.

(i)

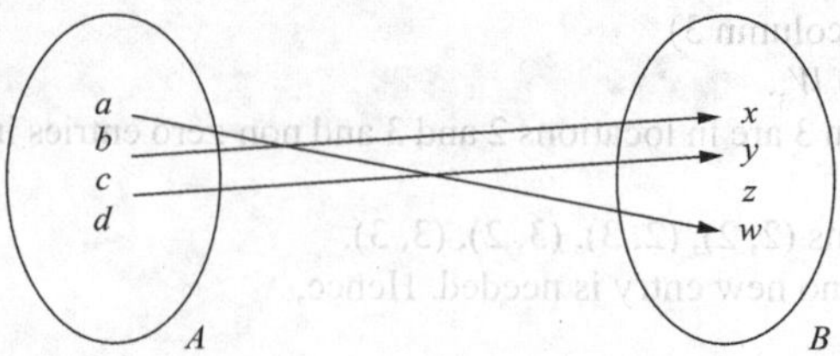

(ii)

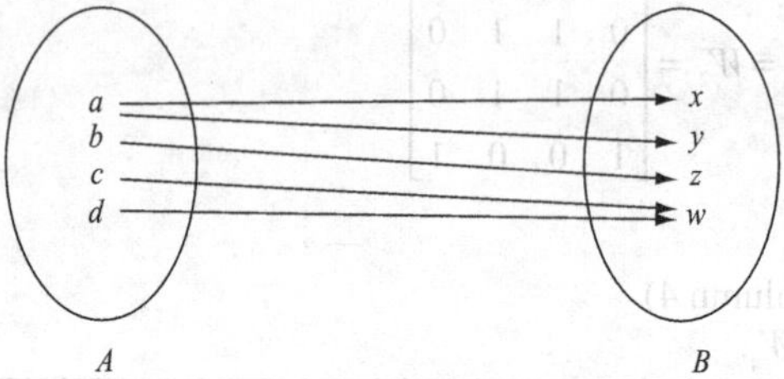

(iii)

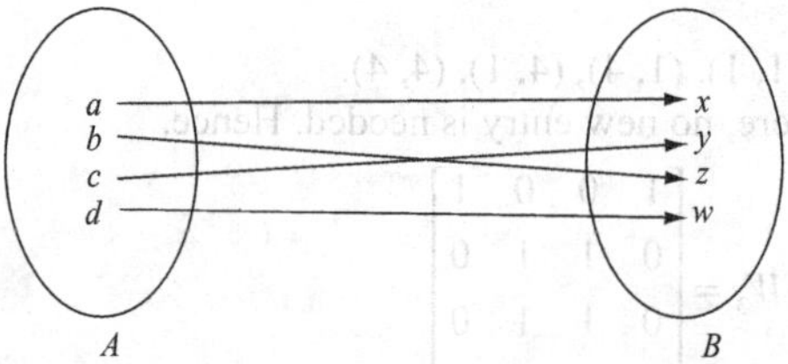

5. Determine which of the following relations from $A = \{a, b, c, d\}$ to $B = \{1, 2, 3, 4, 5\}$ is a function
   (i) $R = \{(a, 1), (b, 2), (c, 3), (d, 5)\}$
   (ii) $R = \{(a, 1), (a, 2), (b, 2), (c, 4), (d, 5)\}$
   (iii) $R = \{(a, 5), (b, 4), (c, 3)\}$
   (iv) $R = \{(a, 1), (d, 4), (b, 3), (c, 5)\}$

6. (i) If $A = \{3, 4, 5, 6\}$ and $B = \{a, b, c, d\}$, then find whether the relation $f = \{(3, a), (3, b), (4, c), (5, d)\}$ is a function.
   (ii) Let $A = \{-1, 0, 1, 2, 3\}$ and let the function $f: A \rightarrow R$ be defined by $f(x) = 2x + 1$. Find the range of f.

7. Let $A = \{a, b, c, d\}$ and $B = \{1, 2, 3\}$ be two given sets. Determine whether the relations $g: A \rightarrow B$ are functions. If it is a function, give its domain and range.
   (i) $g = \{(a, 1), (b, 1), (c, 3)\}$
   (ii) $g = \{(a, 2), (b, 3), (c, 1)\}$
   (iii) $g = \{(a, 3), (b, 2), (c, 1), (d, 2)\}$
   (iv) $g = \{(a, 3), (b, 1), (c, 1), (d, 3)\}$

8. If $f: R \rightarrow R$ is defined by $f(x) = e^x, x \in R$, prove that that the function f is one-one and into.

9. Determine whether the following functions $f: R \to R$ are one-one/many-one/into/onto.
   (i) $f(x) = x + 2$ (ii) $f(x) = x^2$ (iii) $f(x) = x^3$
   (iv) $f(x) = 5x$ (v) $f(x) = 3x + 2$ (vi) $f(x) = x^2 + x$

10. Examine the nature of functions $f: Z \to Z$ defined by
    (i) $f(x) = x + 5$ (ii) $f(x) = 2x + 3$ (iii) $f(x) = x^2 + 1$

11. Find whether the following functions from A to B are one-one/many-one/into/onto.
    (i) $A = \{1, 2, 3, 4\}$, $B = \{1, 2, 3, 4\}$
    $f = \{(1, 1), (2, 3), (3, 4), (4, 2)\}$
    (ii) $A = \{a, b, c, d\}$, $B = \{1, 2, 3, 4, 5\}$
    $f = \{(a, 2), (b, 2), (c, 1), (d, 5)\}$

12. Define $f: A \to A$ where $A = \{1, 2, 3, 4\}$ such that
    (i) f is one-one onto (ii) f is many-one into

13. Show that the functions $f: R \to R$ and $g: R \to R$, where $f(x) = \cos x$ and $g(x) = -\sin x$ are neither one-one nor onto. Find domain and co-domain so that it is one-one onto.

14. Let $f: N \to A$ be defined by $f(x) = 2x$, where A is the set of even natural numbers. Prove that f is one-one onto and hence find f^{-1}.

15. Let $g: R \to R$ be defined by $g(x) = ax + b$, $a, b \in R$. Prove that g is one-one onto and hence find g^{-1}.

16. Let $A = \{a, b, c, d\}$ and $B = \{x, y, z, u\}$ be two sets and functions f and g defined from A to B as follows. Determine whether f^{-1} and g^{-1} exists.
    (i) $f = \{(a, x), (b, u), (c, z), (d, y)\}$
    (ii) $g = \{(a, u), (b, x), (c, y), (d, z)\}$

17. Let f and g be two functions from R to R defined by $f(x) = x^2$ and $g(x) = x + 2$. Find fog and gof. Show that $fog \neq gof$. Also find $gof(1)$ and $fog(1)$.

18. If functions f and g from R to R are defined by
    (i) $f(x) = \sin x$ and $g(x) = x^2$
    (ii) $g(x) = x^2 + 3$ and $f(x) = 2x + 2$
    Show that $fog \neq gof$ in both the cases.

19. Let $A = B = C = R$ and let $f: A \to B$ and $g: B \to C$ be defined by $f(a) = a + 3$ and $g(b) = 2b + 1$ Find
    (i) $gof(5)$ (ii) $fog(5)$ (iii) $gof(-5)$
    (iv) $fog(-5)$ (v) $(gof)\,x$ (vi) $(fog)\,x$

20. Let $A = B = C = R$. Define $f: A \to B$ by $f(x) = 3x + 1$ and $g: B \to C$ by $g(y) = y/4$. Verify $(gof)^{-1} = f^{-1}og^{-1}$.

21. Find: (i) $\lfloor 8.5 \rfloor$ (ii) $\lfloor -8.5 \rfloor$ (iii) $\lfloor -5 \rfloor$

22. Find: (i) $\lceil 8.5 \rceil$ (ii) $\lceil -8.5 \rceil$ (iii) $\lceil -5 \rceil$

23. Find: (i) 34 mod (5) (ii) 40 mod (8) (iii) −35 mod (11)

24. Evaluate: (i) $\log_2 8$ (ii) $\log_2 16$ (iii) $\log_2 64$

25. Evaluate: (i) $\log_{10} 100$ (ii) $\log_{10} 1000$ (iii) $\log_{10} 0.001$

EXERCISE – B

1. Let f be a function that describes the number of operations to carry out an algorithm. The number of items to be processed is represented by n. Describe what happens to the number of steps in each function if the number of items is doubled.
   (i) $f(n) = 5001$ (ii) $f(n) = 2n$
   (iii) $f(n) = 3n^2$ (iv) $f(n) = 2{\cdot}5\, n^3$
2. If $f(n) = n!$, show that $O(f(n)) = O(n^n)$.
3. Show that $g(n) = 1 + 2 + 3 + \ldots + n$ is $O(n^2)$.
4. Show that $f(n) = 5n + \lg n$ is $O(n)$.
5. Show that $g(n) = n^2 (7n^2 - 2n)$ is $O(n^4)$.
6. Show that the functions f and g have the same order for $f(n) = 2n^2 + 5n$ and $g(n) = n^2 + 10n$.
7. Show that the function $f(n) = 5n^4 - 2n^3$ and $g(n) = n^4$ are of the same order.
8. Show that (i) $3n + 5 = \Omega(n)$, (ii) $5n^2 + 3n - 2 = \Omega(n^2)$.
9. If $f(n) = 25n^3 + 5n + 15$, then show that $f(n) = \Omega(n^3)$.
10. Show that $f(n) = 25n^5 - 2n^4 + 5n^3 + 2n + 1$ is $O(n^5)$.
11. Show that $f(n) = \dfrac{n(n+1)}{2}$ is $\Theta(n^2)$.
12. Show that $f(n) = 2^{n+1}$ is $O(2^n)$.
13. Show that $f(n) = 2^{2n}$ is not $O(2^n)$.
14. Show that $f(n) = \Omega(\lg n)$ for $f(n) = 25 + 2 \lg n$.
15. Show that $1^2 + 2^2 + 3^2 \ldots + n^2$ is $O(n^3)$.
16. Show that $1^k + 2^k + 3^k + \ldots + n^k$ is $O(n^{k+1})$, for $k \geq 1$.
17. Using Warshall's algorithm, find the transitive closure R^∞ of the relation $R = \{(1, 1), (1, 2), (2, 3), (3, 3)\}$ on the set $A = \{1, 2, 3\}$.
18. Using Warshall's algorithm, find the connectivity relation R^∞ of the relation $R = \{(1, 1), (2, 1), (2, 2), (2, 3), (3, 3)\}$ on the set $A = \{1, 2, 3\}$.
19. Let $A = \{1, 2, 3\}$ and let $R = \{(1, 1), (1, 2), (1, 3), (2, 3), (3, 1), (3, 2)\}$. Compute transitive closure by using Warshall's algorithm.
20. Using Warshall's algorithm, compute transitive closure of relation R on the set $A = \{1, 2, 3, 4\}$ whose matrix is

$$M_R = \begin{bmatrix} 1 & 1 & 0 & 0 \\ 1 & 0 & 1 & 1 \\ 0 & 1 & 0 & 1 \\ 0 & 1 & 1 & 0 \end{bmatrix}$$

21. Using Warshall's algorithm, compute transitive closure of the relation $R = \{(1, 1), (1, 2), (1, 4), (2, 1), (2, 3), (3, 2), (3, 3), (4, 3), (4, 4)\}$ on the set $A = \{1, 2, 3, 4\}$.

22. Let $A = \{a_1, a_2, a_3, a_4, a_5\}$ and R be a relation on A whose matrix is

$$M_R = \begin{bmatrix} 1 & 0 & 0 & 1 & 0 \\ 0 & 1 & 0 & 0 & 0 \\ 0 & 0 & 0 & 1 & 1 \\ 1 & 0 & 0 & 0 & 0 \\ 0 & 1 & 0 & 0 & 1 \end{bmatrix} = W_0$$

Using Warshall's algorithm show that $W_1 = W_2 = W_3$

4

Mathematical Logic and Methods of Proofs

LEARNING OBJECTIVES

After reading this chapter, you will be conversant with:

- Introduction
- Statement (Proposition)
- Propositional Variables, Simple and Compound Propositions (or Statements)
- Basic Logical Operations
- Tautology and Contradiction
- Logically Equivalent or Equivalent Propositions
- Logical Arguments
- Predicates
- Methods of Proof

4.1 INTRODUCTION

Logic is the discipline that deals with the principles and techniques of reasoning, that are required to recognise valid arguments. Logical reasoning was first developed by the Greek philosopher Aristotle (384–322 BC) who is known as the father of logic. Logic plays an important role in every area of all mathematical and automated reasoning. It is used in mathematics to prove theorems. It also has practical applications in computer programming, in artificial intelligence and in the design of computer systems. One component of logic is the *propositional calculus*, which deals with statements that are either true or false. Another component is *predicate calculus*, which deals with predicates that are propositions containing variables.

In this chapter, the basic rules of mathematics that deal with logical reasoning are introduced.

4.2 STATEMENT (PROPOSITION)

A statement (or a proposition) is a declarative sentence that is either true or false but not both, but is meaningful. The truthfulness or fallacy (falsehood) of a statement is called its truth value, which are denoted by T and F, respectively.

For definiteness of statement, Aristotle has given the following laws:

1. *The Law of Identity*: Under this law, the symbols will have the same meaning throughout the specified problem.
2. *The Law of Excluded Middle*: For every statement or proposition p, either p is true or p is false and there is no possibility of intermediate exclusion.
3. *The Law of Contradiction*: No statement is true and false simultaneously.

Consider the following sentences:

1. There are eight primes less than 20.
2. Aristotle was a great philosopher.
3. Earth revolves round the Sun.
4. $5 + 6 = 11$
5. Fifty one is an even natural number.
6. $x - 3 = 10$
7. Where are you going?
8. Take your medicine.
9. Oh! A great batting by Sachin Tendulkar!
10. What are rational numbers?

The sentences (1), (2), (3), (4) and (5) are statements as they are true except (5) which is false. (6) is not a statement (or proposition) since it is neither true nor false. It can be a proposition if the variable x is assigned some value. (7) and (10) are interrogative sentences, (8) is a command and (9) is an exclamatory sentence. Thus (6), (7), (8), (9) and (10) are not statements (or propositions).

Note 4.1 There can be a declarative sentence that is neither true nor false. A well-known example is the following:

The liar says, 'I always tell a lie'.

Let us assume that the sentence is true. Since he is a liar, it cannot be true. On the contrary, if we assume that the sentence is false, then it will imply that he is not a liar, which is not true. Thus, the above sentence is neither true nor false and so is not a statement or proposition.

4.3 PROPOSITIONAL VARIABLES, SIMPLE AND COMPOUND PROPOSITIONS (OR STATEMENTS)

In logic, letters p, q, r, … denote propositional variables, that is, variables that can be used for statements.

For example, one can write p: The sun rises in the east, q: Delhi is the capital of India, r: Indian Institutes of Technology are the leading technical institutions of India. These variables are assigned letters T and F according as they are true or false, respectively.

4.3.1 Simple Propositions

The following propositions are simple propositions:

1. The sun rises in the east.
2. Einstein was a genius.
3. Newton was a great mathematician.
4. Roses are red.
5. The value of π is 3.14159265
6. The value of e is 2.71828

4.3.2 Compound Proposition

A compound proposition is formed by two or more simple propositions by using logical connectives And, Or and Not.

1. Let p: It is raining.
   q: It is very cloudy.
   The compound statement of the two simple propositions is p and q: It is raining and it is very cloudy.

2. Let p: $25 = 5^2$
   q: $3.15 > \pi$
   The compound statement is p and q: $25 = 5^2$ and $3.15 > \pi$

3. Let p: Ram is playing.
   q: Ram is listening to music.

   The compound statement is p or q: Ram is playing or he is listening to music.

The truth value of a compound statement depends only upon the truth values of simple statements, which are being combined and also the types of connectives being used.

4.4 BASIC LOGICAL OPERATIONS

There are three basic logical operations, which are used to combine propositions. These are conjunction, disjunction and negation, which correspond to English words and, or and not, respectively, and are denoted by symbols $\wedge$, $\vee$ and ~.

1. Negation	p' or $\sim p$	read as no p
2. Conjunction	$p \wedge q$	read as p and q
3. Disjunction	$p \vee q$	read as p or q

where p and q are propositions.

4.4.1 Negation

If p is a statement, then negation of p is a statement not p. The negation of statement. "x equal to δ" is the statement "x does not equal to δ". Thus, negation of "$x = \delta$"is $x \neq \delta$.

The truth table of $\sim p$ relative to p is as follows: Strictly speaking, negation is not connective as it does not combine two statements. In fact, negation is a unary operation for the statements. Further, double negation is positive. Table 4.1 shows truth table for negation.

Table 4.1 Truth table for negation (~)

p	~p
T	F
F	T

Table 4.2 Truth table for double negation

p	~p	~(~p)
T	F	T
F	T	F

Table 4.2 shows that double negation of any statement is the original statement.

Example: Give negation of the following statements:

(a) p: $3 + 7 > 2$ then $\sim p$: $3 + 7 \leq 2$

(b) q: It is a sunny day, then $\sim q$: It is not a sunny day.

(c) r: It is raining, then $\sim r$: It is not raining.

(d) t: Ram is a Hindu god, then $\sim t$: Ram is not a Hindu god.

4.4.2 Conjunction

The conjunction of two statements p and q is the compound statement p and q and is denoted by $p \wedge q$.

To define the truth value of $p \wedge q$, we need to consider the following four cases:

1. p is true, q is true.
2. p is true, q is false.
3. p is false, q is true.
4. p is false, q is false.

All the above cases can be summarised and presented in the form of a table where T and F denote that the statement is true or false. Table 4.3 shows the truth table for $p \wedge q$.

Table 4.3 Truth table for $p \wedge q$

p	q	$p \wedge q$
T	T	T
T	F	F
F	T	F
F	F	F

Consider the following four statements corresponding to four values that the statements p and q assume.

1. p: Delhi is the capital of India (T), q: $3 + 5 = 8$ (T)
2. p: Delhi is the capital of India (T), q: $3 + 5 = 7$ (F)
3. p: Delhi is the capital of England (F), q: $3 + 5 = 8$ (T)
4. p: Delhi is the capital of England (F), q: $3 + 5 = 7$ (F)

In the language of sets, and ($\wedge$ operation) is a binary operation on the set of statements. The compound statement $p \wedge q$ is true when both p and q are true, otherwise it is false.

Example: Form the conjunction of statements p and q for each of the following statements:

(a) p: Ashok is hungry. q: Vinay is sleeping.
(b) p: Plato was a great philosopher. q: Kanpur is a city with high pollution.
(c) p: Varanasi is a city of temples. q: The river Ganga originates from Gangotri.

Solution:

(a) $p \wedge q$: Ashok is hungry and Vinay is sleeping.
(b) $p \wedge q$: Plato was a great philosopher and Kanpur is a city with high pollution.
(c) $p \wedge q$: Varanasi is a city of temples and the river Ganga originates from Gangotri.

4.4.3 Disjunction

A second way of combining two statements p and q is by using the connective OR denoted by $p \vee q$ and read as p or q. This is true only when one or both the statements are true. $p \vee q$ is false when both are false (See Table 4.4).

Table 4.4 Truth table of $p \vee q$

p	q	$p \vee q$
T	T	T
T	F	T
F	T	T
F	F	F

1. Let p: 2 is a prime number and q: $\sqrt{3}$ is a rational number, then $p \vee q$: 2 is a prime number or $\sqrt{3}$ is a rational number. Since p is true, the disjunction $p \vee q$ is true although q is false.
2. Let p: Ganga is a mountain.
   q: Mumbai is the capital of India.

Then $p \vee q$: Ganga is a mountain or Mumbai is the capital of India is the false disjunction because both p and q are false.

4.4.4 Negation of Compound Statements

In the negation of a compound statement, the logical connective changes from and ($\wedge$) to or ($\vee$) and vice versa. That is,

(a) $\sim(p \vee q) = (\sim p) \wedge (\sim q)$

(b) $\sim(p \wedge q) = (\sim p) \vee (\sim q)$

These laws are known as De Morgan's laws.

Table 4.5 Truth table of negation of $p \vee q$

Column	1	2	3	4	5	6	7
	p	q	$p \vee q$	$\sim p$	$\sim q$	$(\sim p) \wedge (\sim q)$	$\sim(p \vee q)$
	T	T	T	F	F	F	F
	T	F	T	F	T	F	F
	F	T	T	T	F	F	F
	F	F	F	T	T	T	T

In Table 4.5, the columns 6 and 7 are identical and hence the result (a) is proved.

Table 4.6 Truth table of negation of $p \wedge q$

Column	1	2	3	4	5	6	7
	p	q	$p \wedge q$	$\sim p$	$\sim q$	$(\sim p) \vee (\sim q)$	$\sim(p \wedge q)$
	T	T	T	F	F	F	F
	T	F	F	F	T	T	T
	F	T	F	T	F	T	T
	F	F	F	T	T	T	T

In Table 4.6, the columns 6 and 7 are identical and hence the result (b) is proved.

Algebra of Propositions: Propositions satisfy various laws that are given below.

1. $p \vee p \equiv p,$	$p \wedge p \equiv p$	Idempotent laws
2. $p \vee q \equiv q \vee p,$	$p \wedge q \equiv q \wedge p$	Commutative laws
3. $(p \vee q) \vee r \equiv p \vee (q \vee r),$	$(p \wedge q) \wedge r \equiv p \wedge (q \wedge r)$	Associative laws
4. $p \vee (q \wedge r) = (p \vee q) \wedge (p \vee r),$	$p \wedge (q \vee r) \equiv (p \wedge q) \vee (p \wedge r)$	Distributive laws
5. $\sim(\sim p) = p$		Involution law
6. $\sim(p \vee q) = \sim p \wedge \sim q,$	$\sim(p \wedge q) = \sim p \vee \sim q$	De Morgan's laws

Some illustrative Examples

Example 1 Translate the following propositions into their symbolic forms.

(a) Aditya is a musician and Ms Rita is a photographer.

(b) Krishna will attend the class or will go to picnic.

(c) Ramesh is a serious student and also a good player.

(d) Rishabh will join a job or will go for higher studies.

Solution:

(a) Let p: Aditya is a musician.
q: Ms Rita is a photographer.
Then the symbolic form is $p \wedge q$.

(b) Let p: Krishna will attend the class.
q: Krishna will go to picnic.
Then the symbolic form is $p \vee q$.

(c) Let p: Ramesh is a serious student.
q: Ramesh is a good player.
Then the symbolic form is $p \wedge q$.

(d) Let p: Rishabh will join a job.
q: Rishabh will go for higher studies.
Then the symbolic form is $p \vee q$.

Example 2 Let p: Vivek Arya is a doctor.
q: Sushil Gupta is an engineer.
be two statements. Translate the following into logical sentences:

(a) $p \vee q$ (b) $p \wedge q$ (c) $\sim(p \wedge q)$ (d) $(p \vee \sim q)$ (e) $\sim p \vee q$ (f) $p \wedge \sim q$

Solution:

(a) Vivek Arya is a doctor or Sushil Gupta is an engineer.

(b) Vivek Arya is a doctor and Sushil Gupta is an engineer.

(c) It is not true that Vivek Arya is a doctor and Sushil Gupta is an engineer.

(d) Vivek Arya is a doctor or Sushil Gupta is not an engineer.

(e) Vivek Arya is not a doctor or Sushil Gupta is an engineer.

(f) Vivek Arya is a doctor and Sushil Gupta is not an engineer.

Example 3 Let p: Mathematics is difficult in general
q: Discrete Mathematics is interesting
be two statements, Give verbal sentence that describes each of the following statements:

(a) $\sim q$ (b) $p \vee q$ (c) $p \wedge q$ (d) $q \vee \sim p$ (e) $\sim p \vee q$ (f) $\sim(p \wedge q)$.

Solution:

(a) Discrete Mathematics is not interesting.

(b) Mathematics is difficult in general or Discrete Mathematics is interesting.

(c) Mathematics is difficult in general and Discrete Mathematics is interesting.

(d) Discrete Mathematics is interesting or Mathematics is not difficult in general.

(e) Mathematics is not difficult in general or Discrete Mathematics is interesting.

(f) It is not true that Mathematics is difficult in general and Discrete Mathematics is interesting.

Example 4 Determine the truth values of each of the following statements:

(a) $3 + 2 = 6$ and $5 + 8 = 13$, (b) $8 + 2 = 10$ and $2 + 5 = 6$

(c) $2 + 2 = 4$ and $3 + 5 = 8$, (d) $2 + 2 = 5$ and $3 + 1 = 5$

Solution: The statement 'p and q' is true only when both statements are true. Thus, (a) False (b) False (c) True (d) False.

Example 5 Let a, b and c be any three real numbers and let p: $a < b$, q: $b < c$ and r: $a < c$. Represent the following sentences symbolically:

1. $(a \geq b)$ or $(b < c)$
2. $(b \geq c)$ or $(a < c)$
3. $(a < b)$ and $[(b \geq c)$ or $(c > a)]$
4. $(a \geq b)$ or $[(b < c)$ and $(c > a)]$

Solution:

1. $\sim p \vee q$
2. $\sim q \vee r$
3. $p \wedge (\sim q \vee r)$
4. $\sim p \vee (q \wedge r)$

4.4.5 Conditional Statements

Many mathematical statements are implications: that is, statement of the form 'p implies q'. The statement p is called the hypothesis and the statement q the conclusion. The symbol $(\rightarrow)$ is read 'implies'.

Example: Statement: 'n is an integer $\rightarrow$ $2n + 1$ is an odd integer'. In the statement, 'n is an integer' is the hypothesis and '$2n + 1$ is an odd integer' is the conclusion.

Implication may also appear without explicit use of the word 'implies'.

The above statement can also be expressed as follows:

1. If n is an integer, then $2n + 1$ is an odd integer.

or

2. n is an integer only if $2n + 1$ is an odd integer.

In a compound statement using the connective 'If…, Then…' is called a conditional statement or implication.

To construct the truth table of the implication $p \rightarrow q$, consider the following example:

$p \rightarrow q$: If you wash my car, then I will pay you Rs. 30.

If you wash my car (p is true) and if I pay you Rs. 30 (q is true).

Then the implication is true.

If you wash my car (p is true) and if I do not pay you Rs. 30 (q is false),

Then promise is isolated, and hence, the implication is false.

If you do not wash my car (p is false), then I may give you Rs. 30 being generous or not and thus q may be true or false. In this case promise has not been tested and so has not been violated. Hence, implication has not been proved false, that is, if p is false, $p \to q$ is true by default.
Thus, $p \to q$ is considered false only if p is true and q is false (see Table 4.7).

Table 4.7 Truth table for $p \to q$

p	q	$p \to q$
T	T	T
T	F	F
F	T	T
F	F	T

Write the implication $p \to q$ for each of the following statements:

1. p: ΔABC is equilateral — q: ΔABC is isosceles.
2. p: If you work hard with sincerity, — q: you will attain great position.
3. p: If $a + b = c$, — q: then $a = c - b$
4. p: If it rains, — q: then there will be bumper crop.
5. p: If you do the assigned work, — q: then you will be rewarded.

Solution:

1. If ΔABC is equilateral, then it is isosceles.
2. If you work hard with sincerity, then you will attain great position.
3. If $a + b = c$, then $a = c - b$.
4. If it rains, then there will be bumper crop.
5. If you do the assigned work, then you will be rewarded.

Note 4.2 Consider the statements:

If it is raining, then $10 + 15 = 25$.

The above example shows that in logic, conditional statements are used in more general sense than is customary. In English, when we say $p \to q$, then we assume the same sort of cause-and-effect relationship between p and q.

4.4.6 Converse, Contrapositive and Inverse

The converse of the implication $p \to q$ (p implies q) is the implication $q \to p$.

Example: Let p and q be two given statements, that is,
p: The triangle is equilateral.
q: The triangle is isosceles.

The implication $p \to q$ states that if the triangle is equilateral, then it is isosceles.
The converse $q \to p$ states that if the triangle is isosceles, then it is equilateral.
The contrapositive of the implication $p \to q$ is $\sim q \to \sim p$, that is, if not q then not p.
If the triangle is not isosceles, then it is not equilateral. The inverse of $p \to q$ is $\sim p \to \sim q$, that is, if not p then not q.
If the triangle is not equilateral, then it is not isosceles (see Table 4.8).

Table 4.8 Truth table of inverse

p	q	$\sim p$	$\sim q$	$p \to q$	$\sim p \to \sim q$
T	T	F	F	T	T
T	F	F	T	F	T
F	T	T	F	T	F
F	F	T	T	T	T

Now given any simple conditional statement, we can obtain the following:
Statement: If $p \to q$
Inverse: The inverse of $p \to q$ is $\sim p \to \sim q$.
Converse: The converse of $p \to q$ is $q \to p$.
Contrapositive: The contrapositive of $p \to q$ is $\sim q \to \sim p$ (see Table 4.9).

Table 4.9 Truth table of converse, contrapositive and inverse

Column	1	2	3	4 Converse	5	6	7 Contrapositive	8 Inverse
	p	q	$p \to q$	$q \to p$	$\sim p$	$\sim q$	$\sim q \to \sim p$	$\sim p \to \sim q$
	T	T	T	T	F	F	T	T
	T	F	F	T	F	T	F	T
	F	T	T	F	T	F	T	F
	F	F	T	T	T	T	T	T

4.4.7 NAND, NOR and XOR Operators

The logical operators NAND (not AND) and NOR (not OR) are defined as follows:

$$p \text{ NAND } q \equiv \sim(p \wedge q) \text{ and } p \text{ NOR } q \equiv \sim(p \vee q)$$

The truth tables for each of the above operators are given in Tables 4.10 and 4.11.

Table 4.10 Truth table for p NAND q $(or\ p \uparrow q)$

p	q	$p \wedge q$	$\sim p \wedge q$
T	T	T	F
T	F	F	T
F	T	F	T
F	F	F	T

Table 4.11 Truth table for p NOR q $(or\ p \downarrow q)$

p	q	$p \vee q$	$\sim(p \vee q)$
T	T	T	F
T	F	T	F
F	T	T	F
F	F	F	T

XOR Operator: If p and q are two statements, then XOR of p and q denoted by $p \oplus q$ is a true statement when either p or q is true but not both and vice versa. The truth values of $p \oplus q$ are shown in Table 4.12.

Table 4.12 Truth table for $p \oplus q$

p	q	$p \oplus q$
T	T	F
T	F	T
F	T	T
F	F	F

4.4.8 Biconditional Statements

If p and q are statements, then the compound statement p if and only if q denoted by $p \leftrightarrow q$ is read as p implies q and is implied by q is called an equivalence or biconditional statement. The equivalence $p \leftrightarrow q$ can also be stated that p is a necessary and sufficient condition for q (see Table 4.13).

Table 4.13 Truth table for $p \leftrightarrow q$

p	q	$p \leftrightarrow q$
T	T	T
T	F	F
F	T	F
F	F	T

4.4.9 Negation of Biconditional Statement

If p and q are any two statements, then

$$\sim(p \leftrightarrow q) = \sim p \leftrightarrow q = p \leftrightarrow \sim q$$

The validity of the statement can be seen from the truth Table 4.14.

Table 4.14 Truth table for negation of $p \leftrightarrow q$

p	q	$p \leftrightarrow q$	$\sim(p \leftrightarrow q)$	$\sim p$	$\sim p \leftrightarrow q$	$\sim q$	$p \leftrightarrow \sim q$
T	T	T	F	F	F	F	F
T	F	F	T	F	T	T	T
F	T	F	T	T	T	F	T
F	F	T	F	T	F	T	F

4.5 TAUTOLOGY AND CONTRADICTION

A statement that is true for all possible values of propositional variables is called a tautology. A statement that is always false is called a contradiction.

Example 1 Prove that the statement $p \vee {\sim}p$ is a tautology and $p \wedge {\sim}p$ is a contradiction.

Table 4.15 Truth table of $p \vee {\sim}p$

p	${\sim}p$	$p \vee {\sim}p$
T	F	T
F	T	T

Table 4.16 Truth table of $p \wedge {\sim}p$

p	${\sim}p$	$p \wedge {\sim}p$
T	F	F
F	T	F

Table 4.15 shows that $p \vee {\sim}p$ is a tautology, and Table 4.16 shows that $p \wedge {\sim}p$ is a contradiction.

Example 2 Construct the truth table of the statement $(p \rightarrow q) \leftrightarrow ({\sim}q \rightarrow {\sim}p)$ and show that it is a tautology.

Table 4.17 Truth table $(p \rightarrow q) \leftrightarrow ({\sim}q \rightarrow {\sim}p)$

p	q	$p \rightarrow q$	${\sim}q$	${\sim}p$	${\sim}q \rightarrow {\sim}p$	$(p \rightarrow q) \leftrightarrow ({\sim}q \rightarrow {\sim}p)$
T	T	T	F	F	T	T
T	F	F	T	F	F	T
F	T	T	F	T	T	T
F	F	T	T	T	T	T

Table 4.17 shows that $(p \rightarrow q) \leftrightarrow ({\sim}q \rightarrow {\sim}p)$ is a tautology as it is true for all values of propositional variables.

Example 3 Show that the binary operation $\vee$ over the set of statements is commutative. Also, show by truth table that the statement $(p \vee q) \leftrightarrow (q \vee p)$ is a tautology.

Table 4.18 Truth table for $(p \vee q) \leftrightarrow (q \wedge p)$

p	q	$(p \vee q)$	$(q \vee p)$	$(p \vee q) \leftrightarrow (q \vee p)$
T	T	T	T	T
T	F	T	T	T
F	T	T	T	T
F	F	F	F	T

The last column in Table 4.18 shows that $(p \vee q) \leftrightarrow (q \vee p)$ is a tautology as it is true for all values of propositional variables. Further, the columns for $(p \vee q)$ and $(q \vee p)$ are identical and hence $(p \vee q) \equiv (q \vee p)$, that is, $\vee$ is commutative.

Example 4 Prove that the conditional statement $p \rightarrow q$ is equivalent to $\sim p \vee q$.

Table 4.19 Truth table for $p \rightarrow q$ and $\sim p \vee q$

Column	1	2	3	4	5
	p	q	$p \rightarrow q$	$\sim p$	$\sim p \vee q$
	T	T	T	F	T
	T	F	F	F	F
	F	T	T	T	T
	F	F	T	T	T

In Table 4.19, the columns (3) and (5) are identical and hence the result.

4.5.1 Contingency

A statement that can be true or false depending upon the truth values of its propositional variables is called a contingency.

Example 5 Show that the statement $(p \rightarrow q) \wedge (p \vee q)$ is a contingency.

Table 4.20 Truth table for $(p \rightarrow q) \wedge (p \wedge q)$

p	q	$p \rightarrow q$	$p \vee q$	$(p \rightarrow q) \wedge (p \vee q)$
T	T	T	T	T
T	F	F	T	F
F	T	T	T	T
F	F	T	F	F

The last column of Table 4.20 shows that $(p \rightarrow q) \wedge (p \vee q)$ is a contingency.

Example 6 Show that the statement $(p \wedge \sim q) \vee (\sim p \wedge q)$ is a contingency.

Table 4.21 Truth table for $(p \wedge \sim q) \vee (\sim p \wedge q)$

p	q	$\sim p$	$\sim q$	$p \wedge \sim q$	$\sim p \wedge q$	$(p \wedge \sim q) \vee (\sim p \wedge q)$
T	T	F	F	F	F	F
T	F	F	T	T	F	T
F	T	T	F	F	T	T
F	F	T	T	F	F	F

The last column of Table 4.21 shows that $(p \wedge \sim q) \vee (\sim p \wedge q)$ is a contingency.

Example 7 Show that the statement $\sim(p \wedge q) \vee (\sim p \wedge \sim q)$ is a contingency.

Table 4.22 Truth table for $\sim(p \wedge q) \vee (\sim p \wedge \sim q)$

p	q	$\sim p$	$\sim q$	$p \wedge q$	$\sim(p \wedge q)$	$\sim p \wedge \sim q$	$\sim(p \vee q) \vee (\sim p \wedge \sim q)$
T	T	F	F	T	F	F	F
T	F	F	T	F	T	F	T
F	T	T	F	F	T	F	T
F	F	T	T	F	T	T	T

The last column of Table 4.22 shows that $\sim(p \wedge q) \vee (\sim p \wedge \sim q)$ depends upon the components associated with the statement and hence is a contingency.

4.6 LOGICALLY EQUIVALENT OR EQUIVALENT PROPOSITIONS

Two compound propositions p and q are said to be logically equivalent or simply equivalent if they have identical truth values for all possible truth values of their component parts and are written as $p \equiv q$.

Example 1 Verify that $p \rightarrow q \equiv \sim p \vee q$

Table 4.23

p	q	$p \rightarrow q$	$\sim p$	$\sim p \vee q$
T	T	T	F	T
T	F	F	F	F
F	T	T	T	T
F	F	T	T	T

In Table 4.23, the columns for $p \rightarrow q$ and $\sim p \vee q$ are identical, and hence the result.

Example 2 Verify that $p \rightarrow q \equiv \sim q \rightarrow \sim p$.

Table 4.24

p	q	$p \rightarrow q$	$\sim q$	$\sim p$	$\sim q \rightarrow \sim p$
T	T	T	F	F	T
T	F	F	T	F	F
F	T	T	F	T	T
F	F	T	T	T	T

The columns in Table 4.24 for $p \rightarrow q$ and $\sim q \rightarrow \sim p$ are identical and hence the result.

Example 3 Prove that $\sim(p \rightarrow q) \equiv p \wedge \sim q$.

Table 4.25

p	q	$p \rightarrow q$	$\sim(p \rightarrow q)$	$\sim q$	$p \wedge \sim q$
T	T	T	F	F	F
T	F	F	T	T	T
F	T	T	F	F	F
F	F	T	F	T	F

In Table 4.25, the columns for $\sim(p \rightarrow q)$ and $p \wedge \sim q$ are identical and hence the result.

Some Illustrative Examples

Example 1 If $5x + 1 = 11$, then $x = 2$. Find its (a) inverse, (b) converse and (c) contrapositive.

Solution: Let p: $5x + 1 = 11$

q: $x = 2$

Then

(a) Inverse : $\sim p \rightarrow \sim q$, that is, if $5x + 1 \neq 11$, then $x \neq 2$.

(b) Converse: $q \rightarrow p$, that is, if $x = 2$, then $5x + 1 = 11$.

(c) Contrapositive: $\sim q \rightarrow \sim p$, that is, if $x \neq 2$, then $5x + 1 \neq 11$.

Example 2 If the morning is good, then I go for a walk. Find its (i) inverse, (ii) converse and (iii) contrapositive.

Solution: Let p: The morning is good.

q: I go for a walk.

(a) *Inverse*: The inverse of $p \rightarrow q$ is $\sim p \rightarrow \sim q$, that is, if the morning is not good, then I do not go for a walk.

(b) *Converse*: The converse of $p \rightarrow q$ is $q \rightarrow p$, that is, if I go for a walk, then morning is good.

(c) *Contrapositive*: The contrapositive of $p \rightarrow q$ is $\sim q \rightarrow \sim p$, that is, if I do not go for a walk, then the morning is not good.

Example 3 State the inverse and converse of the following implications.

(a) If I am fine, then I go to gym.

(b) If I am the director of an institute, then I will go by car.

(c) If I am late, then I take matro to work.

Solution:

(a) *Inverse*: If I am not fine, then I do not go to gym.

Converse: If I go to gym, then I am fine.

(b) *Inverse*: If I am not the director of an institute, then I do not go by car.

Converse: If I go by car, then I am the Director of an institute.

(c) *Inverse*: If I am not late, then I do not take matro to work.

Converse: If I take matro to work, then I am late.

Example 4 Determine the contrapositive of the following statements.

(a) If you work hard, then you can earn money.

(b) If demand increases, then cost inflation takes place.

(c) If production increases, then cost decreases.

Solution:

(a) If you cannot earn good money, then you cannot work hard.

(b) If cost inflation does not take place, then demand does not increase.

(c) If cost does not decrease, then production does not increase.

Example 5 If p, q and r are three statements, then prove that

(a) $p \wedge (q \vee r) \equiv (p \wedge q) \vee (p \wedge r)$

(b) $p \vee (q \wedge r) \equiv (p \vee q) \wedge (p \vee r)$

Solution: (a)

p	q	r	$q \vee r$	$p \wedge (q \vee r)$	$p \wedge q$	$p \wedge r$	$(p \wedge q) \vee (p \wedge r)$	$p \wedge (q \vee r) \leftrightarrow (p \wedge q) \vee (p \wedge r)$
T	T	T	T	T	T	T	T	T
T	T	F	T	T	T	F	T	T
T	F	T	T	T	F	T	T	T
T	F	F	F	F	F	F	F	T
F	T	T	T	F	F	F	F	T
F	T	F	T	F	F	F	F	T
F	F	T	T	F	F	F	F	T
F	F	F	F	F	F	F	F	T

Since the truth values $p \wedge (q \vee r)$ and $(p \wedge q)$ v $(p \wedge r)$ are identical,

$$p \wedge (q \vee r) \equiv (p \wedge q) \vee (p \wedge r)$$

which shows that the operation of conjunction is distributive with respect to disconjunction. It can be observed that if the two compound statements are equivalent, then their biconditional statement or double implication ($\leftrightarrow$) is a tautology.

(b) To prove $p \vee (q \wedge r) = (p \vee q) \wedge (p \vee r)$

p	q	r	$q \wedge r$	$p \vee q$	$p \vee r$	$P \equiv p \vee (q \wedge r)$	$Q \equiv (p \vee q) \wedge (p \vee r)$	$P \rightarrow Q$
T	T	T	T	T	T	T	T	T
T	T	F	F	T	T	T	T	T
T	F	T	F	T	T	T	T	T
T	F	F	F	T	T	T	T	T
F	T	T	T	T	T	T	T	T
F	T	F	F	T	F	F	F	T
F	F	T	F	F	T	F	F	T
F	F	F	F	F	F	F	F	T

Since the truth values of $P = p \vee (q \wedge r)$ and $Q = (p \vee q) \wedge (p \vee r)$ are identical,

$$p \vee (q \wedge r) = (p \vee q) \wedge (p \vee r)$$

which shows that the operation of disjunction is distributive with respect to conjunction. Further, if the two compound statements p and q are equivalent, then $p \leftrightarrow q$ is a tautology.

4.7 LOGICAL ARGUMENTS

An argument is a finite collection of statements $p_1, p_2, \ldots, p_n$ called premises (or hypotheses), which lead to another statement q called conclusion. Such an argument is valid if $p_1 \wedge p_2 \wedge \ldots \wedge p_n \rightarrow q$ is true, we say that q logically follows from $p_1, p_2, \ldots, p_n$.

4.7.1 Validity of an Argument

An argument is valid if and only if the conjunction of the premises implies the conclusion. In this case, $p_1 \wedge p_2 \wedge \ldots \wedge p_n \rightarrow q$ is a tautology.

When q logically follows from $p_1, p_2, \ldots, p_n$, it is written

$$\begin{array}{l} p_1 \\ p_2 \\ \vdots \\ \underline{p_n} \\ \therefore q \end{array}$$

This means that if p_1 is true, p_2 is true,…, p_n is true, the q is true.

Example 1 Prove the validity of the argument

$$\frac{\begin{array}{c}p\\ p\to q\end{array}}{\therefore q}, \text{ i.e., } p\wedge(p\to q)\to q \text{ is a tautology.}$$

p	q	p → q	p ∧ (p → q)	p ∧ (p → q) → q
T	T	T	T	T
T	F	F	F	T
F	T	T	F	T
F	F	T	F	T

Since $p \wedge (p \to q) \to q$ is tautology. The argument is valid.

Fundamental Rules of Inference – Few rules of conclusion are the following:

- The valid argument p is called the rule of detachment.

$$\frac{\begin{array}{c}p\\ p\to q\end{array}}{\therefore q}$$

It is also known as *Modus Ponens*, a Latin term meaning 'method of reasoning'.

A valid argument ~q is known as *Modus Tollens*.

$$\frac{\begin{array}{c}\sim q\\ p\to q\end{array}}{\therefore \sim p}$$

Modus Tollens is a Latin term meaning 'method of denying' (as the conclusion is denial).

4.7.2 Law of Syllogism

A fundamental principle of logical reasoning states that if $p \to q$ and $q \to r$, then $p \to r$.

$$\text{i.e., } \frac{\begin{array}{c}p\to q\\ q\to r\end{array}}{\therefore p\to r}$$

p	q	r	p → q	q → r	(p → q) ∧ (q → r)	p → r	(p → q) ∧ (q → r) ∧ (p → r)
T	T	T	T	T	T	T	T
T	T	F	T	F	F	F	T
T	F	T	F	T	F	T	T
T	F	F	F	T	F	F	T
F	T	T	T	T	T	T	T
F	T	F	T	F	F	T	T
F	F	T	T	T	T	T	T
F	F	F	T	T	T	T	T

The last column shows that $(p \rightarrow q) \wedge (q \rightarrow r) \rightarrow (p \rightarrow r)$ is a tautology.

- Constructive dilemma

$$\begin{array}{c} (p \rightarrow q) \wedge (r \rightarrow s) \\ p \wedge r \\ \hline \therefore q \vee s \end{array}$$

- Disjunctive syllogism

$$\begin{array}{c} p \wedge q \\ \sim p \\ \hline \therefore q \end{array}$$

- Absorption

$$\begin{array}{c} p \rightarrow q \\ \hline \therefore p \rightarrow p \wedge q \end{array}$$

4.7.3 Methods to Test the Validity of an Argument

An argument is valid if the conjunction of its premises implies the conclusion. However, a valid argument does not necessarily lead to a true conclusion.

In order to test the validity of an argument, we proceed as follows:

1. Construct a truth table of all premises and conclusion.
2. Identify the rows called critical rows in which all premises are true. In case no such row is found, the argument is invalid.
3. In each critical row, find whether the conclusion is true. In such a case, the argument is valid, otherwise invalid. However, if at least one critical row contains false conclusion, then the argument is invalid.

Example 1 Determine the validity of the following arguments:

(a) $\begin{array}{c} p \rightarrow q \\ p \\ \hline \therefore q \end{array}$ (b) $\begin{array}{c} p \rightarrow q \\ \sim q \\ \hline \therefore \sim p \end{array}$

Solution:

p	q	~q	p → q	~p
T	T	F	T	F
T	F	T	F	F
F	T	F	T	T
F	F	T	T	T

(a) In the first critical row, $p \rightarrow q$ and p are true and q is true and hence the argument is valid.

(b) In the fourth critical row, $p \rightarrow q$ and $\sim q$ are true and $\sim p$ is true and hence the argument is valid.

Example 2 Examine the validity of the following arguments.

(a) $p \vee q$, $\sim q$, $\therefore p$

(b) $p \rightarrow q$, $q \rightarrow p$, $\therefore p \vee q$

(c) $p \vee q$, $p \rightarrow \sim q$, $p \rightarrow r$, $\therefore r$

Solution:

(a) Construct the truth table for the premises and conclusion.

		Premises		Conclusion
p	q	$\sim q$	$p \vee q$	p
T	T	F	T	T
T	F	T	T	T
F	T	F	T	F
F	F	T	F	F

Since critical row 2 contains true conclusion, the argument is valid.

(b)

		Premises		Conclusion
p	q	$p \rightarrow q$	$q \rightarrow p$	$p \vee q$
T	T	T	T	T
T	F	F	T	T
F	T	T	F	T
F	F	T	T	F

Since critical row 4 does not contain true conclusion, the argument is invalid.

(c)

				Premises			Conclusion
p	q	r	$\sim q$	$p \vee q$	$p \rightarrow \sim q$	$p \rightarrow r$	r
T	T	T	F	T	F	T	T
T	T	F	F	T	F	F	F
T	F	T	T	T	T	T	T
T	F	F	T	T	T	F	F
F	T	T	F	T	T	T	T
F	T	F	F	T	T	T	F
F	F	T	T	F	T	T	T
F	F	F	T	F	T	T	F

Since one of the critical rows, that is, row 6 does not contain true conclusion, the argument is invalid.

Example 3 Test the validity of the argument.

I will be a physicist or I will be a mathematician

I will not be a mathematician

∴ I will be a physicist

Solution:

Let p: I will be a physicist.

q: I will be a mathematician.

In the symbolic form, the argument is

$$\begin{array}{l} p \vee q \\ \underline{\sim q} \\ \therefore p \end{array}$$

		Premises		Conclusion
p	q	$p \vee q$	$\sim q$	p
T	T	T	F	T
T	F	T	T	T
F	T	T	F	F
F	F	F	T	F

The critical row 2 contains true conclusion and hence the argument is valid.

Example 4 Test the validity of the following argument:

If two sides of a triangle are equal, then the opposite angles are equal

Two sides of a triangle are not equal

∴ The opposite angles are not equal

Solution:

Let p: Two sides of a triangle are equal.

q: The opposite angles are equal.

In symbolic form, the argument is

$$\begin{array}{l} p \rightarrow q \\ \underline{\sim p} \\ \therefore \sim q \end{array}$$

		Premises		Conclusion
p	q	$p \rightarrow q$	$\sim p$	$\sim q$
T	T	T	F	F
T	F	F	F	T
F	T	T	T	F
F	F	T	T	T

The critical row 3 does not contain true conclusion and hence the argument is not valid.

Example 5 Test the validity of the argument.

If the morning is fine, I go for a walk
I do not go for a walk
∴ The morning is not fine

Solution: We translate the arguments into symbolic form.
Let p: The morning is fine.
q: I go for a walk.

Then we have
$$\begin{array}{l} p \to q \\ \sim q \\ \hline \therefore \sim p \end{array}$$

The truth table is

Premises				Conclusion
p	q	~q	p → q	~p
T	T	F	T	F
T	F	T	F	F
F	T	F	T	T
F	F	T	T	T

The critical row 4 contains true conclusions and hence the argument is valid.

Example 6 Test the validity of the argument:

If the morning is fine, I go for a walk
The morning is not fine
∴ I do not go for a walk

Solution:
We translate the problems in the symbolic form as in Example 5.

$$\begin{array}{l} p \to q \\ \sim p \\ \hline \therefore \sim q \end{array}$$

The truth table is

		Premises		Conclusion
p	q	p → q	~p	~q
T	T	T	F	F
T	F	F	F	T
F	T	T	T	F
F	F	T	T	T

Since the critical row 3 does not contain true conclusion, hence the argument is invalid.

Example 7 Is the following argument valid?

If income tax rates are lowered, income tax collection increases

Income tax collection increases

∴ Income tax rates are lowered

Solution:

Let p: Income tax rates are lowered

q: Income tax collection increases

Then the argument is of the form

$$\begin{array}{c} p \to q \\ q \\ \hline \therefore p \end{array}$$

The truth table is

	Premises		Conclusion
p	q	$p \to q$	p
T	T	T	T
T	F	F	T
F	T	T	F
F	F	T	F

The critical row 3 does not contain a true conclusion and hence the argument is not valid.

Note 4.3 Another approach to the solution is to verify whether the statement $(p \to q) \wedge q$ implies p. A truth table will show that it is not the case.

Many of the statements that come across in mathematics, computer science, engineering and everyday life are not covered by propositions. A generalisation of this is the predicate as it contains proposition with variables and constants. Predicates have acquired importance due to their application in database and expert systems.

4.8 PREDICATES

From a grammatical point of view, simple declaration sentences involve a subject and a predicate. The word or words that express the property of the subject is called predicate.

It is convenient to represent predicates by capital letters and subject by small letters so that sentences reflect subject–predicate relationship.

Consider the following sentences:

S. Ramanujan was a great mathematician and Karl Friedrich Gauss was a great mathematician.

Here in these two sentences 'was a great mathematician' is the predicate as it describes the property of its subjects

S. Ramanujan and Karl Friedrich Gauss.

If $P(x)$ denotes the property and x the variable denoting subject (i.e., object), then the above sentences can be written as $P(x)$ where P stands for predicate 'was a great mathematician' and x the subject Ramanujan or Gauss. Since $P(x)$ just involves one variable, it is called a unary predicate.

Note 4.4 $P(x)$ is not a proposition but an expression because it involves a variable x that cannot assign any truth value to $P(x)$. However, if x is replaced by a particular object, it becomes a proposition. The set of all values that x can assume is called the universe of discourse.

A predicate may contain two or more variables. A predicate containing two variables x and y denoted by $P(x, y)$ is called a binary predicate. If a predicate contains n variables, it is called an n-ary predicate.

From the above discussion, it is clear that $P(x)$ becomes a proposition function in the sense that for a particular value of x, it becomes a statement. Similarly, $P(x, y)$ will become a proposition for particular values of x and y.

Examples of Binary Predicate

1. If x and y are real numbers, then their product xy is also a real number. Here, 'the product x, y is also a real number' is the binary predicate.
2. If x and y are rational numbers, then $x + y$ is a rational number. Here, '$x + y$ is a rational number' is the binary predicate.
3. For every integer x, there is an integer y such that $x + y = 0$ has the binary predicate 'there is an integer y such that $x + y = 0$'.

4.8.1 Propositional Functions and Quantifiers

Let A be a given set. A propositional function $p(x)$ (or an open sentence) defined on A is true or false, for each $x \in A$, that is, $p(x)$ becomes a statement whenever $x \in A$. The variable x is an open statement and is called a free variable.

The set A is called the domain of $p(x)$ and the set T_p of all elements of A for which $p(x)$ is true is called the truth set of $p(x)$. Thus,

$$T_p = \{x: x \in \text{A}, p(x) \text{ is true}\}$$

Example: Find the truth set of the following propositional functions $p(x)$ defined on the set of natural numbers N.

(a) Let $p(x)$ be '$x + 5 > 10$'
Its truth set $T_p = \{x: x \in N \text{ and } x > 5\}$
$= \{6, 7, 8, \ldots\}$
That is, the set of all natural numbers greater than 5.

(b) Let $p(x)$ be '$x + 8 < 7$'
Its truth set is $T_p = \{x: x \in N \text{ and } x + 8 < 7\}$
Or $T_p = \Phi$ (as there is no natural number satisfying the property)

(c) Let $p(x)$ be '$x + 4 > 3$'
Its truth set $T_p = \{x: x \in N \text{ and } x + 4 > 3\}$
$= \{1, 2, 3, \ldots\} = N$
Thus, $p(x)$ is true for every element of N.

4.8.2 Universal Quantifier

Predicates can be changed into propositions by assigning specific value to their valuables.

Let $p(x)$ be the propositional function defined on the set A. Consider the expression 'For every x in A, $p(x)$ is a true statement'.

i.e., $(\forall\, x \in A), p(x)$ is true
or for all x, $p(x)$ is true

Then the symbol $\forall$, which reads 'for every' or 'for all', is called the universal quantifier.

In this case $T_p = \{x: x \in A \text{ and } p(x) \text{ is true}\} = A$

That is, truth set of $p(x)$ is the entire set A.

Example: The function $p(x) = x + 2 > 2$ over N has universal quantifier as the function is true for every natural number.

4.8.3 Existential Quantifier

Let $p(x)$ be a propositional function defined on a set A. Consider the expression

'There exists x in A such that $p(x)$ is a true statement' written as $(\exists\, x \in A)\, p(x)$, then the symbol $\exists$, which reads 'there exists' or 'for some' or 'for at least one', is called existential quantifier. This means that the propositional function is true for at least one value of x.

i.e., $\{x: x \in A \text{ and } p(x) \text{ is true}\} \neq \Phi$

When the quantifiers are used, one should specify the universe of discourse. If the universe of discourse is changed, the truth value can change.

Example 1. The proposition $(\exists\, x \in N)\,(n + 6 < 8)$ is true
Since $\{n: n + 6 < 8\} = \{1\} \neq \Phi$

Example 2. The proposition $(\exists\, x \in N)\,(x + 5 < 3)$ is false
Since $\{n: n + 5 < 3\} = \Phi$

Example 3. Let $p(x)$: $x^2 = 5$. If the universe of discourse is the act of integer, the $(\exists\, x \in Z)\, p(x)$ is false.
If the universe of discourse is the set of real number, then $(\exists\, x \in \mathrm{R})\, p(x)$ is true.

Example 4. Let the universe of discourse be the set of integers Z. Define proposition

$$p(x): x^2 = x^3$$

Find the truth values of the statements:

(a) $(\forall\, x \in Z)\, p(x)$ (b) $(\exists\, x \in Z)\, p(x)$

Solution:

(a) It is false because for $x = 2$, $2^2 = 2^3$ is false.

(b) It is true because there exists integers 0 and 1 for which $p(x)$ is true.

4.9 METHODS OF PROOF

Proofs in mathematics play a key role in its development and also guarantee the correctness of the results and theorems provided they are proved using logical reasoning.

4.9.1 Direct Proof

Given a set of hypotheses H_1, H_2,…, H_n, which can lead to a conclusion C, means $H_1 \wedge H_2 \wedge \ldots \wedge H_n \rightarrow C$ is a tautology. Such a proof is called direct proof.

Example 1 Prove that the product of two odd integers is an odd integer.

Proof: Let m and n be two odd integers, then they can be written as $m = 2p + 1$ and $n = 2q + 1$ where p and q are integers.

The product $= mn = (2p + 1)(2q + 1)$
$= 4pq + 2(p + q) + 1$
$= 2[2pq + p + q] + 1$, which is an odd integer.

Example 2 Prove that between two rational numbers, there is always a rational number.

Proof: Let m and n be any two given rational numbers.

Since m is a rational number, let $m = \frac{a}{b}$ when $b \neq 0$ and a and b are integers.

Also, n is a rational number, then $n = \frac{c}{d}$, $d \neq 0$, and c and d are integers.

The number between m and n is

$$\frac{m+n}{2} = \frac{1}{2}\left[\frac{a}{b} + \frac{c}{d}\right] = \frac{ad + bc}{2bd}$$

Since $ad + bc$ and bd are integers and $bd \neq 0$, $\frac{m+n}{2}$ is a rational number.

Example 3 The sum of two odd integers is an even integer.

Proof: Let a and b be two odd integers, then

$$a = 2p + 1 \text{ and } b = 2q + 1$$

where p and q are integers.

The sum $a + b = 2p + 1 + 2q + 1$
$= 2p + 2q + 2$
$= 2(p + q + 1)$

Which is an even integer as $p + q + 1$ is an integer.

Example 4 If n is any integer, then $n^2 + n$ is an even integer.

Proof: $n^2 + n = n(n + 1)$ is the product of two consecutive integers, and hence is an even integer.

Example 5 Prove that $9n^2 + 3n - 2$ is even where n is an integer.

Proof:

Case I: n is even.

The product of even integer with any integer is even and so $9n^2$ and $3n$ are even. Thus, $9n^2 + 3n - 2$ is even.

Case II: n is odd.

The product of odd integers is odd. Since n is odd, $9n^2$ and $3n$ are odd. The sum of two odd integers is even. Hence, $9n^2 + 3n - 2$ is even.

Rules of Inference: Since predicates when variables are specified become propositions, all the rules of inference, that is, rules for validity of propositions are also applicable to predicates.

Universal Modus Ponens: $\forall x$, if $P(x)$ then $Q(x)$

For a particular value a, $P(a)$ $\quad \therefore Q(a)$

The argument form consists of two premises and a conclusion.

Universal Modus Tollends: $\forall x$, if $P(x)$ the $Q(x)$

For a particular value $\sim Q(a)$ $\quad \therefore \sim P(a)$

Example 1 Write the following argument using quantifiers, variables and predicates. Prove the validity of the argument. If a number is even, then its square is even.

Solution:

$\forall x$ if x is even, then x^2 in even

$P(x)$: x is even

$Q(x)$: x^2 is even. Let m be an even number, then argument is as follows:

$\forall x$ if $P(x)$ then $Q(x)$

$P(m)$ for a particular m

$\therefore Q(m)$ The argument is valid by Universal *Modulus Ponens*.

Example 2 Write the following argument using quantifiers, variables and predicates. Prove the validity of the argument.

All healthy persons take fruit juice in breakfast.

Ankit does not take fruit juice in breakfast.

Ankit is not a healthy person.

Solution:

$\forall x$ if x is a healthy person, then x takes fruit juice in breakfast.

Let $P(x)$: x is a healthy person

$Q(x)$: x takes fruit juice in breakfast

Let a stand for Ankit, $\forall x$, if $P(x)$ then $Q(x)$.

Let $\sim Q(a)$ $\quad \therefore \sim P(a)$

By Universal *Modus Tollends*, the argument is valid.

4.9.2 Indirect Proof

Proofs that are not direct are called indirect proofs.

There are two kinds of indirect proofs:

1. Proof by contrapositive
2. Proof by contradiction

Proof by Contrapositive: To prove $p \to q$ indirectly by (contrapositive), we have to show that $\sim q \to \sim p$.

Example 1 Let n be an integer. Prove that if n^2 is odd, then n is odd.

Proof Let p: n^2 is odd

q: n is odd

We have to prove $p \to q$. We shall prove the contrapositive $\sim q \to \sim p$.

Let us assume that n is not odd and so n is even. Hence, $n = 2m$ where m is an integer.

Now $n^2 = (2m)^2 = 4m^2 = 2(2m^2)$

Which shows that n^2 is even. We have thus shown that if n is even, then n^2 is even, which is the contrapositive of the given statement. Hence, the given statement is proved.

Example 2 Let n be an integer. Prove that if n^2 is even, then n is even.

Proof: Let p: n^2 is even and q: n is even.
We have to prove $p \to q$. We shall prove the contrapositive $\sim q \to \sim p$.
Let us assume that n is not even and so n is odd.
Hence, $n = 2m + 1$ where m is an integer.

$$\text{Now } n^2 = (2m+1)^2 = 4m^2 + 4m + 1$$
$$= 2(2m^2 + 2m) + 1$$
$$= 2r + 1$$
$$\text{where } r = 2m^2 + 2m$$

We have thus shown that if n is odd, then n^2 is odd, which is the contrapositive of the given statement.

Hence, the given statement is proved.

Example 3 Prove that if $m + n \geq 73$, then $m \geq 37$ or $n \geq 37$, where m and n are positive integers.

Proof: The contrapositive of the given statement is not ($m \geq 37$ or $n \geq 37$) $\to$ not ($m + n \geq 73$) or (not $m \geq 37$) and (not $n \geq 37$) $\to$ not ($m + n \geq 73$) ($\therefore \sim(p \vee q) = \sim p \wedge \sim q$).
Now left-hand side gives

$$m \leq 36 \text{ and } n \leq 36$$

$$m + n \leq 36 + 36$$

$$m + n \leq 72 \to \text{not } (m + n \geq 73)$$

Proof by Contradiction: Sometimes a direct proof of a statement is not simple and straight forward and we do not know how to begin. In such a case, we assume just the negative (or opposite) of the given statement. If this assumption leads to a contradiction, then our assumption is false. Hence, the given statement is true. Such a proof is called a proof by contradiction.

Example 1 Prove that there is no rational number p/q whose square is 2, that is, show that $\sqrt{2}$ is an irrational number.

Proof: Let us assume that there is a rational number p/q whose square is 2. That is, $(p/q)^2 = 2$ where p and q are integers and have no common factor, that is, p and q are in lowest terms and $q \neq 0$.

We now have $p^2 = 2q^2$, which shows that p^2 is even. This implies that p is even.

Let $\quad p = 2m$.

Hence, $\quad p^2 = 4m^2 = 2q^2$

or $\quad q^2 = 2m^2$

which shows that q^2 is even and so q is even. Now p and q are even and have a common factor 2, which contradicts the assumption that p and q have no common factor. Hence, our assumption that $\sqrt{2}$ is a rational number is wrong.

Thus, $\sqrt{2}$ is irrational.

Remember that a real number, which is not rational, is called an irrational number. The number π and e, which are not root of any algebraic equation, are called transcendental numbers.

Example 2 If a is a non-zero rational number and b is an irrational number, then prove that ab is irrational.

Proof: Let us assume that ab is rational and hence

$ab = \dfrac{p}{q}$, where p and q are integers and have no common factor and $q \neq 0$.

Since a is rational, we have $a = \dfrac{m}{n}$ for m and n are integers having no common factor and $n \neq 0$.

Hence, $ab = \dfrac{p}{q}$ leads to $\dfrac{m}{n}b = \dfrac{p}{q}$ or $b = \dfrac{np}{mq}$, $mq \neq 0$

Since n, p, m and q are integers, b is rational.

This contradicts the assumption that b is irrational. Hence, our assumption that ab is rational is wrong. Thus, ab is irrational.

Example 3 Prove that there is no largest integer.

Proof: Let us assume that there is a largest integer n. But $(n + 1)$ is also an integer, which is larger than n. Hence, our assumption that there is a largest integer is wrong. Therefore, there is no largest integer.

4.9.3 Proof by Counter Example

Example 1 Prove or disprove that $n^2 + 41n + 41$ is a prime for every $n \in N$.

To disprove a result, we need only one example for which the implication is false.

Proof: Let $P(n) = n^2 + 41n + 41$ is not prime for every $n \in N$.
Putting $n = 6$, we have

$$P(6) = 6^2 + 41 \times 6 + 41$$
$$= 36 + 246 + 41 = 323$$
$$= 19 \times 17$$

which shows that it is a composite number.
Hence, $P(n) = n^2 + 41n + 41$ is not prime for every $n \in N$.

Example 2 Prove that $n^2 - n + 41$ is not prime for all $n \in N$.

Proof: Let $P(n) = n^2 - n + 41$
Putting $n = 41$

$$P(41) = (41)^2 - 41 + 41 = (41)^2 = 41 \times 41$$

which is not prime. Hence, $n^2 - n + 41$ is not prime for all $n \in N$.

Example 3 Prove or disprove the statement that if x and y are real numbers $x^2 = y^2 \leftrightarrow x = y$.

Proof: The given statement can be written as

$$\forall x\ \forall y\ P(x, y)$$

In order to prove the result, it is required to provide steps, each of which is true for every x and y.

To disprove the result, we need to find only one example whose implication is false. Since $(-7)^2 = (7)^2$ but $7 \neq -7$, the result is false.

4.9.4 Principle of Mathematical Induction

Mathematical induction is one of the most important techniques for proving mathematical statements that cannot be proved easily by direct methods. It is usually employed for results involving positive integers.
Let $P(n)$ be a statement defined for all integers $n \geq n_0$.
The given statement is true if we can prove that

1. $P(n_0)$ is true.
2. If $P(k)$ is true for some $k > n_0$, then $P(k + 1)$ must be true.

Then, $P(n)$ is true for all $n \geq n_0$.

This result is called the Principle of Mathematical Induction. The step (i) is called the basis step while (ii) is called the induction step. The step (ii) is referred to as the induction hypothesis. This step involves showing that the implication $P(k) \rightarrow P(k + 1)$ is a tautology, that is, if $P(k)$ is true, then $P(k + 1)$ is true. Hence, the conclusion $P(n)$ is true.

Example 1 Prove by mathematical induction that the sum of first n odd natural numbers is n^2.

Proof: Let $P(n) = 1 + 3 + 5 + \ldots + (2n - 1) = n^2$ for $n \geq 1$

Clearly $P(1) = 1^2 = 1$ is true.

Now, suppose that the statement is true for $n = k, k \geq 1$.

$$P(k) = 1 + 3 + 5 + \ldots + (2k - 1) = k^2$$

Hence,

$$P(k + 1) = 1 + 3 + 5 + \ldots + (2k - 1) + (2k + 1)$$
$$= k^2 + 2k + 1 = (k + 1)^2$$

which shows that the statement is true for $n = k + 1$.

Hence, by the principle of mathematical induction $P(n)$, that is, $1 + 3 + 5 + \ldots + (2n - 1) = n^2$ is true for all natural numbers n.

Example 2 Prove that for any natural number $n \geq 1$

$$1^2 + 2^2 + 3^2 + \cdots + n^2 = \frac{n(n+1)(2n+1)}{6}$$

Proof: Let $P(n) = 1^2 + 2^2 + 3^2 + \cdots + n^2 = \dfrac{n(n+1)(2n+1)}{6}$

When $n = 1$, the sum of integers from 1^2 to 1^2 is 1.

Also,

$$\frac{1(1+1)(2.1+1)}{6} = 1$$

Hence, the statement is true for $n = 1$, that is, $P(1)$ is true.

Now, suppose that statement is true for $n = k, k \geq 1$.

Then $P(k)$: $1^2 + 2^2 + 3^2 + \ldots + k^2 = \dfrac{k(k+1)(2k+1)}{6}$

Now, for $n = k + 1$, we have

$$1^2 + 2^2 + 3^2 + \ldots + k^2 + (k+1)^2$$

$$= \frac{k(k+1)(2k+1)}{6} + (k+1)^2$$

$$= (k+1)\left[\frac{k(2k+1)}{6} + k + 1\right]$$

$$= (k+1)\left[\frac{k(2k+1) + 6(k+1)}{6}\right]$$

$$= \frac{(k+1)\left[2k^2 + 7k + 6\right]}{6} = \frac{(k+1)\left[2k^2 + 4k + 3k + 6\right]}{6}$$

$$= \frac{(k+1)\left[2k(k+2) + 3(k+2)\right]}{6} = \frac{(k+1)(k+2)(2k+3)}{6}$$

which shows that the statement is true for $n = k + 1$, that is, $P(k + 1)$ is true. Hence, by the principle of mathematical induction, the statement is true for all integers $n \geq 1$.

Example 3 Prove that for any integer $n \geq 1$, $2^{2n} - 1$ is divisible by 3.

Proof: Let $P(n)$, $2^{2n} - 1$ be divisible by 3.
When $n = 1$, $2^{2.1} - 1 = 2^2 - 1 = 4 - 1 = 3$ is divisible by 3.
Now suppose that for $n = k$, $k \geq 1$, the statement is true, i.e., $2^{2k} - 1$ is divisible by 3.

Let $\quad 2^{2k} - 1 = 3m$, then $2^{2k} = 3m + 1$.

$$\text{Now} \quad 2^{2(k+1)} - 1 = 2^{2k+2} - 1 = 2^2.2^{2k} - 1 = 4.2^{2k} - 1$$
$$= 4(3m + 1) - 1$$
$$= 12m + 3 = 3(4m + 1)$$

which is divisible by 3. This shows that the statement is true for $n = k + 1$. Hence, by the principle of mathematical induction $2^{2n} - 1$ is divisible by 3.

Example 4 Prove by the principle of mathematical induction

$$1^3 + 2^3 + 3^3 + \ldots + n^3 = \frac{n^2(n+1)^2}{4}, \; n \geq 1$$

Proof: Let $P(n)$: $1^3 + 2^3 + 3^3 + \ldots + n^3 = \dfrac{n^2(n+1)^2}{4}$

When $n = 1$, the sum of integers from 1^3 to 1^3 is 1.

Also,

$$\frac{1^2(1+1)^2}{4} = \frac{1.4}{4} = 1$$

Hence, the statement is true for $n = 1$.
Now, suppose that the statement is true for $n = k$, $k \geq 1$.
That is, $P(k)$ is true.

Then $P(k)$: $1^3 + 2^3 + 3^3 + \ldots + k^3 = \dfrac{k^2(k+1)^2}{4}$

Now, for $n = k + 1$, we have $1^3 + 2^3 + 3^3 + \ldots + k^3 + (k + 1)^3$

$$= \frac{k^2(k+1)^2}{4} + (k+1)^3$$
$$= (k+1)^2[\frac{k^2}{4} + k + 1]$$
$$= (k+1)^2(\frac{k^2 + 4k + 4}{4})$$
$$= \frac{(k+1)^2(k+2)^2}{4}$$

which shows that $P(k+1)$ is true.

Hence, by the principle of mathematical induction, the statement is true for all integers $n \geq 1$.

Example 5 Prove by the principle of mathematical induction

$$n! > 2^n \text{ for all } n \geq 4$$

Proof: Let $P(n)$: $n! > 2^n$

In this problem, $n_0 = 4$

$4! = 1.2.3.4 = 24 > 2^4$ is true and thus $P(n_0)$, that is, $P(4)$ is true. Now suppose that the statement is true for $n = k$, $k \geq 4$, that is, $k! > 2^k$.

We have to prove that the statement is true for $n = k + 1$, that is, we must prove that

$$(k+1)! > 2^{k+1}$$

Now

$$(k+1)! = (k+1)\,k!$$
$$>(k+1)2^k$$

Since for $k \geq 4$, $k + 1 > 2$

Hence,

$$(k+1)2^k > 2.2^k = 2^{k+1}$$

Hence, $(k+1)! > 2^{k+1}$ as desired, that is, $P(k+1)$ is true. Thus, by the principle of mathematical induction, $n! > 2^n$ for all integers $n \geq 4$.

Example 6 Prove by the principle of mathematical induction

$$P(n):\ 1+a+a^2+\ \ldots + a^{n-1} = \frac{a^n-1}{a-1},\ (a>1) \text{ for all natural numbers.}$$

Proof: $P(1)$ is true since for $n = 1$, LHS = RHS = 1.

Now, suppose that the statement is true for $n = k$, $k > 1$.

That is,

$$P(k):\ 1+a+a^2+\ldots+a^{k-1} = \frac{a^k-1}{a-1}$$

For $n = k + 1$, we have $1 + a + a^2 + \ldots + a^{k-1} + a^k$

$$= \frac{a^k-1}{a-1} + a^k = \frac{a^k-1+a^k(a-1)}{a-1}$$

$$= \frac{a^k-1+a^{k+1}-a^k}{a-1} = \frac{a^{k+1}-1}{a-1}$$

which shows that $P(k+1)$ is true.

Hence, by the principle of mathematical induction, $p(n)$ is true for all positive integral values of n.

Example 7 Prove by the principle of mathematical induction that $2n^3 + 3n^2 + n$ is divisible by 6 for every integer $n \geq 1$.

Proof: Let $P(n)$: $2n^3 + 3n^2 + n$ be divisible by 6

When $n = 1$, $2n^3 + 3n^2 + n = 2(1)^3 + 3(1)^2 + 1 = 2 + 3 + 1 = 6$ is clearly divisible by 6 and thus $P(1)$ is true.

Now suppose that, $P(k)$ is true and hence $2k^3 + 3k^2 + k$ is divisible by 6. Then $2k^3 + 3k^2 + k = 6m$ for some integer m.

Now, we must show that $P(k + 1)$ is true, that is, $2(k + 1)^3 + 3(k + 1)^2 + k + 1$ is divisible by 6. We now have $2(k + 1)^3 + 3(k + 1)^2 + k + 1$

$$
\begin{aligned}
&= 2(k^3 + 3k^2 + 3k + 1) + 3(k^2 + 2k + 1) + k + 1 \\
&= (2k^3 + 3k^2 + k) + 6(k^2 + 2k + 1) \\
&= 6m + 6(k^2 + 2k + 1) \\
&= 6(m + k^2 + 2k + 1)
\end{aligned}
$$

which is divisible by 6. Hence, by the principle of mathematical induction, the given statement is true for every integer $n \geq 1$.

Example 8 Prove by the principle of mathematical induction that $11^{n+2} + 12^{2n+1}$ is divisible by 133.

Proof: Let $P(n)$: $11^{n+2} + 12^{2n+1}$ be divisible by 133.
When $n = 1$, $11^{n+2} + 12^{2n+1}$ is equal to $11^3 + 12^3 = 1331 + 1728$
That is, 3059, which is divisible by 133, and thus $P(1)$ is true.
Let us assume that $P(n)$ is true for $n = k$, $k \geq 1$. That is, $P(k)$ is true and hence $11^{k+2} + 12^{2k+1}$ is divisible by 133.

$$
\begin{aligned}
P(k + 1) &= 11^{k+3} + 12^{2k+3} \\
&= 11(11^{k+2}) + 12^2(12^{2k+1}) \\
&= 11(11^{k+2}) + 144(12^{2k+1}) \\
&= 11(11^{k+2}) + (11 + 133)(12^{2k+1}) \\
&= 11(11^{k+2} + 12^{2k+1}) + 133(12^{2k+1}) \\
&= 11(133m) + 133(12^{2k+1}) \\
&= 133(11m + 12^{2k+1})
\end{aligned}
$$

which is divisible by 133. Hence, by the principle of mathematical induction, the given statement is true for every integer $n \geq 1$.

EXERCISE – A

1. Define a statement (i.e., proposition). Give two examples of simple and compound propositions.
2. Define logical connectives, that is, conjunction, disjunction and negation and give their truth tables.

3. Which of the following sentences are propositions?
   (a) Close the gate.
   (b) How are you?
   (c) What is your name?
   (d) Haridwar is a holy city of India.
   (e) Taj Mahal was built by Emperor Shah Jahan in the memory of his wife.
   (f) May god bless you.
   (g) This sentence is false.

4. Which of the following are propositions?
   (a) Earth moves round the Sun.
   (b) The earth is not flat.
   (c) $x = 3$ is a solution of $x^2 = 9$
   (d) $3 + 5 = 7$
   (e) You must be honest.

5. (a) Express the following statements through appropriate symbols.
   (i) Ayush is honest and simple.
   (ii) Ayush is honest but not simple.
   (iii) Ayush is neither honest nor simple.

   (b) State the following compound statements in words taking p for politician and q for scholar.
   (i) $p \wedge q$
   (ii) $p \vee q$
   (iii) $\sim p \vee \sim q$
   (iv) $\sim(p \wedge q)$

6. (a) Determine the truth value of each of the following statements:
   (i) $1 + 3 = 2$ and $2 + 4 = 6$
   (ii) $1 + 2 = 3$ and $3 + 5 = 8$
   (iii) $2 + 8 = 10$ and $4 + 5 = 2$
   (iv) $10 + 5 = 15$ and $1 + 1 = 2$

   (b) Let p: Gopal is an engineer, q: Ashok is a reputed doctor. Give simple verbal sentence that describes the following statements:
   (i) $\sim p$
   (ii) $p \wedge q$
   (iii) $p \vee q$
   (iv) $p \vee \sim q$

7. Explain the terms tautology and fallacy in relation to truth table.

8. Find which of the following is a tautology or a fallacy (i.e., contradiction).
   (a) $p \vee \sim p$
   (b) $p \wedge \sim p$

9. Let t be a tautology and p an arbitrary proposition. Find the truth value of
   (a) $\sim p \vee t$
   (b) $p \wedge \sim t$

10. Find which of these is a tautology or contradiction.
(a) $p \wedge q \rightarrow p$
(b) $p \vee \sim(p \wedge q)$
(c) $(p \wedge q) \wedge \sim(p \vee q)$

11. Construct the truth table for
(a) $\sim (p \wedge q)$
(b) $(p \rightarrow q) \wedge (q \rightarrow p)$

12. Show that the statement $p \rightarrow (p \rightarrow q)$ is a contingency.

13. Show that the statement $(p \rightarrow (q \wedge r)) \rightarrow \sim(p \rightarrow q)$ is a contingency.

14. Prove that each of the following is a contradiction:
(a) $p \wedge (\sim p)$
(b) $p \leftrightarrow \sim p$

15. Prove that the following statements are tautology:
(a) $p \wedge (p \rightarrow q) \rightarrow q$
(b) $(p \vee q) \wedge (\sim q) \rightarrow p$

16. Test the validity of the following arguments:

(a) $\dfrac{\begin{array}{c} p \vee q \\ \sim p \end{array}}{\therefore q}$
(b) $\dfrac{\begin{array}{c} p \vee q \\ \sim q \end{array}}{\sim p}$

(c) $\dfrac{\begin{array}{c} p \vee q \\ \sim q \end{array}}{p}$
(d) $\dfrac{\begin{array}{c} p \rightarrow q \\ p \rightarrow r \end{array}}{p \rightarrow (q \wedge r)}$

17. Prove the validity of the following argument: If a man is a bachelor, he is unhappy and therefore bachelor man is unhappy.

18. Let p: man is bachelor, q: man is unhappy. Prove the validity of the argument

$$\dfrac{\begin{array}{c} p \rightarrow q \\ q \end{array}}{\therefore p}$$

19. Prove the validity of the following argument. If a man is a bachelor, he is unhappy. If a man is unhappy, he dies young. Therefore, bachelors die young.

20. Consider the following conditional statements: If a cyclone destroys my house or if a tsunami destroys my house, then the insurance company will pay me.
Write
(a) Converse of p
(b) Contrapositive of p

21. Negate the following statements:
(a) All students live in hostels.
(b) In a university, every faculty member is PhD.
(c) It is raining.
(d) It is an unusual summer time.

22. Prepare a truth table to get the negation of the statement 'Anand is rich and humble'.

23. Write an English sentence corresponding to each of the following:
   (a) $\forall x, P(x)$
   (b) $\forall x, Q(x)$
   (c) $\forall x, y, R(x, y)$ using $P(x)$ is odd, $Q(x)$ where x is a perfect square and $R(x, y)$: $x + y$ is even.

24. Given the following open statements:

   $p(x)$: $x > 0$ $q(x)$: x is even
   $r(x)$: x is prime $s(x)$: x is divisible by 2
   $t(x)$: x is odd

   Write the following statements in symbolic form:
   (a) There exists a positive integer which is even.
   (b) If x is odd, then it is not divisible by 2.
   (c) If x is prime, then it is not divisible by 2.
   (d) No odd integer is divisible by 2.
   (e) Every even integer is divisible by 2.
   (f) There exists a negative integer.

EXERCISE – B

1. Which of the following sentences are propositions?
   (a) 19 is a prime number.
   (b) Submit your project for evaluation.
   (c) Are you honest.
   (d) 407 is the product of prime numbers 11 and 37.
   (e) $x \in \{y, z, u, v\}$
   (f) $1 \in \{1, 2, 3, 4\}$
   (g) May God shower His blessings upon you.

2. Determine the truth value of each of the following statements:
   (a) $9 + 9 = 18$ and $5 + 4 = 9$
   (b) $2 + 8 = 10$ and $3 + 9 = 13$
   (c) $3 + 7 = 9$ and $5 + 6 = 11$
   (d) $1 + 3 = 5$ and $13 + 8 = 20$

3. Write the negation of the following prepositions:
   (a) It is very cold today.
   (b) Ganga is a sacred river of India.
   (c) Cow milk is regarded as the best milk for health.
   (d) No one wants to buy my old car.
   (e) Some people in the village have no house.

4. Write the given statements in symbolic form for the prepositions:

   p: Ram is a wise man.

   q: Krishna is a politician.

   (a) Ram is a wise man and Krishna is a politician.
   (b) Ram is not a wise man and Krishna is a politician.
   (c) Ram is a wise man and Krishna is not a politician.
   (d) It is not true that Ram is a wise man and Krishna is a politician.
   (e) Either Ram is not wise or Krishna is a politician.
   (f) Either Ram is wise and Krishna is not a politician.

5. Write the given statements in symbolic form for the propositions.

   p: Nokia is a good mobile.

   q: Nokia is a cheap mobile.

   (a) Nokia is a good and cheap mobile.
   (b) Nokia is not good but cheap mobile.
   (c) Nokia is good and costly mobile.
   (d) Nokia is neither good nor cheap mobile.

6. Let p: It is very cold.

   q: The temp is 1°C.

   Express the meaning of following statements in simple sentences:

   (a) $\sim p$ (b) $p \vee q$ (c) $p \wedge q$ (d) $\sim(p \wedge q)$

7. Given p and q are true and r is false, find the truth value of the following statements:

   (a) $p \vee (q \vee r)$ (b) $p \wedge (q \vee r)$ (c) $\sim p \vee (q \wedge r)$

8. Find the truth table of the following propositions:

   (a) $\sim p \wedge q$
   (b) $p \vee \sim q$
   (c) $\sim p \wedge \sim q$
   (d) $\sim(p \vee q) \vee (\sim p \wedge \sim q)$

9. Let p: x is even.

   q: x is divisible by 3.

   Write the meaning of the following propositions in simple sentences:

   (a) $\sim q$ (b) $p \vee q$ (c) $\sim p \wedge \sim q$

10. Translate each statement into symbols using variables, quantifiers and predicate symbols.

    Let $P(x)$: x can speak English.

    $Q(x)$: x can speak French.

    (a) There is a mathematician who can speak both English and French.
    (b) There is a mathematician who can speak English but not French.
    (c) Every mathematician can speak either English or French.

5

Combinatorics

LEARNING OBJECTIVES

After reading this chapter, you will be conversant with:

- Introduction
- Basic Principle of Counting
- Permutations
- Ordered and Unordered Partitions
- Circular Permutations
- Combinations
- Derangements
- The Pigeonhole Principle
- Elements of Probability
- Multiplication Theorem (Independent Events)
- Baye's Theorem
- Concept of a Random Variable
- Binomial Distribution
- Poisson Distribution

5.1 INTRODUCTION

Combinatorics is the branch of mathematics, which is concerned with the study of techniques of counting. Combinatorial problems arise in many areas of pure mathematics and geometry. It has acquired great importance due to its applications in the analysis of algorithms, computer science, optimization, probability theory, and statistical physics.

In this chapter, counting techniques of sum, multiplication, pigeonhole principle, permutation, combination, and probability theory are discussed.

5.2 BASIC PRINCIPLE OF COUNTING

There are two basic principles of counting: Addition and Multiplication.

5.2.1 Addition Principle

In many problems, we are interested in the number of ways in which an event can happen or a task can be performed.

If $E_1, E_2, \ldots, E_n$ are n mutually exclusive events and E_1 can happen in m_1 ways, E_2 can happen in m_2 ways, E_3 can happen in m_3 ways, … and E_n can happen in m_n ways, then E_1 or E_2 or E_3 or … E_n can happen in $m_1 + m_2 + \ldots + m_n$ ways.

Example 1 A television (TV) showroom contains 10 sets of Sony and 8 sets of Samsung. Find the number of ways in which a customer can select a TV.

Solution: Since there are 10 TV sets of Sony, the customer can select any one of them and hence there are 10 ways of selection. Similarly, there are 8 ways of selecting a Samsung TV. Hence, the total number of ways in which a customer can select a TV set is $10 + 8 = 18$.

Example 2 Suppose E_1 is the event of choosing a prime number less than 15 and E_2 is the event of choosing an odd number less than 10. Find the number of ways in which E_1 or E_2 can happen.

Solution: The event E_1 can happen if the number is any one of the numbers $\{2, 3, 5, 7, 11, 13\} = 6$ numbers. The event E_2 can happen if the number is odd, that is, $\{1, 3, 5, 7, 9\} = 5$ numbers.

Hence, E_1 or E_2 can happen in $6 + 5 – 3 = 8$ ways (3 is subtracted as 3 cases are common to E_1 and E_2).

Example 3 From a deck of playing cards, find the number of ways in which we can draw (1) an ace or a king, (2) a card of diamond or a spade, (3) an ace or a card of hearts.

Solution:

(1) Since there are 4 cards of king and 4 cards of ace, there are $4 + 4 = 8$ ways in which an ace or a king can be drawn from the deck.
(2) Since there are 13 cards of diamond and 13 cards of spade, there are $13 + 13 = 26$ ways in which a card of diamond or a spade can be drawn from the deck.
(3) Since there are 13 cards of heart and 4 cards of ace and a card ace of hearts, there are $13 + 4 - 1 = 16$ ways in which a card of ace or a heart can be drawn from the deck.

Example 4 In how many ways can we get a sum 7 when two distinguishable dice are rolled?

Solution: There are 6 ways in which a sum 7 appears when a pair of dice in rolled.

$$(1, 6), (2, 5), (3, 4), (4, 3), (5, 2), (6, 1)$$

In the above notation (a, b), a denotes the number of dots on one die and b denotes the number of dots on the other die.

5.2.2 Multiplication Principle

Suppose that an event E_1 can happen in n_1 ways and another independent event E_2 can happen in n_2 ways, then both the events E_1 and E_2 can happen in $n_1 n_2$ ways.

The multiplication rule can also be formulated in terms of choices of tasks. Suppose that two tasks T_1 and T_2 are to be performed independently in the sequence T_1T_2. If T_1 can be performed in n_1 ways and T_2 can be performed in n_2 ways, then T_1T_2 can be performed in n_1n_2 ways.

For example, suppose that the task T_1 can be performed in two ways and the task T_2 can be performed in four ways.

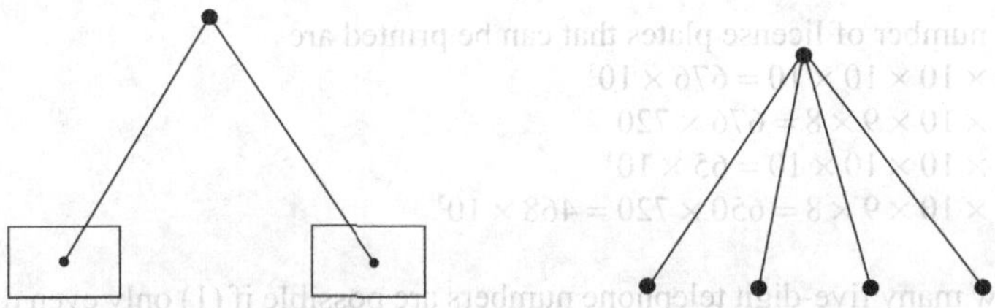

$n_1 = 2$ ways of performing T_1 $n_2 = 4$ ways of performing T_2.

Figure 5.1(a) **Figure 5.1(b)**

Explanation: Each of the two ways of performing T_1 can be associated with any of the four ways of performing T_2 and hence T_1T_2 can be performed in $2 \times 4 = 8$ ways.

Some More Examples

Example 1 A stadium has 3 entry and 4 exit gates. Find the number of ways in which a person can enter and leave the stadium.

Solution: Since there are 3 entry gates, a person can enter through any one of them and so there are 3 ways of entering the stadium. The person can leave the stadium through any of the 4 exit gates. As each gate of entry can be associated with each gate of exit, there are $3 \times 4 = 12$ ways in which a person can enter and leave the stadium.

Example 2 Three persons enter into a hotel that has 5 air-conditioned family suites. Find the number of ways in which they can occupy the suites.

Solution: The first person can occupy any one of the 5 suites. The second person can occupy any one of the remaining 4 suites. Similarly, the third person can occupy any one of the remaining 3 suites. Hence, the total number of ways of occupying the suites is $5 \times 4 \times 3 = 60$.

Example 3 A bank password consists of two letters of English alphabet followed by two digits. How many different passwords are there.

Solution: There are 26 letters in the English alphabet and 10 digits from 0 to 9. The sequence of two digits can be formed in $10 \times 10 = 100$ ways as both places of digits can have any one of the 10 digits. The two-letter sequence can be obtained in 26×26 ways. Hence, the total number of passwords $= 26 \times 26 \times 10 \times 10 = 67{,}600$.

Example 4 The transport department is interested in giving licence plates consisting of two letters of English alphabet followed by three digits.

How many distinct plates can be printed if

(1) repetition of letters and digits are allowed.
(2) only letters can be repeated.
(3) only digits can be repeated.
(4) neither letters nor digits can be repeated.

Solution: The number of license plates that can be printed are

(1) $26 \times 26 \times 10 \times 10 \times 10 = 676 \times 10^3$
(2) $26 \times 26 \times 10 \times 9 \times 8 = 676 \times 720$
(3) $26 \times 25 \times 10 \times 10 \times 10 = 65 \times 10^4$
(4) $26 \times 25 \times 10 \times 9 \times 8 = 650 \times 720 = 468 \times 10^3$

Example 5 How many five-digit telephone numbers are possible if (1) only even digits are used and (2) only odd digits are used.

Solution:

(1) The even digits are 2, 4, 6, and 8. Each of the 5 places can be filled in 4 ways to form a five-digit number. Hence, there can be $4 \times 4 \times 4 \times 4 \times 4 = 4^5$ different numbers.
(2) The odd digits are 1, 3, 5, 7, and 9. Each of the 5 places can be filled in 5 ways. Hence, there can be $5 \times 5 \times 5 \times 5 \times 5 = 5^5$ different numbers.

5.3 PERMUTATIONS

Let $A = \{a_1, a_2, \ldots, a_n\}$ be a set of n elements.

Any arrangement of all the elements of the set A in a line in some order is called a permutation of the objects.

Any arrangement of $r \le n$ of the objects is called a permutation of n objects taken r at a time. This is the same as the number of ways of filling up r places by r objects, taken out of n.

Since any one of the n objects $a_1, a_2, \ldots, a_n$ can be placed in the first position, the number of different ways of filling up the first position is n. Following this, the second position can be filled up in $(n - 1)$ ways; the third position in $(n - 2)$ ways. Continuing in this manner, the rth position can be filled in $(n - (r - 1)$, that is, $(n - r + 1)$ ways. Hence, by the multiplication principle, the number of permutations of n objects taken r at a time denoted by ${}^np_r$ or $p(n, r)$ is

$$
\begin{aligned}
{}^np_r \text{ or } p(n,r) &= n(n-1)(n-2) \ldots (n-r+1) \\
&= \frac{n(n-1)(n-2)\ldots(n-r+1)(n-r)!}{(n-r)!}
\end{aligned}
$$

$$
{}^np_r = \frac{n!}{(n-r)!}
$$

When $r = n$, we have ${}^np_r = n(n - 1)\ldots = n!$, that is, the number of all arrangements of n distinct objects.

Suppose there are four distinct objects, for example, flower pots, identified as a, b, c, and d.

The different orders of arrangement of these four objects by taking three at a time are

(1) *abc*,	*bca*,	*cab*,	*acb*,	*cba*,	*bac*
(2) *abd*,	*bda*,	*dab*,	*adb*,	*dba*,	*bad*
(3) *acd*,	*cda*,	*dac*,	*adc*,	*dca*,	*cad*
(4) *bcd*,	*cdb*,	*dbc*,	*bdc*,	*dcb*,	*cbd*

Thus, the above 24 arrangements of the four objects *a*, *b*, *c*, and *d* obtained by taking three at a time are each called a permutation. It may be seen that six given in (1) are all formed of same *a*, *b*, and *c* in different orders. Similarly, six given in (2) are formed of *a*, *b*, and *d* in different orders and so on.

Example 1 How many numbers can be formed taking only 3 out of the 6 digits: 1, 2, 3, 4, 5, and 6?

Solution: Since each arrangement of 3 digits gives a new number, the different numbers that can be formed is evidently the number of permutations of 6 things taken 3 at a time, that is, ${}^6p_3 = 6 \times 5 \times 4 = 120$.

Example 2 In how many different ways can a true–false test consisting of 9 questions be answered?

Solution: Since each question has two answers, that is, either true or false, answering 9 questions amounts filling 9 places in two ways, that is, $2 \times 2 \times 2 \times 2 \times 2 \times 2 \times 2 \times 2 \times 2 = 2^9 = 2^4 \times 2^5 = 16 \times 32 = 512$.

Example 3

(1) How many three-digit numbers can be formed from the digits 0, 1, 2, 3, 4, 5, 6 if each of the digits can be used only once.

(2) How many of these are odd digit numbers?

Solution:

(1) There are 7 digits, that is, 0, 1, 2, 3, 4, 5, 6.
The number of three-digit numbers $= {}^7p_3 = 7 \times 6 \times 5 = 210$.
This includes numbers whose first digit is 0, which are ${}^6p_2 = 6 \times 5 = 30$.
Hence, three-digit numbers are $210 - 30 = 180$.

(2) The odd digits will have 1, 3, 5 at the unit place.
The number of such odd digit numbers $= 6 \times 5 \times 3 = 90$.
These include numbers whose first digit in 0, i.e., $5 \times 3 = 15$.
Hence, number of odd digit numbers $= 90 - 15 = 75$.

Example 4 How many seven-digit telephone numbers are possible if

(1) only even digits can be used?

(2) the first three digits are 111.

Solution:

(1) The even digits are 2, 4, 6, and 8. Each of the seven places can be filled in four ways by even digits. Hence, different numbers that, are possible, are $4 \times 4 \times 4 \times 4 \times 4 \times 4 \times 4 = (4)^7$

(2) Since first three digits out of seven are fixed, we have to fill up four digits. The possible ways are $4 \times 4 \times 4 \times 4 = 256$

5.3.1 Permutation with Repetitions

The number of permutations of n objects of which n_1 are alike, n_2 are alike,..., n_k are alike is given by,

$$P(n, n_1, n_2, \ldots, n_k) = \frac{n!}{n_1!n_2!\ldots n_k!}$$

The proof can be understood by a particular example. Suppose we want to form all possible seven-letter words using the letters from the word UNUSUAL. There are 7! permutations of letters $U_1NU_2SU_3AL$ where three Us are distinguished. Let us observe the following 6 permutations:

$$U_1U_2U_3NSAL,\ U_2U_3U_1NSAL,\ U_3U_1U_2NSAL$$
$$U_1U_3U_2NSAL,\ U_3U_2U_1NSAL,\ U_2U_1U_3NSAL$$

When the subscripts are removed, they are all the same word. The 6 come from the fact that there are 3! = 3 × 2 × 1 = 6 ways of placing the three Us in the first three positions in the permutation. This is true for each set of three positions in which Us can appear. Hence, there are

$$P(7,3) = \frac{\lfloor 7}{\lfloor 3} = \frac{1.2.3.4.5.6.7}{1.2.3} = 840$$ different seven-letter words that can be formed.

Example 1 In how many ways can the letters of the word **attention** be rearranged?

Solution: There are 9 letters in the word **attention** in which 3 (t's) are of one kind, 2 (n's) of another and 4 other different letters.

Hence, the number of all arrangements $\frac{9!}{3!2!} = 30,240.$

Example 2 Find the number of different signals each consisting of 7 flags hung in a vertical line that can be formed from a set of 3 red flags, 2 blue flags, and 2 white flags.

Solution: We are required to find the arrangements of 7 objects (flags) in which 3 are alike, 2 are alike, and the other 2 are alike.

The number of signals in

$$P(7, 3, 2, 2) = \frac{7!}{3!2!2!} = \frac{7\times6\times5\times\cancel{4}\times\cancel{3}\times\cancel{2}\times1}{\cancel{3}.\cancel{2}.1.\cancel{2}.1.\cancel{2}.1} = 210$$

Example 3 A train consists of 12 carriages of which 5 are AC-I, 4 AC-II and rest AC-III. Find the number of ways in which these carriages can be arranged.

Solution: The problem can be regarded to find the number of arrangements of 12 objects of which 5 are alike, 4 are alike, and 3 are alike.

Required number of permutations is

$$P(12; 5, 4, 3) = \frac{12!}{5!4!3!}$$
$$= \frac{\cancel{5}!\cancel{6}.7.\cancel{8}.\cancel{9}^3.10.11.12}{\cancel{5}!\cancel{24}.\cancel{6}}$$
$$= 7.3.10.11.12 = 27,720$$

Example 4 Four married couples have bought 8 seats in a row for a concert. In how many ways can they be seated?

(1) With no restrictions
(2) If each couple is to sit together
(3) If all the men sit together to the right of all women

Solution:

(1) Since there is no restriction, the number of ways in which 4 couples can occupy the seats is 8! = 1.2.3.4.5.6.7.8 = 40,320.

(2) HW | HW | HW | HW Four couples can occupy seats in 4! ways and each couple can sit in two ways (HW or WH). Hence, the total number of ways of sitting = 4!.2.2.2.2 = 24 × 16 = 384.

(3) WWW W | HHH H Wives can occupy seats in 4! = 24 ways and husbands can occupy seats in 4! = 24 ways. Hence, the number of ways in which they can occupy seats is 24 × 24 = 576

Example 5 Find the number of ways in which four physics books, 3 chemistry books, 3 mathematics books, and 2 books on philosophy can be arranged on a shelf so that all books of the same subject are together.

Solution: The books are arranged on the shelf in four units. The first unit can be filled by any one of the four subjects, the second by any three remaining subjects, the third by any two remaining one subjects, and the last unit by the remaining one subject. Hence, there are 4.3.2.1 = 24 ways to arrange books according to subject matter.

Now physics books can be arranged in 4! ways, chemistry books can be arranged in 3! ways, mathematics book can be arranged in 3! ways, and books on philosophy can be arranged in 2! ways.

Thus, all the arrangements so that the books of the same subject are together is

$$\begin{aligned} &24 \times 4! \times 3! \times 3! \times 2! \\ &= 24 \times 24 \times 6 \times 6 \times 2 \\ &= 41{,}472 \end{aligned}$$

Example 6 In how many different orders can 3 men and 3 women be seated in a row of 6 seats if

(1) any one may sit in any of the seats.
(2) men occupy the first three seats and women occupy the last three seats.
(3) the first and the last seats must be filled by men.
(4) all members of same sex are seated in adjacent seats.
(5) men and women are seated alternately.

Solution:

(1) Since there are 3 men and 3 women who can sit in any of the seats, it amounts to arranging 6 persons. The total number of ways is 6! = 720.

(2) Men occupy first three seats, which is possible 3! = 6 ways. Similarly, women occupy the last three seats, which is possible in 3! = 6 ways. Hence, the total number of ways = 6 × 6 = 36 ways.

(3) The first seat of men can be occupied in 3 ways and the last seat in 2 ways. Now 4 persons, that is, 3 women and 1 man can occupy seat in 4! ways. Hence, the total number of ways is 3.2.4! = 144.

(4) All members of the same sex are seated in adjacent seats is possible in $2.3!3! = 2 \times 6 \times 6 = 72$ ways.
The factor 2 arises due to the fact that in one case men occupy first three seats and then women and another one in which women occupy first three seats then men.

(5) Men and women are seated alternately in $2(3!)^2 = 72$.

Example 7 A state flag is to be designed with 6 vertical strips red, blue, white, and saffron colours. Find the number of ways in which this can be done so that no two strips have the same colours.

1	4 Colours
2	3 Colours
3	3 Colours
4	3 Colours
5	3 Colours
6	3 Colours

Solution: As there are four colours, one can choose the first strip in 4 ways. The second strip can have any one of the remaining 3 colours. The third strip can be of any three colours excluding the colour of the second strip, similarly the fourth strip can have 3 colours and so on. Hence, the number of ways the flag can be designed so that no two adjacent strips have the same colour is $4 \times 3 \times 3 \times 3 \times 3 \times 3$.

$$\text{i.e., } 4 \times 3^5 = 4 \times 243$$
$$= 972 \text{ ways}$$

5.4 ORDERED AND UNORDERED PARTITIONS

5.4.1 Ordered Partitions

Let a box A contain 10 tickets numbered 1, 2, 3,…, 10. It is required to find the number of ways in which one can draw first three tickets, then four tickets, and lastly three tickets. This essentially amounts to finding the number of ordered partitions

$$\{A_1, A_2, A_3,\}$$

of the set A of 10 tickets into cells A_1 containing three tickets, A_2 containing four and A_3 containing three tickets.

These are called ordered partitions since we can distinguish between

$$\{(1, 2, 3), (4, 5, 6, 7), (8, 9, 10)\} \text{ and } \{(8, 9, 10), (4, 5, 6, 7), (1, 2, 3)\}$$

each determining the same partition of A. There are ${}^{10}c_3$ ways of drawing first three tickets for cell A_1; following this, there are ${}^{7}c_4$ ways of drawing the next four tickets for cell A_2 (as 7 tickets are left after first draw of 3 tickets) and finally there are three tickets in the box and there are ${}^{3}c_3$ ways of drawing tickets for the cell A_3.

Hence, there are

$$^{10}c_3 \times {}^{7}c_4 \times {}^{3}c_3 = \frac{\lfloor 10}{\lfloor 3 \lfloor 7} \cdot \frac{\lfloor 7}{\lfloor 4 \lfloor 3} \cdot \frac{\lfloor 3}{\lfloor 3 \lfloor 0} \tag{5.1}$$

$$= \frac{10.9.8.7.6.5.4.3.2.1}{1.2.3.1.2.3.4.1.2.3} \quad (0! = 1)$$

$$= 10.7.6.5.2 = 4200$$

different ordered partitions of A into cells A_1, A_2 and A_3.

It can be observed from (5.1) above that each numerator after first is cancelled by the second term in the denominator and we have $\frac{\lfloor 10}{\lfloor 3 \lfloor 4 \lfloor 3}$ ways.

This result can be generalized for any number of partitions.

THEOREM 5.1 Let a set A contain n elements and let it be partitioned into K cells $A_1, A_2, \ldots, A_k$ containing $n_1, n_2, \ldots, n_k$ elements. The number of different ordered partitions of A is

$$\frac{\lfloor n}{\lfloor n_1 \lfloor n_2 \ldots \lfloor n_k}$$

where $n_1 + n_2 + \ldots + n_k = n$

5.4.2 Unordered Partitions

Very often, it is required to partition a set A into a collection of subsets $A_1, A_2, \ldots, A_k$ where the subsets are unordered. The number of unordered partitions is equal to

$$\frac{\lfloor n}{\lfloor n_1 \lfloor n_2 \ldots \lfloor n_k} \frac{1}{\lfloor k}$$

where k denotes the number of parts of A, which have the same number of elements.

Example: Find the number of ways in which a group of 12 players can be partitioned into three teams A_1, A_2, A_3 so that each team contains four players.

Solution:

Method I:

Let one of the players be selected. Then there are 11 players left and ${}^{11}c_3$ are the number of ways in which three players can be selected and thus A_1 team is formed. Now, let a player be selected who is not in team A_1, then there are ${}^{7}c_3$ ways to choose to form team A_2. The remaining four players will constitute team A_3.

Thus, the total number of ways to partition the players is

$$
\begin{aligned}
{}^{11}c_3 \times {}^{7}c_3 = \frac{11.\cancel{10}^5.\cancel{9}^3}{1.\cancel{2}.\cancel{3}} \times \frac{7.\cancel{6}.5}{1.\cancel{2}.\cancel{3}} &= 11 \times 15 \times 35 \\
&= 11 \times 525 \\
&= 5775
\end{aligned}
$$

Method II:

The partition $\{A_1, A_2, A_3\}$ can be arranged in $\lfloor 3 = 6$ ways as an ordered partition.

There are ordered partitions $\frac{\lfloor 12}{\lfloor 4 \lfloor 4 \lfloor 4} = 34,650$

Hence, the number of unordered partitions is

$$\frac{\lfloor 12}{\lfloor 4 \lfloor 4 \lfloor 4} \cdot \frac{1}{\lfloor 3} = \frac{34,650}{6} = 5775$$

5.5 CIRCULAR PERMUTATIONS

The different orders of arrangements along a circle or along a closed curve that are obtained by taking some or all the objects is called a circular permutation.

Two arrangements along a circle are different only if they are different with respect to their relative positions.

Let *A*, *B*, *C*, *D*, *E*, *F* denote the number of objects along a circle. Consider the following arrangements.

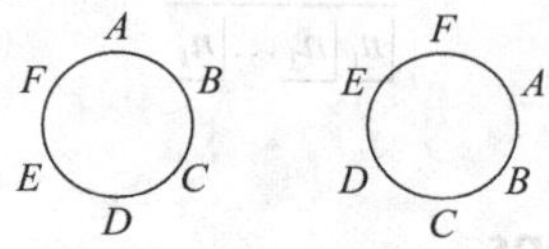

(a) (b)

The two arrangements (a) and (b) are not different because if we travel around the circle in the same direction, the letters are in the same relative position and precisely in the same order (although the letters do not occupy the same absolute position in each).

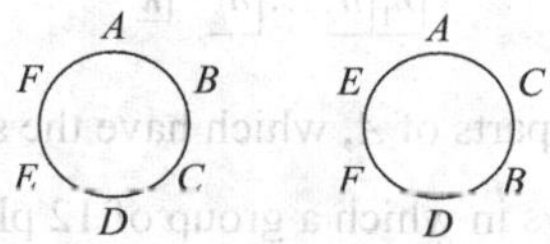

(c) (d)

In the arrangements (c) and (d), the relative positions are different and hence the arrangements are different.

Hence, if we want to arrange objects around a circle, we obtain different orders of arrangements by putting one of the objects in fixed position and changing the position of the rest in all possible ways.

Hence, if *n* objects are arranged around a circle, the total number of arrangements is $(n-1)!$ for if one of them is placed in a fixed position, the remaining $(n-1)$ objects can be arranged in $(n-1)!$ ways.

Example 1 In how many ways can 4 mathematicians and 5 physicists sit around a table so that no two mathematicians are together.

Solution: Let the physicists be seated first, then they can sit in $(5-1)^1$ or 41 ways.

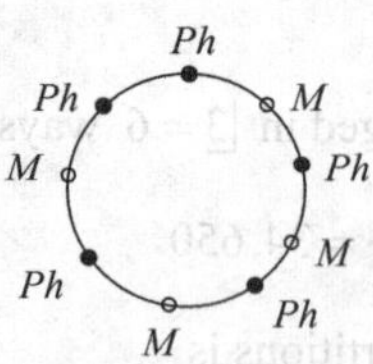

There are 5 locations for the first mathematician, 4 for the second, 3 for the third, and 2 for the fourth.

Hence, the required number of ways $= 4!(5 \times 4 \times 3 \times 2)$
$= 24 \times 120 = 2880$

Example 2 In how many ways can 10 adults and 5 children stand on a circle so that no two children are next to each other?

Solution: The 10 adults can stand on the circle in $(10 - 1)! = 9!$ ways. There are 10 locations for the first child, 9 for the second, 8 for the third, 7 for the fourth and 6 for the fifth.

The required number of ways $= 9!(10 \times 9 \times 8 \times 7 \times 6)$
$= 9!({}^{10}P_5)$

5.6 COMBINATIONS

Given a set of n distinct objects, we are interested in the groups of r objects in which the order does not matter. Such a collection is called a combination of n objects taken r at a time and is denoted by $c(n, r)$ or ${}^n c_r$ or $\binom{n}{r}$.

Suppose that we are given four distinct objects designated by a, b, c, and d. Then the combinations of a, b, c, and d taken three at a time are *abc*, *acd*, *abd*, and *bcd*. The combinations taken two at a time are *ab*, *ac*, *ad*, *bc*, *bd*, and *cd*.

It may be noted that the following combinations *abc*, *bca*, *cab*, *acb*, *cba*, and *bac* are all equal. Each denotes the set $\{a, b, c\}$.

THEOREM 5.2 Let $A = \{a_1, a_2, \ldots, a_n\}$ be a set of n distinct objects. We are concerned with the number of ways in which r objects can be taken from these n objects. Let this number be ${}^n c_r$. If we take any one of the combinations, its r objects can be arranged in $r!$ ways. Hence, the total number of arrangements from all the combinations is

$${}^n c_r \, r!$$

These arrangements can be obtained by considering that r places can be filled by n objects. Hence,

$${}^n p_r = {}^n c_r \, r!$$

$$or \quad \frac{n!}{(n-r)!} = {}^n c_r \, r!$$

$$\therefore \; {}^n c_r = \frac{n!}{r!(n-r)!} = \frac{n(n-1)(n-2)\ldots(n-r+1)}{r!}$$

Example 1 A valid computer password consists of 5 characters, the first of which is a letter chosen from the set $\{A, B, C, D, E\}$ and the remaining 4 characters are letters chosen from English alphabet or a digit. Find the number of different passwords that are possible.

Solution: Since there are 5 letters, a letter can be chosen in ${}^5 c_1$, that is, 5 ways. The remaining 4 characters are to be chosen from 26 letters and 10 digits, namely 36 characters. As repetition is allowed, the four characters can be chosen in $36 \times 36 \times 36 \times 36$, that is, 1,679,616 ways.

Hence, the total number of passwords is $5 \times 1{,}679{,}626 = 8{,}398{,}080$ ways.

Example 2 In how many ways can a committee of four professors and three students be selected from eight professors and seven students?

Solution: The number of ways in which 4 professors can be selected out of 8 is ${}^{8}c_4$. The number of ways in which 3 students can be selected out of 7 students is ${}^{7}c_3$.

Hence, the number of ways of forming a committee of 4 professors and 3 students is

$$
\begin{aligned}
& {}^{8}c_4 \times {}^{7}c_3 \\
&= \frac{8 \times 7 \times 6 \times 5}{1.2.3.4} \times \frac{7.6.5}{1.2.3} \\
&= 70 \times 35 = 2450
\end{aligned}
$$

Example 3 A computer manufacturer concerned with an advertising campaign is considering 5 magazines, 4 newspapers, 3 television stations, and 3 radio stations. In how many ways can 5 advertisements be run if

(1) all 5 are to be magazines.
(2) 2 are to be in magazines, 1 is to be in newspapers and 1 is to be on radio and on the television.

Solution:

(1) There are 5 magazines, which are all to be run and hence there is only one way.
(2) The required number is

$$
\begin{aligned}
& {}^{5}c_2 \cdot {}^{4}c_1 \cdot {}^{3}c_1 \cdot {}^{3}c_1 \\
&= \frac{5 \times 4}{1.2}.4.3.3 = 360
\end{aligned}
$$

Example 4 In how many ways can a committee of 6 persons be selected from a group of 10 persons if one person is to be designated as chairman of the committee.

Solution: As there are 10 persons, the chair can be designated as chairman as any one can be designated.

After the chairman, there are 9 persons from whom 5 members are selected. The number of ways of selection of 5 persons is

$$
{}^{9}c_5 = {}^{9}c_4 = \frac{9 \times 8 \times 7 \times 6}{1.2.3.4} = 126
$$

Hence, by multiplication rule, the number of ways of selecting the committee with chairman is

$$
10 \times {}^{9}c_5 = 1260
$$

Example 5 There are 10 men and 4 women. Find the number of ways in which a committee of 5 persons can be formed if the committee is to have (1) exactly 3 men and (2) at least 2 women.

Solution:

(1) The committee of 5 persons has exactly 3 men and hence the rest will be women, that is, 2 women. The number of ways in which 3 men can be selected out of 10 is ${}^{10}c_3$ and 2 women is ${}^{4}c_2$.
Hence, the committee consisting of exactly 3 men can be formed in

$$^{10}c_3 \times {}^{4}c_2 = \frac{10.\cancel{9}^3.8}{1.\cancel{2}.\cancel{3}} \times \frac{\cancel{4}.3}{\cancel{2}} = 720 \text{ ways}$$

(2) The committee consisting of at least 2 women can be formed as follows:

(i) 2 women and 3 men in ${}^{4}c_2 \times {}^{10}c_3 = 6 \times 120 = 720$
(ii) 3 women and 2 men in ${}^{4}c_3 \times {}^{10}c_2 = 4 \times 45 = 180$
(iii) 4 women and 1 men in ${}^{4}c_4 \times {}^{10}c_1 = 1 \times 10 = 10$

Hence, a committee of 5 persons consisting of at least 2 women can be formed in $(720 + 180 + 10) = 910$ ways.

Example 6 A candidate can choose 8 questions out of 10 questions to answer in an examination.

(1) Find the number of ways in which the candidate can answer.
(2) Find the number of ways to answer if the first two must be answered.

Solution:

(1) The number of ways of choosing 8 questions out of 10 is

$$^{10}c_8 = {}^{10}c_{10-8} = {}^{10}c_2 = \frac{\cancel{10}^5 \times 9}{2} = 45$$

(2) Since the first 2 questions must be answered, the candidate has to select 6 questions out of remaining 8.

$$\text{The number of ways} = {}^{8}c_6 = {}^{8}c_{8-6} = {}^{8}c_2 = \frac{\cancel{8}^4 \times 7}{\cancel{2}} = 28$$

Example 7 A box of 15 electronic components contains 3 defective components but is not known which are those. Find (1) the number of ways in which 3 components can be selected and (2) the number of ways in which selection will consists of at least one defective component.

Solution:

(1) The number of ways in which 3 components can be selected out of 15 is

$$^{15}c_3 = \frac{\cancel{15}^5 \times \cancel{14}^7 \times 13}{1.\cancel{2}.\cancel{3}} = 455.$$

(2) The selection of at least 1 defective out of 3 can happen in the following ways:

(i) 1 defective and 2 nondefective $= {}^3c_1 \times {}^{12}c_2 = 3 \times 66 = 198$ ways

(ii) 2 defective and 1 nondefective $= {}^3c_2 \times {}^{12}c_1 = 3 \times 12 = 36$ ways

(iii) All 3 defective $= {}^3c_3 = 1$ way

Hence, the total number of ways = 198 + 36 + 1 = 235 ways.

Example 8 Find the number of triangles that can be formed by selecting points from a set of 13 points of which 5 are collinear.

Solution: When we take all 13 points, the number of triangles formed $= {}^{13}c_3$.

As 5 points are collinear, they do not form any triangle.

$$= {}^{13}c_3 - {}^5c_3$$

$$\therefore \text{ Total number of triangles } = \frac{13 \times 12 \times 11}{1.2.3} - 10 = 286 - 10$$

$$= 276$$

Example 9 Find the number of lines that can be drawn through 10 points on a circle. As the points are not collinear, the number of lines $= {}^{10}c_2 = \dfrac{10 \times 9}{2} = 45.$

5.6.1 Combinations of Objects not all Different

The total number of combinations by taking some or all at a time from n distinct objects is

$$ {}^nc_1 + {}^nc_2 + {}^nc_3 + \cdots + {}^nc_n$$

$$= 2^n - 1$$

$$\left\{ \begin{array}{l} (1+x)^n = 1 + {}^nc_1 x + {}^nc_2 x^2 + \cdots + {}^nc_n x^n \\ \text{For } x = 1, 2^n - 1 = {}^nc_1 + {}^nc_2 + \cdots + {}^nc_n \end{array} \right\}$$

Example 1 An MBA student has to secure minimum pass marks in each of the 10 papers for his final semester examination. Find the numbers of ways in which the student can fail.

Solution: In each paper, a student can score minimum marks or not. That is, there are two ways: either he passes or fails. The total number of ways in which a student can pass or fail is (2^{10}). But this number includes the case in which student passes in all the subjects.

Hence, the number of ways in which the student can fail is

$$= 2^{10} - 1$$

$$= 1024 - 1 = 1023 \text{ ways}$$

5.7 DERANGEMENTS

Definition 5.1 A derangement of n distinct symbols (or objects), which have some natural order, is a permutation in which no symbol is in its correct position. The number of derangements of n symbols is denoted by D_n.

Example: There are four houses on a particular street. A mischievous postman arrives with one letter addressed to each house. In how many ways can he deliver the letters, one to each house, so that no letter is delivered to the correct house.

Such a question is concerned with the derangement of four objects.

Example: The derangement of two symbols 1, 2 is 1, that is, the derangement 2,1 and so. $D_2 = 1$.

The derangement of three symbols 1, 2, 3 are 2, 3, 1 and 3, 1, 2, hence $D_3 = 2$.

There are nine arrangements of symbols 1, 2, 3, 4

2 1 4 3	2 4 1 3	2 3 4 1
3 4 2 1	3 4 1 2	3 1 4 2
4 3 2 1	4 3 1 2	4 1 2 3

Therefore, $D_4 = 9$. Hence, there are 9 ways in which the mischievous postman can deliver four letters each to a wrong address.

General Formula for D_n: Let us consider the derangement of n symbols 1, 2, 3,…, n. Let A_1 be a set of permutations of 1, 2, 3, 4 in which the number 1 is in the first position; A_2 the number of permutations of 1, 2, 3, 4 in which 2 is in the second position. Define similarly A_3 and A_4.

Since there are 4! permutation of symbols 1, 2, 3, 4, the number of derangements of four symbols is

$$D_4 = 4! - |A_1 \cup A_2 \cup A_3 \cup A_4| \tag{5.2}$$

Now

$$|A_1 \cup A_2 \cup A_3 \cup A_4| = \sum_{i=1}^{4} |A_i| - \sum_{i<j} |A_i \cap A_j| + \sum_{i<j<k} |A_i \cap A_j \cap A_k| - |A_1 \cap A_2 \cap A_3 \cap A_4| \tag{5.3}$$

Now $|A_1| = 3!$ as 1 is in the first position and 2, 3, 4 can be in any of the next three positions, similarly $|A_2| = |A_3| = |A_4| = 3!$ The set $A_1 \cap A_2$ contains those permutations of 1, 2, 3, 4 in which 1 and 2 are in correct positions. There are just two positions such as permutations 1 2 3 4 and 1 2 4 3. Thus, $|A_i \cup A_j| = 2!$ for each of the $4_{C_2} = 6$ terms in the second sum of equation (5.3).

Similarly, there are ${}^4c_3 = 4$ terms of $|A_i \cap A_j \cap A_k|$ equal to 1.

Therefore, $\quad |A_1 \cup A_2 \cup A_3 \cup A_4| = 4(3!) - 6(2!) + 4(1!) - 1$

Hence, from equation (5.1), we have

$$\begin{aligned} D_4 &= 4! - \left(4(3!) - \binom{4}{2}2! + \binom{4}{3}1! - 1\right) \\ &= 4! - 4! + \frac{4!}{2!\,2!}2! - \frac{4!}{3!\,1!}1! + 1 \\ &= 4!\left(1 - \frac{1}{1!} + \frac{1}{2!} - \frac{1}{3!} + \frac{1}{4!}\right) \\ &= 24\left(1 - 1 + \frac{1}{2} - \frac{1}{6} + \frac{1}{24}\right) = 2^4 - 2^4 + 12 - 4 + 1 = 9 \end{aligned}$$

Hence, the number of derangements of $n > 1$ ordered symbols is

$$D_n = n!\left(1 - \frac{1}{1!} + \frac{1}{2!} - \frac{1}{3!} + \frac{1}{4!} - \ldots + (-1)^n \frac{1}{n!}\right)$$

For example, for $n = 5$

$$D_5 = 5!\left(1 - \frac{1}{1!} + \frac{1}{2!} - \frac{1}{3!} + \frac{1}{4!} - \frac{1}{5!}\right)$$

$$= \cancel{5}! - \cancel{5}! + 5.4.3 - 5.4 + 5 - 1$$

$$= 60 - 20 + 5 - 1 = 44$$

Example 1 A postman delivers letters to six houses in a block completely at random. He delivers one letter to each house. Find the number of ways in which the following can be accomplished:

(1) no letter arrives at the right house.
(2) all letters arrive at the right house.
(3) at least one letter arrives at the right house.

Solution:

(1) The six letters can be regarded as 6 symbols. No letter arrives at the right house is equal to the number of derangement of 6 symbols, namely

$$D_6 = 6!\left(1 - \frac{1}{1!} + \frac{1}{2!} - \frac{1}{3!} + \frac{1}{4!} - \frac{1}{5!} + \frac{1}{6!}\right)$$

$$= \cancel{6}! - \cancel{6}! + 6.5.4.3 - 6.5.4 + 6.5 - 6 + 1$$

$$= 360 - 120 - 6 + 1 + 30$$

$$= 391 - 126 = 265$$

(2) There is only 1 way in which all letters arrive at the right house.
(3) The total number of ways in which 6 letters can be arranged is $6! = 720$.

No letter arrives at the right house is $D_6 = 265$. Hence, the number of ways in which at least one letter arrives at the right house is $6! - D_6 = 720 - 265 = 455$.

5.8 THE PIGEONHOLE PRINCIPLE

If n pigeon holes are occupied by $(n + 1)$ or more pigeons, then at least one pigeonhole is occupied by more than one pigeon.

If n pigeon objects are assigned to m pigeonholes (categories having a particular characteristics) and $m < n$, then at least one pigeonhole contains two or more pigeons. This principle can also be expressed in terms of set theoretic notation.

Let A and B be finite sets with $|A| > |B|$, then no function $f: A \rightarrow B$ can be one to one.

For example, no function $\{a, b, c, d\} \rightarrow \{1, 2, 3\}$ can be one to one.

Similarly, no function $\{1, 2, 3, 4\} \rightarrow \{x, y\}$ can be one to one. The statement that no function is one to one essentially means that if B is a set of bird houses and A is a set of pigeons, then the function f will assign at least two pigeons to the same house.

The pigeonhole principle is quite helpful in proving interesting results.

Example 1 Let 8 persons be selected at random from any group, then at least two of them will have been born on the same day of the week.

Solution: Consider the days of the week as pigeonholes and the birthdays of 8 persons as the pigeons. Thus, there are 8 pigeons and 7 pigeonholes. Hence, according to the pigeonhole principle at least two persons will have the same day of their birth.

Example 2 If 13 people are assembled in the same room, then at least two of them have their birthday in the same month.

Solution: Consider the birth months as pigeons and the calendar months as the pigeonholes. Thus, there are 13 pigeons and 12 pigeonholes. According to the pigeonhole principle, there must be at least two people who have their birthdays in the same month.

Example 3 If there are n distinct pairs of shoes in a closet, show that if you choose $(n + 1)$ single shoes at random from the closet you are certain to have a pair.

Solution: The n distinct pairs may be regarded as pigeonholes. The $(n + 1)$ single shoes correspond to $(n + 1)$ pigeons. Hence, there must be at least one pigeonhole with two shoes and thus you certainly have at least one pair of shoes.

5.8.1 Generalized Pigeonhole Principle

If n pigeonholes are occupied by $kn+1$ or more pigeons where k is a positive integer, then at least one pigeonhole is occupied by $k + 1$ or more pigeons.

Example 1 Find the minimum number of persons in a group such that 4 of them are born in the same month.

Solution: Here, pigeonholes are months $n = 12$ and pigeons are persons.

Let
$$k + 1 = 4$$
$$\therefore k = 3$$

Hence, the number of persons (i.e., pigeons) $= nk + 1$
$$= 12.3 + 1 = 37$$

Example 2 Find the minimum number of integers to be selected from the set $\{1, 2, 3, 4, 5, 6, 7, 8, 9\}$ so that (1) the sum of two integers is even and (2) the sum of two integers is 10.

Solution:

(1) Consider the subsets $A = \{1, 3, 5, 7\}$ and $B = \{2, 4, 6, 8\}$ of the given set. By taking one from A and one from B, the sum is not even. Hence, one more digit is required from A or B so that the sum of two integers selected is even. Hence, required number of integers = 3.

The subsets A and B can be regarded as pigeonholes and the integers as pigeons. Hence, by pigeonhole principles, the minimum number of integers = 2 + 1 = 3.

(2) Consider the subsets (1, 9), (2, 8), (3, 7), (4, 6), (5, 5).

These subsets are equivalent to pigeonholes. Hence, the number of pigeons (i.e., integers) to make the sum 10 is 5 + 1 = 6.

Example 3 A bag contains orange, blue, and white socks. Find the minimum number of socks that one needs to choose in order to get two pairs (four) of socks of the same colour.

Solution: Here, colours are pigeonholes and hence $n = 3$, and socks are pigeons and hence

$$k + 1 = 4$$
$$\therefore k = 3$$

Thus, $nk + 1 = 3.3 + 1 = 10$ socks (pigeons) are required so that 4 of them are of the same colour.

5.8.2 Extended Pigeonhole Principle

If m pigeonholes are occupied by n pigeons and $n > m$, then one of the pigeonholes must contain at least $\left\lceil \frac{n}{m} \right\rceil$ objects. The ceiling function $\lceil x \rceil$ means the least integer, which is greater than or equal to x.

$$\lceil 2.5 \rceil = 3,\ \lceil 1.2 \rceil = 2,\ \lceil -2.8 \rceil = -2$$

Example 1 If 34 chairs are positioned around 6 tables in a room, find the minimum number of chairs that will surround some table.

Solution: There are 34 chairs (pigeons) and 6 tables (pigeonholes). Hence, according to pigeonhole principle, the minimum number of chairs around some table is $\left\lceil \frac{34}{6} \right\rceil = 6$.

Example 2 Show that if 35 dictionaries in a library contain a total of 71,578 pages, then one of the dictionaries must have at least 2046 pages.

Solution: Let the pages be pigeons and the dictionaries be the pigeonholes. Then by extended pigeonhole principle, one dictionary must contain at least $\left\lceil \frac{71,578}{35} \right\rceil = 2046$ pages.

5.9 ELEMENTS OF PROBABILITY

The theory of probability is another area where counting techniques are widely used. Historically, probability theory has its origin from games of chance or gambling such as roulette and cards. As of now, probability plays an important role in a large number of decision-making problems arising in industry, different branch of engineering such as space and missiles, information theory, social problems of mortality and insurance. It helps in the development of probabilistic decision models.

5.9.1 Definition

In our day-to-day experience, we come across two types situations: one in which outcomes are always same under similar situations and another in which outcomes are not the same even if essentially similar (or same) situations are brought about. For example, if a particle is subjected to the same set of forces, its motion is always the same. If a body is thrown vertically upwards with a given velocity, it will always rise to the same height. But if a coin is tossed or a die is thrown, the outcome is always not the same. If we call such outcomes as events, then the first type of events can be termed as 'predictable events' and the second type of events as the 'unpredictable events'. The probability of an event provides a measure of certainty of that event. Obviously, it varies from event to event.

- *Trial*: An experiment although repeated essentially under same conditions does not produce the same result every time. Such an experiment is called a trial.
- *Sample Space*: The set of all possible outcomes of a trial (or experiment) is called a sample space and is generally denoted by S.
- *Outcome*: Each element of a sample space is called an outcome. It is also called a sample point.
- *Elementary Event*: Each single element of a sample space is called an elementary event.
- *Event*: An event E is a subset of sample space. This subset, which corresponds to the event, is also called event space.

Example:

1. Tossing of a coin is a trial as it can lead to appearance of head or tail on the uppermost face.
2. Throwing of a die is a trial as it can lead to appearance of its uppermost face with one or two … or six dots.
3. Drawing a card from a pack of playing cards is a trial and leads to 52 cases.
4. Picking an article from a lot is a trial and can lead to two cases, that is, the article is defective or nondefective.
5. The sample space S of possible outcomes when a coin is tossed can be written as $S = \{H, T\}$, where H and T correspond to 'head' and 'tail', respectively.
6. The sample space S of possible outcomes when two coins are tossed simultaneously can be given by, $S = \{HH, HT, TH, TT\}$
7. The sample space S of the experiment of tossing a die consists of six outcomes that can be written as: $S = \{1, 2, 3, 4, 5, 6\}$.
8. Let three items be selected at random from a manufacturing process. Each item is inspected and classified as defective (D) and nondefective (N). The sample space S of the possible outcomes can be written as
   $S = \{NNN, NND, NDN, NDD, DNN, DND, DDN, DDD\}$
9. In tossing a die, the sample space S is written as $S = \{1, 2, 3, 4, 5, 6\}$. Let E_1 denote the event as the appearance of face with odd number of dots and E_2 denote the event as the appearance of the face with even number of dots. Let E_3 denote the event as the appearance of the face with dots which are prime numbers, and E_4 denote the event as the appearance of face with dots, which are perfect squares. Then $E_1 = \{1, 3, 5\}$, $E_2 = \{2, 4, 6\}$, $E_3 = \{2, 3, 5\}$, and $E_4 = \{1, 4\}$.

Favourable Outcomes: Favourable outcomes to an event E are defined as those that entail the happening of the event E.

Example 1 In tossing a pair of coins, let the event E be defined as the appearance of atleast one head.

The sample space $S = \{HH, HT, TH, TT\}$

The outcomes that are favourable to the event E are HT, TH, and HH.

Hence, $E = \{HT, TH, HH\}$.

Example 2 In tossing of a die, if the appearance of an odd number of dots on the upper most face is define as the event E, then there are three favourable outcomes to this event, namely when 1 or 3 or 5 dots appear on the uppermost face of the die. Thus,

$$E = \{1, 3, 5\}$$

Mutually Exclusive Outcomes: Two outcomes are said to be mutually exclusive when the happening of one precludes or prevents the happening of the other in a given trial.

If there are n outcomes, they are mutually exclusive when happening of one rules out happening of others, that is, when one happens, others cannot happen.

Example 1 In tossing of a coin, the event of appearance of head or tail on the upper face are mutually exclusive.

Example 2 In the experiment of tossing a die, all the outcomes, that is, appearance of face with one or two or … six dots on the uppermost face are mutually exclusive.

Mathematically, a set of events $\{E_1, E_2, \ldots, E_n\}$ are mutually exclusive of $E_i \cap E_j = \phi (i \neq j)$.

Equally Likely Outcomes: Equally likely outcomes are defined as those in which no outcome can be considered more likely to happen then the other, that is, all outcomes have equal chances of occurrence.

Example 1 In tossing of a perfectly symmetrical coin, the appearance of head and tail are equally likely outcomes.

Example 2 In tossing of a perfectly symmetrical die, all the six cases of appearance of 1, 2, 3, 4, 5, and 6 on the uppermost face are equally likely outcomes.

Complementary Event: Let E be any subset of sample space S, that is, E be an event. The event

$$\overline{E} = S - E$$

which contains all points of S not belonging to E is called complementary event to E.

***Mathematical Definition of Probability* (Classical Definition):** Let an experiment result in a finite sample space S consisting of N equally likely and mutually exclusive outcomes. If the subset corresponding to the event E has exactly M outcomes, then the probability of the event E denoted by $p(E)$ is

$$p(E) = \frac{M}{N} = \frac{\text{number of favourable outcomes}}{\text{total number of outcomes}} = \frac{n(E)}{n(S)}$$

Impossible Event: If $E = \phi$, then $p(\phi) = 0$, as no outcome is favourable to event E. That is, probability of an impossible event is always 0.

Sure Event: If $E = S$, then $p(S) = p(E) = 1$.

An event E whose probability is 1 is called a Sure Event.

Example 1 A card is drawn at random from a pack of playing cards. Find the probability that the cards will be

(1) Spade (2) Diamond (3) Ace (4) Black in colour.

Solution:

(1) Let A denote the event that the card is a spade, then

$$p(A)=\frac{\text{Favourable no. outcomes}}{\text{Total no. of outcomes}}=\frac{13}{52}=\frac{1}{4}$$

(2) Let D denote the event that the card is a diamond, then

$$p(D)=\frac{13}{52}=\frac{1}{4}$$

(3) Let C denote the event that the card is an ace, then

$$p(C)=\frac{4}{52}=\frac{1}{13}$$

(4) Let B denote the event that the card is black in colour, then

$$p(B)=\frac{26}{52}=\frac{1}{2}$$

Example 2 Find the probability of the sum 7 of numbers appearing on the uppermost face in a single throw of a pair of dice.

Solution: The sample space for the throw of a pair of dice consists of 36 cases, that is,

$$S=\{(1, 1), (1, 2), (1, 3), \ldots, (6, 6)\}$$

If E denotes the event of the sum to be 7, then

$$E=\{(1, 6), (2, 5), (3, 4), (4, 3), (5, 2), (6, 1)\}$$

That is, the number of outcomes favourable to event E is 6.

Required probability $p(E)=\dfrac{6}{36}=\dfrac{1}{6}$

Example 3 A coin is tossed three times. Find the probability of obtaining (1) two heads; (2) at least two heads.

Solution: The sample space is $S=$ {HHT, HHH, HTH, HTT, THH, THT, TTH, TTT}

(1) Let E_1 denote the event of obtaining two heads, then $E_1=$ {HHT, HTH, THH}. The sample space S, which consists of all possible outcomes has eight outcomes and those favourable to event E_1 are 3. Here

$$p(E_1)=\frac{3}{8}$$

(2) Let E_2 denote the event of obtaining at least two heads, then

$$E_2=\{\text{HHH, HHT, HTH, THH}\}$$

Hence, four outcomes are favourable to event E_2. Hence,

$$p(E_2)=\frac{4}{8}=\frac{1}{2}$$

5.9.2 Notations

$p(E)$ will denote the probability of happening of the event E.

$p(\overline{E})$ will denote the probability of *not* happening of the event E or probability of failure of E. $\overline{E}$ is called the complementary event of E.

$p(E_1 \cup E_2)$ will denote the probability of happening of E_1 or E_2 (or both) or the probability of happening of at least one of the two events E_1 or E_2.

$p(E_1 \cup E_2 \cup \ldots \cup E_n)$ will denote the probability of happening of at least one of the events $E_1, E_2, \ldots, E_n$.

$p(E_1 \cap E_2)$ will denote the probability of happening of E_1 and E_2 both.

$p(E_1 \cap E_2 \cap \ldots \cap E_n)$ will denote the probability of happening of E_1 and E_2 and $E_3 \ldots$ and E_n.

$p(E_1/E_2)$ will denote the probability of happening of E_1 when E_2 has already happened. It is called conditional probability of E_1 under E_2.

THEOREM 5.3 The probability $p(E)$ of an event always lies between 0 and 1.

Proof: By definition $p(E) = \frac{m}{n}$, where n is the number of elements in sample space and m is the number of elements in the event space. Clearly $0 \le m \le n$ and hence $0 \le p(E) \le 1$.

THEOREM 5.4 $p(E) + p(\overline{E}) = 1$

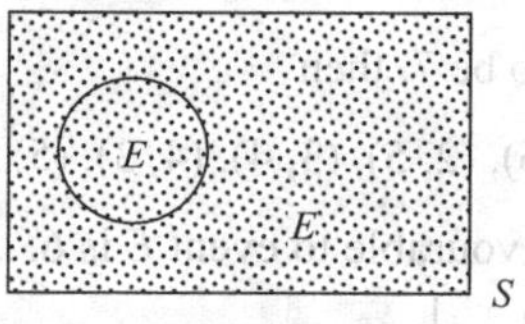

Figure 5.2

Proof: Let E be any event and $\overline{E}$ its complementary event, then

$$E \cup \overline{E} = S$$

and hence $\quad p(E \cup \overline{E}) = p(S) = 1$

or $\quad p(E) + p(\overline{E}) = 1$

This also shows that if E is any event, then $E \cup \overline{E}$ is a sure event.

THEOREM 5.5 If A and B are any two events, then $p(A \cup B) = p(A) + p(B) - p(A \cap B)$.

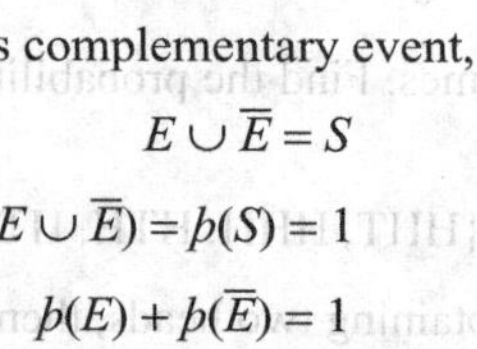

Figure 5.3

Proof: Consider the Venn diagram shown in Fig. 5.3.

The $p(A \cup B)$ is the sum of the probabilities of sample points in $A \cup B$. Now, $p(A) + p(B)$ is the sum of the probabilities of A and B. Thus, we have added the probabilities of $A \cap B$ twice. Since these probabilities are added to obtain the probability of $A \cup B$. Hence, we must subtract the probability of $A \cap B$ once. This gives

$$p(A \cup B) = p(A) + p(B) - p(A \cap B) \qquad \text{(i)}$$

Corollary 1: If A and B are mutually exclusive, $A \cap B = \phi$, then

$$p(A \cap B) = p(\phi) = 0$$

and we have
$$p(A \cup B) = p(A) + p(B) \qquad \text{(ii)}$$

Corollary 2: If the events $A_1, A_2, A_3, \ldots, A_n$ are mutually exclusive, then

$$p(A_1 \cup A_2 \cup A_3 \cup \ldots \cup A_n) = p(A_1) + p(A_2) + \ldots + p(A_n) \qquad \text{(iii)}$$

THEOREM 5.6 (Extension of Theorem 5.5 for three events)

If A_1, A_2, and A_3 are any three events, then

$$p(A_1 \cup A_2 \cup A_3) = p(A_1) + p(A_2) + p(A_3) - p(A_1 \cap A_2) - p(A_2 \cap A_3) - p(A_3 \cap A_1) + p(A_1 \cap A_2 \cap A_3) \qquad \text{(i)}$$

$$= \sum_{i=1}^{3} p(A_i) - \sum_{\substack{i,j=1 \\ i \neq j}}^{3} p(A_i \cap A_j) + p(A_1 \cap A_2 \cap A_3) \qquad \text{(ii)}$$

Proof: Let D denote the event $A_2 \cup A_3$, then

$$\begin{aligned} p(A_1 \cup A_2 \cup A_3) &= p(A_1 \cup D) = p(A_1) + p(D) - p(A_1 \cap D) \qquad \text{(by Theorem 5.5)} \\ &= p(A_1) + p(A_2 \cup A_3) - p\{A_1 \cap (A_2 \cup A_3)\} \\ &= p(A_1) + p(A_2) + p(A_3) - p(A_2 \cap A_3) - [p(A_1 \cap A_2) \cup (A_1 \cap A_3)] \\ &= p(A_1) + p(A_2) + p(A_3) - p(A_2 \cap A_3) - p(A_1 \cap A_2) - p(A_1 \cap A_3) + p(A_1 \cap A_2 \cap A_3) \\ &= \sum_{i=1}^{3} p(A_i) - \sum_{\substack{i,j=1 \\ i \neq j}}^{3} p(A_i \cap A_j) + p(A_1 \cap A_2 \cap A_3) \end{aligned}$$

The general law for n events $A_1, A_2, \ldots, A_n$, which may be proved by induction is

$$p(A_1 \cup A_2 \ldots \cup A_n) = \sum_{i=1}^{n} p(A_i) - \sum_{\substack{i,j=1 \\ i \neq j}}^{n} p(A_i \cap A_j) + \sum_{\substack{i,j,k=1 \\ i \neq j \neq k}}^{n} p(A_i \cap A_j \cap A_k) - \ldots$$
$$+ (-1)^{n-1} p(A_1 \cap A_2 \ldots \cap A_n) \qquad \text{(iii)}$$

Corollary 1: If $A_1, A_2, A_3, \ldots, A_n$ are mutually exclusive events, then $p(A_i \cap A_j) = 0$, $p(A_i \cap A_j \cap A_k) = 0, \ldots, p(A_1 \cap A_2 \cap A_n) = 0$

Consequently, we have [from relation (iii)]

$$p(A_1 \cup A_2 \cup A_3 \ldots \cup A_n) = p(A_1) + p(A_2) + \ldots + p(A_n) \qquad \text{(iv)}$$

This is known as Additive Law of Probability.

THEOREM 5.7 If A and B are any two events, then $p(A-B) = p(A)-p(A \cap B)$

Proof: Let S be a sample space and A and B any two events. The event A can be decomposed into two mutually exclusive events $A-B$ and $A \cap B$, that is,

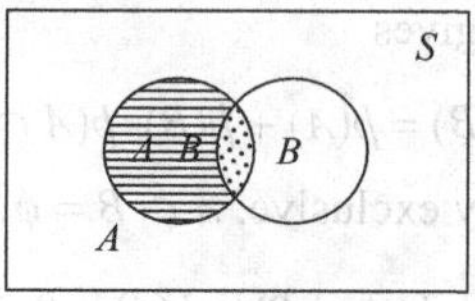

$A \cap B$

$$A = (A-B) \cup (A \cap B)$$

$$\therefore p(A) = p(A-B) + p(A \cap B) \text{ or } p(A-B) = p(A)-p(A \cap B)$$

THEOREM 5.8 If A and B are any two events, then $p(A \cup B) = p(A) + p(B)-p(A \cap B)$

Proof: Let S be a sample space and A, B any two events. Then $A \cup B$ can be decomposed into two mutually exclusive events $A-B$ and B.

Thus, $$A \cup B = (A-B) \cup B$$

$$\therefore p(A \cup B) = p(A-B) + p(B) = p(A) + p(B)-p(A \cap B). \quad \text{by Theorem 5.7}$$

5.9.3 Emperical or Statistical Definition of Probability

If trials are repeated a large number of times n (say), under essentially the same conditions and an event E happens m times, then the probability of event E is given by,

$$p(E) = \lim_{n\to\infty} \frac{m}{n} = p$$

It is assumed that limit is finite and unique.

Remark: The two definitions (mathematical and statistical) of probability are apparently different. In the mathematical definition, the probability of an event is the relative frequency of favourable cases to the total number of cases while in statistical definition, it is the limit if the relative frequency of the happening of the event.

Some Objections to Classical Definition of Probability

The deficiencies of classical definition of probability can be enumerated as follows:

1. The probability of an event can be obtained if all the outcomes of a trial (experiment) are equally likely. It fails to answer when the outcomes are not equally likely. For example, if we know that a coin is biased in favour of heads, then what is the probability of head.
2. The probability of an event being the ratio of the number of points in the event space to that of the sample space of a trial is always a rational number. It fails to accommodate irrational numbers.
3. The definition fails when the total number of points in sample space, i.e., the total number of outcomes as a result of a trial are infinite.

5.9.4 Axiomatic Definition of Probability

The Russian mathematician A.N. Kolmogorov proposed an axiomatic definition of probability, which takes into account all the deficiencies of the classical definition. The definition is as follows:

Let S be a sample space. With each event A in S, we assign a non-negative real number $p(A)$ satisfying the following axioms:

1. $0 \le p(A) \le 1$
2. $p(S) = 1$ and $p(\phi) = 0$
3. If $A_1, A_2, \ldots, A_k$ are mutually exclusive events, then

$p(A_1 \cup A_2 \cup A_3 \ldots \cup A_k) = p(A_1) + p(A_2) + \ldots + p(A_k)$ then $p(A)$ is called the probability of the event A.

Example 1 Let $E = \{E_1, E_2, \ldots, E_n\}$ be an event of a finite sample space S consisting of outcomes, which are not necessarily equally likely. Let $P(E_i)$ denote the probability of occurrence of E_i, then $p(E) = \sum_{i=1}^{n} p(E_i)$.

Example 2 In case of a loaded die, the probability of obtaining a prime number is twice that of obtaining a non-prime number. Find the probability of obtaining an odd number when such a die is rolled.

Solution: When a die is rolled, its sample space consists of six points, that is, $S = \{1, 2, 3, 4, 5, 6\}$. The prime numbers are 2, 3, and 5. The probability of obtaining a prime number is twice that of a non-prime, that is,

p (prime number) = $2p$(non-prime number).

Since the sum of all probabilities is 1, we have

or $\quad 3p(\text{non-prime}) + 3.2p(\text{non-prime}) = 1$

or $\quad 9p(\text{non-prime}) = 1$

$$\therefore\ p(\text{non-prime number}) = \frac{1}{9} \text{ and } p(\text{prime number}) = \frac{2}{9}$$

Required probability of an odd number

$$= p(1) + p(3) + p(5)$$

$$= \frac{1}{9} + \frac{2}{9} + \frac{2}{9} = \frac{5}{9}$$

Example 3 In an examination, the probability of A passing in physics is $\frac{2}{3}$ and that of mathematics is $\frac{4}{9}$. If the probability of passing both courses is $\frac{2}{5}$, find the probability of A passing at least one of the courses.

Solution: Let P denote the event of 'passing in physics' and M the event of 'passing in mathematics'.

Then required probability $= p(P \cup M) = p(P) + p(M) - p(P \cap M)$

$$= \frac{2}{3} + \frac{4}{9} - \frac{2}{5} = \frac{30 + 20 - 18}{45} = \frac{32}{45}$$

Example 4 In a street of 250 houses, after a bomb blast, 45 were found to have broken drains, 180 had broken windows, and 30 had both broken drains and broken windows. Find the probability that a house selected at random is both sound in windows and drains.

Solution: Let D denote the event that 'drain is broken' and W denote the event that 'window is broken'. Then $p(D) = \frac{45}{250}$, $p(W) = \frac{180}{250}$, $p(D \cap W) = \frac{30}{250}$

Now, $p(D \cup W) = p(D) + p(W) + p(D \cap W) = \frac{45 + 180 - 30}{250} = \frac{195}{250} = \frac{39}{50}$

Required probability of both window and drain to be sound

$$= 1 - p(D \cup W) = 1 - \frac{39}{50} = \frac{11}{50}$$

Example 5 A survey among 100 housewives about the two deodorants Dove and Yardley shows that 50 like Dove, 30 like Yardley, and 10 like both. A housewife is selected at random from the group surveyed. Find the probability that she likes neither Dove nor Yardley.

Solution: Let Y denote the event that housewife likes 'Yardley' and D denote the event that she likes 'Dove'.

Then, the probability that she likes either is

$$p(Y \cup D) = p(Y) + p(D) - p(D \cap Y)$$

$$= \frac{30}{100} + \frac{50}{100} - \frac{10}{100} = \frac{70}{100} = \frac{7}{10}$$

Hence, the probability that she likes neither is

$$p(\overline{Y} \cap \overline{D}) = 1 - p(Y \cup D) = 1 - \frac{7}{10} = \frac{3}{10}$$

5.9.5 Multiplication Theorem for Probability

Conditional Probability

The probability that an event A will occur, knowing that a certain other event B has already occurred, is called the conditional probability of A given B. It is denoted by $p(A/B)$ and is defined as

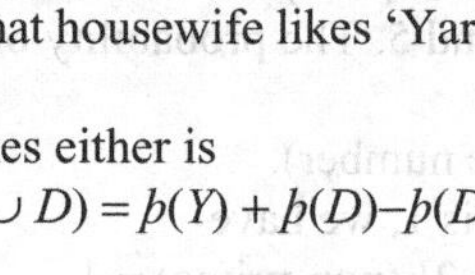

$$p(A/B) = \frac{p(A \cap B)}{p(B)} \tag{5.4}$$

Proof: As pictured in Venn diagram $p(A/B)$, is the relative probability A with respect to reduced space B. Let S be the sample space and $(A \cap B)$ contain k sample points. Also, let $n(A) = m + k$ and $n(B) = n + k$, then

$$p(A/B) = \frac{k}{n+k} = \frac{\frac{k}{n(s)}}{\frac{n+k}{n(s)}} = \frac{p(A \cap B)}{p(B)}$$

where $n(s)$ is the number of sample points in sample space S.

Example 1 A pair of fair dice is tossed. Find the probability of appearance of 3 on one die when the sum is 7.

Solution: Let A denote the event that one die shows 3 and B denote the event that sum is 7.

Then required probability $= \dfrac{p(A \cap B)}{p(B)}$

Now $B = \{(1, 6), (2, 5), (3, 4), (4, 3), (5, 2), (6, 1)\}$ consists of 6 elements. Two of the elements (3, 4) and (4, 3) belong to A.

$$\therefore p(A/B) = \frac{2}{6} = \frac{1}{3}$$

Example 2 The probability that a regular scheduled flight depart on time is 0.85 and the probability that it arrives on time is 0.80. The probability that it arrives and departs on time is 0.75. Find the probability that a plane (1) arrives on time given that it departed on time and (2) departs on time given that it arrived on time.

Solution: Let D denote the event of departure and A the event of arrival on time. We are given $p(D) = 0.85$, $p(A) = 0.80$, $p(A \cap D) = 0.75$

Hence, the required probability is

$$(1)\ p(A/D) = \frac{p(A \cap D)}{p(D)} = \frac{0.75}{0.85} = \frac{15}{17} = 0.88$$

$$(2)\ p(D/A) = \frac{p(A \cap D)}{p(A)} = \frac{0.75}{0.80} = \frac{15}{16} = 0.94$$

Example 3 The probability that an automobile being filled with gasoline will also need an oil change is 0.25; the probability that it needs a new oil filter is 0.35; and the probability that both the oil and filter need change is 0.15. (1) If the oil had to be changed, find the probability that a new oil filter is needed. (2) If a new oil filter is needed, find the probability that oil has to be changed.

Solution: Let O denote the event that oil change will take place and N the event of change of new oil filter.

We are given $p(O) = 0.25, p(N) = 0.35, p(O \cap N) = 0.15$

(1) Required probability is $p(N/O) = \dfrac{p(N \cap O)}{p(O)} = \dfrac{0.15}{0.25} = \dfrac{3}{5} = 0.6$

(2) Required probability is $p(O/N) = \dfrac{p(O \cap N)}{p(N)} = \dfrac{0.15}{0.35} = \dfrac{3}{7} = 0.43$

Example 4 In a married couple, the probability that the husband watches a particular TV serial is 0.35 and the probability that the wife watches the serial is 0.45. The probability that the husband watches the serial, given that his wife does, is 0.6. Find the probability that (1) the married couple watches the serial; (2) at least one of the couple watches the serial; (3) the wife watches the serial given that her husband does.

Solution: Let M denote the event that husband watches the serial and W denote the event that wife watches the serial.

We are given $p(M) = 0.35, p(W) = 0.45, p(M/W) = 0.6$

Since
$$p(M/W) = \frac{p(M \cap W)}{p(W)}$$
$$\therefore p(M \cap W) = p(W)p(M/W)$$

(1) $p(\text{married couple watches the serial}) = p(M \cap W) = p(W)p(M/W) = 0.45 \times 0.6 = 0.27$

(2) Required probability $= p(M \cup W) = p(M) + p(W) - p(M \cap W) = 0.35 + 0.45 - 0.27 = 0.80 - 0.27 = 0.53$

(3) Required probability $= p(W/M) = \dfrac{p(W \cap M)}{p(M)} = \dfrac{0.27}{0.35} = \dfrac{27}{35} = 0.77$

5.9.6 Independent Events

Consider an experiment of drawing two cards in succession from a pack of playing cards. Let the events be defined as follows:

A: The first drawn card is a king

B: The second drawn card is a king

Here, two cases will arise: (1) When the first card is replaced before the second card is drawn. (2) When the first card is not replaced before the second card is drawn.

Now in case (i) $p(A) = \dfrac{4}{52}$, as there are 52 cards and 4 cards of king. Since the first card is replaced before the second is drawn, the pack again has 52 cards, and hence $p(B/A) = \dfrac{4}{52} = p(B)$, that is, the probability of drawing a king.

In such a case, events A and B are said to be Independent.

Further, in case (ii) $p(A) = \dfrac{4}{52}$, $p(B/A) = \dfrac{3}{51}$ because the first drawn card of king has not been replaced, and hence, the pack has only 51 cards having 3 cards of king. In such a case events A and B are said to dependent.

Definition 5.2 Two events A and B are said to be independent if the probability of each is unaffected whether the other happens or not. Thus, events A and B are independent if $p(A/B) = p(A)$ and $p(B/A) = p(B)$.

5.10 MULTIPLICATION THEOREM (INDEPENDENT EVENTS)

Multiplying equation (5.4) by $p(B) \neq 0$, we have

$$p(B \cap A) = p(A \cap B) = p(B)\, p(A/B) \tag{5.5}$$

as events $A \cap B$ and $B \cap A$ are equivalent, we can also write

$$p(A \cap B) = p(A)\, p(B/A) \tag{5.6}$$

That is, if in an experiment, the events A and B can both occur, then

$$p(A \cap B) = p(A)\, p(B/A)$$

This is known as multiplication theorem. If the events A and B are independent,

$$p(A \cap B) = p(A)\, p(B) \tag{5.7}$$

The theorem can be extended to three or more events. If A, B, and C are three events, then

$$p(A \cap B \cap C) = p(A)\, p(B/A)\, p(C/A \cap B)$$

If A, B, and C are mutually independent events, then

$$p(A \cap B \cap C) = p(A)\, p(B)\, p(C)$$

It may, however, be noted that mutual independence does not mean simply pairwise independence. For mutual independence, each event must be independent not only of each other but also of any combination of others. Thus, the events A, B, and C are mutually independent if and only if

$$p(A/B) = p(A),\ p(B/C) = p(B),\ p(A/C) = p(A), \text{ and } p(C/A \cap B) = p(C)$$

Example 1 Suppose we have a box containing 15 fuses, of which 5 are defective. Find the probability that (1) none is defective; (2) exactly one is defective; (3) at least one is defective; and (4) all are defective when three fuses are selected in random.

Solution: Let D denote the event that the fuse is defective and N denote the event that the fuse is non-defective.

Total number of outcomes of selecting 3 fuses from the box containing 15 fuses $= {}^{15}C_3 = \dfrac{\cancel{15}^5 \times \cancel{14}^7 \times 13}{1.\cancel{2}.\cancel{3}} = 455$

(1) Probability of none defective means all the three are selected from the non-defective fuses

$$= \frac{{}^{10}C_3}{{}^{15}C_3} = \frac{\cancel{10}^2 \times \cancel{9}^3 \times 8}{\cancel{15}_3 \times \cancel{14}_7 \times 13} = 24/91$$

(2) Probability of one defective $= \dfrac{{}^5c_1 \times {}^{10}c_2}{{}^{15}c_3} = \dfrac{\cancel{5} \times 45 \times \cancel{2} \times \cancel{3}}{\cancel{15} \times \cancel{14}_7 \times 13} = \dfrac{45}{91}$

(3) Probability of at least one defective = p(1 defective) + p(2 defective) + (3 defective)

$= 1 - p(0 \text{ defective}) = 1 - \frac{24}{91} = \frac{67}{91}.$

(4) Probability of 3 defective $= \frac{{}^5C_3}{{}^{15}C_3} = \frac{5 \times 4^2 \times 3}{15 \times 14_7 \times 13} = \frac{2}{91} = \frac{2}{91}$

Example 2 A box contains 20 electronic components, of which 5 are defective. If two components are selected at random in succession, then find the probability that both are defective. (1) When the first component is not replaced before the second draw and (2) when the first component is replaced.

Solution: Let A denote the event that the first component is defective and B the event that the second component is defective. Thus, $A \cap B$ is the event so that both components are defective.

(1) $p(A \cap B) = p(A)\, p(B/A) = \left(\frac{5}{20}\right) \cdot \frac{4}{19} = \frac{1}{19}$

(2) Since the first drawn component is replaced before the second drawn, the events A and B are independent. Hence,

$$p(A \cap B) = p(A)\, p(B/A) = p(A)\, p(B) = \left(\frac{5}{20}\right)\left(\frac{5}{20}\right) = \frac{1}{4} \cdot \frac{1}{4} = \frac{1}{16}$$

Example 3 A problem of design is given to three architects whose chances of designing are $\frac{1}{4}, \frac{1}{2}$, and $\frac{3}{4}$, respectively. Find the probability that the design problem will be solved.

Solution: Let A, B and C denote the events that the architects will be able to design the problem. Then

$$p(A) = \frac{1}{4}, \quad p(B) = \frac{1}{2}, \quad p(C) = \frac{3}{4}$$

$$\therefore p(\bar{A}) = \frac{3}{4}, \quad p(\bar{B}) = \frac{1}{2}, \quad p(\bar{C}) = \frac{1}{4}$$

The probability that all the three architects will fail to design

$= p(\bar{A} \cap \bar{B} \cap \bar{C})$ (as A, B, and C are independent $\bar{A}$, $\bar{B}$, and $\bar{C}$ are also independent)

$= p(\bar{A})\, p(\bar{B})\, p(\bar{C})$

$= \frac{3}{4} \cdot \frac{1}{2} \cdot \frac{1}{4} = \frac{3}{32}$

$\therefore$ Required probability $= 1 - \frac{3}{32} = \frac{29}{32}.$

Example 4 The probability that a new airport will get an award for its design is 0.24; the probability that it will get an award for the efficient use of material is 0.28. The probability that it will get both the awards is 0.15. Find the probability that it will get (1) at least one award and (2) only one of the two awards.

Solution: Let A and B be two events defined as follows:

A: To get an award for design
B: To get an award for efficient use of material

Then $p(A) = 0.24$, $p(B) = 0.28$, and $p(A \cap B) = 0.15$.

(1) The probability of at least one award $= p(A \cup B) = p(A) + p(B) – p(A \cap B)$
$= 0.24 + 0.28 – 0.15 = 0.37$

(2) The probability of getting only one of the two awards $= p\{(\bar{A} \cap B) \text{ U } (A \cap \bar{B})\}$
$= p(\bar{A} \cap B) + p(A \cap \bar{B}) = p(A) + p(B) – 2\,p(A \cap B) = 0.24 + 0.28 – 0.30 = 0.52$
$– 0.30 = 0.22$

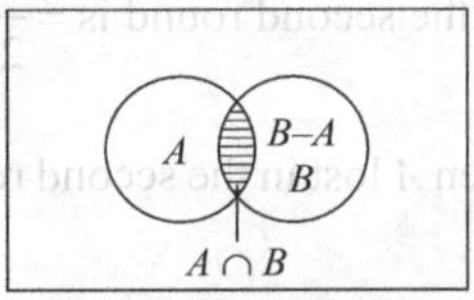

Note 5.1 $\bar{A} \cap B = B – A$ and $B = (B – A) \cup (A \cap B) \Rightarrow p(B) = p(B – A) + p(A \cap B) \Rightarrow p(\bar{A} \cap B)$ $= p(B – A) = p(B) – p(A \cap B)$. Similarly, $p(A \cap \bar{B}) = p(A – B) = p(A) – p(A \cap B)$.

Example 5 Three machines A, B and C manufacture, respectively, 35 per cent, 55 per cent, and 10 per cent of the total production. The percentage of defective items produced by three machines are 4, 2, and 1, respectively. If an item is chosen at random, then find the probability that it is defective.

Solution: Let A, B, and C denote the events that the items are produced by machines A, B, and C, respectively. That is,

A: an item is produced by machine A
B: an item is produced by machine B
C: an item is produced by machine C

Let D denote the event that the item is defective. Now $p(D/A) = 0.04$, $p(D/B) = 0.02$, $p(D/C) = 0.01$.

Required Probability

$= p(D) = p\{(D \cap A) \cup (D \cap B) \cup (D \cap C)\}$ $\quad \therefore P \cap Q = Q \cap P$ for any two sets P and Q

$= p(A \cap D) + p(B \cap D) + p(C \cap D)$

$= p(A)\,p(D/A) + p(B)\,p(D/B) + p(C)\,p(D/C)$

$= 0.35 \times 0.04 + 0.55 \times 0.02 + 0.10 \times 0.01 = 0.0140 + 0.0110 + 0.0010 = 0.026$

Some More Illustrative Examples

Example 1 *A* and *B* toss a coin alternately on the understanding that the first to throw a head wins. Find their respective chances of winning.

Solution: The probability of *A* winning in the first throw is ½ and that of *A* losing is ½.

The probability of *B* throwing a head is $\frac{1}{2}$. Hence, the probability that *A* loses and *B* wins is $= \frac{1}{2} \times \frac{1}{2} = \frac{1}{4}$.

The probability that *A* and *B* lose in the first round is $= \frac{1}{2} \times \frac{1}{2} = \frac{1}{4}$.

The probability that *A* wins (when *A* and *B* both have lost in the first round) $= \frac{1}{2} \times \frac{1}{4} = \frac{1}{8}$

and the probability that *A* loses the second round is $= \frac{1}{2} \times \frac{1}{4} = \frac{1}{8}$.

The probability that *B* wins (when *A* lost in the second round) $= \frac{1}{2} \times \frac{1}{8} = \frac{1}{16}$ and the probability that *B* loses now is $= \frac{1}{2} \times \frac{1}{8} = \frac{1}{16}$.

The probability of *A* winning (when *A* and *B* both lost in the first two rounds) $= \frac{1}{2} \times \frac{1}{16} = \frac{1}{32}$ and so on.

The total probability of *A* winning $= \frac{1}{2} + \frac{1}{8} + \frac{1}{32} + \cdots = \frac{1}{2}\left(1 + \frac{1}{4} + \frac{1}{4^2} + \cdots\right)$

$$= \frac{1}{2} \cdot \frac{1}{1 - \frac{1}{4}} = \frac{1}{\not{2}} \cdot \frac{\not{4}}{3} = \frac{2}{3}$$

Hence, the probability of *B* winning $= 1 - \frac{2}{3} = \frac{1}{3}$.

Example 2 A system consists of three components *A*, *B*, and *C*. The system will function if all the three components are working. The probability that the components *A*, *B*, and *C* will function for one year are 0.80, 0.85, and 0.95, respectively. Find the probability that the system will fail to work before the end of the year.

Solution: The probability that the system will function for a year is

$$p(A \cap B \cap C) = p(A)\, p(B)\, p(C)$$
$$= 0.80 \times 0.85 \times 0.95 = 0.646$$

The probability that the system will fail to work before the end of the year is

$$p(\bar{A} \cup \bar{B} \cup \bar{C}) = 1 - p(A \cap B \cap C) = 1 - 0.646 = 0.354$$

Example 3 The probability that A can solve a given problem in mathematics is 70 per cent and B can solve is 80 per cent. Find the probability that a problem selected at random will be solved if they both A and B try.

Solution: Let A denote the event that the problem is solved by A and B denote the event that the problem is solved by B.

The problem can be solved if

(1) A succeeds and B fails
(2) A fails and B succeeds
(3) both A and B succeed

The required probability is

$$p\left((A\cap\overline{B})\cup(\overline{A}\cap B)+(A\cap B)\right)$$
$$= p(A\cap\overline{B})+p(\overline{A}\cap B)+p(A\cap B)$$
$$= p(A)p(\overline{B})+p(\overline{A})p(B)+p(A)p(B)$$
$$= \frac{7\not{0}}{10\not{0}}\times\frac{2\not{0}}{10\not{0}}+\frac{3\not{0}}{10\not{0}}\times\frac{8\not{0}}{10\not{0}}+\frac{7\not{0}}{10\not{0}}\times\frac{8\not{0}}{10\not{0}}$$
$$= \frac{1}{100}[14+24+56]$$
$$= 0.94$$

Example 4 The probability that a teacher will give an unannounced test during a class is 0.2. If a student is absent twice, what is the probability that he will miss at least one test.

Solution: Let A denote the event that the teacher gives first unannounced test and B denote the event that the teacher will give second unannounced test.

$$\text{Here } p(A) = 0.2, p(B) = 0.2$$

The probability of missing both the tests is

$$= p(\overline{A}\cap\overline{B}) = p(\overline{A})p(\overline{B}) = (1-0.2)\,(1-0.2)$$
$$= 0.8\times 0.8$$
$$= 0.64$$

∴ Required probability of missing at least one test

$$= 1-0.64 = 0.36$$

Example 5 The probability that a doctor makes correct diagnosis of a patient is 0.8. Given that the doctor makes incorrect diagnosis, the probability that the patient files a law suit against him, is 0.95. Find the probability that the doctor makes an incorrect diagnosis and the patient files a suit against him.

Solution: Let A denote the event that the doctor makes a correct diagnosis and L the event that the patient files law suit against the doctor.

The $p(A) = 0.8,\ p(\overline{A}) = 0.2 p\left(L/\overline{A}\right) = 0.95$

Required probability = $(\bar{A} \cap L) = p(\bar{A}).p\left(L / \bar{A}\right) = 0.2 \times 0.95$

$$= 0.19$$

Example 6 Two candidates A and B appear in interviews for two vacancies in the same post. The probability of A's selection is 0.35 and that of B is 0.6. What is the probability that (1) only one is selected; (2) both will be selected; (3) none is selected.

Solution: Let E_1 denote the event that the candidate A is selected and E_2 denote the event that the candidate A is selected.

Then $$p(E_1) = 0.35, p(E_2) = 0.6$$

$$\therefore p(\bar{E}_1) = 0.65, p(\bar{E}_2) = 0.4$$

(1) The probability that one is selected

$$\begin{aligned} &= p[(E_1 \cap \bar{E}_2) \cup (\bar{E}_1 \cap E_2)] \\ &= p(E_1 \cap \bar{E}_2) + p(\bar{E}_1 \cap E_2) \\ &= p(E_1)P(\bar{E}_2) + p(\bar{E}_1)P(E_2) \\ &= 0.35 \times 0.4 + 0.65 \times 0.6 \\ &= 0.14 + 0.39 = 0.53 \end{aligned}$$

(2) The probability that both will be selected

$$\begin{aligned} &= p(E_1 \cap E_2) = p(E_1)p(E_2) \\ &= 0.35 \times 0.6 = 0.21 \end{aligned}$$

(3) The probability that none is selected

$$\begin{aligned} &= p(\bar{E}_1 \cap \bar{E}_2) = p(\bar{E}_1)p(\bar{E}_2) \\ &= 0.65 \times 0.4 = 0.26 \end{aligned}$$

Example 7 Three newspapers A, B, and C are published in a city. It is estimated from a survey that 20 per cent read A, 16 per cent read B, 14 per cent read C, 8 per cent read A and B, 5 per cent read A and C, 4 per cent read B and C and 2 per cent read all the newspapers. Find the probability that a person chosen at random (1) reads A but not B (2) does not read any paper.

Solution: It is given that $p(A) = \frac{20}{100}, p(B) = \frac{16}{100}, p(C) = \frac{14}{100}, p(A \cap B) = \frac{8}{100},$

$$p(B \cap C) = \frac{4}{100}, p(C \cap A) = \frac{5}{100}, p(A \cap B \cap C) = \frac{2}{100}$$

(1) $p(A \cap \bar{B}) = p(A - B) = p(A) - p(A \cap B) = \frac{20}{100} - \frac{8}{100} = 0.12$

(2) $p(\bar{A} \cap \bar{B} \cap \bar{C}) = 1 - p(A \cup B \cup C)$

$$= 1-[p(A)+p(B)+p(C)-\{p(A\cap B)+p(B\cap C)+p(C\cap A)\}+p(A\cap B\cap C)]$$

$$= 1-\left[\left\{\frac{20+16+14}{100}-\frac{8+4+5}{100}+\frac{2}{100}\right\}\right]$$

$$= 1-\frac{52-17}{100} = 1-\frac{35}{100} = 1-0.35 = 0.65.$$

Example 8 A can hit two times in three shots; *B* three times in four shots and *C* four times in five shots. Find the probability that at least two shots hit the target if they fire a volley.

Solution: Here $p(A)=\frac{2}{3}, p(B)=\frac{3}{4}, p(C)=\frac{4}{5}\ \therefore p(\bar{A})=\frac{1}{3}, p(\bar{B})=\frac{1}{4}, p(\bar{C})=\frac{1}{5}$

The probability that at least two will hit the target is

$$p[(\bar{A}\cap B\cap C)\cup(A\cap \bar{B}\cap C)\cup(A\cap B\cap \bar{C})\cup(A\cap B\cap C)$$

$$= p(\bar{A}\cap B\cap C)+p(A\cap \bar{B}\cap C)+p(A\cap B\cap \bar{C})+p(A\cap B\cap C)$$

$$= p(\bar{A})\,p(B)\,p(C)+p(A)\,p(\bar{B})\,p(C)+p(A)\,p(B)\,p(\bar{C})+p(A)\,p(B)\,p(C)$$

$$= \frac{1}{3}\cdot\frac{3}{4}\cdot\frac{4}{5}+\frac{2}{3}\cdot\frac{1}{4}\cdot\frac{4}{5}+\frac{2}{3}\cdot\frac{3}{4}\cdot\frac{1}{5}+\frac{2}{3}\cdot\frac{3}{4}\cdot\frac{4}{5}=\frac{1}{60}[12+8+6+24]=\frac{50}{60}=\frac{5}{6}.$$

Example 9 Find the probability that a leap year selected at random will contain

(1) 53 Mondays and Tuesdays
(2) 53 Mondays or Tuesdays

Solution: There are 366 days in a leap year. It contains 52 weeks and 2 days. These two days may be the following 7 pairs:

1. Sunday and Monday
2. Monday and Tuesday
3. Tuesday and Wednesday
4. Wednesday and Thursday
5. Thursday and Friday
6. Friday and Saturday
7. Saturday and Sunday

Let *A* denote the event of 53rd Monday and B of 53rd Tuesday. Then

$$p(A)=\frac{2}{7}\quad\text{and}\quad p(B)=\frac{2}{7}$$

as Monday and Tuesday appear in two outcomes.

(1) p (53 Mondays and 53 Tuesdays) $= p(A\cap B)=\frac{1}{7}$ (Monday and Tuesday occur only in one outcome)

(2) p (53 Mondays or 53 Tuesdays)

$$= p(A \cup B) = p(A) + p(B) - p(A \cap B) = \frac{2}{7} + \frac{2}{7} - \frac{1}{7} = \frac{3}{7}$$

Example 10 A bag contains 2 white and 3 black balls and a second bag contains 3 white and 5 black balls.

(1) A bag is selected at random and one ball is drawn from it. What is the probability that the ball is black?

(2) If one ball is drawn from each bag, what is the probability that both are of the same colour?

Solution: Let E_1 denote the event that bag no. 1 is selected and E_2 the event that bag no. 2 is selected. Let W denote the event that the ball drawn is *white* and B the event that the ball drawn is *black*, then

$$p(E_1) = \frac{1}{2}, p(E_2) = \frac{1}{2}, p(B/E_1) = \frac{3}{5}, p(B/E_2) = \frac{5}{8}$$

(1) $p(\text{ball drawn is black}) = p(B) = p(B \cap E_1) \cup (B \cap E_2)$

$$= p(E_1 \cap B) + p(E_2 \cap B)$$

$$= p(E_1)\, p(B/E_1) + p(E_2)\, p(B/E_2)$$

$$= \frac{1}{2}.\frac{3}{5} + \frac{1}{2}.\frac{5}{8} = \frac{1}{2}\left(\frac{3}{5} + \frac{5}{8}\right) = \frac{1}{2}\frac{(25+24)}{40} = \frac{49}{80}$$

(2) The probability of drawing a white ball from first bag is 2/5 and the probability of drawing a white ball from second bag is 3/8. Hence, the probability of drawing a white balls from both bags $= \frac{2}{5} \times \frac{3}{8} = \frac{6}{40} = \frac{3}{20}$.

Similarly, the probability of drawing black balls from both bags $= \frac{3}{5} \times \frac{5}{8} = \frac{15}{40}$.
Hence, the probability that both balls are of the same colour.
i.e., (either white or black)

$$= p(W \cup B)$$

$$= p(W) + p(B) = \frac{6}{40} + \frac{15}{40} = \frac{21}{40}$$

Example 11 An anticraft gun can take a maximum of four shots at any enemy plane moving away from it. The probabilities of hitting at the first, second, third, and fourth shots are 0.4, 0.3, 0.2, and 0.1, respectively. What is the probability that a shot from the gun hits the plane?

Solution: Let E_i be the event that ith shot from the gun hits the enemy plane ($i = 1, 2, 3, 4$).

Then $\quad p(E_1) = 0.4, p(E_2) = 0.3, p(E_3) = 0.2, \text{and } p(E_4) = 0$

Required probability is

$$p(E_1 \cup E_2 \cup E_3 \cup E_4) = 1 - p\left(\overline{E_1} \cap \overline{E_2} \cap \overline{E_3} \cap \overline{E_4}\right)$$

$$= 1 - p(\overline{E_1})\, p(\overline{E_2})\, p(\overline{E_3})\, p(\overline{E_4})$$

$$= 1 - (1 - 0.4)(1 - 0.3)(1 - 0.2)(1 - 0.1)$$
$$= 1 - 0.6 \times 0.7 \times 0.8 \times 0.9$$
$$= 1 - 0.3024$$
$$= 0.6976$$

Example 12 A five-digit number is formed by digits 1, 2, 3, 4, 5 without repetition. Find the probability that the number is divisible by 4.

Solution: Total number of five-digit numbers = 5! = 120

Now a number is divisible by 4 if the last two digits (i.e., digits at unit and 10 place) are divisible by 4. Therefore, the last two digits can be 12, 24, 32, 52 and thus there are four ways. Corresponding to each of these, there are 3! = 6 ways of filling the remaining three digits.

Hence, the total number of favourable ways = 6 × 4 = 24

$\therefore$ The required probability $= \dfrac{24}{120} = \dfrac{1}{5}$

Example 13 If four whole numbers taken at random are multiplied, find the probability that the last digit in the product is divisible by 1, 3, 7, or 9.

Solution: The total number of digits that can be in the last place (i.e., at unit's place) is 10 as the digits 0, 1, 2, 3,…, 9 can occur in unit's place.

Clearly, if the last digit in any of the four numbers is 0, 2, 4, 5, 6, or 8, then the last digit in the product cannot be 1, 3, 7, or 9.

Thus, in order that the last digit in the product may be 1, 3, 7, or 9, it is necessary that the *last digit* in each number must be 1, 3, 7, or 9. The favourable cases are 4.

The probability p that the last digit is 1, 3, 7, 9 is $p = \dfrac{4}{10} = \dfrac{2}{5}$.

Hence, the required probability $= p^4 = \left(\dfrac{2}{5}\right)^4 = \dfrac{16}{625}$.

5.11 BAYE'S THEOREM

Baye's theorem finds its application in revising the original probability of events after gaining additional information.

Let $A_1, A_2, \ldots, A_n$ be n mutually exclusive and exhaustive events and let an event B follow these events. Now, if on actual observation B is found to have occurred, then the probability that it was preceded by a particular event A_k is

$$p(A_k / B) = \frac{p(A_k)\, p(B/A_k)}{\sum_{i=1}^{n} p(A_i)\, p(B/A_i)}$$

Proof: The event B can happen either when A_1 happens or A_2 happens … or A_n happens, and hence

$$B = (A_1 \cap B) \cup (A_2 \cap B) \cup \ldots (A_i \cap B) \ldots \cup (A_n \cap B)$$

$$p(B)=\sum_{i=1}^{n} p(A_i \cap B) = \sum_{i=1}^{n} p(A_i)\, p(B/A_i)$$

Now

$$p(A_k \cap B) = p(A_k)\, p(B/A_k) = p(B)\, p(A_k/B)$$

$$\therefore\ p(A_k/B) = \frac{p(A_k)\, p(B/A_k)}{p(B)} = \frac{p(A_k)\, p(B/A_k)}{\sum_{i=1}^{n} p(A_i)\, p(B/A_i)}$$

On account of the fact that the probability of A_k is determined after the event B has happened, $p(A_k/B)$ is called posterior probability.

Example 1 A production company has three units A_1, A_2, and A_3, which produce 35 per cent, 40 per cent, and 25 per cent, respectively, of the production. Experience shows that 5 per cent items produced by A_1 are defective, 2 per cent items produced by A_2, and 3 per cent items produced by A_3 are defective. An item is selected at random and found to be defective. Find the probability that it was produced by unit A_2.

Solution: Let A_1, A_2, and A_3 denote the events that items are provided by units A_1, A_2, and A_3, respectively. Let D denote the event that the item is defective. Then

$$p(A_1) = \frac{35}{100} = 0.35, p(A_2) = 0.40, p(A_3) = 0.25, \text{ and}$$

$$p(D/A_1) = 0.05, p(D/A_2) = 0.02, p(D/A_3) = 0.03$$

Required probability is $p(A_2/D) = \dfrac{p(A_2)\, p(D/A_l)}{\sum_{i=1}^{3} p(A_i) p(D/A_i)}$

$$= \frac{0.40 \times 0.02}{0.35 \times 0.05 + 0.40 \times 0.02 + 0.25 \times 0.03}$$

$$= \frac{0.0080}{0.0175 + 0.0080 + 0.0075}$$

$$= \frac{0.0080}{0.0330} = \frac{8}{33}$$

$$= 0.242$$

Example 2 Police plan to enforce speed limits by using radar traps at four different locations within the city limits. The radar traps at each of the locations L_1, L_2, L_3 and L_4 are operated at 50 per cent, 40 per cent, 30 per cent, and 20 per cent of the time. A person who is speeding on his way to work has probabilities of 0.3, 0.2, 0.4, and 0.1, respectively, of passing through on his way to work. What is the probability that he passed through radar trap located at L_4?

Solution: Let L_1, L_2, L_3, and L_4 denotes the events of passing through its location L_1, L_2, L_3, and L_4, respectively. Let S denote the event of receiving a speeding ticket. Then

$$p(L_1) = \frac{50}{100} = 0.5, p(L_2) = \frac{40}{100} = 0.4, p(L_3) = \frac{30}{100} = 0.3, p(L_4) = \frac{20}{100} = 0.2, \text{ and}$$

$$p(S/L_1) = 0.3, p(S/L_2) = 0.2, p(S/L_3) = 0.4, p(S/L_4) = 0.1$$

$$\text{Required probability} = p(L_4/S) = \frac{p(L_4)\,p(S/L_4)}{\sum_{i=1}^{4} p(L_i)\,p(S/L_i)}$$

$$= \frac{0.2 \times 0.1}{0.5 \times 0.3 + 0.4 \times 0.2 + 0.3 \times 0.4 + 0.2 \times 0.1}$$

$$= \frac{0.02}{0.15 + 0.08 + 0.12 + 0.02}$$

$$= \frac{0.02}{0.37} = \frac{2}{37}$$

$$= 0.054$$

5.12 CONCEPT OF A RANDOM VARIABLE

If we toss two coins together, then the sample space

$$S = \{\text{HH, HT, TH, TT}\}$$

Consists of four points. The probability of each out come is $\frac{1}{4}$. Let the number of appearance of heads be denoted by a variable X, which can assume values 0, 1, and 2.

Now

$$p(X = 0) = \text{probability of 0 head} = \{\text{TT}\} = \frac{1}{4}$$

$$p(X = 1) = \text{probability of 1 head} = \{\text{HT, TH}\} = \frac{2}{4} = \frac{1}{2}$$

$$p(X = 2) = \text{probability of 2 head} = \{\text{HH}\} = \frac{1}{4}$$

This can be arranged in the following tabular form

No of Heads	Probabilities
0	$\frac{1}{4}$
1	$\frac{1}{2}$
2	$\frac{1}{4}$
Total	1.0

This is similar to a frequency distribution except that instead of frequencies of events, we have their probabilities. Such a distribution is called probability distribution and the variable X whose value depends upon chance is called a Random variable.

Consider the sample space giving detailed description of each possible outcome when three electronic components are tested. This can be written as

$$S = \{\text{NNN, NND, DNN, DND, NDN, NDD, DDN, DDD}\}$$

where N denotes (non-defective) and D denotes (defective).

Let the variable X denote the number of defective items.

When three electronic components are tested, then X can assume values 0, 1, 2, and 3.

Now $$p(X=0) = p\{\text{NNN}\} = \frac{1}{8},\ p(X=1) = p\{\text{NND, NDN, DND}\} = \frac{3}{8}$$

$$p(X=2) = p\{\text{NDD, DND, DDN}\} = \frac{3}{8},\ p(X=3) = p\{\text{DDD}\} = \frac{1}{8}$$

The sum of all probabilities is one, the variable X is a random variable.

Definition 5.3 A random variable is a function defined on a sample space which associates a real number with each element of the sample space. It is usually denoted by letters X, Y, Z, etc.

5.12.1 Probability Distribution

Let a discrete random variable X assume values $x_1, x_2, x_3, \ldots, x_i, \ldots, x_n$ with probabilities $p_1, p_2, p_3, \ldots, p_i, \ldots, p_n$, then the set of values of x_i together with their probabilities constitutes what is known as a probability distribution. The distribution is usually given by a table.

x_1	x_2		x_i	…	x_n
p_1	p_2		p_i	…	p_n

where $p(X = x_i) = p_i$

Example 1 A box contains 10 items of which three are defective. Find the probability distribution of the number of defective items if a sample of three items is selected at random.

Solution: The sample space consists of ${}^{10}c_3$, that is, 120 different samples of size 3. Let X denote the number of defective items, then $X = \{0, 1, 2, 3\}$

7 Good
3 Defective

$$p(X = 0) = \frac{{}^7c_3}{{}^{10}c_3} = \frac{35}{120} = \frac{7}{24}$$

$$p(X = 1) = \frac{{}^7c_2 \times {}^3c_1}{{}^{10}c_3} = \frac{63}{120} = \frac{21}{40}$$

$$p(X = 2) = \frac{{}^7c_1 \times {}^3c_2}{{}^{10}c_3} = \frac{21}{120} = \frac{7}{40}$$

$$p(X = 3) = \frac{{}^3c_3}{{}^{10}c_3} = \frac{1}{120}$$

The probability distribution of the random variable X is

X:	0	1	2	3
$p(X)$:	$\frac{35}{120}$	$\frac{63}{120}$	$\frac{21}{120}$	$\frac{1}{120}$

Example 2 A shipment of 8 computers to an educational institution contains 3 that are defective. If a department is given 2 of these computers by random choice, find the probability distribution for the number of defectives.

Solution: The sample space contains ${}^8C_2$, that is, 28 different samples of size 2. The random variable X, which denotes the number of defective computers, can assume values 0, 1 and 2, that is $X = \{0, 1, 2\}$.

$$p(X = 0) = \frac{{}^5c_2}{{}^8c_2} = \frac{10}{28},\ p(X = 1) = \frac{{}^3c_1 \times {}^5c_1}{{}^8c_2} = \frac{15}{28}$$

5 Non-defective
3 Defective

$$p(X = 2) = \frac{{}^3c_2}{{}^8c_2} = \frac{3}{28}$$

Thus, the probability distribution of random variable X is

X	0	1	2
$p(X)$	$\frac{10}{28}$	$\frac{15}{28}$	$\frac{3}{28}$

5.13 BINOMIAL DISTRIBUTION

5.13.1 Repeated Independent Trials with Two Outcomes

Consider an experiment of repeated trials in which each trial has only *two* possible outcomes. Such a situation always arises when we deal with testing of items. Each trial indicates whether the item is defective or non-defective. We can define the two outcomes as a 'success' or 'failure'.

These trials are independent as the outcome of any trial does not depend upon the outcome of the previous trial just as in tossing of a coin or drawing a card in succession from a park of cards if the card is replaced every time before the next draw. These independent repeated trials are called Bernoulli trials named after swiss mathematician Jacob Bernoulli. Further the probability of success or that of failure remains constant in each trial.

Let a trial be repeated so that we have a series of n trials. Let p be the Probability of success and q that of failure in each Bernoulli trial so that $p + q = 1$.

It is required to find the probability of r successes in n trials. If S denotes the success and F the failure, then the probability of r successes and $(n–r)$ failures in a given order say

$$SFSSFS \ldots$$

is

$$pqppqp \ldots = p^r q^{n-r}$$

as the number of p's is equal to the number of successes, that is, r and the number of q's is equal to the number of failures, that is, $n–r$.

In order to find the required probability, we have to count all different orders, in which r successes and $(n–r)$ failures can occur. This is in fact all possible arrangements of n letters of which r are S and $(n–r)$ are F and is equal to $\dfrac{n!}{r!\,(n-r)!}$ i.e., ${}^nc_r$

Hence, by addition theorem, the probability of getting r successes in a trial is

$$ {}^nc_r\, p^r\, q^{n-r}$$

which can be denoted $b(r: n, p)$

Thus $p(r \text{ successes}) = b(r: n, p) = {}^nc_r p^r q^{n-r}$ $r = 0, 1, 2, 3, \ldots, n$

If the number X of successes is a random variable, then

$$p(X = r) = b(r; n, p) = {}^nc_r\, p^r\, q^{n-r}$$

We notice that the probabilities of	0	1	2	3	r	n
	q^n	${}^nc_1 p q^{n-1}$	${}^nc_2 p^2 q^{n-2}$	${}^nc_3 p^3 q^{n-3}$	${}^nc_r p^r q^{n-r}$	p^n

which are successive terms of the binomial expansion of $(q + p)^n$, and hence, the above probability distribution is known as 'binomial distribution'.

It can be noted that the sum of all the probabilities is 1

$$q^n + {}^nc_1\, pq^{n-1} + {}^nc_2\, p^2 q^{n-2} + \ldots + {}^nc_r\, p^r q^{n-r} + \ldots + p^n$$

$$= (q + p)^n$$

$$= 1^n$$

$$= 1$$

5.13.2 Mean and Variance

Associated with any random variable X, there are two important parameters, that is, mean and variance. Mean denoted by μ gives expectation of X while variance σ^2 provides a measure of dispersion of X.

The expectation of X is

$$\mu = E(X) = \sum_{i=1}^{n} x_i p_i = \sum_{r=0}^{n} r.\,{}^{n}c_r p^r q^{n-r}$$

$$= \sum_{r=0}^{n} r.\frac{\underline{|n}}{\underline{|r}\,\underline{|n-r}} p^r q^{n-r}$$

$$= \sum_{r=1}^{n} \frac{n\,\underline{|n-1}\,.p.p^{r-1}\,q^{n-r}}{\underline{|r-1}\,\underline{|n-r}}$$

$$= np\sum_{r=1}^{n} \frac{\underline{|n-1}}{\underline{|r-1}\,\underline{|n-r}} p^{r-1} q^{n-r}$$

$$= np\,(q+p)^{n-1}$$

$$= np.1$$

$$= np$$

Hence, mean of binomial distribution $b(r\colon n, p)$ is np

$$\text{Now variance} = \sigma^2 = E(X-\mu)^2 = \sum_{i=0}^{n} p_i(x_i-\mu)^2 = \sum_{i=1}^{n} p_i(x_i^{\,2} - 2\mu x_i + \mu^2)$$

$$= \sum_{i=0}^{n} p_i x_i^{\,2} - 2\mu \sum_{i=1}^{n} p_i x_i + \mu^2$$

$$= \sum_{i=0}^{n} p_i x_i^{\,2} - 2\mu.\mu + \mu^2 = \sum_{i=1}^{n} p_i x_i^{\,2} - \mu^2$$

$$= \sum_{r=0}^{n} r^2.\,{}^{n}c_r p^r q^{n-r} - n^2 p^2$$

$$= \sum_{r=0}^{n} [r(r-1)+r]\ {}^{n}c_r\, p^r q^{n-r} - n^2 p^2$$

$$= \sum_{r=0}^{n} [r(r-1).\,{}^{n}c_r p^r q^{n-r} + \sum_{r=0}^{n} r.\,{}^{n}c_r p^r q^{n-r} - n^2 p^2$$

$$= \sum_{r=0}^{n} r(r-1)\frac{\underline{|n}}{\underline{|r}\,\underline{|n-r}} p^r q^{n-r} + np - n^2 p^2$$

$$= \sum_{r=2}^{n} [n(n-1)\frac{\underline{|n-2}\,.p^2}{\underline{|r-2}\,\underline{|n-r}}.p^{r-2} q^{n-r} + np - n^2 p^2$$

$$= n(n-1)p^2 \sum_{r=2}^{n} \frac{\underline{|n-2}}{\underline{|r-2}\,\underline{|n-r}}.p^{r-2} q^{n-r} + np - n^2 p^2$$

$$= n(n-1)p^2(q+p)^{n-2} + np - n^2p^2$$
$$= n(n-1)p^2.1 + np - n^2p^2$$
$$\therefore \sigma^2 = n^2p^2 - np^2 + np - n^2p^2 = np(1-p) = npq$$

Example 1 The probability that a particular type of component will survive a given shock test is $\frac{2}{3}$. Find the probability that 2 of the next 5 components tested will survive.

Solution: Let shock test of the component be regarded as a trial and its survival as the success. If p denotes the probability of success and q that of failure in a single trial, then

Number of tests $n = 5, p = \frac{2}{3}, q = \frac{1}{3}$

Required probability $= p$ (2 successes)

$$= b\left(2:5, \frac{2}{3}\right) = {}^5c_2\left(\frac{2}{3}\right)^2\left(\frac{1}{3}\right)^3$$

$$= 10 \times \frac{4}{3\times3\times3\times3\times3} = \frac{40}{243}.$$

Example 2 The probability that a patient will survive from a delicate heart operation is 0.9. What is the probability that exactly 5 of the next 7 patients having this operation will survive?

Solution: Let the heart operation of a patient be regarded as a trial and his survival as the success.

If n denotes the number of trials and p probability of success and q that of failure in a single trial, then

$$n = 7; p = 0.9, q = 1 - p = 0.1$$

Required probability $= p$ (5 successes) $= b\,(5{:}7, 0.9)$

$$= {}^7c_5\,(0.9)^5\,(0.1)^2$$

$$= \frac{7\times \not{6}3}{\not{2}} \times 0.5905 \times 0.01$$

$$= 0.1240$$

Example 3 If 10 per cent of components produced by a machine are defective, determine the probability that out of 5 components selected at random (1) none; (2) one; (3) at most one component will be defective.

Solution: Let selection of a component be regarded as a trial and getting a defective component as the success

If n denotes the number of trials and p the probability of success and q that of failure, then

$$n = 5, p = \frac{10}{100} = 0.1, q = 1 - p = 0.9$$

(1) $p(0 \text{ success}) = b(0: 5, 0.1) = (0.9)^5 = 0.5905$

(2) $p(1 \text{ success}) = b(1: 5, 0.1) = {}^5c_1(0.1)(0.9)^4 = 0.3281$

(3) The probability of at most 1 success $= p(0 \text{ success}) + p(1 \text{ success})$

$$= 0.5905 + 0.3281$$
$$= 0.9186.$$

Example 4 The incidence of an occupational disease in an industry is that the workers have 20 per cent chance of suffering from it.

Find the probability that out of 6 workers selected at random

(1) 4 will catch the disease.

(2) At least 4 will catch the disease.

Solution: Let the selection of a worker be regarded as a trial and suffering from the disease as the success.

If n denotes the number of trials and p probability of success and q that of failure, then

$$n = 6,\ p = \frac{20}{100} = \frac{1}{5},\ q = 1 - p = 1 - \frac{1}{5} = \frac{4}{5}$$

(1) $p\,(4 \text{ successes}) = b\left(4 : 6, \frac{1}{5}\right) = {}^6c_4 p^4 q^2$

$$= {}^6c_4\left(\frac{1}{5}\right)^4\left(\frac{4}{5}\right)^2$$
$$= {}^6c_4\left(\frac{1}{5}\right)^4\left(\frac{4}{5}\right)^2$$
$$= 15 \times (0.2)^4(0.8)^2$$
$$= 15 \times 0.0016 \times 0.64$$
$$= 0.01536$$

(2) $p(\text{at least 4 successes}) = p(4 \text{ successes}) + p(5 \text{ successes}) + p(6 \text{ successes})$

$$= {}^6c_4\left(\frac{1}{5}\right)^4\left(\frac{4}{5}\right)^2 + {}^6c_5\left(\frac{1}{5}\right)^5\left(\frac{4}{5}\right) + \left(\frac{1}{5}\right)^6$$
$$= \left(\frac{1}{5}\right)^4 [{}^6c_2(0.8)^2 + {}^6c_1(0.2)(0.8) + (0.2)^2]$$
$$= 0.0016\,[15 \times 0.64 + 6 \times 0.16 + 0.04]$$
$$= 0.0016\,[9.60 + 0.96 + 0.04]$$
$$= 0.01696$$

Example 5 The probability of a man hitting a target is $\frac{1}{4}$. If he fires 7 times, find the probability of his hitting the target at least twice.

Solution: Let fireing be regarded as a trial and hitting the target as the success, then

$$n=7,\ p=\frac{1}{4},\ q=1-p=1-\frac{1}{4}=\frac{3}{4}$$

$$\begin{aligned}
\text{Prob (At least 2 successes)} &= p(2)+p(3)+\ldots+p(7\text{ successes})\\
&= 1-p\text{ (0 success)}-p\text{ (1 successes)}\\
&= 1-q^7-{}^7c_1pq^6\\
&= 1-\left(\frac{3}{4}\right)^7-7\times\frac{1}{4}\times\left(\frac{3}{4}\right)^6\\
&= 1-\left(\frac{3}{4}\right)^6\left(\frac{3}{4}+\frac{7}{4}\right)=1-\left(\frac{3}{4}\right)^6\times\frac{5}{2}=1-(0.75)^6\,(2.5)\\
&= 1-0.1780\times2.5=1-0.4449\\
&= 0.5551
\end{aligned}$$

Example 6 The probability that a bomb dropped from a plane will strike the target is $\frac{1}{5}$. If 6 bombs are dropped, find the probability that (1) exactly two will strike the target; (2) at least two will strike the target.

Solution: Let dropping of a bomb be regarded as a trial and striking the target as the success. Then $n=6,\ p=\frac{1}{5},\ q=1-p=\frac{4}{5}$.

(1) $p(\text{2 successes})=b\left(2:\ 6,\frac{1}{5}\right)={}^6c_2\left(\frac{1}{5}\right)^2\left(\frac{4}{5}\right)^4=15\times(0.2)^2\times(0.8)^4=0.2458$

(2) $p(\text{at least 2 success})=1-p(\text{0 success})-p(\text{1 success})=1-\left(\frac{4}{5}\right)^6-{}^6c_1\times\frac{1}{5}\times\left(\frac{4}{5}\right)^5$

$$\begin{aligned}
&=1-\left(\frac{4}{5}\right)^5\left(\frac{4}{5}+\frac{6}{5}\right)=1-2\times\left(\frac{4}{5}\right)^5\\
&=1-2\times(0.8)^5\\
&=1-2\times3277\\
&=1-0.6554\\
&=0.3446
\end{aligned}$$

5.14 POISSON DISTRIBUTION

We come across several situations in which the probability p of an event is very small but the number of trials n is so large that the event occurs several times as can be seen from the following cases.

1. The number of typographical errors by an expert typist in typing the manuscript of a book is fairly high.
2. The probability of a person suffering from a disease such as blood cancer is very small, but in a big city in a year the number of persons suffering from the disease is sufficiently high.
3. Individually being born blind is a rare event and hence its probability is very small, but the number of persons born blind in a big city in a year is quite high.

Thus, evidently Poisson distribution is a limiting case of binomial distribution when p(or q) is very small and n is large so that the average number of successes np (or nq) is a finite constant m (say).

In the binomial distribution, the probability of r successes is given by,

$$b(r, n, p) = {}^nc_r\, p^r q^{n-r},\ p+q=1$$

Let $\lim\limits_{x\to\infty} np = m$

$$\text{Now, } b(r, n, p) = \frac{n(n-1)(n-2)\dots(n-r+1)}{r!} p^r \left(1-\frac{m}{n}\right)^{n-r} \tag{5.8}$$

Taking limit of (5.8) as $n \to \infty$, we have

$$\lim_{x\to\infty} b(r, n, p) = \lim_{x\to\infty} \frac{\left(1-\frac{1}{n}\right)\left(1-\frac{2}{n}\right)\dots\left(1-\frac{r-1}{n}\right)}{r!}(np)^r\left(1-\frac{np}{n}\right)^{n-r}$$

$$= \lim_{x\to\infty} \frac{\left(1-\frac{1}{n}\right)\left(1-\frac{2}{n}\right)\dots\left(1-\frac{r-1}{n}\right)}{r!}\,\frac{(np)^r\left(1-\frac{np}{n}\right)^n}{\left(1-\frac{np}{n}\right)^r}$$

$$= \lim_{x\to\infty} \frac{\left(1-\frac{1}{n}\right)\left(1-\frac{2}{n}\right)\dots\left(1-\frac{r-1}{n}\right)}{r!}\,\frac{m^r\left(1-\frac{m}{n}\right)^n}{\left(1-\frac{m}{n}\right)^r}$$

$$= \frac{m^r e^{-m}}{r!} \qquad \because \lim_{x\to\infty}\left(1-\frac{m}{n}\right)^n = e^{-m}$$

$$\text{and } \lim_{x\to\infty}\left(1-\frac{m}{n}\right)^r = (1-0)^r = 1$$

Thus, the probabilities of

	0	1	2	......	r
Successes are	e^{-m}	me^{-m}	$\frac{m^2e^{-m}}{2!}$	......	$\frac{m^re^{-m}}{r!}$.

The above probability distribution of the number of successes is called Poisson distribution.

Definition 5.4 The probability distribution of the random variable X is called Poisson distribution if

$$p(X=r) = \frac{e^{-m}m^r}{r!} \qquad r = 0, 1, 2, \ldots$$

where m is known as parameter of the distribution.

If N is the total frequency of a Poisson distribution, then the frequency of $X = r$ is $N.\frac{e^{-m}m^r}{r!}$, $r = 0, 1, 2, \ldots$

The sum of all probabilities of a Poisson distribution is

$$= \sum_{r=0}^{\infty} \frac{e^{-m}m^r}{r!} = e^{-m}\sum_{r=0}^{\infty}\frac{m^r}{r!} = e^{-m}e^{m} = e^0 = 1$$

5.14.1 Mean and Variance

If X is the PoissonVariate, then

$$\mu = \text{Mean} = E(X) = \sum_{i=0}^{\infty} x_i p_i = \sum_{r=0}^{\infty} r.\frac{e^{-m}m^r}{r!}$$

$$= me^{-m}\sum_{r=1}^{\infty}\frac{m^{r-1}}{(r-1)!} = me^{-m}[1 + m + \frac{m^2}{2!} + \frac{m^3}{3!} + \ldots]$$

$$= me^{-m}.e^{m} = m$$

Variance $= \sigma^2 = E(X - \mu)^2$
$= E(X^2) - \mu^2$

Now

$$E(X^2) = \sum_i x_i^2 p_i = \sum_{r=0}^{\infty} r^2 \frac{e^{-m}m^r}{r!}$$

$$= \sum_{r=0}^{\infty}\{r(r-1)+r\}e^{-m}\frac{m^r}{r!}$$

$$= e^{-m}\left[\sum_{r=0}^{\infty} r(r-1)\frac{m^r}{r!} + \sum_{r=0}^{\infty} r\cdot\frac{m^r}{r!}\right]$$

$$= e^{-m}\left[m^2\sum_{r=2}^{\infty}\frac{m^{r-2}}{(r-2)!} + m\sum_{r=1}^{\infty}\frac{m^{r-1}}{(r-1)!}\right]$$

$$= e^{-m}\left[m^2e^m + me^m\right] = m^2 + m$$

$\therefore$ Variance $= \sigma^2 = m^2 + m - m^2 = m$

5.14.2 Another Approach to Poisson Distribution

Let the random variable X denote the number of telephone calls during time t at a telephone switch board. We assume that (1) the calls are independent and (2) the probability of a call in time dt is $\lambda\, dt$ where λ is a real number.

Let $P_x(t)$ be the probability of x calls in time t, then $P_x(t + dt)$ will denote the probability of x calls in time $t + dt$. Now two mutually exclusive cases will arise.

1. x calls in time t and no call in time dt
2. $(x - 1)$ calls in time t and one call in time dt

(The possibility of more than one call in time dt is neglected)

$$\therefore\ P_x(t + dt) = P_x(t)(1-\lambda dt) + P_{x-1}(t)\lambda dt$$

or
$$\frac{P_x(t+dt) - P_x(t)}{dt} = \lambda[P_{x-1}(t) - P_x(t)]$$

Taking limit as $dt \to 0$, we get

$$\frac{d}{dt}P_x(t) = \lambda[P_{x-1}(t) - P_x(t)] \qquad (5.9)$$

Let its solution be
$$P_x(t) = \frac{(\lambda t)^x}{x!}f(t) \qquad (5.10)$$

Substituting (5.10) in (5.9), we get

$$\frac{(\lambda t)^x}{x!}f'(t) + x\frac{(\lambda t)^{x-1}}{x!}\lambda f(t) = \lambda\left[\frac{(\lambda t)^{x-1} f(t)}{(x-1)!} - \frac{(\lambda t)^x}{x!}f(t)\right]$$

$$\therefore\ \frac{f'(t)}{f(t)} = -\lambda$$

This gives $f(t) = Ce^{-\lambda t}$, where C is an arbitrary constant.

Putting $t = 0$, we have $C = f(0)$.

From relation (5.10), we have $P_0(0) = f(0) = 1$ as the probability of no call in zero time is 1 $\therefore C = 1$.

Hence, $f(t) = e^{-\lambda t}$

Substituting for $f(t)$ in (5.10), we have

$$P_x(t) = \frac{(\lambda t)^x}{x!}e^{-\lambda t} = \frac{m^x e^{-m}}{x!} \qquad \text{where } m = \lambda t$$

which shows that the number of calls in time t is a Poisson variate. If X is the random variable representing the number of calls in time t, then

$$p(X = x) = \frac{m^x e^{-m}}{x!} \quad x = 0, 1, 2,\ldots$$

Since $m = \lambda t$ is the mean of the distribution, λ denotes mean arrival rate of telephone calls per second.

The following are further examples of Poisson variates.

1. The number of buses reaching bus depot in time t.
2. The number of aeroplanes arriving at an airport in time t.
3. The number suicides in a city in a year.
4. The number of persons joining a queue at a counter in time t.
5. The number of radio-active particles passing through a counter in time t.

Example 1 Fit a Poisson distribution to the following data and calculate the theoretical frequencies

x:	0	1	2	3	4
f:	192	100	24	3	1

Total frequency $N = 192 + 100 + 24 + 3 + 1 = 320$

Mean $$m = \frac{1}{N}\sum_i f_i x_i = \frac{1}{320}[192\times 0 + 100\times 1 + 24\times 2 + 3\times 3 + 4\times 1]$$

$$= \frac{1}{320}[0 + 100 + 48 + 9 + 4] = \frac{161}{320} = 0.5031$$

Required Poisson distribution $= N.\dfrac{e^{-m}m^r}{r!} \qquad r = 0, 1, 2, 3, 4$

$$\therefore \text{Frequency}(x = r) \qquad = 320\frac{e^{-0.5031}(0.5031)^r}{r!}$$

$$= 320\times 0.6047\frac{(0.5031)^r}{r!}$$

$$= 193.489\frac{(0.5031)^r}{r!}, r = 0, 1, 2, 3, 4$$

Putting $r = 0, 1, 2, 3, 4$ we get the theoretical frequencies.

$r = 0$, frequency $= 193.489 \simeq 193$

$r = 1$, frequency $= 193.489\dfrac{(0.5031)^1}{1!} = 97.344 \simeq 97$

$r = 2$, frequency $= 193.489\dfrac{(0.5031)^2}{2!} = 24.487 \simeq 24$

$r = 3$, frequency $= 193.489\dfrac{(0.5031)^3}{3!} = 4.106 \simeq 4$

$r = 4$, frequency $= 193.489\dfrac{(0.5031)^4}{4!} = 0.516 \simeq 1$

Hence, the theoretical frequencies are

x:	0	1	2	3	4
Theoretical f:	193	97	24	4	1

Example 2 Experience shows that a box of 400 components of a company has 1 per cent defective components. Find the probability that such a box has (1) no defective; (2) one component defective; and (3) at most 3 components defective.
Given $e^{-4} = 0.0183$.

Solution: Let selection of a component from the box be regarded as a trial and getting a defective component as the success. If p denotes the probability of success, then $p = \frac{1}{100} = 0.01$ which is small, as the number of trials $n = 400$ is large and so the problem can be solved by Poisson distribution.

The parameter $m = np = 400 \times 0.01 = 4$

Required Poisson distribution $p(x = r) = \frac{e^{-m} m^r}{r!}$ $\quad r = 0, 1, 2, \ldots$

$$= \frac{e^{-4} 4^r}{r!} \quad r = 0, 1, 2, \ldots$$

(1) $p(0 \text{ defective}) = p(x = 0) = e^{-m} = e^{-4} = 0.0183$

(2) $p(1 \text{ defective}) = p(x = 1) = e^{-m} = 4e^{-4} = 4 \times 0.0183$
$= 0.0732$

(3) $p(\text{at most 3 defective}) = p(x \leq 3)$

$$= \sum_{r=0}^{3} \frac{e^{-4} 4^r}{r!}$$

$$= e^{-4}\left(1 + 4 + \frac{4^2}{2!} + \frac{4^3}{3!}\right)$$

$$= 0.0183\left(1 + 4 + 8 + \frac{32}{3}\right)$$

$$= 0.0183 \times \frac{71}{3} = 0.0183 \times 23.6667$$

$$= 0.4331$$

Example 3 A manufacturer of cotter pins knows that 5 per cent of his product is defective if he sells cotter pins in boxes of 100 and guarantees that not more than 10 pins will be defective. What is the probability that a box will fail to meet the guaranteed quality.

Solution: The number of trials $n = 100$

The probability that a cotter pin is defective, i.e., the probability of success $p = \frac{5}{100}$

$\therefore$ Parameter for Poisson distribution $m = np = 100 \times \frac{5}{100} = 5$

$\therefore$ Probability of r success $= \frac{e^{-5}5^r}{r!}$ $r = 0, 1, 2, \ldots$

$$P(\text{more than 10 defective}) = 1 - \sum_{r=0}^{10} \frac{e^{-5}5^r}{r!}$$

Example 4 The probability of getting no misprint is e^{-2}. Find the probability that a page contains more than 2 errors.
(Given $e^{-2} = 0.135$)

Solution: The required Poisson distribution is

$$p(r\,\text{errors}) = \frac{e^{-2}2^r}{r!} \qquad r = 0, 1, 2, \ldots$$

Since p (0 error) $= e^{-2} = e^{-m} \Rightarrow m = 2$, the parameter of Poisson distribution

Hence $p(\text{more than 2 errors}) = p(2\text{ errors}) + p(3\text{ errors}) + \ldots$

$$= 1 - p(0\text{ error}) - p(1\text{ error}) = 1 - e^{-2} - 2e^{-2}$$

$$= 1 - 3e^{-2} = 1 - 3 \times 0.1353 = 1 - 0.4059$$

$$= 0.5941.$$

Example 5 If the probability of a bad reaction from certain vaccine is 0.001. Determine the probability that one of 2000 individuals (1) none; (2) 1; (3) more than 1; (4) more than 2; (5) exactly 3 will get a bad reaction, given $e^{-2} = 0.1353$.

Solution: Let getting a bad reaction be regarded as success, then $p = 0.001$ and the selection of an individual being regarded as a trial, then the number of trials $n = 2000$.

Hence, parameter of Poisson distribution $m = 2000 \times 0.001$
$= 2$

Hence, the required Poisson distribution is

$$p(r\,\text{success}) = \frac{e^{-2} \cdot 2^r}{r!} \qquad r = 0, 1, 2, \ldots$$

(1) $p(0\text{ success}) = e^{-2} = 0.1353$

(2) $p(1\text{ success}) = 2e^{-2} = 2 \times 0.1353 = 0.2706$

(3) $p(\text{more than 1 success}) = p(2\text{ success}) + p(3\text{ success}) + \ldots$
$= 1 - p(0\text{ success}) - p(1\text{ success})$
$= 1 - e^{-2} - 2e^{-2}$

$$= 1 - 3e^{-2} = 1 - 3 \times 0.1353$$
$$= 1 - 0.4059 = 0.5941.$$

(4) $p(\text{more than 2 successes}) = p(3 \text{ successes}) + p(4 \text{ successes}) + \ldots$

$$= 1 - p(0 \text{ success}) - p(1 \text{ success}) - p(2 \text{ successes})$$

$$= 1 - e^{-2} - 2e^{-2} - \frac{2^2}{2} e^{-2}$$

$$= 1 - 5e^{-2} = 1 - 5 \times 0.1353$$
$$= 1 - 0.6765 = 0.3235$$

(5) $p(3 \text{ successes}) = \frac{2^3}{3!} e^{-2} = \frac{8^4}{6_3} \times 0.1353 = \frac{4 \times 0.1353}{3}$

$$= \frac{0.5412}{3} = 0.1804.$$

Example 6 Suppose the number of telephone calls an operator received from 8:30 a.m. to 8:40 a.m. follow Poisson distribution with mean 2. Find the probability that the operator will receive (1) no call; (2) 1 call in that time interval tomorrow, given $e^{-2} = 0.1353$.

Solution: Poisson distribution has mean $m = 2$.

Hence, the required Poisson distribution is

$$p(r \text{ calls in the given time}) = \frac{e^{-m} m^r}{r!} = \frac{e^{-2} . 2^r}{r!}, \qquad r = 0, 1, 2, \cdots$$

(1) $p(\text{no call}) = p(r = 0) = e^{-2} = 0.1353$

(2) $p(1 \text{ call}) = p(r = 1) = 2e^{-2} = 2 \times 0.1353 = 0.2706$

EXERCISE – A

1. A license plate contains two distinct letters of English alphabet followed by four digits with first digit non-zero. Find the number of distinct license plates that can be printed.
2. Find the number of ways in which four persons can be accommodated in three different hotels.
3. Find the numbers of ways in which three balls coloured white, black, and red can be placed in 10 boxes.
4. In how many ways can three men and six women be arranged in a row of 3 men always stranding next to each other.
5. Find the number of 4 digit numbers that contain no repeated digits.
6. How many seven letter words can be formed by using the letters of the word BENZENE.
7. Find the number of 7 bit messages that can be sent using four 1's and three 0's.

8. A bit is either 0 or 1. *A* byte is a sequence of 8 bits. Find (i) the number of bytes that can be formed; (ii) find the number of bytes that begin with 11 and end with 00.
9. How many different numbers can be formed by taking only three out of 5 digits 1, 2, 3, 4, 5?
10. In how many ways can 5 letters A, B, C, D, E be arranged so that A and B occupy contiguous positions.
11. Find the number of ways in which 10 persons working in an industry can be divided into groups containing four, three, and three.
12. Find the number of ways in which three examinations can be scheduled in a five-day period so that no two examinations are scheduled for the same day.
13. Three are 20 true or false questions in an examination paper. Find the number of ways, a student can answer the question if he or she can also choose to leave the answer blank.
14. Find the number of ways in which three examinations can be scheduled in a six-day period, there being no restriction on the number of examination scheduled each day.
15. Find the number of committees of 4 that can be formed from a select group of 10 members.
16. Find the number of ways in which a team consisting of 4 teachers and 3 students can be selected from 8 teachers and 5 students.
17. Find the number of different passwords consisting of 5 characters, the first of which is a letter from the set $\{A, B, C, D, E\}$ and the remaining are digits.
18. Find the number of ways in which a committee of 5 persons can be selected from a group of 12 persons if one person is to be designated as chairman of the committee.
19. An industry has 10 positions vacant for a particular section. Find the number of different batches of candidates that can be selected out of 15 applicants. Also, find the number of times a particular candidate can be selected.
20. A bag contains 6 blue and 5 red and 5 white marbles. Find the number of ways in which 4 marbles can be drawn from the bag if (i) they have any colour; (ii) two must be blue and one must be red and one white; (iii) they must all be of the same colour.
21. In an examination, a student is required to answer 10 questions out of 13. (i) How many choices has he; (ii) how many choices if first two questions are compulsory; (iii) how many if he must answer exactly 3 out of first 5 questions; (iv) how many if he must answer at least 3 of the first 5 questions.
22. A group of 15 persons sit around a circular table. Find the total number of ways in which they can sit. Also, find the number of ways of sitting if two particular persons always sit together.
23. Find the number of ways in which 5 boys and 5 girls can be seated around a circular table so that no two girls sit together.
24. Find the number of ways in which 10 students can be *partitioned* into teams containing five, three, and two.

5. A pair of dice is tossed and the two numbers appearing on the top are noted. Find the probability for each of the following events:
   (i) $A = \{\text{the sum is } 11\}$
   (ii) $B = \{1 \text{ appears on first die}\}$
   (iii) $C = \{\text{the sum is 7 or less}\}$

6. Find the probability of appearance of at least one head in the toss of 3 fair coins.

7. In a quiz contest, three persons A, B, and C participated and their chances of standing first are $\frac{1}{3}, \frac{1}{7}$, and $\frac{1}{8}$, respectively. Find the probability of either of them winning the contest.

8. Four cards are drawn from a pack of 52 one after the other without replacement. Find the probability that they are all kings.

9. A group of children has 10 boys and 3 girls. If 4 children are selected at random, find the probability that (i) they are all boys; (ii) 2 are boys and 2 girls.

0. Three television sets are selected at random from a lot of 15, of which 5 are defective. Find the probability that (i) one is defective; (ii) none is defective; (iii) at least one is defective.

1. A problem of design is given to three architects A, B, and C whose chances of designing are $\frac{1}{4}, \frac{1}{3}$, and $\frac{2}{5}$, respectively. Find the probability of design to be solved.

2. A group of 15 persons sit around a circular table. Find the probability that two particular persons always sit together.

3. Let A and B be events with $p(A) = \frac{1}{4}$, $p(B)\ \frac{1}{5}$, $p(A \cup B) = \frac{2}{5}$ (i) Find $p(A/B)$ and $p(B/A)$; (ii) A and B independent.

4. A pair of fair dice is tossed. If the numbers appearing are different, find the probability that (i) the sum is even; (ii) the sum exceeds 10.

nditional Problem

5. In an University examination, 30 per cent failed in physics, 20 per cent failed in mathematics, and 10 per cent failed in both physics and mathematics. Find the probability that
   (i) he failed in physics if he failed in mathematics
   (ii) he failed in mathematics if he failed in physics
   (iii) he failed in physics or mathematics
   (iv) he failed neither in physics nor in mathematics

ye's Theorem

6. In a factory, an item is manufactured by three machines A, B, and C, which produce 30, 25, and 45 of the total; of their output 3, 4, and 5 per cent are defective. An item is selected at random from the produce and is found to be defective. Find the probability that the defective item was produced by (i) machine A; (ii) machine B; and (iii) machine C.

25. (i) In how many ways can 15 students be partitioned into *three* groups so tha has 5 students.
    (ii) In how many ways can 15 students be selected for groups A, B, and C so t 5 students.

Pigeonhole Principle

26. The possible remainders when a positive integer is divided by 8 are 0, 1, 2, The remainders are pigeonholes. Assign each number (pigeon) to the pigeonh by its remainder when the number is divided by 8. Since there are 9 pigeons, of them must have the same remainder.

27. The 11 digits can be regarded as pigeons and there are 10 pigeon holes, that is 2, 3, 4, 5, 6, 7, 8, 9, hence in 11 digits number, at least two digits must repeat.

28. In English alphabet, there are 26 letters, which can be regarded as pigeon holes word (27 pigeons), at least two letters are same.

29. Let the pages be pigeons and the dictionaries the pigeon holes. Let each page to the dictionary in which it appears. By extended pigeon hole principle, on must contain at least $\left[\frac{61326}{30}\right]$, that is, 2045 pages.

30. Let the cars be taken as pigeons and colours as the pigeon holes, then by exte hole principle at least $\left[\frac{50}{6}\right]$, that is, 9 cars will have the same colour.

EXERCISE – B

Probability

1. Let A and A be two events such that $p(A) = 0.6$, $p(B) = 0.5$, $p(A \cap B) = 0$ probability that
   (i) A does not occur (ii) B does not occur
   (iii) A or B occurs (iv) neither A nor B occurs

2. In a race, the probability of A winning is $\frac{1}{5}$ and the probability of B winning is probability of either of them winning.

3. A card is drawn from a deck of 52 cards. Find the probability that
   (i) the card is an ace
   (ii) the card is an spade
   (iii) the card is a face card (king, queen or jack)
   (iv) the card is a face and a diamond
   (v) the card is an ace or a heart

4. Two cards are drawn from a pack of 52. Find the probability that (i) both are ki is a club and the other is a spade.

EXERCISE – C

Binomial Distribution

1. If 10 per cent of the workers in an industry are allergic to its chemical pollution, determine the probability that out of 5 workers selected at random (i) two are allergic: (ii) none is allergic; (iii) at most 1 is allergic.
2. A survey of city shows that 20 per cent of the adults have high blood pressure. A sample of four adults is selected at random. Find the probability that (i) exactly one of them has high blood pressure; (ii) all have high blood pressure; (iii) at least two have high blood pressure.
3. The probability that an entering student will graduate is 0.4. Determine the probability that out of 5 students (i) none; (ii) one; (iii) at least one will graduate.
4. An insurance salesman sells policies to 5 men, all of identical age and good health. According to actuarial tables, the probability that a man of this age will be alive 30 years hence is 0.6. Find the probability that in 30 years (i) all men; (ii) at least 3 men; and (iii) only 2 men will be alive.
5. The incidence of an occupational disease in an industry is such that the workers have 20 per cent chance of suffering from it. What is the probability that out of 6 workers, 4 or more will catch the disease.
6. The probability of a man hitting a target is $\frac{1}{4}$. If he fires 7 times, find the probability of his hitting the target at least twice.
7. The probability that a patient recovers from a delicate operation is 0.9. Find the probability that of the 7 patients having this operation (i) 5; (ii) at least 5 will recover.
8. What is the probability of 5 occurrences of a two year flood in a 10 year period.
9. What is the probability that a storm with return period of 20 years will occur once in a 10-year period.
10. The probability that a fluorescent light has a useful life of 500 hours in $\frac{3}{4}$. Find the probability that at least 6 will have a useful life of 500 hours among 8 such lights.

EXERCISE – D

Poisson Distribution

1. Fit Poisson distribution to the following data and calculate theoretical frequencies given

   $e^{-0.5} = 0.61$

Deaths	0	1	2	3	4
Frequency	122	60	15	2	1

2. Suppose that 120 misprints are distributed randomly throughout a book of 100 pages. Find the probability that a given page contains (i) no misprints; (ii) 2 misprints, given $e^{-1.2} = 0.3012$.
3. Find the probability that at most 4 defective fuses will be found in a box of 200 fuses if experience shows that 2 per cent of such fuses are defective, given $e^{-4} = 0.0183$.
4. An insurance company found that only 0.01 per cent of the population is involved in a certain type of accident each year. If its 1000 policy holders were randomly selected from the population, find the probability that not more than 2 of its clients are involved in such an accident next year, given $e^{-0.1} = 0.9048$.
5. If 2 per cent of the fluorescent lights manufactured by a company are defective, find the probability that in a sample of 100 lights (i) 1; (ii) 2; (iii) at least 1; (iv) at most 1 will be defective, given $e^{-2} = 0.1353$.
6. A manufacturer knows that the condenser he makes contains on the average 1 per cent defective. He packs them in boxes of 100. What is the probability that a box picked at random will contain 4 or more faulty condensers?
7. A car hire firm has two cars which it hires out day by day. The number of demands for a car on each day is distributed as a Poisson distribution with mean 1.5. Calculate the probable number of days in a year on which (i) neither car is on demand and (ii) a car demand is refused (given $e^{-1.5} = 0.2231$).
8. In a book of 400 pages, a proof reader finds no error in 230 pages, in 120 pages one error on each page, in 45 pages two errors on each page and in 5 pages 3 errors on each page. Fit a Poisson distribution to the data and calculate the theoretical frequencies.
9. The distribution of the number of road accidents per day in a city is Poisson with mean 4. Find the number of days out of 100 days when there will be
   (i) no accident
   (ii) at least 2 accidents
   (iii) at most 3 accidents
   (iv) between 2 and 5 accidents.
10. In a certain factory turning razor blades, there is a small chance of (1/500) for any blade to be defective. The blades are in packets of 10. Use Poisson distribution to calculate the approximate number of packets containing (i) no defective; (ii) one ; and (iii) two defective blades in a consignment of 10,000 packets.

6

Recurrence Relations and Generating Functions

LEARNING OBJECTIVES

After reading this chapter, you will be conversant with:

- Introduction
- Sequences
- Linear Recurrence Relation with Constant Coefficients
- *E* and Δ Operators Method
- Method of Generating Functions

6.1 INTRODUCTION

Recurrence relations, also called difference equation, arise in all situations in which a sequential relation exists between various values of the independent variable. This leads to the study of discrete functions that arise in physical problems. Such types of relations arise in control theory, electrical networks, and radar-tracking devices, which receive discrete pulses from the target being tracked and many other fields.

6.2 SEQUENCES

A sequence is a function whose domain is the set of natural numbers and whose range is the set of real numbers and is denoted by $a_1, a_2, a_3, \ldots, a_n, \ldots$ $(a \in R)$.

Thus, a sequence is a list of elements that are real numbers in order: a first element, a second element, and so on. If the list steps after n stops, $n \in N$, then the sequence is called a finite sequence otherwise an infinite sequence.

1. 1, 1, 0, 0, 1, 1 is a finite sequence.
2. 1, 3, 5, 7, 9, 10, 12, 14 is a finite sequence.
3. $1, \frac{1}{2}, \frac{1}{3}, \frac{1}{4}, \ldots$ is an infinite sequence.
4. 1, 3, 5, 7,… is an infinite sequence.
5. The sequence $\frac{1}{2}, \frac{1}{4}, \frac{1}{8}, \ldots$ can also be written as $a_n = 2^{-n}, n \geq 1$
6. The sequence 1, −1, 1, −1,… can be written as $a_n = (-1)^{n+1}, n \geq 1$
7. The sequence 1, −1, 1, −1,… can be written as $b_n = (-1)^n, n = 0, 1, 2\ldots$

Note 6.1

1. In a sequence, the elements may all be different or some of them may be repeated.
2. Some authors write the sequence starting with $n = 0$ instead of $n = 1$.

6.2.1 Recurrence Relation

Consider the sequence

$$2, 5, 8, 11, \ldots$$

which can be written as

$$a_n = a_{n-1} + 3,\ a_1 = 2,\ n \geq 2 \tag{6.1}$$

A relation of type (6.1) is called a recurrence relation.

To define a sequence recursively, a recurrence relation includes the information about the beginning of the sequence. This information is called the initial condition or conditions for the sequence.

Definition 6.1 A recurrence relation for the sequence $\{a_n\}$ is an equation in terms of one or more of the previous terms of the sequence $a_1, a_2, \ldots, a_{n-1}$ and the initial conditions.

(1) The recurrence relation
$a_n = a_{n-1} + 2$ with initial condition $a_1 = 5$ and $n \geq 2$ defines the sequence

$$5, 7, 9, 11, 13, \ldots$$

(2) The recurrence relation
$f_n = f_{n-1} + f_{n-2}, n > 2$, with initial conditions $f_1 = 1, f_2 = 2$ defines Fibonacci sequence 1, 1, 2, 3, 5, 8,…

(3) The recurrence relation

$${}^nc_r = {}^{n-1}c_r + {}^{n-1}c_{r-1},\ {}^nc_0 = 1,\ {}^nc_n = 1$$

defines binomial coefficients.

Backtracking: It is a technique used for finding an explicit formula for a sequence defined by a recurrence relation, which is illustrated by the following examples.

Example 1 The recurrence relation $a_n = a_{n-1} + 3$, with $a_1 = 2$, defines the sequence 2, 5, 8, 11,…. To find an explicit formula, we backtrack the value of a_n by substituting the definition of a_{n-1}, $a_{n-2}, \ldots$ until a pattern is clear.

Solution:

$$a_n = a_{n-1} + 3 = (a_{n-2} + 3) + 3 = a_{n-2} + 2.3$$
$$= a_{n-3} + 3 + 2.3 = a_{n-3} + 3.3$$
$$\ldots\ldots\ldots\ldots\ldots\ldots\ldots\ldots\ldots$$
$$= a_{n-(n-1)} + (n-1).3 = a_1 + (n-1).3$$
$$= 2 + 3n - 3 = 3n - 1$$
$$\therefore a_n = 3n - 1, n \geq 1$$

Example 2 Find the explicit formula for the sequence defined by the recurrence relation

$$a_n = 3\,a_{n-1} + 1, a_1 = 5$$

Solution:

$$a_n = 3a_{n-1} + 1$$
$$= 3[3a_{n-2} + 1] + 1$$
$$= 3^2\,a_{n-2} + 3 + 1$$
$$= 3^2[3a_{n-3} + 1] + 3 + 1$$
$$= 3^3\,a_{n-3} + 3^2 + 3 + 1$$
$$= 3^3[3a_{n-4} + 1] + 3^2 + 3 + 1$$
$$= 3^4\,a_{n-4} + 3^3 + 3^2 + 3 + 1$$
$$\vdots$$
$$= 3^{n-1}\,a_{n-(n-1)} + 3^{n-2} + 3^{n-3} + \ldots + 3^2 + 3 + 1$$
$$a_n = 3^{n-1}a_1 + (3^{n-2} + 3^{n-3} + \ldots + 3^2 + 3 + 1)$$
$$a_n = 5.3^{n-1} + \frac{3^{n-1}-1}{(3-1)} = 5.3^{n-1} + \frac{1}{2}(3^{n-1}-1)$$

or $$a_n = \left(5 + \frac{1}{2}\right).3^{n-1} - \frac{1}{2} = \frac{11}{2}.3^{n-1} - \frac{1}{2}$$

$$a_n = \frac{1}{2}(11.3^{n-1} - 1)\; n \geq 1$$

Note 6.2 A recurrence relation is also called a difference equation and these two names will be used interchangeably.

A recurrence relation may also be defined as an equation relating the values of the function $y = f(x)$ and one or more of its differences Δy, $\Delta^2 y$,..., where $\Delta y_n = y_{n+1} - y_n$: the symbol Δ is known as forward difference operator.

For example

$f(x+4) - 2f(x+3) + 5(x+2) - f(x+1) + 3f(x) = 0$ is a difference equation that can be written as

$$y_{n+4} - 2y_{n+3} + 5y_{n+2} - y_{n+1} + 3y_n = 0$$

or $$a_{n+4} - 2a_{n+3} + 5a_{n+2} - a_{n+1} + 3a_n = 0$$

6.3 LINEAR RECURRENCE RELATION WITH CONSTANT COEFFICIENTS

A recurrence relation is said to be linear if each subscripted term appears in first degree. The terms such as a_n, a_{n-1} do not appear.

A linear recurrence relation of order k with constant coefficients is expressed as

$$c_0a_{n+k} + c_1a_{n+k-1} + c_2a_{n+k-2} + \ldots + c_ka_n = f(n) \tag{6.2}$$

where $f(n)$ is some function of n and $c_0, c_1, c_2, \ldots, c_k$ are constants. The order of a linear recurrence relation is the difference between the largest and the smallest subscript appearing in the relation. If $f(n) = 0$, then the recurrence relation is called homogeneous.

1. The relation $a_n - 2a_{n-1} + 3a_{n-2} = 0$ is a linear homogeneous recurrence relation of order 2.
2. The recurrence relation $a_n = 2a_{n-1} - 5a_{n-3}$ is a linear homogeneous relation of order 3.
3. The recurrence relation $a_n = a_{n-1}^2 + a_{n-2}$ is not a linear homogenous relation.
4. $a_n = 2a_{n-1} + 5a_{n-2} + n^2 + 3$ is a linear non-homogenous recurrence relation of order 2.

Solution of a Recurrence Relation: An explicit formula for a_n, which satisfies the recurrence relation with initial conditions is called the solution of the recurrence relation.

General Solution: The general solution of a recurrence relation of order k is a solution that involves k arbitrary constants.

Particular Solution: A particular solution is a solution obtained from the general solution by assigning particular values to arbitrary constants.

Note 6.3 The recurrence relation (6.2) can also be written as

$$c_0a_n + c_1a_{n-1} + c_2a_{n-2} + \ldots + c_ka_{n-k} = f(n) \tag{6.3}$$

Three methods of solving recurrence relations are discussed in this chapter.

1. Characteristic root method
2. Finite difference method using E and Δ operators
3. Generating function method

6.3.1 Characteristic Root Method

Characteristic Equation of the Recurrence Relation: Consider a linear homogeneous equation (6.2) with $f(n) = 0$ of order k with constant coefficients, that is,

$$c_0a_{n+k} + c_1a_{n+k-1} + c_2a_{n+k-2} + \ldots + c_ka_n = 0 \text{ of order } k \text{ with } c_0 \neq 0 \text{ and } c_k \neq 0.$$

Let us assume that its solution is of the form $a_n = Am^n$.

Substituting for a_n in the left hand side of (6.2), we get

$$Am^n(c_0m^k + c_1m^{k-1} + c_2m^{k-2} + \ldots + c_K) = 0$$

$$\text{or}\quad c_0m^k + c_1m^{k-1} + c_2m^{k-2} + \ldots + c_K = 0 \tag{6.4}$$

Equation (6.4) is called the characteristic equation, which being a polynomial equation of degree k will have k roots $m_1, m_2, \ldots, m_k$.

Since we are interested in the non-trivial solution, the solution corresponding to $m = 0$ is disregarded.

The general solution of the recurrence relation (6.2) will be the linear combination of the terms.

When all the roots are real and distinct, the solution is $a_n = A_1 m_1^n + A_2 m_2^n + \ldots + A_k m_k^n$ where $A_1, A_2, \ldots, A_k$ are constants determined by initial conditions.

Now different cases arise:

Case I: Roots are real and equal.

If two roots m_1 and m_2 are equal, then $(A_1 + nA_2)\ m_1^n$ is the homogeneous solution of recurrence relation.

Case II: If three roots m_1, m_2, and m_3 are equal, then $(A_1 + A_2 n + A_3 n^2)\ m_1^n$ is the homogeneous solution of the recurrence relation.

In general, if the root m_1 is repeated r times, then $(A_1 + A_2 n + A_3 n^2 + \ldots + A_r n^{r-1})\ m_1^n$ is the homogeneous solution of recurrence relation.

Case III: The roots are non-repeated complex roots.

Let $\alpha \pm i\beta$ be the root of the characteristic equation, then $(\alpha + i\beta)^n$ and $(\alpha - i\beta)^n$ are solutions. Hence, we have $A_1(\alpha + i\beta)^n + A_2(\alpha - i\beta)^n$ as the homogeneous solution.

Case IV: The roots are repeated complex roots.

Let $\alpha \pm i\beta$ be the complex root repeated twice, then the homogenous solution of the recurrence relation is $(A_1 + A_2 n)(\alpha + i\beta)^n + (A_3 + A_4 n)(\alpha - i\beta)^n$, where A_1, A_2, A_3, and A_4 are constants.

Example 1 Solve the recurrence relation $a_{n+2} - 3a_{n+1} + 2a_n = 0$.

Solution: Let the solution be $a_n = A\ m^n$.

Its substitution in the given equation gives the characteristic equation

$$Am^n(m^2 - 3m + 2) = 0$$

This gives $m^2 - 3m + 2 = 0$

$$\text{or} \quad (m-1)(m-2) = 0 \Rightarrow m = 1, 2$$

Hence, the required solution is

$$a_n = A_1(1)^n + A_2(2)^n$$

Example 2 Solve the recurrence relation

$$a_n = 5a_{n-1} - 6a_{n-2}\ n \geq 2, \text{ given } a_0 = 1, a_1 = 4$$

Solution:

Let the solution be $a_n = Am^n$.

Its substitution in the given recurrence relation gives

$$Am^{n-2}(m^2 - 5m + 6) = 0$$

This gives $(m^2 - 5m + 6) = 0$

$$\text{or} \quad (m-2)(m-3) = 0 \Rightarrow m = 2, 3$$

Hence, the required solution is

$$a_n = A_1 2^n + A_2 3^n$$

Applying the initial conditions $a_0 = 1$ and $a_1 = 4$, we have

$$1 = A_1 + A_2$$

$$4 = A_1.2 + A_2.3$$

$$\therefore\ 2A_1 + 3A_2 = 4$$

$$\text{or}\quad 2A_1 + 3(1 - A_1) = 4$$

$$\text{or}\quad 2A_1 - 3A_1 = 4 - 3 = 1$$

$$\text{or}\quad A_1 = -1$$

$$\therefore\ A_2 = 2$$

Hence, the solution is

$$a_n = -2^n + 2(3^n).$$

Example 3 Solve the recurrence relation $a_n = 6a_{n-1} - 9a_{n-2}$, $n \geq 2$, with initial condition $a_0 = 1$, $a_1 = 4$.

Solution: The characteristic equation is

$$m^2 - 6m + 9 = 0 \Rightarrow (m - 3)^2 = 0 \Rightarrow m = 3, 3$$

Hence, the solution is

$$a_n = (A_1 + A_2 n)3^n$$

Applying initial condition, we have

$$a_0 = 1 = A_1$$

$$a_1 = 4 = (A_1 + A_2.1)3$$

$$\text{or}\quad 4 = (1 + A_2)3 = 3 + 3A_2$$

$$\therefore\ A_2 = \frac{1}{3}$$

Hence, the solution is

$$a_n = \left(1 + \frac{1}{3}n\right)3^n = 3^n + n.3^{n-1}$$

Example 4 For the Fibonacci sequence (1, 1, 2, 3, 5, 8,...) defined by recurrence relation

$a_n = a_{n-1} + a_{n-2}$, $n \geq 2$, with initial conditions $a_1 = 1$, $a_2 = 1$, show that

$$a_n = \frac{1}{\sqrt{5}}\left[\left(\frac{1+\sqrt{5}}{2}\right)^n - \left(\frac{1-\sqrt{5}}{2}\right)^n\right]$$

is the explicit formula for the sequence.

Solution: The given equation is

$$a_n - a_{n-1} - a_{n-2} = 0 \tag{i}$$

Let us assume solution in the form $a_n = Am^n$. Its substitution in the given Equation (i) gives

$$Am^{n-2}\ (m^2 - m - 1) = 0$$

$$\text{or} \quad m^2 - m - 1 = 0,$$

which is its characteristic equation whose solution is $m = \dfrac{1 \pm \sqrt{1+4}}{2} = \dfrac{1 \pm \sqrt{5}}{2}$

The two distinct roots are $\dfrac{1+\sqrt{5}}{2}$ and $\dfrac{1-\sqrt{5}}{2}$ say α, β

Clearly $\alpha + \beta = 1$, $\alpha\beta = -1$

Hence, the general solution is $a_n = A_1\left(\dfrac{1+\sqrt{5}}{2}\right)^n + A_2\left(\dfrac{1-\sqrt{5}}{2}\right)^n$ (ii)

where A_1 and A_2 are arbitrary constants to be determined from the initial conditions $a_1 = 1$ and $a_2 = 1$.

Putting $n = 1$ and 2 in (ii), we have

$$1 = \left(\frac{1+\sqrt{5}}{2}\right)A_1 + \left(\frac{1-\sqrt{5}}{2}\right)A_2 = \alpha A_1 + \beta A_2$$

$$\text{and} \quad 1 = \left(\frac{1+\sqrt{5}}{2}\right)^2 A_1 + \left(\frac{1-\sqrt{5}}{2}\right)^2 A_2 = \alpha^2 A_1 + \beta^2 A_2$$

$$\therefore \quad A_1 = \frac{\alpha}{1+\alpha^2} \text{ and } A_2 = \frac{\alpha}{1+\beta^2}$$

Its solution is $A_1 = \dfrac{1}{\sqrt{5}}$, $A_2 = -\dfrac{1}{\sqrt{5}}$

Hence, the required solution is

$$a_n = \frac{1}{\sqrt{5}}\left[\left(\frac{1+\sqrt{5}}{2}\right)^n - \left(\frac{1-\sqrt{5}}{2}\right)^n\right]$$

The nth term of Fibonacci sequence is

$$a_n = \frac{1}{\sqrt{5}}\left(\frac{1+\sqrt{5}}{2}\right)^n - \frac{1}{\sqrt{5}}\left(\frac{1-\sqrt{5}}{2}\right)^n$$

Example 5 Solve recurrence relation $a_n - 7a_{n-2} + 6a_{n-3} = 0$ with initial condition $a_0 = 4$, $a_1 = 2$, $a_2 = 18$.

Solution: The characteristic equation of the given recurrence relation is

$$m^3 - 7m + 6 = 0$$

$$\text{or} \quad (m-1)(m-2)(m+3) = 0 \quad \therefore m = 1, 2, -3$$

The general solution is $$a_n = A_1(1)^n + A_2(2)^n + A_3(-3)^n \qquad \text{(i)}$$
where A_1, A_2, and A_3 are arbitrary constants to be determined by the initial conditions.

Putting $n = 0$, 1, and 2 in solution (i), we get

$$a_0 = 4 = A_1 + A_2 + A_3$$

$$a_1 = 2 = A_1 + 2A_2 - 3A_3$$

$$a_2 = 18 = A_1 + 4A_2 + 9A_3$$

$$\therefore A_1 = 1, A_2 = 2, A_3 = 1$$

The solution is given by,

$$a_n = (1)^n + 2(2)^n + (-3)^n$$

$$\text{or} \quad a_n = 1 + 2^{n+1} + (-1)^n 3^n$$

Example 6 Solve the recurrence relation

$$a_n = a_{n-1} + 8a_{n-2} - 12a_{n-3}$$

Solution: The given equation can be written as

$$a_n - a_{n-1} - 8a_{n-2} + 12a_{n-3} = 0$$

The characteristic equation is

$$m^3 - m^2 - 8m + 12 = 0$$

$$\text{or} \quad (m-2)^2(m+3) = 0 \quad \therefore m = 2, 2, -3$$

The general solution is

$$a_n = (A_1 + nA_2)2^n + A_3(-3)^n$$

Example 7 Solve the recurrence relation

$$a_n + 4a_{n-2} = 0 \text{ given } a_0 = 1, a_1 = 0$$

Solution: The characteristic equation is

$$m^2 + 4 = 0 \quad \therefore m = 2i, -2i$$

Hence, the general solution is

$$a_n = A_1(2i)^n + A_2(-2i)^n$$

$$= A_1.2^n\left(\cos\frac{\pi}{2} + i\sin\frac{\pi}{2}\right)^n + A_2.2^n\left(\cos\frac{\pi}{2} - i\sin\frac{\pi}{2}\right)^n$$

$$= 2^n.A_1\left(\cos\frac{n\pi}{2} + i\sin\frac{n\pi}{2}\right) + 2^n.A_2\left(\cos\frac{n\pi}{2} - i\sin\frac{n\pi}{2}\right)$$

$$a_n = 2^n(A_1 + A_2)\cos\frac{n\pi}{2} + 2^n(iA_1 - iA_2)\sin\frac{n\pi}{2}$$

$$a^n = 2^n\left(c_1\cos\frac{n\pi}{2} + c_2\sin\frac{n\pi}{2}\right)$$

Putting $n = 0$ and 1 for initial condition, $a_0 = 0$, $a_1 = 1$, we get

$$a_0 = 0 = c_1$$

$$a_1 = 1 = 2c_2\sin\frac{\pi}{2} \Rightarrow c_2 = \frac{1}{2}$$

Hence, the required solution, which is obtained by substituting for C_1 and C_2 in the general solution, is

$$a_n = 2^n\frac{1}{2}\sin\frac{n\pi}{2}$$

$$\text{or}\quad a_n = 2^{n-1}\sin\frac{n\pi}{2}$$

6.3.2 Non-homogeneous Recurrence Relation

Consider the non-homogenous linear recurrence relation

$$c_0a_n + c_1a_{n-1} + c_2a_{n-2} + \ldots + c_ka_{n-k} = f(n) \tag{6.5}$$

Its solution consists of two parts:

1. Homogenous solution, which is obtained by solving (6.5) with right-hand member $f(n) = 0$, which has been discussed in 6.3.1.
2. Particular solution, which is the solution that satisfies the given recurrence relation (6.5) with $f(n)$ on the right-hand side (RHS).

Hence, the required solution = homogenous solution + particular solution

Particular Solution: There are two methods for finding the particular solution of non-homogenous recurrence relations:

1. Methods of undetermined coefficients
2. E and Δ operator method

Method of Undetermined Coefficients: This method is useful when the function $f(n)$ consists of special forms. We consider solution according to the form of $f(n)$ containing a number of unknown coefficients that are determined by the initial conditions.

The forms of solution for various $f(n)$ are given as follows:

Form of $f(n)$	Form of corresponding particular solution	Form of $f(n)$	Form of corresponding particular solution
α(a constant)	A	$(\alpha_0 n^k + \alpha_1 n^{k-1} + \ldots + \alpha_k)\beta^n$ where β is not a root of characteristic equation	$(A_0 n^k + A_1 n^{k-1} + \ldots + A_k)\beta^n$
$\alpha\beta^n$ (α, β are constants)	$A\beta^n$ *		
$\alpha \cos nc$	$A \cos nc + B \sin nc$	$(\alpha_0 n^k + \alpha_1 n^{k-1} + \ldots + \alpha_k)\beta^n$ where β is a characteristic root of multiplicity r	$n^r (A_0 n^k + A_1 n^{k-1} + \ldots + A_k)\beta^n$
$\alpha \sin nc$	$A \cos nc + B \sin nc$		
$\alpha_0 n^k + \alpha_1 n^{k-1} + \ldots + \alpha_k$ (polynomial of degree k)	$A_0 n^k + A_1 n^{k-1} + \ldots + A_k$		
αn^k	$A_0 n^k + A_1 n^{k-1} + \ldots + A_k$		

*$nA\beta^n$ if β is a root of the characteristic equation.

Some Illustrative Examples

(1) When $f(n)$ has constant value, that is, $f(n) = \alpha$

Example 1 Solve the recurrence relation

$$a_{n+2} - 5a_{n+1} + 6a_n = 4, \text{ given } a_0 = 1, a_1 = 2$$

The characteristic equation of the homogeneous equation

$$a_{n+2} - 5a_{n+1} + 6a_n = 0 \qquad \text{(i)}$$

is $\quad m^2 - 5m + 6 = 0 \quad \text{or} \quad (m-2)(m-3) = 0 \Rightarrow m = 2, 3$

Hence, the homogenous solution is

$$a_n^h = c_1 2^n + c_2 3^n \qquad \text{(ii)}$$

where c_1 and c_2 arbitrary constants.

The RHS of the given recurrence relation is constant and hence for the particular solutions we assume $a_n = A$.

Substituting for a_n in the given equation

$$A - 5A + 6A = 4$$

or $\quad 2A = 4 \quad \therefore A = 2 \quad \therefore$ Particular solution $a_n^p = 2$

Hence, the solution that consists of homogenous solution and the particular solution is given by,

$$a_n = c_1 2^n + c_2 3^n + 2 \qquad \text{(iii)}$$

Putting $n = 0, 1$ for applying the initial conditions $a_0 = 1, a_1 = 2$
We get

$$1 = c_1 + c_2 + 2 \Rightarrow c_1 + c_2 = -1$$

and

$$2 = 2c_1 + 3c_2 + 2 \Rightarrow 2c_1 + 3c_2 = 0$$

$$\therefore \quad c_1 = -3, c_2 = 2$$

Hence, the required solution is $a_n = a_n^h + a_n^p$

$$a_n = c_1 2^n + c_2 3^n + 2 = (-3)2^n + (2)3^n + 2$$

(2) When $f(n)$ is of the form $f(n) = \alpha^n$

Example 2 Solve $a_n = 2a_{n-1} + 3a_{n-2} + 5^n, n \geq 2$ given $a_0 = 1, a_1 = 5$

Solution: The homogenous equation is

$$a_n - 2a_{n-1} - 3a_{n-2} = 0$$

The characteristic equation is

$$m^2 - 2m - 3 = 0$$

$$\text{or} \quad (m + 1)(m - 3) = 0 \Rightarrow m = -1, 3$$

Hence, the homogenous solution is

$$a_n^h = C_1(-1)^n + C_2(3)^n$$

For the particular solution, we assume

$$a_n^p = A.5^n$$

Its substitution in the given equation gives

$$A.5^n = 2A.5^{n-1} + 3A.5^{n-2} + 5^n$$

$$\text{or} \quad 25A = 10A + 3A + 25 \quad \text{(cancelling } 5^{n-2} \text{ from both sides)}$$

$$\text{or} \quad 12A = 25 \therefore A = \frac{25}{12}$$

$$\therefore \quad a_n^p = \left(\frac{25}{12}\right)5^n$$

Hence, the required solution is

$$a_n = a_n^h + a_n^p$$

$$a_n = c_1(-1)^n + c_2(3)^n + \frac{25}{12}(5)^n$$

The initial conditions give

$$a_0 = 2 = c_1 + c_2 + \frac{25}{12} \quad \therefore c_1 + c_2 = 2 - \frac{25}{12} = -\frac{1}{12}$$

$$\text{and } a_1 = 5 = -c_1 + 3c_2 + \frac{25}{12}.5 \quad \therefore \quad -c_1 + 3c_2 = 5 - \frac{125}{12} = \frac{60-125}{12}$$

$$\text{or } -c_1 + 3c_2 = -\frac{65}{12}$$

Hence, we have $4c_2 = -\frac{1}{12} - \frac{65}{12} = -\frac{66}{12}$

$$\therefore c_2 = -\frac{\not{66}\ \not{22}\ 11}{4\not{12} \times \not{4}\ 2} = -\frac{11}{8}$$

$$\therefore c_1 = -\frac{1}{12} - c_2 = \frac{11}{8} - \frac{1}{12} = \frac{33-2}{24} = \frac{31}{24}$$

Therefore, the required solution is

$$a_n = \frac{31}{24}(-1)^n - \frac{11}{8}(3^n) + \frac{25}{12}\cdot(5^n)$$

(3) When $f(n)$ is a polynomial

Example 3 Solve $a_n - 6a_{n-1} + 9a_{n-2} = (n+1)^2 \; n \geq 2$

Solution: The characteristic equation of the homogeneous equation

$$a_n - 6a_{n-1} + 9a_{n-2} = 0$$

is

$$m^2 - 6m + 9 = 0$$

$$\text{or} \quad (m-3)^2 = 0 \quad \therefore m = 3, 3$$

Hence, the homogeneous solution is

$$a_n^h = (c_1 + c_2 n)3^n$$

The particular solution corresponding to the term $(n+1)^2$ on the RHS of the given equation is $A_0 + A_1 n + A_2 n^2$ [since $f(n)$ is a polynomial of degree 2].

$$\text{Let } a_n^p = A_0 + A_1 n + A_2 n^2$$

Substituting in the given equation

$$(A_0 + A_1 n + A_2 n^2) - 6[A_0 + A_1(n-1) + A_2(n-1)^2] + 9[A_0 + A_1(n-2) + A_2(n-2)^2] = 1 + 2n + n^2$$

or

$$(A_0 - 6A_0 + 9A_0 + 6A_1 - 6A_2 - 18A_1 + 36A_2) + (A_1 - 6A_1 + 12A_2 + 9A_1 - 36A_2)n + (A_2 - 6A_2 + 9A_2)n^2 = 1 + 2n + n^2$$

This gives

$$\left.\begin{aligned} 4A_0 - 12A_1 + 30A_2 &= 1 \\ 4A_1 - 24A_2 &= 2 \\ 4A_2 &= 1 \end{aligned}\right\} A_2 = \frac{1}{4},\ A_1 = 2,\ A_0 = \frac{35}{8}$$

and

The particular solution a_n^p is

$$a_n^p = \frac{35}{8} + 2n + \frac{1}{4}n^2$$

Hence, the required solution is

$$a_n = a_n^h + a_n^p$$

$$= (c_1 + c_2 n)3^n + \frac{1}{4}n^2 + 2n + \frac{35}{8}$$

(4) When $f(n)$ involves β^n when β is a characteristic root

Example 4 Find the particular solution of the recurrence relation

$$a_{n+2} - 4a_{n+1} + 4a_n = 3n + 2^n$$

Solution: The characteristic equation is

$$m^2 - 4m + 4 = 0$$

or

$$(m - 2)^2 = 0$$

$$\therefore m = 2, 2$$

The homogeneous solution $a_n^h = (c_1 + nc_2)2^n$

The RHS of the recurrence relation contains two terms 3^n and 2^n and hence for the particular solution, we assume $a_n^p = A_1 + A_2 n + A_3 n^2 2^n$. Here, $A_3 n^2 2^n$ is taken in the solution because 2 is the characteristic root of multiplicity 2.

Substituting for a_n^p in the given recurrence relation, we get

$$[A_1 + A_2(n + 2) + A_3(n + 2)^2 2^{n+2}] - 4[A_1 + A_2(n + 1) + A_3(n + 1)^2 2^{n+1}] + 4[A_1 + A_2 n + A_3 n^2 2^n] = 3n + 2^n$$

or

$$(A_1 - 2A_2) + A_2 n + [4(n + 2)^2 - 8(n + 1)^2 + 4n^2]A_3 2^n = 3n + 2^n$$

This gives $A_1 - 2A_2 = 0$, $A_2 = 3$, $\therefore A_1 = 2A_2 = 6$

$$8A_3 = 1 \qquad \therefore A_3 = \frac{1}{8}$$

Hence, the required solution is

$$a_n = a_n^h + a_n^p$$

$$\text{or} \quad a_n = (c_1 + c_2 n)2^n + \frac{1}{8}n^2 2^n + 3n + 6$$

Example 5 Solve $a_{n+2} - 6a_{n+1} + 8a_n = 2^n + 6n$

Solution: The characteristic equation is

$$m^2 - 6m + 8 = 0 \text{ or } (m - 2)(m - 4) = 0 \qquad \therefore m = 2, 4$$

Hence, the homogeneous solution is

$$a_n^h = c_1 2^n + c_2 4^n \tag{i}$$

For the particular integral, we assume

$$a_n^p = A_1 n 2^n + A_2 n + A_3 \tag{ii}$$

Since 2^n occurs on the RHS of the given recurrence relation and 2 is a characteristic root of multiplicity one, 2^n is multiplied by n.

Substitution of (ii) in the given recurrence relation gives

$$[A_1(n + 2)2^{n+2} + A_2(n + 2) + A_3] - 6[A_1(n + 1)2^{n+1} + A_2(n + 1) + A_3]$$
$$+ 8[A_1 n 2^n + A_2 n + A_3] = 2^n + 6n$$

Comparison of terms from both sides, gives

$$3A_3 - 4A_2 = 0,\ 3A_2 = 6 \quad \therefore A_2 = 2$$

$$\text{or} \quad A_3 = \frac{4}{3}A_2 = \frac{4}{3}.2 = \frac{8}{3}$$

$$[4(n+2)A_1 - 12(n+1)A_1 + 8nA_1] = 1$$

$$\text{or} \quad -4A_1 = 1 \quad \therefore A_1 = -\frac{1}{4}$$

Hence, the required solution is

$$a_n = a_n^h + a_n^p = c_1 2^n + c_2 4^n - \frac{1}{4}n2^n + 2n + \frac{8}{3}$$

$$\text{or} \quad a_n = (c_1 - \frac{1}{4}n)2^n + c_2 4^n + 2n + \frac{8}{3}$$

6.4 *E* AND Δ OPERATORS METHOD

These methods are based on the concept of finite differences that deal with functions $y = f(x)$ tabulated for equally spaced values of independent variable x.

Forward Difference Operator Δ: Let the independent variable x assume equally spaced values $a, a + h, a + 2h, \ldots, a + nh$. The corresponding values of the function $y = f(x)$ are $f(a), f(a + h), f(a + 2h), \ldots, f(a + nh)$.

The differences

$$f(a+h) - f(a), f(a+2h) - f(a+h), \ldots, f(a+nh) - f(a+\overline{n-1}\,h)$$

are called the first forward differences of the function $y = f(x)$.

These are denoted by $\Delta f(a), \Delta f(a + h), \ldots, \Delta f(a + \overline{n-1}\,h)$.

The operator Δ is called the forward difference operator.

Definition 6.2 Forward difference operator Δ is defined by,

$$\Delta f(x) = f(x + h) - f(x)$$

that is, each value of the function is subtracted from the next tabulated value. The operator Δ can be applied repeatedly on the given function $f(x)$. Thus, $\Delta\{\Delta f(x)\}$ can be obtained and is denoted by $\Delta^2 f(x)$

$$\begin{aligned}\Delta^2 f(x) &= \Delta\{\Delta f(x)\} = \Delta\{f(x + h) - f(x)\}\\ &= \Delta f(x + h) - \Delta f(x)\\ &= f(x + 2h) - 2f(x + h) + f(x)\end{aligned}$$

This shows that the difference of first-order difference is the second-order difference. Similarly, third and higher order differences can be obtained. The nth-order forward difference is denoted by $\Delta^n f(x)$.

Shift Operator *E*: The operator E known as shift operator is defined by,

$$Ef(x) = f(x + h)$$

$$E^2 f(x) = f(x + 2h), E^{-2} f(x) = f(x - 2h), \ldots, E^p f(x) = f(x + ph)$$

That is, the function is evaluated at a point shifted p times the interval h.

6.4.1 Interrelation Between Operators *E* and Δ

$$\begin{aligned}\Delta f(x) &= f(x + h) - f(x)\\ &= Ef(x) - f(x)\\ \Delta f(x) &= (E - 1) f(x)\end{aligned}$$

Hence,

$$\Delta = E - 1 \quad \text{or} \quad E = 1 + \Delta$$

6.4.2 Differences of a Polynomial Function

1. Let $f(x) = a$, a is constant then $\Delta f(x) = \Delta a = a - a = 0$
   It can also be explained as

$$\Delta f(x) = \Delta\, ax^0$$
$$= a\,\Delta x^0 = a[(x + h)^0 - x^0] = a[1 - 1] = 0$$

2. Let $f(x) = ax$, a is constant

$$= a\Delta x = a[x + h - x] = ah \text{ (constant)}$$

and $\qquad \Delta^2 f(x) = \Delta^2 ax = \Delta(\Delta ax) = \Delta(ah) = 0 \text{ from } 1$

Thus, the first forward difference of a polynomial of degree 1 is constant and second and higher order differences are 0.

3. Let

$$f(x) = ax^2$$
$$= a\,\Delta x^2$$
$$= a\,[(x + h)^2 - x^2] = a[2xh + h^2]$$
$$\Delta^2 f(x) = \Delta^2\, ax^2 = \Delta(\Delta ax^2)$$
$$= \Delta\, a(2xh + h^2)$$
$$= 2ah\,\Delta x + a\,\Delta h^2$$
$$= 2a.h^2 + 0 = 2ah^2 \text{ (constant)}$$

Thus, the second forward differences of a polynomial of degree 2 are constant and third and higher differences are 0.

The above discussion suggests that nth-order differences of a polynomial of degree n is constant and we have the following theorem.

THEOREM 6.1 The nth-order differences of a polynomial of degree n are constant and all higher order differences are 0.

Proof: Let $f(x)$ be a polynomial of degree n, then

$$f(x) = a_0 x^n + a_1 x^{n-1} + a_2 x^{n-2} + \ldots + a_{n-1} x + a_n$$
$$\Delta f(x) = f(x + h) - f(x)$$
$$= a_0\,[(x + h)^n - x^n] + a_1[(x + h)^{n-1} - x^{n-1}] + \ldots + a_{n-1}\,[x + h - x]$$
$$= a_0\,[\,x^n + n.hx^{n-1} + \ldots + h^n - x^n\,] + a^1\,[\,x^{n-1} + (n-1)hx^{n-2}$$
$$+ \ldots + h^{n-1} - x^{n-1}] + \ldots + a_{n-1}\,h$$
$$\Delta f(x) = a_0\,nhx^{n-1} + \text{terms of lower degree in } x.$$

Thus, the first difference of a polynomial of degree n is a polynomial of degree $(n - 1)$.

The second-order difference

$$\Delta^2 f(x) = \Delta\,\Delta f(x)$$
$$= \Delta[a_0\,nhx^{n-1} + \text{terms of lower degree in } x]$$
$$= a_0\,nh\,[(x + h)^{n-1} - x^{n-1}] + \text{terms of lower degree in } x$$

$$\Delta^2 f(x) = a_0 n(n-1)h^2 x^{n-2} + \text{terms of lower degree in } x$$

Thus, the second-order difference is a polynomial of degree $(n-2)$.

Proceeding in this manner, we get

$$\Delta^n f(x) = a_0\, n(n-1)(n-2) \ldots 1.h^n = n!\, a_0\, h^n$$

which is a constant and hence $(n+1)$th and higher order differences will be 0.

Representation of Differences in Terms of Function Values: Let the function $y = f(x)$ assume value $y_0, y_1, y_2, \ldots, y_n$ for equally spaced values $x_0, x_1, x_2, \ldots, x_n$ of the independent variable x.

$$\Delta y_i = y_{i+1} - y_i$$

$$\Delta^2 y_i = \Delta(y_{i+1} - y_i) = \Delta y_{i+1} - \Delta y_i = y_{i+2} - 2y_{i+1} + y_i$$

$$\Delta^3 y_i = \Delta(\Delta^2 y_i) = \Delta(y_{i+2} - 2y_{i+1} + y_i)$$

$$= \Delta y_{i+2} - 2\Delta y_{i+1} + \Delta y_i$$

$$= y_{i+3} - y_{i+2} - 2y_{i+2} + 2y_{i+1} + y_{i+1} - y_i$$

$$\Delta^3 y_i = y_{i+3} - 3y_{i+2} + 3y_{i+1} - y_i$$

In general,

$$\Delta^n y_i = y_{i+n} - {}^n c_1\, y_{i+n-1} + {}^n c_2\, y_{i+n-2} - \ldots + (-1)^n y_i$$

The nth-order differences can also be obtained as follows:

$$\Delta^n y_i = (E-1)^n y_i$$

$$= [E^n - {}^n c_1\, E^{n-1} + {}^n c_2\, E^{n-2} - \ldots + (-1)^n] y_i$$

$$\Delta^n y_i = y_{i+n} - {}^n c_1\, y_{i+n-1} + {}^n c_2\, y_{i+n-2} - \ldots + (-1)^n y_i$$

When differences are taken about y_0, that is, $i = 0$, then

$$\Delta^n y_0 = y_n - {}^n c_1\, y_{n-1} + {}^n c_2\, y_{n-2} - \ldots + (-1)^n y_0$$

In particular,

$$\Delta^2 y_0 = y_2 - 2y_1 + y_0$$

$$\Delta^3 y_0 = y_3 - 3y_2 + 3y_1 - y_0$$

$$\Delta^3 y_1 = y_4 - 3y_3 + 3y_2 - y_1$$

and so on.

6.4.3 Factorial Notation

A product of the form $x(x-1)(x-2) \ldots (x-r+1)$ is denoted by $[x]^r$ and is called a factorial.

In particular, $[x] = x$, $[x]^2 = x(x-1)$, $[x]^3 = x(x-1)(x-2)$, and so on.

In general, $[x]^n = x(x-1)(x-2) \ldots (x-n+1)$

In case the interval of differencing is h

$$[x]^n = x(x-h)(x-2h) \ldots (x - \overline{n-1}\, h)$$

The factorial notation finds its application in the theory of finite differences. The result of differencing $[x]^r$ is analogous to that of differentiating x^r.

Example 1 Express $y = 2x^3 - 3x^2 + 3x - 10$ in factorial notation and show that $\Delta^3 y = 12$.

Solution:

$$y = 2x^3 - 3x^2 + 3x - 10$$

$$\text{or} \quad y = 2[x]^3 + 3[x]^2 + 2[x] - 10 \text{ (in factorial notation)}$$

$$\therefore \Delta y = 2.3.[x]^2 + 3.2[x] + 2$$

$$\Delta^2 y = 2.3.2[x] + 3.2.1 = 12[x] + 6$$

$$\therefore \Delta^3 y = 12$$

Application of Operators *E* and Δ in the Solution of Difference Equations

Consider the non-homogeneous linear difference equation

$$a_{n+k} + c_1\, a_{n+k-1} + c_2\, a_{n+k-2} + \ldots + c_k\, a_n = f(n) \tag{6.6}$$

of order k. This can be written as

$$(E^k + c_1\, E^{k-1} + c_2\, E^{k-2} + \ldots + c_k)\, a_n = f(n) \tag{6.7}$$

Rules for Finding Homogeneous Solution (Complementary Function)

Consider the equation (6.7) with its right-hand member equal to 0. That is

$$(E^k + c_1\, E^{k-1} + c_2\, E^{k-2} + \ldots + c_k)\, a_n = 0 \tag{6.8}$$

The symbolic coefficient equated to 0 gives

$$E^k + c_1\, E^{k-1} + c_2\, E^{k-2} + \ldots + c_k = 0$$

This equation being a polynomial of degree k will have k roots and is called the auxiliary equation.

1. First-order Homogeneous Linear Difference Equation

Consider the equation $a_{n+1} - \lambda a_n = 0$, when λ is any constant. (6.9)

Writing in the symbolic form, we have

$$(E - \lambda)a_n = 0$$

$$\text{or} \quad Ea_n = \lambda a_n$$

Putting $n = 0, 1, 2, \ldots$ successively, we get

$$Ea_0 = \lambda a_0 \quad \therefore\; a_1 = \lambda a_0$$

$$Ea_1 = \lambda a_1 \quad \therefore\; a_2 = \lambda a_1 = \lambda(\lambda a_0) = \lambda^2 a_0$$

$$Ea_2 = \lambda a_2 \quad \therefore\; a_3 = \lambda a_2 = \lambda(\lambda^2 a_0) = \lambda^3 a_0$$

$$a_n = \lambda^n a_0, \text{ where } a_0 \text{ is some constant}$$

Thus, the solution of $(E - \lambda)\, a_n = 0$ is $a_n = c\lambda^n$, where c is an arbitrary constant.

2. Second-order Homogeneous Linear Different Equation
Consider the second-order equation

$$a_{n+2} + \alpha a_{n+1} + \beta a_n = 0 \tag{6.10}$$

Writing in the symbolic form

$$(E^2 + \alpha E + \beta)a_n = 0 \tag{6.11}$$

The auxiliary equation is

$$E^2 + \alpha E + \beta = 0 \tag{6.12}$$

which being quadratic will have two roots. Now three cases will arise:

Case I: Equation (6.12) has real and distinct roots, say λ_1 and λ_2. Equation (6.11) can be written as

$$(E-\lambda_1)(E-\lambda_2)a_n = 0$$

The solution of equation $(E - \lambda_1)\, a_n = 0$ and $(E - \lambda_2)\, a_n = 0$ is the required solution

$$a_n = c_1\lambda_1^n + c_2\lambda_2^n$$

where c_1 and c_2 are arbitrary constants.

Case II: Let the roots of (6.12) be real and equal, that is, $\lambda_1 = \lambda_2$, then Equation (6.11) becomes

$$(E - \lambda_1)^2\, a_n = 0 \tag{6.13}$$

Define $a_n = \lambda_1^n\, b_n$, where b_n is a new dependent variable, then (6.13) becomes

$$\left(E^2 - 2\lambda_1 E + \lambda_1^2\right).\lambda_1^n b_n = 0$$

or $$\lambda_1^{n+2}\, b_{n+2} - 2\lambda_1^{n+2}\, b_{n+1} + \lambda_1^{n+2}\, b_n = 0$$

or $$b_{n+2} - 2b_{n+1} + b_n = 0$$

or $$\Delta^2 b_n = 0$$

which shows that second-order difference of b_n is 0 and hence b_n is a linear function of n.

Let $b_n = c_1 + c_2 n$, where c_1 and c_2 are arbitrary constants.
Hence, the required solution is

$$a_n = \lambda_1^n\, b_n = \left(c_1 + c_2 n\right)\lambda_1^n \tag{6.14}$$

Case III: Let the roots of (6.12) be complex, that is, $\alpha \pm i\beta$, then the solution is

$$\begin{aligned} a_n &= c_1(\alpha + i\beta)^n + c_2(\alpha - i\beta)^n \\ &= r^n[c_1(\cos n\theta + i \sin n\theta) + c_2(\cos n\theta - i\sin n\theta)] \end{aligned}$$

or $\quad a_n = r^n[A_1\cos n\theta + A_2\sin n\theta]$ where $r = \sqrt{\alpha^2 + \beta^2}$ $\tan\theta = \beta/\alpha$

where A_1 and A_2 are arbitrary constants.

The Particular Integral (Bool's Operational Method)

Consider the non-homogeneous linear difference equation

$$a_{n+k} + c_1 a_{n+k-1} + c_2 a_{n+k-2} + \ldots + c_k a_n = f(n) \tag{6.15}$$

which can be written in the symbolic form

$$(E^k + c_1 E^{k-1} + c_2 E^{k-2} + \ldots + c_k) a_n = f(n) \tag{6.16}$$

or $$\phi(E) a_n = f(n) \tag{6.17}$$

where $$\phi(E) = E^k + c_1 E^{k-1} + c_2 E^{k-2} + \ldots + c_k \tag{6.18}$$

Then, the particular integral of equation (6.17) is

$$a_n^p = \frac{1}{\phi(E)} f(n) \tag{6.19}$$

Different cases arise depending upon the function $f(n)$.

Case I: When $f(n) = A$, where A is a constant

Since $EA = A$,

$$\begin{aligned}\phi(E)A &= (E^k + c_1 E^{k-1} + c_2 E^{k-2} + \ldots + c_k)A \\ &= A + c_1 A + c_2 A + \ldots + c_k A \\ &= (1 + c_1 + c_2 + \ldots + c_k)A \\ &= \phi(1)A\end{aligned}$$

This implies $E = 1$

Hence, the particular integral

$$a_n^p = \frac{1}{\phi(E)} A = \frac{1}{\phi(1)} A \text{ provided } \phi(1) \neq 0$$

Case II: When $f(n) = A\beta^n$, where A and β are constants.

$$\begin{aligned}\phi(E)(A\beta^n) &= (E^k + c_1 E^{k-1} + c_2 E^{k-2} + \ldots + c_k)A\beta^n \\ &= A(\beta^{n+k} + c_1 \beta^{n+k-1} + c_2 \beta^{n+k-2} + \ldots + c_k \beta^n) \\ &= A(\beta^k c_1 \beta^{k-1} + c_2 \beta^{k-2} + \ldots + c_k)\beta^n\end{aligned}$$

$$\therefore \phi(E)(A\beta^n) = A\phi(\beta)\beta^n$$

Which implies $E = \beta$. Thus, to obtain the particular integral, we put $E = \beta$.

Hence, the particular integral

$$a_n^p = \frac{1}{\phi(E)} A\beta^n = \frac{1}{\phi(\beta)} A\beta^n \text{ provided } \phi(\beta) \neq 0$$

Special Cases

1. When $A = 1$, $a_n^p = \dfrac{1}{\phi(E)}\beta^n = \dfrac{1}{\phi(\beta)}\beta^n, \phi(\beta) \neq 0.$
2. When $\phi(\beta) = 0$

   then $(E - \beta)$ is a factor of Equation (6.17).

Now consider

$$(E - \beta)a_n = A\beta^n$$

The particular integral is $a_n^p = \dfrac{1}{(E-\beta)} A\beta^n = A\,n\,\beta^{n-1}$

3. For the relation

$$(E - \beta)^2 a_n = A\beta^n$$

The particular integral is $a_n^p = \dfrac{1}{(E-\beta)^2} A\beta^n = \dfrac{An(n-1)}{2!}\beta^{n-2}$

4. For the relation

$$(E - \beta)^3 a_n = A\beta^n$$

The particular integral is $a_n^p = \dfrac{1}{(E-\beta)^3} A\beta^n = \dfrac{An(n-1)(n-2)}{3!}\beta^{n-3}$

Case III: When $f(n)$ is a polynomial of degree m in n.

Now,

$$\phi(E) = \phi(1 + \Delta) \quad (\therefore E = 1 + \Delta)$$

Hence,

$$\frac{1}{\phi(E)} f(n) = \frac{1}{\phi(1+\Delta)} f(n) = (b_0 + b_1\Delta + b_2\Delta^2 + \ldots + b_m\Delta^m + \ldots)f(n)$$

By expanding $\phi(1 + \Delta)$ in powers of Δ.

Now,

$$\frac{1}{\phi(E)} f(n) = (b_0 + b_1\Delta + b_2\Delta^2 + \ldots + b_m\Delta^m) f(n)$$

[Retaining terms upto Δ^m as $f(n)$ is a polynomial of degree m]

Hence, the particular integral

$$a_n^p = b_0 f(n) + b_1\Delta f(n) + b_2\Delta^2 f(n) + \ldots + b_m\Delta^m f(n)$$

Case IV: When $f(n)$ is a polynomial of degree m in n and is of the form $\beta^n f(n)$

We have

$$E^k[\beta^n f(n)] = \beta^{n+k} f(n + k)$$

$$= \beta^n . \beta^k E^k f(n)$$

$$\text{or} \quad E^k[\beta^n f(n)] = \beta^n(\beta E)^k f(n)$$

Thus we shall have

$$\frac{1}{\phi(E)}\beta^n f(n) = \beta^n \frac{1}{\phi(E\beta)} f(n)$$

$$= \beta^n \frac{1}{\phi\{\beta(1+\Delta)\}} f(n)$$

$$= \beta^n \left[\phi(\beta + \beta\Delta)\right] f(n)$$

Hence, the particular integral

$$a_n^p = \beta^n \left[\phi(\beta + \beta\Delta)\right]^{-1} f(n)$$

Some Illustrative Examples

Example 1 Find the solution of the recurrence relation

$$a_{n+2} - 8a_{n+1} + 15a_n = 24$$

Solution: The given recurrence equation can be written in the symbolic form as

$$(E^2 - 8E + 15)a_n = 24$$

The auxiliary equation is

$$(E^2 - 8E + 15) = 0$$

$$\text{or} \quad (E - 3)(E - 5) = 0 \quad \therefore \; E = 3, 5$$

Hence, the homogenous solution (i.e., complementary function) is

$$a_n^h = A_1 3^n + A_2 5^n$$

The particular integral $= a_h^p = \dfrac{1}{E^2 - 8E + 15}(24) = \dfrac{24}{\left(1^2 - 8.1 + 15\right)} = \dfrac{24}{8} = 3$

(putting $E = 1$)
Hence, the required solution is

$$a_n = a_n^h + a_n^p = A_1.3^n + A_2.5^n + 3$$

Example 2 Solve the recurrence relation $a_{n+2} - 6a_{n+1} + 5a_n = 10.6^n$

Solution: The given equation in the symbolic form is

$$(E^2 - 6E + 5)a_n = 10.6^n$$

The auxiliary equation is

$$E^2 - 6E + 5 = 0$$

$$\text{or} \quad (E-1)(E-5) = 0 \quad \therefore \; E = 1, 5$$

The homogenous solution is

$$a_n^h = A_1 (1)^n + A_2 (5)^n$$

The particular integral

$$a_n^p = \frac{1}{E^2 - 6E + 5} 10.6^n$$

$$= 10 \frac{1}{6^2 - 6.6 + 5} .6^n \quad (\text{putting } E = 6)$$

$$= 2.6^n$$

Hence, the required complete solution is $a_n = A_1(n)^n + A_2(5)^n + 2.6^n$

Example 3 Solve $a_{n+2} - 6a_{n+1} + 9a_n = 3^n$

Solution: The given equation in the symbolic form is $(E^2 - 6E + 9)a_n = 3^n$

The auxiliary equation is $(E^2 - 6E + 9)$ or $(E - 3)^2 = 0 \therefore E = 3,3$

The homogeneous solution $a_n^p = (A_1 + A_2 n)3^n$

The particular integral $\quad a_n^p = \dfrac{1}{(E-3)^2}.3^n = \dfrac{n(n-1)}{2!}3^{n-2}$

Hence, the required complete solution is $a_n = (A_1 + A_2 n)3^n + \dfrac{n(n-1)}{2!}3^{n-2}$.

Example 4 Solve the recurrence relation

$$a_{n+2} - 9a_n = n^2 + n + 1$$

Solution: The symbolic form is $(E^2 - 9)a_n = n(n - 1) + 2n + 1$

The auxiliary equation is $E^2 - 9 = 0 \quad \therefore E = -3, 3$

Hence, the homogenous solution is $a_n^h = A_1(3)^n + A_2(-3)^n$

The particular integral

$$\begin{aligned}
a_h^p &= \frac{1}{E^2 - 9}\left[[n]^2 + 2[n] + 1\right] \\
&= \frac{1}{(1+\Delta)^2 - 9}\left[[n]^2 + 2[n] + 1\right] \\
&= \frac{1}{\Delta^2 + 2\Delta - 8}\left[[n]^2 + 2[n] + 1\right] \\
&= -\frac{1}{8}\left[1 - \left(\frac{\cancel{2}}{\cancel{8}_4}\Delta + \frac{1}{8}\Delta^2\right)\right]^{-1}\left[[n]^2 + 2[n] + 1\right] \\
&= -\frac{1}{8}\left[1 + \left(\frac{1}{4}\Delta + \frac{1}{8}\Delta^2\right) + \left(\frac{1}{4}\Delta + \frac{1}{8}\Delta^2\right)^2 + \ldots\right]\left[[n]^2 + 2[n] + 1\right] \\
&= -\frac{1}{8}\left[1 + \frac{1}{4}\Delta + \frac{3}{16}\Delta^2 + \ldots\right]\left[[n]^2 + 2[n] + 1\right] \\
&= -\frac{1}{8}\left[[n]^2 + 2[n] + 1 + \frac{1}{4}\{2[n] + 2\} + \frac{3}{\cancel{16}_8}\cancel{2}\right] \\
&= -\frac{1}{8}\left[n(n-1) + 2n + 1 + \frac{1}{\cancel{4}}.\cancel{2}n + \frac{1}{\cancel{4}}\cancel{2} + \frac{3}{8}\right] \\
&= -\frac{1}{8}\left[n^2 - n + 2n + 1 + \frac{1}{2}n + \frac{1}{2} + \frac{3}{8}\right] \\
a_h^p &= -\frac{1}{8}\left[n^2 + \frac{3}{2}n + \frac{15}{8}\right]
\end{aligned}$$

Hence, the required solution is $a_n = a_n^h + a_n^p = A_1(3)^n + A_2(-3)^n - \frac{1}{8}n^2 - \frac{3}{16}n - \left(\frac{15}{64}\right)$

Example 5 Solve the recursive relation $a_{n+2} - 4a_{n+1} + 4a_n = n^2.3^n$

Solution: The symbolic form of the given relation is $(E^2 - 4E + 4)\, a_n = n^2.3^n$

Here, RHS function is of the form $\beta^n f(n)$, where $f(n)$ is a polynomial of finite degree in *n*, that is, of degree 2.

The auxiliary equation is $E^2 - 4E + 4 = 0$ or $(E - 2)^2 = 0 \quad \therefore E = 2, 2$

Hence, the homogeneous solution $a_n^h = (A_1 + A_2 n)2^n$

The particular integral

$$a_n^p = \frac{1}{E^2 - 4E + 4} n^2.3^n = \frac{1}{(E-2)^2} 3^n n^2$$

$$= 3^n \frac{1}{(3E-2)^2} n^2$$

$$= 3^n \frac{1}{[3(1+\Delta)-2]^2} [n(n-1)+n]$$

$$= 3^n \frac{1}{(1+3\Delta)^2} \left[[n]^2 + [n]\right]$$

$$= 3^n (1+3\Delta)^{-2} \left[[n]^2 + [n]\right]$$

$$= 3^n \left[1 - 6\Delta + \frac{(-2)(-3)}{2_1}(3\Delta)^2 - \ldots\right]\left[[n]^2 + [n]\right]$$

$$= 3^n \left[1 - 6\Delta + 27\Delta^2 \ldots\right]\left[[n]^2 + [n]\right]$$

$$= 3^n \left[[n]^2 + [n] - 6\{2[n] + 1\} + 27(2.1)\right]$$

$$= 3^n \left[n(n-1) + n - 12n - 6 + 54\right]$$

$$= 3^n \left[n^2 - 12n + 48\right]$$

Hence, the required solution is

$$a_n = a_n^h + a_n^p = (A_1 + A_2 n)2^n + 3^n (n^2 - 12n + 48)$$

Case V: When $f(n) = \cos kn$ or $\sin kn$

$$\text{The particular integral} = \frac{1}{\phi(E)} \cos kn = \frac{1}{\phi(E)} (\text{real part of } e^{ikn})$$

$$= \text{Real part of } \frac{1}{\phi(E)} e^{ikn}$$

Similarly for $f(x) = \sin kn$

$$\text{The particular integral} = \text{Imaginary part of } \frac{1}{\phi(E)} e^{ikn}$$

Example 1 Solve the recurrence equation $a_{n+2} + k^2 a_n = \cos kn$

The auxiliary equation is $(E^2 + k^2) = 0$

$$\therefore E = \pm ik \quad \text{i.e., } ik, -ik$$

$$\text{or} \quad E = k\left(\cos\frac{\pi}{2} + i\sin\frac{\pi}{2}\right), k\left(\cos\frac{\pi}{2} - i\sin\frac{\pi}{2}\right)$$

The homogeneous solution is

$$a_n^h = A_1 k^n\left(\cos\frac{n\pi}{2} + i\sin\frac{n\pi}{2}\right) + A_2 k^n\left(\cos\frac{n\pi}{2} - i\sin\frac{n\pi}{2}\right)$$

$$or \quad a_n^h = k^n\left(c_1\cos\frac{n\pi}{2} + c_2\sin\frac{n\pi}{2}\right)$$

The particular integral

$$a_n^p = \frac{1}{E^2 + k^2}\cos kn$$

$$= \text{Real part}\left\{\frac{1}{E^2 + k^2}e^{ikn}\right\}$$

$$= \text{Real part}\frac{1}{\left(e^{ik}\right)^2 + k^2}e^{ikn}$$

$$= \text{Real part}\frac{e^{ikn}}{k^2 + e^{2ik}} \times \frac{k^2 + e^{-2ik}}{k^2 + e^{-2ik}}$$

$$= \text{Real part}\frac{k^2 e^{ikn} + e^{ik(n-2)}}{1 + k^2\left(e^{2ik} + e^{-2ik}\right) + k^4}$$

$$= \text{Real part}\frac{k^2\left(\cos kn + i\sin kn\right) + \cos(n-2)k + i\sin(n-2)k}{1 + 2k^2\cos 2k + k^4}$$

$$= \frac{k^2\cos kn + \cos(n-2)k}{1 + 2k^2\cos 2k + k^4}$$

Hence, the required solution is

$$a_n = a_n^h + a_n^p = k^n\left(c_1\cos\frac{n\pi}{2} + c_2\sin\frac{n\pi}{2}\right) + \frac{k^2\cos kn + \cos(n-2)k}{1 + 2k^2\cos 2k + k^4}$$

Example 2 Solve $a_{n+2} + k^2 a_n = \sin kn$

Proceeding as in Example 1, the homogeneous solution is

$$a_n^h = k^n\left(c_1\cos\frac{n\pi}{2} + c_2\sin\frac{n\pi}{2}\right)$$

The particular integral = Imaginary part of $\dfrac{1}{E^2 + k^2} e^{ikn}$

$$= \text{Imaginary part of } \frac{k^2 e^{ikn} + e^{ik(n-2)}}{1 + 2k^2 \cos 2k + k^4}$$

$$= \frac{k^2 \sin kn + \sin(n-2)k}{1 + 2k^2 \cos 2k + k^4}$$

Hence, the required solution is

$$a_n = a_n^h + a_n^p = k^n \left(c_1 \cos \frac{n\pi}{2} + c_2 \sin \frac{n\pi}{2} \right) + \frac{k^2 \sin kn + \sin(n-2)k}{1 + 2k^2 \cos 2k + k^4}$$

6.5 METHOD OF GENERATING FUNCTIONS

The generating functions provide an elegant and powerful method to solve recurrence relations. In this method, the sequences are represented as coefficients of various powers of a variable in a power series. Generating functions transform recurrence relation into an equation whose solution leads to some closed form for the generating function. With the help of this closed form, coefficients of various powers of the variable are known, which is the required solution of the original difference equation.

Definition 6.3 Let $a_0, a_1, a_2, \ldots, a_n, \ldots$ be an infinite sequence, then the power series denoted by

$$G(z) = a_0 + a_1 z + a_2 z^2 + \ldots + a_n z^n + \ldots \tag{6.20}$$

is called the generating function for the sequence $\{a_n\}$. If all the coefficients are 0 after some terms a_k onwards, then $G(z)$ is a polynomial of degree k.

It may be noted that there is an obvious correspondence between generating functions and sequences, that is,

$$a_0 + a_1 z + a_2 z^2 + \ldots + a_n z^n + \ldots \leftrightarrow a_0, a_1, a_2, \ldots, a_n, \ldots$$

That is, given any sequence, we can write the corresponding generating function and vice versa. The variable z is simply used as a device for locating the coefficients, which are really of interest.

The generating function defined by (6.20) is called an ordinary generating function to distinguish from other types of generating functions such as exponential and binomial generating functions.

6.5.1 Properties of Generating Functions

1. Let $\{a_n\}$ and $\{b_n\}$ be two sequences with generating functions $A(z)$ and $B(z)$, respectively. Let $\{c_n\}$ be a sequence, where $c_n = a_n + b_n$, then the generating function $C(z)$ of the sequence $\{c_n\}$ is defined by,

$$C(z) = A(z) + B(z)$$

Example 1 Let $a_n = 3^n$ and $b_n = 5^n$ and $c_n = a_n + b_n$

Solution: The generating function of $\{c_n\}$ is

$$C(z) = A(z) + B(z) = \sum_{n=0}^{\infty} 3^n z^n + \sum_{n=0}^{\infty} 5^n z^n$$

$$= (1 + 3z + (3z)^2 + \ldots) + (1 + 5z + (5z)^2 + \ldots)$$

$$= \frac{1}{1-3z} + \frac{1}{1-5z}$$

$$= \frac{2-8z}{1-8z+15z^2}$$

2. Convolution: The convolution of $\{a_n\}$ and $\{b_n\}$ denoted by,

$$c_n = a_n * b_n$$

is defined by,

$$c_n = a_0 b_n + a_1 b_{n-1} + a_2 b_{n-2} + \ldots + a_k b_{n-k} + \ldots + a_n b_0$$

is the coefficient of z^n in the product

$(a_0 + a_1 z + a_2 z^2 + \ldots + a_{n-1} z^{n-1} + a_n z^n + \ldots)(b_0 + b_1 z + b_2 z^2 + \ldots + b_{n-1} z^{n-1} + b_n z^n + \ldots)$

Thus,

$$C(z) = A(z)\, B(z)$$

Convolution is an associative and commutative operation. It is distributive with respect to addition.

3. If $A(z)$ is the generating function for the sequence $\{a_n\}$, then $\alpha A(z)$ is the generating function for the sequence $\{\alpha a_n\}$.

Example 2 If $a_n = 3^n$ and its generating function is $A(z)$, find the generating function $B(z)$, where $b_n = 5.3^n$.

Solution: $A(z) = \sum_{n=0}^{\infty} a_n z^n = \sum_{n=0}^{\infty} 3^n z^n = 1 + 3z + 3z^2 + \ldots = (1-3z)^{-1}$

Now,

$$B(z) = \sum_{n=0}^{\infty} b_n z^n = \sum_{n=0}^{\infty} 5.3^n z^n = 5\sum_{n=0}^{\infty} 3^n z^n = 5A(z)$$

Some Illustrative Examples

Example 1 Find the generating function of the sequence 1, 1, 1, 1,....

Solution: Let the given sequence be $\{a_n\}$, then

$$a_0 = 1, a_1 = 1, a_2 = 1, \ldots$$

Hence, the generating function is

$$G(z) = a_0 + a_1 z + a_2 z^2 + \dots$$
$$= 1 + z + z^2 + \dots = (1 - z)^{-1}$$

Example 2 Find the generating function of the sequence 1, –1, 1, –1, 1,...

Solution: Let the given sequence be $\{a_n\}$, then

$$a_0 = 1, a_1 = -1, a_2 = 1, a_3 = -1, \dots$$

Hence, the generating function is

$$G(z) = \sum_{n=0}^{\infty} a_n z^n = a_0 + a_1 z + a_2 z^2 + a_3 z^3 + \dots$$
$$= 1 - z + z^2 - z^3 \dots = (1 + z)^{-1}$$

Example 3 Find the generating function of the sequence 1, 1, 1, 1, 1.

Solution: Let the given sequence be $\{a_n\}$, then

$$a_0 = 1, a_1 = 1, a_2 = 1, a_3 = 1, a_4 = 1, a_5 = 0, a_6 = 0 \text{ and all other terms are } 0.$$

Hence, the generating function is

$$G(z) = a_0 + a_1 z + a_2 z^2 + a_3 z^3 + a_4 z^4$$
$$= 1 + z + z^2 + z^3 + z^4$$

Example 4 Find the generating function of the sequence 1, 2, 3,..., *n*,....

Solution: Let the given sequence be $\{a_n\}$, then

$$a_0 = 1, a_1 = 2, a_2 = 3, \dots$$

Hence, the generating function is

$$G(z) = \sum_{n=0}^{\infty} a_n z^n = 1 + 2z + 3z^2 + \dots + nz^{n-1} + \dots$$
$$= (1 - z)^{-2}$$

Example 5 Find the generating function of the sequence 0, 1, 2, 3,...

Solution: Let the given sequence be $\{a_n\}$, then

$$a_0 = 0, a_1 = 1, a_2 = 2, a_3 = 3, \dots$$

Hence, the generating function is

$$G(z) = 0 + 1.z + 2z^2 + 3z^3 + \dots$$
$$= z(1 + 2z + 3z^2 + \dots) = z(1 - z)^{-2}$$

Some Generating Functions

	Sequence $\{a_n\}$	Generating Function G(z)
1.	$a_n = 1$ for all n	$(1-z)^{-1}$
2.	$a_n = n+1$	$(1-z)^{-2}$
3.	$a_n = \dfrac{1}{n!}$	e^z
4.	${}^nc_k p^k q^{n-k}$ where $k = 0, 1, 2, \ldots, n$ and $p+q=1$	$(q+pz)^n$
5.	$1, 3, 3^2, \ldots, 3^n, \ldots$	$(1-3z)^{-1}$
6.	$1, 2, 4, 8, 16, \ldots$	$(1-2z)^{-1}$

6.5.2 Closed Form for Generating Function

Consider the infinite sequence

$1, \alpha, \alpha^2, \alpha^3, \ldots,$ where α is any constant.

The generating function for the sequence is

$$G(z) = 1 + \alpha z + \alpha^2 z^2 + \alpha^3 z^3 + \ldots$$
$$= (1 - \alpha z)^{-1}$$

which is the required closed form.

THEOREM 6.2 If the generating function for the sequence whose general term a_n is known, then the generating function for a_{n+1} and a_{n+2} can also be obtained.

Proof: Let $G(z)$ be the generating function for the sequence whose general term is a_n, then we can write

$$G(z) = a_0 + a_1 z + a_2 z^2 + \ldots + a_n z^n + a_{n+1} z^{n+1} + \ldots$$

$$\therefore \quad \frac{G(z) - a_0}{z} = a_1 + a_2 z + a_3 z^2 + \ldots + a_{n+1} z^n + \ldots$$

which shows that

$\dfrac{G(z) - a_0}{z}$ is the generating function for the sequence whose general term is a_{n+1}.

Similarly, $\dfrac{G(z) - a_0 - a_1 z}{z^2} = a_2 + a_3 z + a_4 z^2 + \ldots + a_{n+2} z^n + \ldots$ is the generating function for the sequence whose general term is a_{n+2}.

Proceeding in this manner

$$\frac{G(z) - a_0 - a_1 z - a_2 z^2 - \ldots - a_{k-1} z^{k-1}}{z^k}$$

generates the sequence

$$a_k + a_{k+1}z + a_{k+2}z^2 + \ldots + a_n z^{n-k} + \ldots$$

$$= \sum_{n=k}^{\infty} a_n z^{n-k}$$

$$= \sum_{n=0}^{\infty} a_{n+k} z^n$$

which shows that the general term is a_{n+k}.

Some Problems Solved by Generating Function

Example 1 Solve the recurrence relation

$$a_n = 3a_{n-1} - 2a_{n-2}, n \geq 2$$

by generating function method given $a_0 = 2$ and $a_1 = 3$

Solution: Multiplying the given recurrence relation by z^n and summing for $n = 2$ to ∞, we have

$$\sum_{n=2}^{\infty} a_n z^n - 3\sum_{n=2}^{\infty} a_{n-1} z^n + 2\sum_{n=2}^{\infty} a_{n-2} z^n = 0$$

or $\quad (G(z) - a_0 - a_1 z) - 3z(G(z) - a_0) + 2z^2 G(z) = 0$

Substituting for a_0 and a_1

$$(2z^2 - 3z + 1)\,G(z) - 2 - 3z + 6z = 0$$

$$\therefore G(z) = \frac{2-3z}{2z^2 - 3z + 1} = \frac{2-3z}{(1-z)(1-2z)} = \frac{1}{1-z} + \frac{1}{1-2z}$$

or $\quad G(z) = \sum_{n=0}^{\infty} 1.z^n + \sum_{n=0}^{\infty} 2^n z^n = \sum_{n=0}^{\infty} (1 + 2^n) z^n$

Thus,

$$a_n = 1 + 2^n$$

Example 2 Solve the recurrence relation $a_{n+2} - 2a_{n+1} + a_n = 2^n$, given $a_0 = 2$, $a_1 = 1$, by the method of generating functions.

Solution: Multiplying the given recurrence relation by z^n and summing for $n = 0$ to ∞, we get

$$\sum_{n=0}^{\infty} a_{n+2} z^n - 2\sum_{n=0}^{\infty} a_{n+1} z^n + \sum_{n=0}^{\infty} a_n z^n = \sum_{n=0}^{\infty} 2^n z^n$$

or $\quad \frac{1}{z^2}[G(z) - a_0 - a_1 z] - \frac{2}{z}[G(z) - a_0] + G(z) = \frac{1}{1-2z}$

Substituting for a_0 and a_1 and multiplying by z^2

$$G(z)-2-z-2z[G(z)-2] + z^2G(z) = \frac{z^2}{1-2z}$$

$$\text{or} \quad (z^2 - 2z + 1)\, G(z) = 2 - 3z + \frac{z^2}{1-2z}$$

$$\text{or} \quad G(z) = \frac{2-3z}{(1-z)^2} + \frac{z^2}{(1-z)^2(1-2z)}$$

$$= \frac{2}{(1-z)^2} - 3.\frac{z}{(1-z)^2} + \frac{1}{1-2z} - \frac{1}{(1-z)^2}$$

$$\text{or} \quad G(z) = \frac{1}{(1-z)^2} - 3.\frac{z}{(1-z)^2} - \frac{1}{(1-2z)}$$

$$\text{or} \quad G(z) = \sum_{n=0}^{\infty} a_n z^n = \sum_{n=0}^{\infty}(n+1)z^n - 3\sum_{n=0}^{\infty} nz^n + \sum_{n=0}^{\infty} 2^n z^n$$

Hence, the required solution is $a_n = (n + 1) - 3n + 2^n = 1 - 2n + 2^n$.

Example 3 Find the generating function for the Fibonacci sequence $\{a_n\}$ defined by,

$$a_n = a_{n-1} + a_{n-2},\ n \geq 2, \quad \text{and} \quad a_0 = 1,\ a_1 = 1$$

Solution: Let $G(z)$ be the generating function, that is, $G(z) = \sum_{n=0}^{\infty} a_n z^n$

Multiplying the given recurrence relation by z^n and summing up for $n = 2$ to $n = \infty$

$$\sum_{n=2}^{\infty} a_n z^n = \sum_{n=2}^{\infty} a_{n-1} z^n + \sum_{n=2}^{\infty} a_{n-2} z^n$$

$$\text{or} \quad G(z) - a_0 - a_1 z = z\{G(z) - a_0\} + z^2G(z)$$

Substituting for a_0 and a_1

$$G(z) - 1 - z - z\{G(z) - 1\} - z^2G(z) = 0$$

$$\text{or} \quad (1 - z - z^2)\, G(z) = 1$$

$$\text{or} \quad G(z) = \frac{1}{1-z-z^2}$$

Example 4 Solve the recurrence relation $a_n = 3a_{n-1} - 2a_{n-2}$, $n \geq 2$ by generating function given $a_0 = 5$ and $a_1 = 3$.

Solution: Let $G(z)$ be the generating function, that is, $G(z) = \sum_{n=0}^{\infty} a_n z^n$, multiplying the given recurrence relation by z^n and summing up from $n = 2$ to $n = \infty$, we get

$$\sum_{n=2}^{\infty} a_n z^n - 3\sum_{n=2}^{\infty} a_{n-1} z^n + 2\sum_{n=2}^{\infty} a_{n-2} z^n = 0$$

$$\text{or} \quad G(z) - a_0 - a_1 z - 3z\{G(z) - a_0\} + 2z^2 G(z) = 0$$

Substituting for a_0 and a_1

$$G(z) - 5 - 3z - 3z\{G(z) - 5\} + 2z^2 G(z) = 0$$

$$\text{or} \quad (1 - 3z + 2z^2)\, G(z) = 5 + 3z - 15z = 5 - 12z$$

$$G(z)\frac{5-12z}{1-3z+2z^2} = \frac{5-12z}{(1-z)(1-2z)} = \frac{7}{1-z} - 2.\frac{1}{1-2z}$$

$$\text{or} \quad G(z) = 7(1 - z)^{-1} - 2(1 - 2z)^{-1}$$

Hence,

$$a_n = 7.1^n - 2.2^n = 7.1^n - 2^{n+1}$$

Example 5 Find the generating function for the following sequences:

(1) 1, 0, 1, 0, 1,…
(2) 0, 0, 3, 3, 3,…
(3) 1, −2, 4, −2,…
(4) 5, −5, 5, −5,…
(5) 1, a, a^2, a^3,…
(6) 0^2, 1^2, 2^2, 3^3,…

Solution:

(1) The required generating function is

$$G(z) = 1 + 0z + z^2 + 0z^3 + z^4 + \ldots$$

$$= 1 + z^2 + z^4 + \ldots = (1 - z^2)^{-1}$$

$$= \frac{1}{1-z^2}$$

(2) The required generating function is

$$G(z) = 0 + 0z + 3z^2 + 3z^3 + 3z^4 + \ldots$$

$$= 3z^2 (1 + z + z^2 + \ldots) = 3z^2 (1 - z)^{-1}$$

$$= \frac{3z^2}{1-z}$$

(3) The required generating function is

$$G(z) = 1 - 2z + 4z^2 - 8z^2 + \dots$$
$$= 1 - 2z + (2z)^2 = \frac{1}{1+2z}$$

(4) The required generating function is

$$G(z) = 5 - 5z + 5z^2 - 5z^3 + \dots$$
$$= 5(1 - z + z^2 - z^3 + \dots)$$
$$= 5(1 + z)^{-1} = \frac{5}{1+z}$$

(5) The required generating function is $G(z) = 1 + az + a^2z^2 + a^3z^3 + \dots = (1 - az)^{-1}$

(6) The required generating function is

$$G(z) = 0^2 + 1^2z + 2^2z^2 + 3^2z^3 + 4^2z^4 + \dots$$
$$= z + 4z^2 + 9z^3 + 16z^4 \dots$$
$$= z\,(1 + 4z + 9z^2 + 16z^3 \dots)$$
$$= z\,(1 + z)\,(1 - z)^{-3}$$

6.5.3 Combinatorial Method

Generating functions provide a convenient method for solving many counting problems. An important class of problems is to count the number of solutions of the equation of the form

$$e_1 + e_2 + \dots + e_n = r \tag{6.21}$$

where r is a constant and e_i are non-negative integers which may be subjected to some constraints.

Example 1 Find the number of solutions of equations (6.21)
Where

$$0 \le e_i \le 1,\ i = 1, 2, \dots, n.$$

Solution: Since e_1 can be 0 or 1, the polynomial $1 + z$ can be used to represent the possibilities and hence $A_1(z) = 1 + z$ can be taken as its generating function.

Similarly, $A_2(z) = 1 + z$ can be taken as the generating function for e_2 and so on.

The product $A_1(z)\,A_2(z) \dots A_n(z) = (1 + z)^n = \sum_{r=0}^{n} {}^nc_r z^r$

The coefficient of z^r, that is, ${}^nc_r$ is the number of solution of equation (6.21), that is, there are ${}^nc_r$ solutions of (6.21) satisfying the constraints $0 \le e_i \le 1$.

Example 2 Find the generating function for a_r, the number of ways of selecting r object, from n objects with any number of repetitions.

Solution: Since each object can be selected any number of times, the generating functions is

$$G(z) = (1 + z + z^2 + \ldots)^n$$

where each term $1 + z + z^2 + \ldots$ corresponds to the number of times the object is selected. To find a_r, we find the coefficient of z^r in $G(z)$.

$$G(z) = \left(\frac{1}{1-z}\right)^n = (1-z)^{-n} = \sum_{r=0} a_r z^r$$

Hence,

$$a_r = \frac{(-1)^r(-n)(-n-1)(-n-2)\ldots(-n-r+1)}{r!}$$

$$= \frac{n(n+1)(n+2)\ldots(n+r-1)}{r!} = {}^{n+r-1}c_r$$

∴ This example demonstrates a practical situation where equation of the form (6.21) given above occurs.

Example 3 A coffee house can serve three types of coffee to its customers. Find the number of possible orders by five customers.

Solution: Let i, j, and k denote the customers, then every order can be represented by,

$$i + j + k = 5$$

where i, j, k are non-negative integers.

Let z^i denote that cold coffee was ordered by i customers. Since i can take values 0, 1, 2, 3, 4, and 5, the polynomial $1 + z + z^2 + z^3 + z^4 + z^5$ represents various possibilities. Similarly, $1 + y + y^2 + y^3 + y^4 + y^5$ and $1 + x + x^2 + x^3 + x^4 + x^5$ represent various possibilities of hot coffee and espresso coffee. Thus, we want the products of the form $z^i y^j x^k$, where $i + j + k = 5$ in $(1 + z + z^2 + z^3 + z^4 + z^5)(1 + y + y^2 + y^3 + y^4 + y^5)(1 + x + x^2 + x^3 + x^4 + x^5)$

Taking $x = y = z$ for the sake of convenience, we want the coefficient of x^5 in the product

$$(1 + x + x^2 + x^3 + x^4 + x^5)^3$$

$$= \left(\frac{1-x^6}{1-x}\right)^3 = \frac{(1-x^6)^3}{(1-x)^3} = (1-x^6)^{-3}(1-x)^{-3}$$

$$= (1-x^6)^3 \sum_{r=0} {}^{3+r-1}c_r x^r$$

In the factor $(1 - x^6)^3$, each term is of degree greater than 5 except the first term, which is 1.

Hence, for x^5, $r = 5$, and the coefficient is ${}^{3+5-1}c_5 = {}^7c_5 = {}^7c_2 = 21$.

Hence, there are 21 solutions.

THEOREM 6.3 The number of all possible arrangements (permutations) of n objects, of which n_1 are of first type, n_2 are of second type,…, n_k items of kth type, is

$$\frac{n!}{n_1!\, n_2! \ldots n_k!}$$

where $n_1 + n_2 + \ldots + n_k = n$

Example 4 Use generating function to find the number of ways to select 8 balls from a box containing 3 green, 3 red, and 3 black balls.

Solution: Let x_1, x_2, and x_3 represent green, red, and black balls. Consider the number of integer solutions of the equation

$$x_1 + x_2 + x_3 = 8$$

The required solution is the coefficient of x^8 of the generating function

$$(1 + x + x^2 + x^3 + \ldots)^3$$

$$= (1 - x)^{-3} = \sum_{r=0} {}^{3+r-1}c_r\, x^r$$

The coefficient of x^8 corresponds to $r = 8$. The coefficient is

$${}^{3+8-1}c_8 = {}^{10}c_8 = {}^{10}c_2 = \frac{\overset{5}{\not{10}} \times 9}{\not{2}} = 45$$

Example 5 Use generating function to find the number of ways of selecting 12 balls from a box containing red, blue, black, and white balls in such a way that one can select

(1) at least 2 balls of each colour
(2) at most 2 white balls
(3) an even number of red balls

Solution: Let x_1, x_2, x_3, and x_4 represent the number of red, blue, black, and white balls. The number of integer solutions should satisfy the equation

$$x_1 + x_2 + x_3 + x_4 = 12$$

(1) For each $x_i \geq 2$, the generating function is

$$(x^2 + x^3 + x^4 + \ldots\ldots)^4$$

$$= x^8(1 + x + x^2 + \ldots)^4 = x^8(1-x)^{-4} = \sum_{r=0} {}^{4+r-1}c_r\, x^{8+r}$$

For the coefficient x^{12}, $r = 4$

$\therefore$ Required number of ways $= 4 + 4 - 1c_4 = 7c_4 = 7c_3 = 35$.

(2) The required number of ways is the coefficient of x^{12} in the generating function

$$(1 + x + x^2)(1 + x + x^2 + x^3 + \ldots)^3$$

$$= (1 + x + x^2)(1-x)^{-3} = (1 + x + x^2)\sum_{r=0} {}^{3+r-1}c_r\, x^r$$

$$= \sum_{r=0} {}^{3+r-1}c_r\, x^r + \sum_{r=0} {}^{3+r-1}c_r\, x^{r+1} + \sum_{r=0} {}^{3+r-1}c_r\, x^{r+2}$$

The coefficient of x^{12} is

$$^{3+12-1}c_{12} + {}^{3+11-1}c_{11} + {}^{3+10-1}c_{10}$$
$$= {}^{14}c_{12} + {}^{13}c_{11} + {}^{12}c_{10} = {}^{14}c_2 + {}^{13}c_2 + {}^{12}c_2$$
$$= \frac{{}^{7}\not{1}4\times 13}{2!} + \frac{13\times \not{1}2^{6}}{2!} + \frac{{}^{6}\not{1}2\times 11}{2!} = 91 + 78 + 66 = 235$$

(3) The required number of ways is the coefficient of x^{12} in the generating function

$$(1 + x^2 + x^4 + x^6 + \ldots)(1 + x + x^2 + \ldots)^3$$
$$= (1 + x^2 + x^4 + x^6 + \ldots)(1-x)^{-3}$$
$$= (1 + x^2 + x^4 + x^6 + x^8 + x^{10} + x^{12} +) \sum_{r=0} {}^{3+r-1}c_r\, x^r$$
$$= \sum_{r=0} {}^{3+r-1}c_r\, x^r + \sum_{r=0} {}^{3+r-1}c_r\, x^{r+2} + \sum_{r=0} {}^{3+r-1}c_r\, x^{r+4} + \sum_{r=0} {}^{3+r-1}c_r\, x^{r+6}$$
$$+ \sum_{r=0} {}^{3+r-1}c_r\, x^{r+8} + \sum_{r=0} {}^{3+r-1}c_r\, x^{r+10} + \sum_{r=0} {}^{3+r-1}c_r\, x^{r+12}$$

$\therefore$ The coefficient of x^{12} is

$$^{3+12-1}c_{12} + {}^{3+10-1}c_{10} + {}^{3+8-1}c_8 + {}^{3+6-1}c_6 + {}^{3+4-1}c_4 + {}^{3+0-1}c_0$$
$$= {}^{14}c_{12} + {}^{12}c_{10} + {}^{10}c_8 + {}^{8}c_6 + {}^{6}c_4 + {}^{4}c_2 + {}^{2}c_0$$
$$= {}^{14}c_2 + {}^{12}c_2 + {}^{10}c_2 + {}^{8}c_2 + {}^{6}c_2 + {}^{4}c_2 + 1$$
$$= \frac{{}^{7}\not{14}\times 13}{\not{2}!} + \frac{{}^{6}\not{12}\times 11}{\not{2}!} + \frac{{}^{5}\not{10}\times 9}{\not{2}!} + \frac{{}^{4}\not{8}\times 7}{\not{2}!} + \frac{{}^{3}\not{6}\times 5^2}{\not{2}!} + \frac{{}^{2}\not{4}\times 3}{\not{2}!} + 1$$
$$= 91 + 66 + 45 + 28 + 15 + 6 + 1$$
$$= 252$$

Example 6 Find the number of solutions of the equation

$$x_1 + x_2 + x_3 = 5$$

where x_1, x_2, x_3 are non-negative integer variables.

Solution: The required solution is the coefficient of x^5 in the generating function

$$(1 + x + x^2 + \ldots)^3$$
$$= (1-x)^{-3} = \sum_{r=0} {}^{3+r-1}c_r\, x^r$$

For x^5, $r = 5$, and so the number of solution is $^{3+5-1}c_5 = {}^{7}c_5 = {}^{7}c_2 = 21$.

Alternate solution Here, $r = 5$ and $n = 3$.

Hence, the number of solutions is $^{n+r-1}c_r = {}^{5+3-1}c_5 = {}^{7}c_5 = {}^{7}c_2 = 21$.

Example 7 Find the number of non-negative integer solutions of the equations

(1) $x_1 + x_2 + x_3 = 3$
(2) $x_1 + x_2 + x_3 + x_4 = 10$

Solution:

(1) $x_1 + x_2 + x_3 = 3$

Here, $n = 3$ and $r = 3$

Hence, the number of non-negative integer solution is

$$^{n+r-1}c_3 = {}^{3+3-1}c_3 = {}^5c_2 = {}^5c_2 = 10.$$

(2) $x_1 + x_2 + x_3 + x_4 = 10$

Here, $r = 10$ and $n = 4$

Hence, the number of non-negative integer solutions is

$$^{n+r-1}c_r = {}^{10+4-1}c_{10} = {}^{13}c_3 = \frac{13 \times \cancel{12}^{2} \times 11}{1.\cancel{2}.\cancel{3}} = 286$$

Example 8 Find the generating function for a_r, that is, the number of non-negative integer solutions of

$$e_1 + e_2 + e_3 + e_4 = r$$

where

$$0 \le e_1 \le 3,\ 0 \le e_2 \le 3,\ 2 \le e_3 \le 5,\ 2 \le e_4 \le 5$$

Solution: Let

$$A_1(x) = 1 + x + x^2 + x^3 = A_2(x)$$
$$A_3(x) = x^2 + x^3 + x^4 + x^5 = A_4(x)$$

The required generating function is

$$A_1(x)A_2(x)A_3(x)A_4(x) = (1 + x + x^2 + x^3)^2(x^2 + x^3 + x^4 + x^5)^2$$

	Cold coffee	Hot coffee	Expresso
1	0	0	5
2	0	1	4
3	0	2	3
4	0	3	2
5	0	4	1

(Continued)

(Continued)

	Cold coffee	Hot coffee	Expresso
6	0	5	0
7	1	0	4
*8	1	1	3
*9	1	2	2
*10	1	3	1
11	1	4	0
12	2	0	3
*13	2	1	2
*14	2	2	1
15	2	3	0
16	3	0	2
*17	3	1	1
18	3	2	0
19	4	1	0
20	4	0	1
21	5	0	0

Note: The number of positive integer solution is six marked *.

6.5.4 Linear Diophantine Equations

Here, we are interested in finding out the number of positive integer solutions of the equation of the type

$$x_1 + x_2 + \ldots + x_n = m \tag{6.22}$$

THEOREM 6.4 The equation

$$x_1 + x_2 + \ldots + x_n = m \; (n \geq m)$$

has exactly ${}^{m-1}c_{n-1}$ positive integer solutions

Proof: The equation

$$x_1 + x_2 + \ldots + x_n = m$$

contains $(n - 1)$ plus signs between variables. If the sum of the variables is m, it will have $(m - 1)$ plus signs. Thus, we have to choose $(n - 1)$ out of $(m - 1)$ plus signs, which is possible in ${}^{m-1}c_{n-1}$ ways.

Thus, the number of positive integer solutions are ${}^{m-1}c_{n-1.}$

Example 1 Find the positive integer solution of the equation $x + y + z = 5$.

Solution: Here, the number of variables is 3, that is, $n = 3$ and $m = 5$.

Hence, the number of positive solutions is ${}^{5-1}c_{3-1}$ or ${}^{4}c_2 = 6$.

Example 2 Find the number of positive integer solutions of the equation $x_1 + x_2 + x_3 + x_4 = 10$.

Solution: Here, the number of variables is 4, that is, $n = 4$ and $m = 10$

Hence, the number of positive integer solutions is ${}^{10-1}C_{4-1}$ or ${}^{9}C_3$, i.e., $\frac{9 \times 8 \times 7}{1.2.3} = 84$.

Note 6.4 Sometimes there is some restriction on the variables.

Example 3 Find the number of positive integer solutions of the equation

$$x + y + z = 8 \text{ where } x > 2, y > 0, z > 0$$

Solution: Here we first convert the problem into positive integer conditions. Since $x > 2$, $x - 2 > 0$ and hence let $u = x - 2 > 0$.

Thus, the problem is reduced to find the number of positive integer solutions of the equation

$$u + 2 + y + z = 8$$
$$\text{or} \quad u + y + z = 6,\ u > 0, y > 0, z > 0$$

As the number of variables is 3, that is, $n = 3$, and $m = 6$.

The number of positive integer solution is ${}^{6-1}c_{3-1}$ or ${}^{5}c_2 = \frac{5 \times 4}{2} = 10$

EXERCISES

1. Find the general term a_n for the following sequences:
   (i) $1, -1, 1, -1, \ldots$
   (ii) $1, \frac{1}{2}, \frac{1}{3}, \frac{1}{4}, \ldots$
   (iii) $1, 2, 3, 4, \ldots$
   (iv) $1, 3, 3^2, 3^3, \ldots$
   (v) $1, 0, 1, 0, \ldots$

2. Find the generating function for the following sequences:
   (i) $1, -1, 1, -1, \ldots$
   (ii) $1, 0, 1, 0, 1, \ldots$
   (iii) $1, 1, 1, 1, \ldots$
   (iv) $1, 1, \frac{1}{2!}, \frac{1}{3!}, \ldots$

3. Which of the following recurrence relations are linear homogeneous. If the relation is linear homogeneous, given its degree.
   (i) $a_n = 2a_{n-1}$
   (ii) $a_n = 2a_{n-1} - 5a_{n-2}$
   (iii) $b_n = 5^n b_{n-1}$
   (iv) $b_n = 2b_{n-1} + 10$
   (v) $c_n = \sqrt{c_{n-1} + c_{n-2}}$

4. Use backtracking method for finding explicit formula for the following recurrence relations:
   (i) $a_n = 3.5\, a_{n-1}$, $a_1 = 0$
   (ii) $a_n = 5a_{n-1} + 1$, $a_1 = 2$
5. Solve the following linear homogeneous recurrence relation by using characteristic root method:
   (i) $a_n = 10a_{n-1}$
   (ii) $a_n = 2a_{n-1} + 3a_{n-2}$
   (iii) $b_n = 3b_{n-1} + 10b_{n-2}$, given $b_0 = 5$, $b_1 = 11$
   (iv) $a_n = a_{n-1} + a_{n-2}$, $n \geq 2$, $a_1 = 1$ and $a_2 = 1$
   (v) $4a_n - 4a_{n-1} + a_{n-2} = 0$, given $a_0 = 1$ and $a_1 = 2$
   (vi) $a_{n+2} = 3a_{n+1} - 2a_n$, given $a_0 = 5$, $a_1 = 8$
6. Solve the following non-homogeneous linear recurrences relations:
   (i) $a_{n+2} - 5a_{n+1} + 6a_n = 5^n + 2n$
   (ii) $a_n - 10a_{n-1} + 25a_{n-2} = 16(n^2 + n)$
   (iii) $a_{n+2} - 2a_{n+1} + a_n = n^2.2^n$
   (iv) $a_{n+2} - 4a_{n+1} + 4a_n = 2^n$
7. Using Bool's operational method, solve the following recurrence relations:
   (i) $3a_{n+1} - 2a_n = 15$
   (ii) $a_{n+2} - 5a_{n+1} + 6a_n = 5^n$
   (iii) $4a_{n+2} - 4a_{n+1} + a_n = n^2 + 5n$
   (iv) $a_{n+2} - 6a_{n+1} + 9a_n = n^2 2^n$
8. Solve the recurrence relations by generating function.
   (i) $a_{n+1} - 3a_n = 2^n$ with $a_0 = 5$
   (ii) $a_{n+2} - 5a_{n+1} + 6a_n = 5$ with $a_0 = a_1 = 1$
9. Apply generating function method to solve the following non-homogeneous recurrence relations:
   (i) $a_n + 2a_{n-1} = 2n + 3$, given $a_0 = 3$
   (ii) $a_n = 5a_{n-1} + 7^{n-1}$ with $a_0 = 1$
   (iii) $a_{n+2} - 3a_{n+1} + 2a_n = 3^n$ given $a_0 = 1$, $a_1 = 3$
10. Find the number of positive integer solution of the linear Diophantine equation

$$x_1 + x_2 + x_3 = 7\; x_1 > 0, x_2 > 0, x_3 > 0$$

11. Find the number of positive integer solutions of the linear Diophantine equation $x_1 + x_2 + x_3 = 11$

$$\text{where } x_1 > 1, x_2 > 1 \quad \text{and} \quad x_3 > 1.$$

12. Find the number of non-negative integer solutions of the equation

$$x_1 + x_2 + x_3 + x_4 = 10$$

7

Algebraic Structures

LEARNING OBJECTIVES

After reading this chapter, you will be conversant with:

- Introduction
- Binary Operation
- Algebraic Structures
- Congruences
- Permutations
- Integral Powers of an Element
- Cyclic Group
- Subgroups
- Coset Decomposition
- Isomorphism and Homomorphism of Groups
- Algebraic Systems With Two Binary Operations
- Ring, Subring and Ideals
- Integral Domain
- Field

7.1 INTRODUCTION

Algebraic structures play an important role in computer science due to their applications in computing techniques dealing with problems in operations research, graph theory, and relational databases. They also have found applications in the study of crystals in physics and in biological science in the analysis of human genes. The important structures that are discussed in this chapter are semi-groups, monoids, groups, subgroups, homomorphism and isomorphism of groups, rings, integral domains, and fields.

7.2 BINARY OPERATION

If we add two integers a and b, we get another integer denoted by $a + b$; if we subtract b from a, we get another integer denoted by $a - b$ and if we multiply a and b, we get another integer denoted by $a.b$. In all these three processes, there is one common property involved: given any two integers, their product under the operation is an integer. In this case, we say that the set is closed under the operation or closure property is satisfied or a binary operation is defined over the set.

Definition 7.1 A map or a function $o : S \times S \rightarrow S$ is called a binary operation over the set S, the function o associates each ordered pair (a, b) in $S \times S$ with a unique element $c \in S$. The o-image of (a, b) is denoted by $a \; o \; b$, which is called product of a and b under the binary operation o.

Some Examples

1. Ordinary addition over the set of integers Z is a binary operation. Thus, o: $Z \times Z \rightarrow Z$, where $a \; o \; b = a + b$, a, $b \in Z$.
2. Ordinary subtraction over the set of integers Z is a binary operation where $a \; o \; b = a - b$, a, $b \in Z$.
3. Ordinary multiplication over the set of integers Z is a binary operation where $a \; o \; b = a.b$, a, $b \in Z$.
4. Division over the set of integers Z is not a binary operation.
5. Subtraction over the set of natural numbers N is not a binary operation.
6. Addition, subtraction, and multiplication are binary operations over the set of real numbers R, over the set of complex numbers C and over the set of rational numbers Q.
7. Addition of vectors is a binary operation over the set of vectors V.
8. Vector product is a binary operation over the set of vectors V.
9. Matrix addition is a binary operation over set of $m \times n$ matrices.
10. Matrix multiplication is a binary operation over the set of square matrices of order n.

7.3 ALGEBRAIC STRUCTURES

A set S with one or more binary operations satisfying specific laws (e.g., commutative, associative or distributive) is called an algebraic structure.

Commutative Binary Operation: A binary operation o over a set S is said to be commutative if $a \; o \; b = b \; o \; a$, $\forall a, b \in S$.

1. Ordinary addition and multiplication over the set of integers Z, the set of real numbers R, and the set of complex numbers C are commutative, but subtraction is not a commutative binary operation.
2. Matrix addition is a commutative binary operation over the set of $m \times n$ matrices, but multiplication is not commutative over the set of square matrices of order n.

Associative Binary Operation: A binary operation o over a set S is said to be associative if for every three elements a, b, $c \in S$,

$$(a \; o \; b) \; o \; c = a \; o \; (b \; o \; c)$$

1. Ordinary addition and multiplication over the set of real numbers R are associative while subtraction is not associative.
2. Matrix addition is associative over the set of all matrices of order $m \times n$. Matrix multiplication is associative over the set of square matrices of order n.
3. Vector addition is an associative binary operation over the set of vectors V.

Distributive Laws: Let a set S have two binary operations, for example addition and multiplication denoted by (+) and (.) defined over it. Then multiplication is said to be distributive with respect to addition if

$$a.(b + c) = a.b + a.c \qquad \text{(Left distributive law)}$$

$$(b + c).a = b.a + c.a \qquad \text{(Right distributive law)}$$

Addition is distributive with respect to multiplication if

$$a + (b.c) = (a + b).(a + c) \qquad \text{(Left distributive law)}$$

$$(b.c) + a = (b + a).(c + a) \qquad \text{(Right distributive law)}$$

1. Ordinary multiplication is distributive with respect to addition over the sets R, C, and Q.
2. Scalar product operation is distributive with respect to vector addition over the set of vectors.
3. Matrix multiplication is distributive with respect to matrix addition over the set of square matrices of order n.

Identity Element: Identity element is a neutral element in the sense that when it is combined with any element, its effect on the outcome is nil, that is, the element remains unchanged. In the number system, 0 and 1 are identity elements for addition and multiplication. The zero vector is the identity element for vector addition over the set of vectors.

Definition 7.2 An element e in a set S on which a binary operation o has been defined is called an identity element if

$$a \; o \; e = e \; o \; a = a, \qquad \forall a \in S$$

If $a \; o \; e = a$, then e is called the right identity; and if $e \; o \; a = a$, then e is called the left identity.

Inverse of an Element: Let S be a set on which a binary operation o has been defined. If corresponding to each element $a \in S$, there exists an element $b \in S$ such that

$$a \; o \; b = b \; o \; a = e$$

then b is called the inverse of a and is denoted by a^{-1}.

If $a \; o \; b = e$, then b is called right inverse of a; and if $b \; o \; a = e$, then b is called left inverse of a.

7.3.1 Semi-Group

A semi-group G is a system consisting of a non-empty set G of elements $a, b, c, \ldots$ and a binary operation o satisfying the postulate:

The binary operation o is associative, that is,

$$(a \; o \; b) \; o \; c = a \; o \; (b \; o \; c), \qquad \forall a, b, c \in G$$

Examples

1. The set of natural numbers N is a semi-group with respect to addition.
2. The set of real numbers R is a semi-group with respect to addition.
3. The set of complex numbers C is a semi-group with respect to addition.
4. The set of square matrices of order n is a semi-group with respect to matrix multiplication.
5. The set of vectors is a semi-group with respect to vector addition.
6. The set of rational numbers Q is not a semi-group with respect to binary operation o defined by $a \ o \ b = a^2 + b^2$.

7.3.2 Monoid

A monoid G is a system consisting of a non-empty set G of elements $a, b, c, \ldots$ and a binary operation o satisfying the following postulates:

M1. The binary operation o is associative, that is,

$$(a \ o \ b) \ o \ c = a \ o \ (b \ o \ c) \qquad \forall a, b, c \in G$$

M2. There exists an element e in G called identity element such that

$$e \ o \ a = a \ o \ e = a, \qquad \forall a \in G$$

Example 1 The set of natural numbers N is a monoid with respect to multiplication.

Solution:

(1) The multiplication of two natural numbers is a natural number, and hence multiplication is a binary operation over N.

(2) The multiplication is associative as

$$(ab)c = a(bc), \qquad \forall a, b, c \in N$$

(3) There exists a natural number 1 such that

$$1.\,a = a.1 = a, \qquad \forall a \in N$$

Thus, there exists identity element in N. Hence, the set of natural numbers N is a monoid with respect to multiplication.

Example 2 The system $\{N, +\}$ is not a monoid with respect to addition as it does not contain identity element zero for addition.

Example 3 Let A be the set of people of different heights. Define binary operation $\otimes$ by,

$$a \otimes b = \text{Taller of } a \text{ and } b,$$

Then $\{a, \otimes\}$ is a monoid.

Solution:

(1) The binary operation $\otimes$ is defined as taller of any two persons of the set belonging to the set A.

(2) The binary operation $\otimes$ is associative as $(a \otimes b) \otimes c = a \otimes (b \otimes c) =$ Tallest of a, b, and c.

(3) The identity element is the shortest person in A.

7.3.3 Group

A group G is a system consisting of a non-empty set G of elements $a, b, c, \ldots$ and a binary operation o satisfying the following postulates:

G1. The binary operation o is associative, that is,

$$(a \; o \; b) \; o \; c = a \; o \; (b \; o \; c), \qquad \forall a, b, c \in G$$

G2. There exists an element e in G called left identity such that

$$e \; o \; a = a, \qquad \forall a \in G$$

G3. For each element $a \in G$, there exists an element $a^{-1} \in G$ such that

$$a^{-1} \; o \; a = e$$

If in addition to above postulates

$$a \; o \; b = b \; o \; a, \qquad \forall a, b \in G$$

then the group G is called an abelian group or commutative group.

Finite and Infinite Groups: A group G consisting of a finite number of elements is called a finite group otherwise an infinite group. The number of elements in the group is called the order of the group.

Examples

1. The set of integers $Z = \{0, \pm 1, \pm 2, \ldots\}$ is an abelian group with respect to addition.
2. The set of non-zero integers Z^* is not a group with respect to multiplication as inverse of an integer is not an integer although identity element 1 exists.
3. The set of non-zero rational numbers Q^* is an abelian group with respect to multiplication.
4. The set of complex numbers C is an abelian group with respect to addition.
5. The set of non-zero complex numbers is an abelian group with respect to multiplication.
6. The set of vectors V is an abelian group with respect to vector addition.
7. The set of matrices of order $m \times n$ is an abelian group with respect to matrix addition.
8. The set $\{1, -1, i, -i\}$, that is, fourth roots of unity is an abelian group with respect to multiplication.

THEOREM 7.1 The identity element in a group is unique.

Proof: Let e and e' be two identity elements in a group G. Then we have

$$ee' = e' \text{ if } e \text{ is the identity}$$

and $$ee' = e \text{ if } e' \text{ is the identity}$$

But ee' is a unique element of G. Therefore, $ee' = e = e'$. Hence, the identity element in a group is unique.

General Properties of a Group: Let $\{G, o\}$ be a group. Henceforth, $a \; o \; b$ will be denoted by ab.

THEOREM 7.2 In a group G,

(1) $ab = ac \Rightarrow b = c$ (Left cancellation law)
(2) $ba = ca \Rightarrow b = c$ (Right cancellation law)

Proof: (1) Let $ab = ac$
Therefore,

$$a^{-1}(ab) = a^{-1}(ac) \Rightarrow (a^{-1}a)b = (a^{-1}a)c$$
$$\Rightarrow eb = ec$$
$$\Rightarrow b = c$$

Similarly, the proof of (2) can be done.

THEOREM 7.3 In a group G, the left identity element is also the right identity.

Proof: Let e be the left identity, then $ea = a \quad \forall a \in G$
Let a^{-1} be the left inverse of a, then $a^{-1}a = e$
Now $a^{-1}(ae) = (a^{-1}a)e = ee = e = a^{-1}a$ which shows that $ae = a$ by left cancellation law. Hence, e is the right identity.

THEOREM 7.4 The left inverse of an element is also its right inverse.

Proof: Let a^{-1} be the left inverse of a, and hence $a^{-1}a = e$, where e is the identity element. Now, $a^{-1}(aa^{-1}) = (a^{-1}a)\,a^{-1} = ea^{-1} = a^{-1} = a^{-1}\,e$. Therefore, $aa^{-1} = e$ by left cancellation law.
$\therefore a^{-1}$ is the right inverse of a.

THEOREM 7.5 The inverse of each element in a group is unique.

Proof: Suppose b and c are two inverse elements of a in group G. Let e be the identity element, then

$$ab = ba = e \text{ and } ac = ca = e$$

Now,

$$ab = e = ac \Rightarrow b = c \quad \text{(by left cancellation law)}$$

and so b and c are same. Hence, inverse of an element in a group is unique.

Note 7.1 The identity element is its own inverse since

$$ee = e \text{ and hence } e^{-1} = e$$

THEOREM 7.6 If the inverse of a is a^{-1}, then inverse of a^{-1} is a, that is,

$$(a^{-1})^{-1} = a$$

Proof: If e is the identity element, then

$$a^{-1}a = e$$

Pre-multiplying by $(a^{-1})^{-1}$, we have

$$(a^{-1})^{-1}\,a^{-1}\,a = (a^{-1})^{-1}e = (a^{-1})^{-1}$$

or $$[(a^{-1})^{-1} a^{-1}]\, a = (a^{-1})^{-1}$$

or $$ea = (a^{-1})^{-1} \Rightarrow a = (a^{-1})^{-1}$$

THEOREM 7.7 REVERSAL LAW

Reversal Law: The inverse of the product of any two elements in a group is the product of their inverses in a reversed order.

Proof: Let G be a group and let $a, b \in G$. Let a^{-1} and b^{-1} be their respective inverses.

Now, $$(ab)(b^{-1}a^{-1}) = a(bb^{-1})\, a^{-1} = aea^{-1} = (ae)\, a^{-1} = aa^{-1} = e$$

Similarly, $$(b^{-1}a^{-1})(ab) = b^{-1}(a^{-1}a)\, b = b^{-1}\, eb = b^{-1}(eb) = b^{-1}b = e$$

Hence, $b^{-1}a^{-1}$ is the inverse of ab, that is,

$$(ab)^{-1} = b^{-1}a^{-1}$$

The reversal rule can be generalized

$$(a, b, c, \ldots k)^{-1} = k^{-1} \ldots c^{-1}b^{-1}a^{-1}$$

THEOREM 7.8 A set G with an associative binary operation is a group if the equations

$$ax = b,\ ya = b,\ a, b \in G$$

have unique solutions in G.

Proof: Let $ya = a$ have unique solution $y = e$, then $ea = a$, $a \in G$.
Let c be an arbitrary element in G, then $ax = c$ has unique solution in G.
Now,

$$ec = e(ax) = (ea)x = ax = c, \qquad \forall c \in G$$

Hence, e is the left identity.
The unique solution of

$$ya = e$$

is the left universe of a in G.

Hence, G is a group.

Some Illustrative Examples

Example 1 Show that if $a^2 = a$, then $a = e$, e being the identity element of the group G.

Solution: $a^2 = a \Rightarrow aa = a$
$\Rightarrow aa = ae$
$\Rightarrow a = e$ (by left cancellation law)

Example 2 Find x if $axa = c$ in a group G.

Solution: We are given $axa = c$

$$\Rightarrow a^{-1}(axa)a^{-1} = a^{-1}ca^{-1}$$
$$\Rightarrow (a^{-1}a) \times (aa^{-1}) = a^{-1}ca^{-1}$$
$$\Rightarrow exe = a^{-1}ca^{-1}$$
$$\Rightarrow x = a^{-1}ca^{-1}$$

Example 3 If G is an abelian group, then prove that $(ab)^n = a^n b^n$, $a, b \in G$, where n is any positive integer.

Solution: Let the result be true for $n = k > 0$, then

$$(ab)^k = a^k b^k$$

Now

$$\begin{aligned}(ab)^{k+1} &= (ab)^k (ab)\\ &= (a^k b^k)(ab)\\ &= a^k(b^k a)b\\ &= a^k(ab^k)b\\ &= (a^k a)(b^k b) \qquad (\because G \text{ is abelian } b^k a = ab^k)\\ &= a^{k+1} b^{k+1}\end{aligned}$$

Hence, the result is true for $n = k + 1$.
For $n = 1$, $ab = ab$ (the result is true).
Hence, $(ab)^n = a^n b^n$, where n is a positive integer.

Example 4 If for every element a in a group G, $a^2 = e$ (the identity element), then G is abelian.

Solution: Let a and b be any two elements in the group G, then $ab \in G$. Hence,

$$\begin{aligned}(ab)^2 = e &\Rightarrow (ab)(ab) = e\\ &\Rightarrow ab = (ab)^{-1}\\ &\Rightarrow ab = b^{-1}a^{-1}\end{aligned}$$

But,

$$a^2 = e \Rightarrow aa = a \Rightarrow a = a^{-1}.$$

Similarly,

$$b^2 = e \Rightarrow b = b^{-1}.$$

Thus, we have $ab = ba$ for every a, b in the group G, and hence group G is abelian.

Example 5 Prove that groups of order 1, 2, 3, and 4 are abelian.

Solution: Let G be a group of order 1, then $G = \{e\}$ and we have $ee = e$.
Now let G be a group of order 2, then $G = \{e, a\}$. The multiplication table will have a blank:

0	e	a
e	e	a
a	a	…

The blank can be filled by a or e. Since there can be no repetition of elements in any row or column, the blank space will be filled by e, and hence, we have

0	e	a
e	e	a
a	a	e

The associative and other properties are now satisfied and the group G is abelian.

Next, let $G = \{a, b, e\}$ be a group of order 3.

We have a multiplication table where we shall have four blanks:

0	e	a	b
e	e	a	b
a	a	…	…
b	b	…	…

Now aa can be either e or b. Since aa cannot be identity, $aa = b$ and hence $ab = e$. Further, $bb \neq e$, and hence $bb = a$.

The multiplication table will be

0	e	a	b
e	e	a	b
a	a	b	e
b	b	e	a

The associative and other properties are satisfied. The table shows that the group is abelian.

Let group G consist of 4 elements, that is, $G = \{e, a, b, c\}$. The identity element is its own inverse.

Let $a^{-1} = a$, $b^{-1} = b$, and $c^{-1} = c$, that is, each element is its own inverse, then G is abelian (Theorem 7.9).

Now let us assume that $b^{-1} = c$, $c^{-1} = b$, and $a^{-1} = a$, then $bc = c^{-1}c = e$, and $cb = b^{-1}b = e$.

As $a^{-1} = a$, we have $aa = e$.

Consider the multiplication (composition table):

0	e	a	b	c
e	e	a	b	c
a	a	e	c	b
b	b	c	a	e
c	c	b	e	a

Now ab can either be b or c. As $ab = b \Rightarrow a = e$ and so $ab = c$ and $ac = b$.

The composition table shows that G is abelian.

Example 6 If G is a group such that

$$(ab)^n = a^n b^n$$

for three consecutive integers, then

$$ab = ba$$

Solution: We are given

$$(ab)^n = a^n b^n$$

$$(ab)^{n+1} = a^{n+1} b^{n+1}$$

and $(ab)^{n+2} = a^{n+2}b^{n+2}$

Now, $(ab)(a^n b^n) = (ab)(ab)^n = (ab)^{n+1}$

or $a(ba^n)b^n = a^{n+1}b^{n+1}$

or $a(ba^n)b^n = a(a^n b)b^n$

$\Rightarrow ba^n = a^n b$ (i) (by cancellation law)

Similarly, $ba^{n+1} = a^{n+1}b$ (ii)

Now, $ba^{n+1} = a^{n+1}b \Rightarrow (ba^n)a = a^{n+1}b$

$\Rightarrow (a^n b)a = a^{n+1}b$ (using (i))

$\Rightarrow a^n(ba) = a^n(ab)$

$\Rightarrow ba = ab$ (by left cancellation law)

THEOREM 7.9 If every element of a group G is its own inverse, then G is abelian.

Proof: Let a and b be any two elements in G, then $a = a^{-1}$ and $b = b^{-1}$.
Now,

$$ab = a^{-1}b^{-1} = (ba)^{-1} = ba$$

Thus, $ab = ba$ which shows that G is abelian.

THEOREM 7.10 If G is a group with identity e and $a^2 = e$ for every $a \in G$, then G is abelian.

Proof: Given $a^2 = e$ for every $a \in G$

$$\therefore\ aa = e$$

Pre-multiplying by a^{-1}, we have

$$a^{-1}(aa) = a^{-1}e$$
$$\Rightarrow (a^{-1}a)a = a^{-1}$$
$$\Rightarrow ea = a^{-1}$$
$$\Rightarrow a = a^{-1}$$

Thus, every element is its own inverse, and hence G is abelian (Theorem 7.9).

Remark: Group theory has important application in coding theory, which is used in communication systems. It helps in transmitting the data without any noise by detecting distortion due to it and then rectifying the messages by using encoder and decoder.

Some Examples:

Example 1 Prove that the set of all nth roots of unity is a group with respect to multiplication.

Solution: We know that

$$(1)^{\frac{1}{n}} = (\cos 2r\pi + i\sin 2r\pi)^{\frac{1}{n}} \quad \text{where } r \text{ is any integer}$$
$$= \cos\frac{2r\pi}{n} + i\sin\frac{2r\pi}{n} \quad \text{By De Moivre's theorem}$$
$$= e^{\frac{2\pi i}{n}}, r = 0, 1, 2\cdots,(n-1)$$

Let G denote the set of nth roots of unity, then

$$G=\left\{1,\ e^{\frac{2r_1\pi i}{n}},\ e^{\frac{4\pi i}{n}},\dots,\ e^{\frac{2(n-2)\pi i}{n}},\ e^{\frac{2(n-1)\pi i}{n}}\right\}$$

Let $\quad a=e^{\frac{2r_1\pi i}{n}}$ and $b=e^{\frac{2r_2\pi i}{n}}$, where $r_1\le n-1,\ r_2\le n-1$

Then, $\quad ab=e^{\frac{2(r_1+r_2)\pi i}{n}}\in G\ \text{ if } r_1+r_2\le n-1$

if $r_1 + r_2 > n - 1$, then $r_1 + r_2 = n + k$, $k \le n - 2$
In this case

$$ab=e^{\frac{2(n+k)\pi i}{n}}=e^{2\pi i}\,.\,e^{\frac{2k\pi}{n}}$$

$$=e^{\frac{2k\pi}{n}}\in G\ \text{ as } k\le n-2$$

Hence, multiplication is a binary operation.

The binary operation of multiplication is associative as the elements of G are complex numbers and multiplication of complex numbers is associative.

The identity of any element is $e^{\frac{2\pi i o}{n}}=1$.

The inverse of any element $e^{\frac{2\pi i r}{n}}$ is $e^{\frac{2(n-r)\pi i}{n}}$.

Further, the operation is commutative, and hence, G is an abelian group.

Example 2 Prove that the set of integers Z forms a group with respect to binary operation (*) defined by $a * b = a + b + 1$.

Solution:

(1) Binary operation (or closure property)
Let a and b be any two integers, then $a + b + 1$ is also an integer, and hence binary operation is defined.

(2) Associative law
Let a, b, c be any three integers, that is, $a, b, c \in Z$, then

$$(a * b) * c = (a + b + 1) * c = a + b + 1 + c + 1 = a + b + c + 2$$

and $\quad a * (b * c) = a * (b + c + 1) = a + b + c + 1 + 1 = a + b + c + 2$

Hence,

$a * (b * c) = (a * b) * c$, i.e., (*) is associative binary operation.

(3) Existence of Identity Element
Let a be any element and e the identity element, then

$$a * e = e * a = a$$
$$\Rightarrow a + e + 1 = a$$
$$\Rightarrow e = -1 \in Z$$

Thus, -1 is the identity element.

(4) Existence of Inverse

For each element $a \in Z$, there must exist an element $b \in Z$ such that

$$b * a = e = -1$$

or $$b + a + 1 = -1$$

or $$b = -a - 2$$

Hence, each element a has its inverse $-a - 2$.

Further, $a * b = a + b + 1 = b + a + 1 = b * a$.

Therefore, the set Z is an abelian group for the given binary operation.

Example 3 Show that the set of positive rational numbers Q^+ forms an abelian group with respect to the binary operation (*) defined by $a * b = \dfrac{ab}{2}$.

(1) Binary operation (closure property)

Let $a, b \in Q^+$, then $a * b = \dfrac{ab}{2} \in Q^+$ as a and b both being positive rational numbers, their product $\dfrac{ab}{2}$ is also a positive rational number and hence binary operation (*) is defined.

(2) Associative law

Let a, b, and $c \in Q^+$, then $(a * b) * c = \left(\dfrac{ab}{2}\right) * c = \dfrac{abc}{4}$

Also $$a * (b * c) = a * \left(\frac{bc}{2}\right) = \frac{abc}{4}$$

Hence, $$a * (b * c) = (a * b) * c.$$

(3) Existence of Identity Element: Let a be any element and e the identity, that is, $a, e \in Q^+$, then

$$e * a = a \Rightarrow \frac{ea}{2} = a$$

$$\Rightarrow e = 2$$

Thus, 2 is the identity element.

(4) Existence of Inverse: Let a be any element in Q^+ and b its inverse, then

$$b * a = e$$

$$\Rightarrow \frac{ba}{2} = 2$$

$$\Rightarrow b = \frac{4}{a} \in Q^+.$$

Hence, the inverse of a is $\dfrac{4}{a}$.

Further, $$a * b = \frac{ab}{2} = \frac{ba}{2} = b * a.$$

Hence, the set Q^+ is an abelian group with respect to the given binary operation.

7.3.4 Modulo (*m*) Operations

Two new types of operations on a set known as modulo (*m*) are discussed due to their applications.

Additional Modulo (*m*): Let *a* and *b* be any two integers and m a fixed positive integer. The addition of *a* and *b* modulo *m* written as $a +_m b$ is defined by,

$$a +_m b = r, 0 \le r < m$$

where *r* is the least non-negative reminder when the ordinary sum $a + b$ is divided by *m*. Thus,

$$15 +_5 8 = 3,\ 11 +_4 6 = 1,\ 9 +_3 8 = 2$$
$$-20 +_5 6 = 1 \text{ as } -20 + 6 = -14 = 5(-3) + 1 = 1$$

Multiplication Modulo (*p*): The multiplication modulo *p* (where *p* is a fixed positive integer) of any two integers *a* and *b* written as $a \times_p b$ is defined by,

$$a \times_p b = r, 0 \le r < p$$

where *r* is the least non-negative remainder when the ordinary product *ab* is divided by *p*. Thus,

$$7 \times_3 5 = 2 \quad \text{as} \quad 7 \times 5 = 3(11) + 2 = 2$$
$$15 \times_8 7 = 1 \quad \text{as} \quad 15 \times 7 = 105 = 8(13) + 1 = 1$$

THEOREM 7.11 Additive group of integers modulo *m*, the set $G = \{0, 1, 2,\ldots, m-1\}$ is an abelian group under the operation of addition modulo *m*.

(1) *Binary Operation*: For all $a, b \in G$,

$$a +_m b = r \in G \quad \text{as} \quad 0 \le r \le m - 1$$

(2) *Associative Law*: Let $a, b, c \in G$, then $(a +_m b) +_m c$ = least non-negative remainder when the sum $(a + b) + c$ is divided by *m* and $a +_m (b +_m c)$ = least non-negative remainder when the $a + (b + c)$ is divided by *m*.

Thus, $$(a +_m b) +_m c = a +_m (b +_m c)$$

(3) *Existence of Identity*: We have $0 \in G$, it is identity as $0 +_m a = a$ for all $a \in G$.

(4) *Existence of Inverse*: The inverse of 0 is 0 itself. For each $a \in G$, its inverse is $(m - a)$, which is in G $(m - a) +_m a = 0$. Thus, $(m - a)$ is the inverse of *a*.

(5) *Commutative Law*: $a +_m b = b +_m a = r$, where *r* is the least non-negative remainder when $a + b$ is divided by *m*. Hence, *G* is an abelian group.

Example 1 Prove that the set $G = \{0, 1, 2, 3, 4\}$ is an abelian group under addition modulo 5.

Solution: The composition table is as follows:

+5	0	1	2	3	4
0	0	1	2	3	4
1	1	2	3	4	0
2	2	3	4	0	1
3	3	4	0	1	2
4	4	0	1	2	3

(1) *Binary Operation* (*or closure property*): Since all the entries in the composition table are elements of the set *G*, the binary operation of ordinary addition modulo 5 is defined.

(2) *Associative Law*: Let *a*, *b*, *c* be any three elements in *G*, then

$$(a +_5 b) +_5 c = a +_5 (b +_5 c)$$

Let 4, 3, 2 be the elements of *G*, then

$$(4 +_5 3) +_5 2 = 2 +_5 2 = 4$$

Also,

$$4 +_5 (3 +_5 2) = 4 +_5 0 = 4$$

Hence, $(4 +_5 3) +_5 2 = 4 +_5 (3 +_5 2)$ (associative law is satisfied)

(3) *Existence of Identity*: The integer 0 is the identity element as $0 +_m a = a +_m 0 = a, \forall a \in G$

(4) *Existence of Inverse*: The table shows that the inverses of 0, 1, 2, 3, 4 are 0, 4, 3, 2, 1, respectively, and thus each element of *G* has its inverse.

(5) *Commutative Law*:

$$a +_5 b = b +_5 a$$

Now $a +_5 b = r$, where *r* is the least negative reminder when $a + b$ is divided by 5

$$= b +_5 a$$

For example, $3 +_5 4 = 4 +_5 3 = 2$

As all the postulates of the abelian group are satisfied, *G* is an abelian group.

Example 2 Multiplicative group of integer modulo *p*.

The set $G = \{1, 2, 3, \ldots, p - 1\}$ of $(p - 1)$ integers where *p* is prime is an abelian group under ordinary multiplication modulo *p*.

Proof:

(1) *Binary Operation* (*closure property*): Let *a* and *b* be any two elements belonging to *G*,

then $\quad 1 \le a \le (p - 1), \quad 1 \le b \le p - 1$

Now, $\quad a \times_p b = r, \quad 1 \le r \le (p - 1)$

where *r* is the least positive remainder when the product *ab* is divided by *p*.

Since *p* is prime, *ab* is not exactly divisible by *p*, and hence the remainder cannot be zero and we shall have $1 \le r \le p - 1$. Hence, $a \times_p b = r \in G$ and hence the set *G* is closed under the operation of ordinary multiplication modulo *p*.

(2) *Associative Law*: Let *a*, *b*, *c* be any three elements of *G*.

Then $\quad a \times_p (b \times_p c) = a \times_p (bc)$

= Least positive remainder when *a*(*bc*) is divided by *p*

= Least positive remainder when (*ab*)*c* is divided by *p*

$= ab \times_p c = (a \times_p b) \times_p c$

Thus associative law is satisfied.

(3) *Existence of Identity*: The element $1 \in G$ is the identity as $1 \times_p a = a$ for every $a \in G$

(4) *Existence of Inverse*: In order to show that every element has inverse in G, let q be any member of set G. consider $1 \times_p q, 2 \times_p q, 3 \times_p q, \ldots, (p-1) \times_p q$. All these are elements of G and are distinct. For if not, let i and j be two unequal integers less than p, then

$$
\begin{aligned}
i \times_p q = j \times_p q &\Rightarrow iq - jq \text{ is divisible by } p \\
&\Rightarrow (i-j)q \text{ is divisible by } p \\
&\Rightarrow \text{either } (i-j) \text{ or } q \text{ is divisible by } p
\end{aligned}
$$

Both $(i-j)$ and q are less than p. Since p is prime, we have $i - j = 0$ as q is less than p.

This implies $i = j$, which contradicts the assumption that i and j are unequal. Thus, all elements are distinct and one of these must be equal to 1.

$$\text{Let } q' \times_p q = 1$$

Then, q' is the inverse of q.

(5) *Commutative Law*:

$$
\begin{aligned}
a \times_p b &= \text{Least positive remainder when } ab \text{ is divided by } p \\
&= \text{Least positive remainder when } ba \text{ is divided by } p \\
&= b \times_p a
\end{aligned}
$$

Hence, the set G is an abelian group.

Example 1 Prove that the set $G = \{1, 2, 3, 4\}$ is an abelian group with respect to ordinary multiplication modulo 5.

Solution: The composition table is as follows:

	1	2	3	4
1	1	2	3	4
2	2	4	1	3
3	3	1	4	2
4	4	3	2	1

(1) *Binary Operation* (*closure property*): Since all the entries in the composition table are elements of G, the binary operation of ordinary multiplication modulo 5 is defined. We also say that closure property is satisfied.

(2) *Associative Law*: For any three elements $a, b, c \in G$, we have

$$
\begin{aligned}
a \times_5 (b \times_5 c) &= a \times_5 (bc) \\
&= \text{Least positive remainder when } a(bc) \text{ is divided by 5} \\
&= \text{Least positive remainder when } (ab)c \text{ is divided by 5} \\
&= (ab) \times_5 c
\end{aligned}
$$

For example, $(2 \times_5 3) \times_5 4 = 1 \times_5 4 = 4$; also $2 \times_5 (3 \times_5 4) = 2 \times_5 2 = 4$

$\therefore (2 \times_5 3) \times_5 4 = 2 \times_5 (3 \times_5 4)$, that is, associative law is satisfied.

(3) *Existence of Identity*: The element $1 \in G$ is the identity as

$$1 \times_5 a = a \qquad \forall a \in G.$$

(4) *Existence of Inverse*: For every element $a \in G$, there exists $b \in G$ such $b \times_5 a = 1$

1 is its own inverse, the inverse of 2 is 3, the inverse of 3 is 2 and 4 is its own inverse.

(5) *Commutative Law*: The operation is commutative as the corresponding rows and columns are identical in the composition table. For example, $2 \times_5 3 = 1 = 3 \times_5 2$.
Hence G is an abelian group.

7.4 CONGRUENCES

Definition 7.3 Two integers a and b are said to be congruent modulo a positive integer m if $a - b$ is divisible by m, that is, there exists an integer k such that $a - b = km$ and is written as $a \equiv b \pmod{m}$. Congruences behave as equalities as can be seen from the following relations:

1. If $a \equiv b \pmod{m}$, then
   (i) $a + x \equiv b + x \pmod{m}$
   (ii) $ax \equiv bx \pmod{m}$
2. If $a \equiv b \pmod{m}$ and $c \equiv d \pmod{m}$, then
   (i) $a + c \equiv b + d \pmod{m}$
   (ii) $a - c \equiv b - d \pmod{m}$
   (iii) $ac \equiv bd \pmod{m}$

7.4.1 Residue Classes

THEOREM 7.12 The congruence modulo m relation in the set of integers has m distinct equivalence classes called residue classes or congruence classes modulo m.

Proof: As congruent modulo relation (mod m) is an equivalence relation, it decomposes the set of integers Z into m disjoint classes putting all integers in the same class, which are congruent modulo m. Since every integer is congruent (modulo m) to its remainder when divided by m, there are m classes corresponding to m possible reminders 0, 1, 2,…, $(m - 1)$. The first class consists of all integers $\equiv 0 \pmod{m}$, that is, intcgers whose reminder is 0 when divided by m. The second class consists of all integers $\equiv 1 \pmod{m}$ and so on, the mth class consists of all integers $\equiv (m - 1) \pmod{m}$. Thus, there are m sets of integers

$$mk,\ mk + 1,\ mk + 2, \ldots,\ mk + (m - 1)$$

where k runs through all integers. These classes are denoted by $\{0\}, \{1\}, \{2\}, \ldots, \{(m - 1)\}$ and are called residue classes modulo m.

Let us consider residue classes mod 5. We shall have five classes:

$$\{0\} = \{0, \pm 5, \pm 10, \pm 15, \ldots\}$$

$$\{1\} = \begin{Bmatrix} 1, & 6, & 11, & 16, \ldots \\ -4, & -9, & -14, & -19, \ldots \end{Bmatrix}$$

$$\{2\} = \begin{Bmatrix} 2, & 7, & 12, & 17, \ldots \\ -3, & -8, & -13, & -18, \ldots \end{Bmatrix}$$

$$\{3\} = \begin{Bmatrix} 3, & 8, & 13, & 18, \ldots \\ -2, & -7, & -12, & -17, \ldots \end{Bmatrix}$$

$$\{4\} = \begin{cases} 4, & 9, & 14, & 19, \ldots \\ -1, & -6, & -11, & -16, \ldots \end{cases}$$

7.4.2 Addition and Multiplication of Residue Classes

Definition 7.4 Let $\{r_1\}$ and $\{r_2\}$ be two residue classes mod m, where $0 \le r_1 \le m - 1$ and $0 \le r_2 \le m - 1$.
The sum is written as

$$\{r_1\} + \{r_2\} = \{r_1 + r_2\}$$

where $r_1 + r_2$ must be reduced to mod m if $r_1 + r_2 \ge m$.

Example 1 Let us consider residue classes modulo 6, that is,

$$\{0\}, \{1\}, \{2\}, \{3\}, \{4\}, \{5\}.$$

Then,

$$\{3\} + \{5\} = \{8\} = \{2\}$$
$$\{2\} + \{4\} = \{6\} = \{0\}$$
$$\{4\} + \{5\} = \{9\} = \{3\}$$

Definition 7.5 The product of residue classes $\{r_1\}$ and $\{r_2\}$ mod m is defined as follows:

$$\{r_1\}.\{r_2\} = \{r_1 r_2\}, \quad 0 \le r_1 \le m - 1, 0 \le r_2 \le m - 1$$

where $r_1\, r_2$ is reduced to mod m where $r_1\, r_2 \ge m$.

THEOREM 7.13 The residue classes modulo m form a group with respect to addition of residue classes.

Proof: Let set S of residue class modulo m be

$$S = \{\{0\}, \{1\}, \{2\}, \ldots, \{r_1\}, \ldots, \{r_2\}, \ldots, \{m - 1\}\}$$

Now,

$$\{r_1\} + \{r_2\} = \{r\}$$

where r is the least non-negative remainder when $r_1 + r_2$ is divided by m.

Clearly $0 \le r < m$, and hence, $\{r\} \in S$. Thus, addition of residue classes is a binary operation.

The addition of residue classes modulo m is associative. The identity element is $\{0\}$ as $\{0\} + \{r\} = \{r\}$.

The additive inverse of $\{r\}$ is $\{m - r\}$ since $\{m - r\} + \{r\} = \{0\}$. Thus, all the postulates of group are satisfied.

THEOREM 7.14 The non-zero residue classes modulo of a positive prime integer p is a group with respect to multiplication of residue classes.

Proof: The set S of non-zero residue classes modulo p is

$$S = \{\{1\}, \{2\}, \ldots, \{r_1\}, \ldots, \{r_2\}, \ldots, \{p - 1\}\}$$

Let $\{r_1\}\,\{r_2\} = \{r\}$

where r is the least positive reminder when r_1r_2 is divided by p.

Since $0 < r < p$ and hence $\{r\} \in S$. Therefore, multiplication is a binary operation over the set S. The multiplication of residue classes is associative. The identity element is $\{1\}$, since

$$\{1\}\,\{r\} = \{r\} \qquad 0 < r < p$$

The inverse of $\{r_1\}$ is $\{r_2\}$ where $r_1r_2 \equiv 1 \pmod p$.
Since the equation $r_1.x \equiv 1 (\text{mod}\, p)$ has solution, $\{r_2\}$ exists.
Thus, all the postulates of group are satisfied.

7.5 PERMUTATIONS

Let $S = \{a_1, a_2, \ldots, a_n\}$ be a finite set having n distinct elements, then a one to one and onto mapping $f: S \to S$ is called a permutation of degree n. Thus, permutation is the operation of re-arranging n distinct objects among themselves. If $(b_1, b_2, \ldots, b_n)$ is some other arrangement, then

$$f = \begin{pmatrix} a_1 & a_2 & a_i & a_n \\ b_1 & b_2 & b_i & b_n \end{pmatrix}$$

defines a permutation where $f(a_i) = b_i$ for all i. The order of columns is immaterial.

Example: Let $S = \{1, 2, 3, 4\}$ be a finite set, then

$$f_1 = \begin{pmatrix} 1 & 2 & 3 & 4 \\ 2 & 3 & 4 & 1 \end{pmatrix}, \quad f_2 = \begin{pmatrix} 1 & 2 & 3 & 4 \\ 3 & 2 & 1 & 4 \end{pmatrix}, \quad f_3 = \begin{pmatrix} 1 & 2 & 3 & 4 \\ 1 & 2 & 3 & 4 \end{pmatrix}$$

are all permutations of degree 4.

Equality of Permutations: Two permutations f_1 and f_2 of degree n are said to be equal if $f_1(a_i) = f_2(a_i)$ $1 \le i \le n$

Consider the following two permutations of degree 4, that is,

$$f_1 = \begin{pmatrix} 1 & 2 & 3 & 4 \\ 3 & 4 & 1 & 2 \end{pmatrix}, \quad f_2 = \begin{pmatrix} 2 & 3 & 4 & 1 \\ 4 & 1 & 2 & 3 \end{pmatrix}$$

Here, $f_1 = f_2$ because

$$f_1(1) = 3 = f_2(1), f_1(2) = 4 = f_2(2), f_1(3) = 1 = f_2(3), \quad \text{and} \quad f_1(4) = 2 = f_2(4)$$

If a set S has n distinct elements, it will have $n!$ distinct arrangements, and thus there will be $n!$ permutations. The set consisting of all permutations of degree n is called a symmetric set of permutations of degree n. If $S = \{1, 2, 3\}$, then the symmetric set of permutations is

$$\left\{ \begin{pmatrix} 1 & 2 & 3 \\ 2 & 3 & 1 \end{pmatrix}, \begin{pmatrix} 1 & 2 & 3 \\ 3 & 1 & 2 \end{pmatrix}, \begin{pmatrix} 1 & 2 & 3 \\ 1 & 2 & 3 \end{pmatrix}, \begin{pmatrix} 1 & 2 & 3 \\ 1 & 3 & 2 \end{pmatrix}, \begin{pmatrix} 1 & 2 & 3 \\ 3 & 2 & 1 \end{pmatrix}, \begin{pmatrix} 1 & 2 & 3 \\ 2 & 1 & 3 \end{pmatrix} \right\}$$

Identity Permutation: A permutation in which each element is replaced by itself is called an identity permutation and is denoted by *I*.

For example, $I = \begin{pmatrix} a_1 & a_2 & a_3 & \dots & a_i & \dots & a_n \\ a_1 & a_2 & a_3 & \dots & a_i & \dots & a_n \end{pmatrix}$ is the identity permutation of degree *n*.

Product of Permutations: Let the two permutations on *n* symbols be

$$f_1 = \begin{pmatrix} a_1 & a_2 & a_3 & \dots & a_i & \dots & a_n \\ b_1 & b_2 & b_3 & \dots & b_i & \dots & b_n \end{pmatrix} \text{ and } f_2 = \begin{pmatrix} b_1 & b_2 & b_3 & \dots & b_i & \dots & b_n \\ c_1 & c_2 & c_3 & \dots & c_i & \dots & c_n \end{pmatrix}$$

then $f_1 f_2 = \begin{pmatrix} a_1 & a_2 & a_3 & \dots & a_i & \dots & a_n \\ c_1 & c_2 & c_3 & \dots & c_i & \dots & c_n \end{pmatrix}$

For f_1 replaces a_1 by b_1 and f_2 replaces b_1 by c_1, and hence $f_1 f_2$ replaces a_1 by c_1. In general, $f_1 f_2$ replaces a_i by c_i, clearly $f_1 f_2$ is a permutation on *n* symbols.

Permutation multiplication is not commutative, in general. This is illustrated by considering two permutations on 4 symbols.

Let $$f_1 = \begin{pmatrix} 1 & 2 & 3 & 4 \\ 3 & 2 & 4 & 1 \end{pmatrix} \text{ and } f_2 = \begin{pmatrix} 1 & 2 & 3 & 4 \\ 4 & 1 & 2 & 3 \end{pmatrix}$$

By re-arranging columns, f_2 can be written as follows:

$$f_2 = \begin{pmatrix} 3 & 2 & 4 & 1 \\ 2 & 1 & 3 & 4 \end{pmatrix}$$

Now $$f_1 f_2 = \begin{pmatrix} 1 & 2 & 3 & 4 \\ 3 & 2 & 4 & 1 \end{pmatrix}\begin{pmatrix} 3 & 2 & 4 & 1 \\ 2 & 1 & 3 & 4 \end{pmatrix} = \begin{pmatrix} 1 & 2 & 3 & 4 \\ 2 & 1 & 3 & 4 \end{pmatrix}$$

And $f_2 f_1 = \begin{pmatrix} 1 & 2 & 3 & 4 \\ 4 & 1 & 2 & 3 \end{pmatrix}\begin{pmatrix} 4 & 1 & 2 & 3 \\ 1 & 3 & 2 & 4 \end{pmatrix}$ (columns f_1 being re-arranged)

$$= \begin{pmatrix} 1 & 2 & 3 & 4 \\ 1 & 3 & 2 & 4 \end{pmatrix}$$

Thus, $$f_1 f_2 \neq f_2 f_1$$

Group of Permutations: The *n*! permutations on *n* symbols $(a_1, a_2, a_3, \dots, a_n)$ form a group with respect to permutation multiplication. This group is called symmetric group and is denoted by S_n.

(1) *Binary Operation*: The multiplication of two permutations on *n* symbols is again a permutation on *n* symbols, and hence multiplication is a binary operation.

(2) Permutation multiplication is associative

The permutation $f_1 = \begin{pmatrix} a_1 & a_2 & \dots & a_n \\ b_1 & b_2 & \dots & b_n \end{pmatrix}$

can be represented by $f_1 = \begin{pmatrix} \alpha \\ \beta \end{pmatrix}$ where α and β denote some arrangement of n symbols. Let the other two permutations after suitable rearrangement of their columns be

$$f_2 = \begin{pmatrix} \beta \\ \gamma \end{pmatrix} \text{ and } f_3 \begin{pmatrix} \gamma \\ \delta \end{pmatrix}$$

Then

$$(f_1 f_2) f_3 = \left\{\begin{pmatrix} \alpha \\ \beta \end{pmatrix}\begin{pmatrix} \beta \\ \gamma \end{pmatrix}\right\}\begin{pmatrix} \gamma \\ \delta \end{pmatrix} = \begin{pmatrix} \alpha \\ \gamma \end{pmatrix}\begin{pmatrix} \gamma \\ \delta \end{pmatrix} = \begin{pmatrix} \alpha \\ \delta \end{pmatrix}$$

And
$$f_1(f_2 f_3) = \begin{pmatrix} \alpha \\ \beta \end{pmatrix}\left\{\begin{pmatrix} \beta \\ \gamma \end{pmatrix}\begin{pmatrix} \gamma \\ \delta \end{pmatrix}\right\} = \begin{pmatrix} \alpha \\ \beta \end{pmatrix}\begin{pmatrix} \beta \\ \delta \end{pmatrix} = \begin{pmatrix} \alpha \\ \delta \end{pmatrix}$$

Hence,
$$(f_1 f_2) f_3 = f_1(f_2 f_3)$$

(3) Existence of identity

The identity element is the identity permutation

$$I = \begin{pmatrix} a_1 & a_2 & a_3 \ldots & a_i & \ldots & a_n \\ a_1 & a_2 & a_3 \ldots & a_i & \ldots & a_n \end{pmatrix}$$

which leaves each symbol unchanged.

(4) Existence of Inverse

Let
$$f = \begin{pmatrix} a_1 & a_2 & a_3 & \ldots & a_i & \ldots & a_n \\ b_1 & b_2 & b_3 & \ldots & b_i & \ldots & b_n \end{pmatrix}$$

then
$$f^{-1} = \begin{pmatrix} b_1 & b_2 & b_3 & \ldots & b_i & \ldots & b_n \\ a_1 & a_2 & a_3 & \ldots & a_i & \ldots & a_n \end{pmatrix}$$

$$\therefore f.f^{-1} = \begin{pmatrix} a_1 & a_2 & a_3 & \ldots & a_i & \ldots & a_n \\ a_1 & a_2 & a_3 & \ldots & a_i & \ldots & a_n \end{pmatrix} = I$$

Note 7.2 For permutation on n symbols, one can use numbers 1, 2, 3,…, n instead of letters a_1, a_2,…, a_n.

Dihedral Groups: The consideration of symmetries of regular polygons leads to certain permutation groups known as dihedral groups.

To construct a dihedral group, let us consider an equilateral triangle in Fig. 7.1 with vertices denoted by 1, 2, and 3 with its centroid at O.

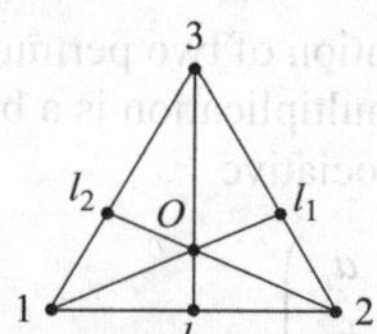

Figure 7.1

The symmetries of this triangle can be described as follows:

There is a counterclockwise rotation f_2 of the triangle about o through 120°. Then f_2 can be expressed as the permutation

$$f_2 = \begin{pmatrix} 1 & 2 & 3 \\ 2 & 3 & 1 \end{pmatrix}$$

The next counterclockwise rotation f_3 about o through 240°, which can be expressed as the permutation

$$f_3 = \begin{pmatrix} 1 & 2 & 3 \\ 3 & 1 & 2 \end{pmatrix}$$

Finally, there is a counterclockwise rotation f_1 about o through 360°, which can be expressed as the permutation

$$f_1 = \begin{pmatrix} 1 & 2 & 3 \\ 1 & 2 & 3 \end{pmatrix}$$

These three symmetries are called rotational symmetries. We may also obtain three additional symmetries of the triangle (known as axial symmetries) g_1, g_2, and g_3 by reflecting about the lines l_1, l_2, and l_3, respectively. These reflections are denoted by the following permutations:

$$g_1 = \begin{pmatrix} 1 & 2 & 3 \\ 1 & 3 & 2 \end{pmatrix}, \quad g_2 = \begin{pmatrix} 1 & 2 & 3 \\ 3 & 2 & 1 \end{pmatrix}, \quad g_3 = \begin{pmatrix} 1 & 2 & 3 \\ 2 & 1 & 3 \end{pmatrix}$$

Thus, $G = \{f_1, f_2, f_3, g_1, g_2, g_3\}$ is a group with respect to the composition of functions ⊛ and $(G, ⊛)$ is called the dihedral group.

With the operation ⊛, we can obtain a multiplication as shown in below table. Entries in the table can be obtained either algebraically or geometrically. Algebraically, the computation $f_2 ⊛ g_2$ is carried out as follows:

$$f_2 ⊛ g_2 = \begin{pmatrix} 1 & 2 & 3 \\ 2 & 3 & 1 \end{pmatrix} ⊛ \begin{pmatrix} 1 & 2 & 3 \\ 3 & 2 & 1 \end{pmatrix} = \begin{pmatrix} 1 & 2 & 3 \\ 1 & 3 & 2 \end{pmatrix} = g_1$$

Geometrically, we proceed as in Figs. 7.2–7.4.

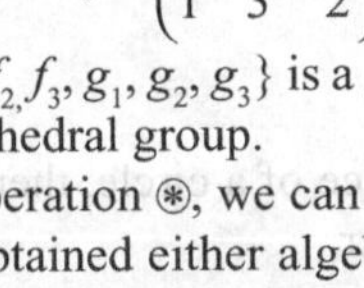

Figure 7.2

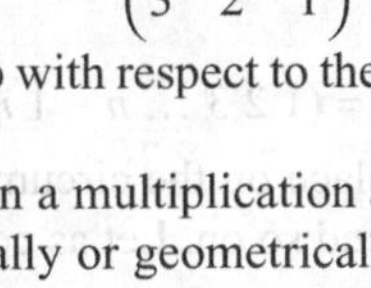

Figure 7.3

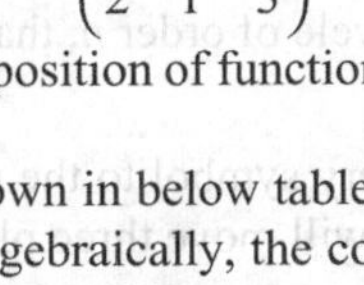

Figure 7.4

Similarly, $g_2 ⊛ g_2 = f_1$, $g_3 ⊛ g_3 = f_1$, $g_1 ⊛ g_1 = f_1$

The multiplication table for the operation ⊛ is shown below:

⊛	f_1	f_2	f_3	g_1	g_2	g_2
f_1	f_1	f_2	f_3	g_1	g_2	g_3
f_2	f_2	f_3	f_1	g_3	g_1	g_2
f_3	f_3	f_1	f_2	g_2	g_3	g_1

(Continued)

(Continued)

g_1	g_1	g_2	g_3	f_1	f_2	f_3
g_2	g_2	g_3	g_1	f_3	f_1	f_2
g_2	g_3	g_1	g_2	f_2	f_3	f_1

7.5.1 Cyclic Permutation

The permutation that replaces n symbols cyclically is called a cyclic permutation (or a cycle) of degree n.
For example, the permutation

$$C = \begin{pmatrix} 1 & 2 & 3 \ldots & n-1 \ldots & n \\ 2 & 3 & 4 \ldots & n & \ldots & 1 \end{pmatrix}$$

is a cycle of degree n. It can also be written as follows:

$$C = (1\ 2\ 3\ \ldots\ n-1\ n)$$

which means that each member is replaced by its successor on the right and the last member by the first.

7.5.2 Powers of Cycle

Consider the cycle of order n, that is,

$$C = (1\ 2\ 3\ \ldots\ n-1\ n)$$

If C moves every symbol to the next place on the circumference of a circle, then C^2 will move two places, C^3 will move three places and so on. Let as consider

$$\begin{aligned} C &= (1\ 2\ 3\ 4\ 5), \text{ then} \\ C^2 &= (1\ 3\ 5\ 2\ 4) \\ C^3 &= (1\ 4\ 2\ 5\ 3) \\ C^4 &= (1\ 5\ 4\ 3\ 2) \\ C^5 &= (1)\ (2)\ (3)\ (4)\ (5) \\ &= I \end{aligned}$$

Alternating Group: The $n!/2$ even permutations form a group with respect permutation multiplication. This group is called alternating group and is denoted by A_n.

THEOREM 7.15 Every permutation can be expressed as a product of disjoint cycles, that is, cycles having no element in common.
This is illustrated by examples.

Example 1 Consider the permutation

$$f = \begin{pmatrix} 1 & 2 & 3 & 4 & 5 & 6 & 7 \\ 2 & 4 & 3 & 5 & 1 & 7 & 6 \end{pmatrix}$$

This can be represented by

$$f = (1\ \ 2\ \ 4\ \ 5)(3)(6\ \ 7) = (1\ \ 2\ \ 4\ \ 5)(6\ \ 7)$$

The cycle of length 1 represents identity permutation, and hence can be omitted.

Example 2 Consider the permutation

$$g = \begin{pmatrix} 1 & 2 & 3 & 4 & 5 & 6 & 7 \\ 2 & 3 & 1 & 5 & 4 & 7 & 6 \end{pmatrix}$$
$$= (1\ 2\ 3)(4\ 5)(6\ 7)$$

Note 7.3 A cycle does not change by changing the place of its elements provided that their relative position is not changed.
For example,

$$(1\ 2\ 3\ 4\ 5) = (2\ 3\ 4\ 5\ 1) = (3\ 4\ 5\ 1\ 2) = (4\ 5\ 1\ 2\ 3) = (5\ 1\ 2\ 3\ 4)$$

7.5.3 Transposition

A cycle of the form (a, b) of length 2, which interchanges the symbols a and b and leaves all other unchanged, is called a transposition. Every cycle and hence every permutation can be expressed as a product of transpositions in several ways.
For example,

$$(1\ \ 2\ \ 3\ \ 4) = (1\ \ 2)(1\ \ 3)(1\ \ 4)$$
$$= (1\ \ 2)(2\ \ 3)(3\ \ 4)(4\ \ 2)(1\ \ 3)$$

It can be noted that in the two representations, the number of transpositions in both cases is odd.

THEOREM 7.16 If a permutation P has been expressed as a product of a transpositions and also as a product of b transpositions, then either both a and b are odd or both a and b are even, that is, $a = b \pmod 2$

Proof: Consider the alternating function A on n distinct symbols $x_1, x_2, \ldots, x_n$ which is a product of $\frac{n(n-1)}{2}$ factors of the form $(x_i - x_j)$, $i < j$. This can be written as follows:

$$A = (x_1 - x_2)(x_1 - x_3)(x_1 - x_4) \ldots (x_1 - x_n)$$
$$(x_2 - x_3)(x_2 - x_4) \ldots (x_2 - x_n)$$
$$\ldots$$
$$(x_{n-2} - x_{n-1})(x_{n-2} - x_n)$$
$$(x_{n-1} - x_n)$$
$$= \prod_{i<j}^{n} (x_i - x_j)$$

If we operate on A by a transposition $t = (x_i, x_j)$, $i < j$, then the following changes will occur in A.

1. All the factors in A, which contain neither x_i nor x_j will remain unaltered.
2. The factor $(x_i - x_j)$ is changed to its negative $(x_j - x_i)$.
3. The factors that contain either x_i or x_j can be grouped into pairs of products $(x_k - x_i)(x_k - x_j)$, where $k \neq i$ or j, which remain unaltered when x_i and x_j are interchanged.

Hence, A operated upon by a transposition t will give $-A$.

Now, the permutation P considered as the product of a transpositions when operated on the alternating function A will give $(-1)^a A$. Similarly, the permutation P when considered as the product of b transposition will give $(-1)^b A$.

Therefore, $$(-1)^a A = (-1)^b A$$

or $$(-1)^a = (-1)^b$$

that is, either both a and b are odd or both are even.

Hence, $a \equiv b \pmod 2$

7.5.4 Even and Odd Permutations

A permutation is called an even or odd according as it can be expressed as a product of even or odd number of transpositions.

THEOREM 7.17(a) Of the $n!$ permutations, $n!/2$ are even and $n!/2$ are odd.

Proof: Let the odd permutations be $O_1, O_2, \ldots, O_p$ and the even permutations be $E_1, E_2, \ldots, E_q$, then $p + q = n!$

If we multiply each permutation by a transposition t, then each of tO_i $(i = 1, 2, \ldots, p)$ will be even and each $tE_j (j = 1, 2, \ldots, q)$ will be odd.

All the permutations tE_j are distinct. For, if not, let

$$tE_r = tE_s \Rightarrow E_r = E_s$$

Similarly, all tO_i are different.

Since an even permutation cannot be equal to an odd permutation, the P odd permutations are q odd permutations tE_j in some order. Similarly, q even permutations are p even permutations tO_i in some order.

Hence, $$p = q = n!/2$$

THEOREM 7.17(b) The set A_n of all even permutations of degree n is a group with respect to permutation multiplication.

Proof: The product of any two even permutations in an even permutation, and hence permutation multiplication is a binary operation over the set A_n. We know that permutation multiplication is associative. If I is the identity permutation of degree n, then I is an even permutation, and hence $I \in A_n$.

Hence, $$If = fI = f, \quad \forall f \in A_n$$

Let f be any even permutation of degree n, that is, $f \in A_n$, then f^{-1} is the inverse of f in the group of permutations of degree n. f^{-1} is an even permutation because

$$ff^{-1} = I \text{ (an even permutation)}$$

Thus, $f \in A_n \Rightarrow f^{-1} \in A_n$,

that is, each element of A_n has its inverse in A_n.

Hence, the set A_n is a group with respect to permutation multiplication.

Note 7.4 The identity permutation I is even. Let the identity permutation be expressed as a product of p transposition. Then the permutation I operating on the alternating function A (Theorem 7.16) on n symbols will change to $(-1)^p A$. Now,

$$IA = A = (-1)^p A \Rightarrow (-1)^p = 1$$

$\therefore p$ is even and thus identity permutation is an even permutation.

7.6 INTEGRAL POWERS OF AN ELEMENT

Let G be a group and a its any element, then by associative law $a \circ a \circ a$ can be grouped in any manner. By general associative law, $a \circ a \circ a \ldots$ to n factors can be grouped in any manner and is denoted by a^n.

Thus

$$a^n = a \circ a \circ a \ldots \text{ to } n \text{ factors.}$$

Further, we define $a^0 = e$ (the identity element in G)

and $a^{-n} = (a^{-1})^n$ where n is a positive integer

THEOREM 7.18 For any element a in a group

(1) $a^r \circ a^s = a^{r+s}$

(2) $(a^r)^s = a^{rs}$

for all integral values of r and s.

Proof: The proof is left as an exercise for students.

Order of an Element: An element a in a group G is said to be of order n, if n is the least positive integer such that $a^n = e$ (the identity element).

If no positive integral power of a is equal to unity, then the element a is said to be of order zero or infinity.

Consider the set of fourth roots of unity

that is, $G = \{1, -1, i, -i\}$

which is a group with respect to ordinary multiplication.

The identity element 1 is of order 1 as $(1)^1 = 1$

The element -1 is of order 2 as $(-1)^2 = 1$

The element i is of order 4 as $(i)^4 = 1$

And the element $-i$ is of order 4 as $(-i)^4 = 1$

7.7 CYCLIC GROUP

A group G is said to be cyclic group if every element of G is of the form a^m, where $a \in G$ and m is an integer.

The element a is called the generator of the group.

Symbolically

$$G = \{a^n : n \in I\} \text{for multiplication}$$

and
$$G = \{na : n \in I\} \text{for addition}$$

If G is a cyclic group generated by a, then we write $G = \{a\}$.

Example 1 The set G of cube roots of unity, that is,

$$G = \{1, \omega, \omega^2\} \text{ is a cyclic group}$$

It has two generators ω and ω^2

Example 2 The set G of fourth roots of unity, that is,

$$G = \{1, -1, i, -i\} \text{is a cyclic group.}$$

It has two generators, i and $-i$.

Example 3 The integral powers of a, that is,

$$\{a^{-3}, a^{-2}, a^{-1}\ a^0 = e, a, a^2, a^3, \ldots\}$$

is a cyclic group with respect to multiplication.

THEOREM 7.19 If $G = \{a\}$ is an infinite cyclic group, then a^{-1} is also a generator.

Proof: Let G be an infinite cyclic group and let a be its generator.

If b is any element in G, then $b = a^m$, where m is an integer or $b = (a^{-1})^{-m}$

Thus, each element is generated by a^{-1}, and hence a^{-1} is also a generator of G.

THEOREM 7.20 Every cyclic group is an abelian group.

Proof: Let G be a cyclic group generated by a.

Let x and y be any two elements in G, then $x = a^m$ and $y = a^n$, where m and n are integers

$$\therefore xy = a^m . a^n = a^{m+n} = a^{n+m} = a^n . a^m = yx$$

which shows that G is abelian.

Some Theorems on Integral Powers of an Element

THEOREM 7.21 If the element a of a group G is of order n, then $a^m = e$ iff $m = nq$

Proof: Let $m > n$ say $m = nq + r$, $\quad 0 \le r < n$

If
$$a^m = e, \text{ then } a^{nq+r} = e$$

$$\Rightarrow (a^n)^q . a^r = e$$

$$\Rightarrow a^r = e$$

Since a is of order n, $a^n = e$, and hence $a^r = e$ is not possible as $r < n$. Hence, $r = 0$ and so $m = nq$

Conversely, let $m = nq$, then $a^m = a^{nq} = (a^n)^q = (e)^q = e$.

THEOREM 7.22 The identity element e is the only element of order 1.

Proof: Obviously $(e)^1 = e$.

THEOREM 7.23 The order of any power of a cannot exceed the power of a.

Proof: Let a be of order n and a^p of order m, then we have to prove that $m \leq n$.

Since a is of order n, $a^n = e$

Now $$(a^p)^n = (a^n)^p = e^p = e$$

But a^p is of order m, and hence $m \leq n$.

THEOREM 7.24 The order of an element a of a group is the same as that of its inverse a^{-1}

Proof: Let the order of a and a^{-1} be n and m, respectively. Since a is of order n,

$$a^n = e \Rightarrow (a^n)^{-1} = e^{-1}$$
$$\Rightarrow (a^{-1})^n = e$$
$$\Rightarrow m \leq n$$

Similarly, a^{-1} is of order m,

$$(a^{-1})^m = e \Rightarrow (a^m)^{-1} = e$$
$$\Rightarrow a^m = e^{-1} = e$$
$$\Rightarrow n \leq m$$

Then, we have $m \leq n$ and $n \leq m$, which implies $m = n$, that is, a and a^{-1} are of the same order.

THEOREM 7.25 The elements a and $x^{-1}ax$ have the same order, a and x being any two elements of a group.

Proof: Let the element a be of order n, then

$$a^n = e$$

Now $$(x^{-1}ax)^2 = (x^{-1}ax)(x^{-1}ax)$$
$$= x^{-1}a(xx^{-1})ax$$
$$= x^{-1}aeax$$
$$= x^{-1}a^2x$$

In general
$$(x^{-1}ax)^n = x^{-1}a^nx = x^{-1}ex = x^{-1}x = e$$

Hence, $x^{-1}ax$ is of order n

THEOREM 7.26 If a is an element of order n and p is prime to n, then a^p is also order n.

Proof: Let a^p be of order m. Since the order of any power of a cannot exceed the order of a,

$$m \leq n \tag{i}$$

Since p is prime to n, that is, $(p, n) = 1$, there exist integers x and y such that $px + ny = 1$. Hence, we can write

$$a = a^{px + ny} = a^{px}.a^{ny} = a^{px}(a^n)^y = a^{px} e^y = a^{px} = (a^p)^x$$

Thus, a is power of a^p and so $n \leq m$. Consequently, $m = n$.

7.8 SUBGROUPS

A non-empty set H of a group $(G, *)$ is called a subgroup of G if $(H, *)$ is a group by itself. Every group G has two trivial subgroups, that is, G itself and the identity group consisting of the identity element only. All other subgroups are called proper subgroups.

THEOREM 7.27 The necessary and sufficient conditions that a non-empty subset H of a group $(G, *)$ is a subgroup are

(1) $a \in H, b \in H \Rightarrow a * b \in H$
(2) $a \in H \Rightarrow a^{-1} \in H$

Proof: The conditions are sufficient. The condition (1) implies that the binary operation (*) in G is also the binary operation in H. The binary operation (*) is associative as it is the binary operation in G.

From condition (2), $a \in H, a^{-1} \in H \Rightarrow a * a^{-1} \in H$

$$\Rightarrow e \in H$$

which shows that identity element of G is also the identity element of H and belongs to H. The condition (2) also shows that every element $a \in H$ has its inverse $a^{-1} \in H$. Hence, H is a group by itself and therefore is a subgroup of G.

The Conditions are Necessary: Let the subset H be a group with respect to the same binary operation (*) of G. Then the condition (1), that is, $a \in H, b \in H \Rightarrow a * b \in H$ necessarily follows. If e is the identity element in G, then $e * a = a \ \forall \ a \in H$ as every element of H is also an element of G. Thus, e is the identity element of H also and must be unique since H being a group has unique identity. Again, let $a \in H$ and a^{-1} be inverse of a in G, then $a^{-1} * a = e$, which shows that a^{-1} is also the inverse of a in H. But this inverse must be unique as H is a group. Therefore, the condition (2) $a \in H \Rightarrow a^{-1} \in H$

Note 7.5 The conditions (1) and (2) of Theorem 7.27 can be combined into a single condition (Theorem 7.28).

THEOREM 7.28 The necessary and sufficient condition for a non-empty subset H of a group $(G, *)$ to be a subgroup is $a \in H, b \in H \Rightarrow ab^{-1} \in H$.

Proof:

The Condition is Necessary: Let H be a group with the binary operation (*), which is for G, then H and G will have the same identity element.
Let $a \in H$ and $b \in H$. b^{-1} also belongs to H as H is a group.
Now, $a \in H, b^{-1} \in H \Rightarrow a * b^{-1} \in H$ (as * is also the binary operation for H).

The Condition is Sufficient: Let H be a non-empty subset of the group G and let $a \in H, b \in H \Rightarrow a * b^{-1} \in H$.

Since H is non-empty, there will exist an element $a \in H$.

Existence of Identity: Taking $b = a$ in the above condition, we have

$$a \in H, a \in H \Rightarrow a * a^{-1} \in H$$
$$\Rightarrow e \in H$$

where e is the identity element of G.

Existence of Inverse: Now let $b \in H$, then $e \in H, b \in H \Rightarrow e * b^{-1} \in H$

$$\Rightarrow b^{-1} \in H$$

Thus,

$$b \in H \Rightarrow b^{-1} \in H.$$

Binary Operation: Further, $a \in H, b^{-1} \in H \Rightarrow a * (b^{-1})^{-1} \in H$

$$\Rightarrow a * b \in H$$
$$a \in H, b \in H \Rightarrow a * b \in H$$

which shows that (*) is the binary operation for H also.

Associative Law: As all the elements of H are also the elements of group G, and hence obey associative law.

Hence, all the axioms for group hold for H, that is, H is a group.

Some Example of Subgroups

Example 1 Let $G = \{1, -1, i, -i\}$ be a multiplicative group. Prove that the subset $H = \{1, -1\}$ is a subgroup.

(1) Binary operation

*	1	−1
1	1	−1
−1	−1	1

Since all the entries in the composition table are elements of H, and hence multiplication binary operation on H also.

(2) Associative law holds.
(3) 1 is the identity element.
(4) Every element is its own inverse.

Hence, H is a group. Since H is a subset of a group G, it is a subgroup.

Example 2 Let G be the additive group of integers. Investigate whether the following subsets of G are subgroup.

(1) The subset H_1 of all odd integers, that is, $\{\pm 1, \pm 3, \pm 5, \ldots\}$
(1) The subset H_2 of all even integers, that is, $\{0, \pm 2, \pm 4, \ldots\}$
(3) The subset $H_3 = \{0, \pm 3, \pm 6, \ldots\}$, that is, all are integral multiples of 3.

Solution:

(1) The set H_1 does not satisfy closure law as the sum of two odd integers is always an even integer, and hence H_1 is not a subgroup of G.

(2) The set H_2 is closed under addition as the sum of two even integers is always even. The operation of addition is associative as

$$(a+b)+c = a+(b+c) \qquad \forall a, b, c \in H_2$$

The identity element is 0 and every element $a \in H_2$ has its inverse $-a \in H$.
Therefore, H_2 is a subgroup of G.

(3) The set H_3 is a subgroup (left as an exercise for students).

Example 3 Let S_3 be a symmetric group on three symbols $\{1, 2, 3\}$. Then the following subsets of S_3 are subgroups:

(1) $H_1 = \{I, (1\ 2)\}$
(2) $H_2 = \{I, (1\ 3)\}$
(3) $H_3 = \{I, (2\ 3)\}$
(4) $A_3 = \{I, (1\ 2\ 3)(1\ 3\ 2)\}$

Solution:
It can be easily verified that subsets H_1, H_2, and H_3 are groups. Each element is its own inverse in H_1, H_2, and H_3.

(4) The set A_3 is the subset of even permutation of S_3 and so is a group. Thus, A_3 is a subgroup of S_3.

Example 4 The multiplicative group G of sixth roots of unity, that is, $G = \left\{ e^{\frac{n\pi i}{3}}, n = 0, 1, 2, 3, 4, 5 \right\}$ has the following two proper subgroups.

$$H_1 = \left\{ e^{\frac{2\pi i}{3}}, e^{\frac{4\pi i}{3}}, 1 \right\}$$

and $$H_2 = \{e^{\pi i}, 1\}$$

left as an exercise for students.

Example 5 Let H_1 and H_2 be two subgroups of a group G, then $H_1 \cap H_2$ is also subgroup of G.

Let $x_1 \in H_1 \cap H_2$ and $x_2 \in H_1 \cap H_2$.
Then, $x_1 \in H_1$ and $x_2 \in H_1$ which implies $x_1 x_2 \in H_1$
Similarly $x_1 x_2 \in H_2$.
Therefore, $x_1 x_2 \in H_1 \cap H_2$
Again $x_1 \in H_1$ and $x_1 \in H_2$
Hence, $x_1^{-1} \in H_1$ and $x_1^{-1} \in H_2 \Rightarrow x_1^{-1} \in H_1 \cap H_2$
Therefore, $H_1 \cap H_2$ is a subgroup of G (Theorem 7.27).
The union of H_1 and H_2, that is, $H_1 \cup H_2$ is not generally a subgroup.
Let $\{Z, +\}$ be the additive group of integers. That is,
Let $Z = \{0, \pm 1, \pm 2, \pm 3, \ldots\}$ be a group.
Let $H_1 = \{0, \pm 2, \pm 4, \pm 6, \ldots\}$ and let $H_2 = \{0, \pm 3, \pm 6, \pm 9, \ldots\}$ be two subgroups of Z.

Then, $H_1 \cup H_2 = \{0, \pm 2, \pm 3, \pm 4, \pm 6, \ldots\}$ is not a group. $2 + 3 = 5$ does not belong to $H_1 \cup H_2$ and thus addition is not a binary operation on $H_1 \cup H_2$.

7.9 COSET DECOMPOSITION

Let G be a group and H any subgroup of G. Let a be any element of G. The collection $\{h_1a, h_2a, \ldots, h_ia, \ldots\}$, where $h_i \in H$ is defined as right coset of H in G and is denoted by Ha. Thus,

$$Ha = \{h_1a, h_2a, \ldots, h_ia, \ldots\}$$

Similarly, the collection $aH = \{ah_1, ah_2, \ldots, ah_i, \ldots\}$ is called left coset of H in G.

Since $\quad eH = He = H,$

H is both a left coset and a right coset of H in G, e being the identity element of G.

If $h \in H$, then $hH = H$ as the product of element of H by h are the elements of H in some order.

If G is a group under addition, then the cosets of H in G are defined by,

$$H + a = \{h_i + a, h_i \in H, a \in G\} \text{ right coset and}$$
$$a + H = \{a + h_i, h_i \in H, a \in G\} \text{ left coset}$$

The number of left (right) cosets of H in G is called the index of H in G and is denoted by (G/H).

THEOREM 7.29 If $a, b \in G$ and $a \neq b$ then $aH = bH \Leftrightarrow a^{-1}b \in H$

Proof: Let $aH = bH$, then $H = a^{-1}bH$, which shows that $a^{-1}b \in H$.
Conversely, let $a^{-1}b \in H$. Now $bH = a(a^{-1}b)H = aH$.

THEOREM 7.30 A subgroup H of a cyclic group G is also cyclic and is generated by a^m, where a is a generator of G and m is the least positive integer such that $a^m \in H$.

Proof: Since a is a generator of G, the subgroup H will contain powers of a, that is, a^p and also its inverse a^{-p} where p is an integer. Let m be the least positive integers such that $a^m \in H$. We have to prove that H is generated by a^m.

Let a^q be any element in H where $q > m$, then by division algorithm

$$q = mr + s \qquad \text{where } o \leq s < m$$

Now $(a^m)^r = a^{mr} \in H$, and hence its inverse $a^{-mr} \in H$ ($\because H$ is a subgroup).

Also, $\quad a^q \in H.$

Hence, $\quad q^{q-mr} \in H \Rightarrow a^s \in H$

This is not possible as m is the least positive integer such that $a^m \in H$. Therefore, $s = 0$ and so $q = mr$.

Hence,

$$a^q = a^{mr} = (a^m)^r$$

Thus, every element of H is in the form $(a^m)^r$, which shows that a^m is a generator of H.

THEOREM 7.31 Lagrange's theorem
The order of a subgroup H of a finite group G is a divisor of the order of the group G.

Proof: Let G be a finite group of order n and its subgroup H of order m. The group G can be decomposed into disjoint left cosets of H in G. The number of these cosets will be finite since G is finite. Since H contains m elements, each coset contains m elements. The total number of elements of these cosets is km which is equal to the number of elements of G, that is, n, k being the number of cosets. Hence, $n = km$ which shows that m, the order of the subgroup H, is a divisor of the order of the group G.

Corollary 1: The order of an element of a finite group is a divisor of the order of the group.

Proof: Let G be a group of order n. Let a be an element of G and its order be m, then the cyclic group $\{a\}$ generated by a is of order m, but the cyclic group $\{a\}$ is a subgroup of G. Hence, its order m is a divisor of n, that is, the order of G.

Corollary 2: Every group of prime order p is cyclic.

Proof: The order of the cyclic subgroup generated by an element $a \neq e$ must divide prime p, and hence is of order p itself. Thus, the group consists of powers of a and so is cyclic.

THEOREM 7.32 Fermat's theorem
If a is an integer and p a prime, then $a^p \equiv a(\text{mod } p)$.

Proof: The set G of non-zero residue classes a prime integer p is a multiplicative group of order $p - 1$, that is, $1, 2, 3, \ldots, p - 1$ and its identity is 1. Let a be any integer. Now two cases will arise, that is, either a is divisible by p or a is not divisible by p.

Case I: Let a be divisible by p, then

$$a \equiv o \text{ (mod } p)$$

Now $a^p - a = a(a^{p-1} - 1)$ is also divisible by p, and hence the theorem is true.

Case II: Let a be not divisible by p, that is, $a \not\equiv 0 \text{ mod}(p)$, then it must be congruent to one of the $(p - 1)$ residues. Hence,

$$a^{p-1} \equiv 1 \text{ (mod } p) \qquad \text{(since } G \text{ is of order } p - 1)$$

$$\text{or} \qquad a^p \equiv a(\text{mod } p)$$

7.9.1 Congruence Relation in Groups

Let H be a subgroup of group G. Define congruence relation in G by $a \equiv b \pmod{H}$, $a, b \in G$. if $b \in aH$, that is, $b = ah_1$ for some $h_1 \in H$

This congruence relation is an equivalence relation in G as can be seen from the following:

1. Reflexivity $a \equiv a \pmod{H}$ since $a = ae \qquad (e \in H)$
2. Symmetry $a \equiv b \pmod{H} \Rightarrow b \equiv a \pmod{H}$
   If $a \equiv b \pmod{H}$, then $b = ah_1 \qquad (h_1 \in H, b \in aH)$
   $\therefore a = bh_1^{-1}$
   Therefore, $a \in bH$, which shows that $b \equiv a \pmod{H}$.
3. Transitivity $a \equiv b \pmod{H}$, $b \equiv c \pmod{H} \Rightarrow a \equiv c \pmod{H}$

Now

$$a \equiv b \pmod{H} \Rightarrow b \in aH, \text{ that is, } b = ah_1 \qquad (h_1 \in H)$$

and

$$b \equiv c \pmod{H} \Rightarrow c \in bH, \text{ that is, } c = bh_2 \qquad (h_2 \in H)$$

Hence,

$$c = bh_2 = (ah_1)\,h_2 = a(h_1\,h_2) = ah_3 \qquad (h_1\,h_2 = h_3 \in H)$$

That is, $a \equiv c \pmod{H}$.

This equivalence relation decomposes group G into disjoint classes, which are left cosets of H in G, which can be obtained as follows:

The first left coset of H in G is H itself. Now if $a \notin H$, then $a \in aH$ and so the second coset of H in G is aH.

Now if $b \notin H$ and $b \notin aH$, then bH is the next coset of H in G. In this manner, we have disjoint cosets of H in G as $H, aH, bH, \ldots$

Example 1 Let $(Z, +)$ be the additive group of integers and let H be a subgroup of integral multiples of a fixed integer m. Then disjoint left cosets of H in Z are

$$km, 1 + km, 2 + km, \ldots, (m-1) + km$$

where $k = 0, \pm 1, \pm 2, \ldots$

These are also residue classes modulo m

$$\{0\}, \{1\}, \ldots, \{m-1\}$$

Hence, coset is also called the residue class.

Example 2 Let S_3 be the symmetric group on three symbols $\{1, 2, 3\}$, then $S_3 = \{I, (1\ 2), (1\ 3), (2\ 3), (1\ 2\ 3), (1\ 3\ 2)\}$

Let H be the subgroup, that is, $H = \{I, (1\ 2)\}$, then its all left cosets are H, $\{(1\ 3), (1\ 3)\,(1\ 2) = (1\ 3\ 2)\}$, $\{(2\ 3), (2\ 3)\,(1\ 2) = (2\ 3\ 1) = (1\ 2\ 3)\}$.

The right cosets of H in S_3 are

$$H, \{(1\ 3), (1\ 2)\,(1\ 3) = (1\ 2\ 3)\}, \{(2\ 3), (1\ 2)\,(2\ 3) = (1\ 3\ 2)\}$$

Hence, we find that except the coset H, which is both a left coset and a right coset, a left coset is not the right coset.

Definition 7.6 The number of left (right) cosets of H in G is called the index of H in G and is denoted by $(G{:}H)$.

For example 2, $(S_3{:}H) = 3$

7.9.2 Normal Subgroups

A subgroup H of a group G is called a normal subgroup if $aH = Ha$ for every $a \in G$, that is, if each left coset is also the right coset.

If G is abelian, then all its subgroups are normal subgroups. However, if G is not abelian, yet it may have a subgroup such that $aH = Ha$ for every $a \in G$.

Note 7.6 Every group G has at least two normal subgroups, namely G itself and the subgroup consisting of the identity element e alone.

Definition 7.7 A normal subgroup can also be defined as follows. A subgroup H of a group G is said to be a normal subgroup of G if for every $a \in G$ and $h \in H$, $ah^{-1}a \in H$

From this definition, it follows that H is a normal subgroup of G.

$$\text{If } aHa^{-1} \subseteq H, \qquad \forall a \in G.$$

THEOREM 7.33 A subgroup H of a group G is normal if and only if $aH^{-1}a = H \qquad \forall a \in G$

Proof: Let $aH^{-1}a = H, \qquad \forall a \in G$

Then, $aH^{-1}a \subseteq H, \qquad \forall a \in G$ (i)

Therefore, H is a normal subgroup of G.

Conversely, let H be a normal subgroup of G.

Then, $$aH^{-1}a \subseteq II, \qquad \forall a \in G$$

Since $$a \in G \Rightarrow a^{-1} \in G, \text{ we have}$$

$$a^{-1} H (a^{-1})^{-1} \subseteq H \qquad \forall a \in G$$

$$\Rightarrow a^{-1}Ha \subseteq H$$

$$\Rightarrow a(a^{-1}Ha)\, a^{-1} \subseteq aHa^{-1}$$

$$\Rightarrow (aa^{-1})\, H\, (aa^{-1}) \subseteq aHa^{-1}$$

$$\Rightarrow H \subseteq aHa^{-1} \qquad \forall a \in G \qquad \text{(ii)}$$

From (i) and (ii), we have $aHa^{-1} = H$.

THEOREM 7.34 The intersection of any two normal subgroups is a normal subgroup.

Proof: Let H_1 and H_2 be two normal subgroups of a group G. Since H_1 and H_2 are subgroups, $H_1 \cap H_2$ is a subgroup of G.

To prove that $H_1 \cap H_2$ is a normal subgroup of group G, let $a \in G$ and h be any element of $H_1 \cap H_2$. Since H_1 is a normal subgroup of G,

$$aha^{-1} \in H_1 \qquad \forall a \in G, h \in H_1$$

Similarly, since H_2 is a normal subgroup of G

$$aha^{-1} \in H_2 \qquad \forall a \in G, h \in H_2$$

Thus,

$$aha^{-1} \in H_1 \cap H_2 \qquad \forall a \in G, h \in H \cap H_2$$

Hence, $H_1 \cap H_2$ is a normal subgroup of G.

Some Solved Examples

Example 1 Prove that every subgroup of an abelian group is normal.

Solution: Let H be a subgroup of an abelian group G.
Then for every $a \in G$ and $h \in H$,

$$aha^{-1} = a(a^{-1}h) \qquad \text{(since } G \text{ is abelian)}$$
$$= (aa^{-1})h = h \in H$$

Thus, $aha^{-1} \in H$, and hence H is a normal subgroup of G.

Example 2 If H is a normal subgroup of a group G, then cosets of H form a group under the operation of coset multiplication defined by $(aH)(bH) = abH$. This group is called quotient group (or factor group) and is denoted by G/H.

Solution:

(1) Coset multiplication is associative:

$$[(aH)(bH)](cH) = (abH)(cH) = (ab)cH$$
$$(aH)[(bH)(cH)] = (aH)(bcH) = a(bc)H$$

Since multiplication in G is associative, $(ab)c = a(bc)$.

(2) Identity element is H as

$$(aH)H = a(HH) = aH \text{ and } H(aH) = H(Ha) = (HH)a = Ha = aH$$

Hence, G/H has identity element $H = eH = He$

and $$(aH)(eH) = aeH = aH$$

Also $$(eH)(aH) = eaH = aH$$

(3) Further, $a^{-1}H$ is the inverse of aH as

$$(a^{-1}H)(aH) = (a^{-1}a)H = eH = eH = H$$

and $$(aH)(a^{-1}H) = (aa^{-1})H = eH = H$$

Thus, G/H is a group under coset multiplication.

Example 3 If H_1 is a normal subgroup of G and H_2 is a subgroup of G such that $H_1 \subseteq H_2 \subseteq G$, then show that H_1 is also a normal subgroup of H_2.

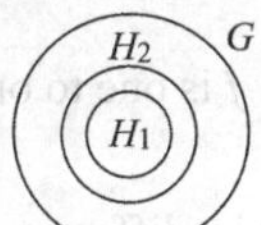

Solution: Since H_1 is a normal subgroup of G and so is a subgroup of G.
Further, H_2 is a subgroup and $H_1 \subseteq H_2$ and so H_1 is a subgroup of H_2.
To show that H_1 is a normal subgroup of H_2, let a be any element in H_2.

Now $a \in H_2 \Rightarrow a \in G$. Since H_1 is normal in G, $aH_1 = H_1 a$. Thus, H_1 is a normal subgroup of H_2 and we have $aH_1 = H_1 a, \quad \forall a \in H_2$.

Example 4 The set G of 2×2 non-singular matrices with real number elements is a group under the operation of matrix multiplication. Let H be the subset of G consisting of matrices of the form

$$\begin{pmatrix} a & 0 \\ b & c \end{pmatrix}$$

Then H is a subgroup of G since H is closed under multiplication and identity matrix $\begin{pmatrix} 1 & 0 \\ 0 & 1 \end{pmatrix} \in H$.

But H is not a normal subgroup of G since

$$\begin{pmatrix} 1 & 3 \\ 2 & 4 \end{pmatrix}^{-1} \begin{pmatrix} 1 & 0 \\ 0 & 1 \end{pmatrix} \begin{pmatrix} 1 & 3 \\ 2 & 4 \end{pmatrix} = \begin{pmatrix} -2 & -9 \\ 1 & 4 \end{pmatrix}$$

does not belong to H as the right upper entry is not zero.

Example 5 Let S_n be the symmetric group on n symbols. Show that the alternating group A_n is a normal subgroup of S_n.

Solution: Let α be any element of S_n and β any element of A_n. We know that A_n is a subgroup of S_n. Since $\beta \in A_n$, β is an even permutation. Now consider $\alpha \beta \alpha^{-1}$, $\alpha \in S_n$, then α can be an odd or even permutation.

If α is odd, then $\alpha \beta \alpha^{-1}$ is even because α^{-1} is also odd. If α is even, then $\alpha \beta \alpha^{-1}$ is even because α^{-1} is also even.

Hence,

$$\alpha \beta \alpha^{-1} \in A_n \qquad \forall \alpha \in S_n, \beta \in A_n.$$

Therefore, A_n is a normal subgroup of S_n.

7.10 ISOMORPHISM AND HOMOMORPHISM OF GROUPS

Let G be a group with binary operation o and G' be another group with binary operation o'. If there exists a map $f: G \to G'$ such that

$$f(a \ o \ b) = f(a) o' f(b),\ a,b \in G; f(a), f(b) \in G'$$

that is, f preserves compositions in G and G', then the mapping f is said to be homomorphism. If f is one to one onto, then G is said to be isomorphic to G' and symbolically written as $G \cong G'$.

Example 1 Let G be the group of real numbers R under addition and G' be the group of positive real numbers R_+ under multiplication. Define map $f: G \to G'$ by,

$$f(a) = 2^a, \qquad a \in R$$

Now f is one to one as $f(a_1) = f(a_2) \Rightarrow 2^{a_1} = 2^{a_2}$

$$\Rightarrow a_1 = a_2$$

That is, different elements have different images.

Further, f is ONTO.
Let α be any positive real number, then there will exist a real number a such that

$$2^a = \alpha \Rightarrow a = \ln_2 \alpha$$

Also, $$f(a+b) = 2^{a+b} = 2^a.2^b = f(a)f(b),$$

which shows that compositions are preserved. Hence, $G \cong G'$

Example 2 Let R be the additive group of real numbers R and R_+ the multiplicative group of positive real numbers, then the mapping $f: R \to R_+$ defined by

$$f(x) = e^x, \qquad \forall\, x \in R$$

is an isomorphism.

Solution: The mapping is one to one as

$$f(x_1) = f(x_2) \Rightarrow e^{x_1} = e^{x_2} \Rightarrow x_1 = x_2$$

f is ONTO. Let y be any positive real number in R_+ then there exists a real number $lny \in R$ such that

$$f(lny) = e^{lny} = y$$

Hence, the mapping is ONTO.
Further,

$$f(x+y) = e^{x+y} = e^x.e^y = f(x) \cdot f(y)$$

which shows that compositions are preserved. Hence, f is an isomorphism.

Example 3 Show that the additive group of integers $G = \{0, \pm 1, \pm 2, \ldots, \pm n, \ldots\}$ is isomorphic to the multiplicative group $= \{2^n: n = 0, \pm 1, \pm 2, \ldots\}$ defined by $f(n) = 2^n$, $n \in Z$.

Solution: It can be easily verified that the mapping f is one to one and ONTO.
Further,

$$f(n_1 + n_2) = 2^{n_1 + n_2} = 2^{n_1} 2^{n_2} = f(n_1) f(n_2)$$

which shows that the compositions are preserved.
Hence, $$G \cong G'.$$

THEOREM 7.35 If the mapping $f: G \to G'$ is an isomorphism, then

(1) $f(e) = e'$, where e is the identity of G and e' is the identity of G'.

(2) If $f(a) = a'$ $(a \in G, a' \in G')$
then $f(a^{-1}) = (a')^{-1}$

Proof:

(1) Let $f(e) = e'$ and a any element in G, then $f(a) \in G'$
Let $f(a) = a', a' \in G'$
Now $a = ae$ (e is the identity in G)

$$\therefore f(a) = f(ae) = f(a)\, f(e)$$
$$\Rightarrow a' = a'e'$$

which shows that $e' = f(e)$ is the identity for G'. That is, identities correspond if $G \cong G'$.

(2) Let $f(a) = a'$

Since $\quad aa^{-1} = e \quad$ (e is the identity in G)

$$\therefore f(aa^{-1}) = f(e)$$

or $\quad f(a)\, f(a^{-1}) = e' \quad$ (since f is an isomorphism)

or $\quad a' f(a^{-1}) = e'$

Consequently, $f(a^{-1})$ is the inverse of a'

that is, $\quad f(a^{-1}) = (a')^{-1}$

which shows that inverses correspond if $G \cong G'$.

Note 7.7 From the above properties, it is clear that if two groups are isomorphic, they can be regarded as the same group from the point of view of their abstract structure. The relation between the elements of G is carried over to the corresponding relation between their images in G'. The concept of isomorphic finds its application in the study of particle physics.

THEOREM 7.36 Cayley's Theorem: Every finite group G is isomorphic to a permutation group.

Proof: Let $G = \{a_1, a_2, \ldots, a_n\}$ be a group of order n. If $b \in G$, then the products $a_1 b, a_2 b, \ldots, a_i b, \ldots, a_n b$ are all distinct. For if not, let

$$a_i b = a_j b \Rightarrow a_i = a_j \qquad \text{(by cancellation law)}$$

Hence, elements $a_i b$ $(i = 1, 2, \ldots, n)$ are the n elements in G in some order. Therefore, $a_i \to a_i b$ is a one-to-one mapping of G onto itself.

That is, $\begin{pmatrix} a_1 & a_2 & a_i & a_n \\ a_1 b & a_2 b & a_i b & a_n b \end{pmatrix}$ is a permutation on n symbols and can be written as $\begin{pmatrix} a_i \\ a_i b \end{pmatrix}$.

Let it be denoted by π_b. Let G' be the set of n permutations π_b where $b = a_i\,(i = 1, 2, \ldots, n)$. Consider the map $f\colon G \to G'$ defined by,

$$f(b) = \begin{pmatrix} a_i \\ a_i b \end{pmatrix} = \pi_b \qquad (b \in G)$$

The mapping f is one to one, since for every $b, c \in G$

$$f(b) = f(c) \Rightarrow \pi_b = \pi_c$$
$$\Rightarrow a_i b = a_i c$$
$$\Rightarrow b = c \text{ (by cancellation law)}$$

The mapping f is ONTO as for every permutation π_d, there exists $d \in G$.

The mapping f preserves compositions as

$$f(bc) = \pi_{bc} = \begin{pmatrix} a_i \\ a_i bc \end{pmatrix}$$

$$= \begin{pmatrix} a_i \\ a_i b \end{pmatrix}\begin{pmatrix} a_i b \\ a_i bc \end{pmatrix}$$

Now $\begin{pmatrix} a_i b \\ a_i bc \end{pmatrix}$ and $\begin{pmatrix} a_i \\ a_i c \end{pmatrix}$ are the same permutations with their columns different, and hence

$$f(b.c) = \begin{pmatrix} a_i \\ a_i b \end{pmatrix}\begin{pmatrix} a_i \\ a_i c \end{pmatrix} = \pi_b \pi_c = f(b).f(c)$$

Therefore, $G \cong G'$.

Example 4 Find the permutation group G' isomorphic to the multiplication group $G = \{1, i, -1, -i\}$.

We can set up the following correspondence between the elements of G and the group G' of permutations on symbols $1, i, -1, -i$.

$$1 \to \begin{pmatrix} 1 & i & -1 & -i \\ 1 & i & -1 & -i \end{pmatrix} = I$$

$$i \to \begin{pmatrix} 1 & i & -1 & -i \\ i & -1 & -i & 1 \end{pmatrix} = (1 \quad i \quad -1 \quad -i)$$

$$-1 \to \begin{pmatrix} 1 & i & -1 & -i \\ -1 & -i & 1 & i \end{pmatrix} = (1 \quad -1)(i \quad -i)$$

$$-i \to \begin{pmatrix} 1 & i & -1 & -i \\ -i & 1 & \mathrm{i} & -1 \end{pmatrix} = (1 \quad -i \quad -1 \quad i)$$

Hence, the permutation group G' isomorphic to G is

$$\{I, (1 \quad i \quad -1 \quad -i), (1 \quad -1)(i \quad -i), (1 \quad -i \quad -1 \quad i)\}$$

7.11 ALGEBRAIC SYSTEMS WITH TWO BINARY OPERATIONS

The algebraic systems with one binary operation, namely semi-group and group, are not adequate to define the set of real numbers that need two binary operations to be defined over a set. One such algebraic system is called a ring.

7.11.1 Rings

Definition 7.8 A system $(R, +)$ consisting of a non-empty set R and two binary operations called addition and multiplication denoted by $(+)$ and $(\cdot)$ is called a ring if the following postulates are satisfied:

R_1: The system $(R, +)$ is an abelian group.
R_2: The system $(R, \cdot)$ is a semi-group.
That is, operation of multiplication is associative

$$(a.\ b).c = a.(b.\ c) \qquad \forall a, b, c \in R$$

R_3: The operation of multiplication is right and left distributive with respect to addition.

$$a.(b + c) = a.\ b + a.\ c \qquad \text{(left distributive law)}$$
$$(b + c).a = b.\ a + c.a \qquad \text{(right distributive law) } \forall a, b, c \in R$$

Since $(R, +)$ is a group, there will exist an identity element for the operation of addition, which is called zero element of the ring and is denoted by 0. Thus,

$$a + 0 = 0 + a = a, \qquad \forall a \in R$$

Also, for every element $a \in R$, there will exist its additive inverse denoted by $-a$ such that

$$a + (-a) = (-a) + a = 0$$

Commutative Ring: A ring in which multiplication is commutative is called a commutative ring, that is, if $a.b = b.a$, $\forall a, b \in R$

Ring with Unity Element: If a ring R contains an element 1 such that for every $a \in R$, $a.1 = 1.a = a$, then the ring is called a ring with unity element 1.

Unity element, if it exists, is the identity element for the operation of multiplication.

Example 1 The set Z of integers $(0, \pm 1, \pm 2, \ldots)$ is a commutative ring with unity element with respect to ordinary addition and multiplication.

Solution: Since the sum and the product of any two integers is an integer, ordinary addition, and multiplication are binary operations over Z.
Further, $\{Z, +\}$ is an abelian group.
and $\{Z, \cdot\}$ is a semi-group.

and multiplication is both left and right distributive with respect to addition. Hence, $(Z, +, \cdot)$ is a ring.

Further multiplication is commutative and 1 is the identity for multiplication, $(Z, +, \cdot)$ is a commutative ring with unity element.

Example 2 The set of even integers with zero, that is, $\{0, \pm 2, \pm 4, \ldots\}$ is a ring with respect the ordinary addition and multiplication. It does not possess unity element.

Example 3 The set Q of rational numbers, the set R of real numbers and the set C of complex numbers are commutative rings with unity elements.

Example 4 Prove that the set of matrices M of order 2×2 with their elements as real numbers (rational numbers or integers) form a ring with respect to addition and multiplication of matrices.

Solution: The sum and product of two matrices of order 2×2 are matrices of order 2×2 and hence addition and multiplication are binary operations.

(1) $(M, +)$ is an abelian group. $\begin{bmatrix} 0 & 0 \\ 0 & 0 \end{bmatrix}$ is the identity element and $\begin{bmatrix} -a & -b \\ -c & -d \end{bmatrix}$ is the inverse of $\begin{bmatrix} a & b \\ c & d \end{bmatrix} \in M$.

(2) $(M, \cdot)$ is a semigroup as matrix multiplication is associative.

(3) Matrix multiplication is distributive with respect to addition over a set of square matrices M.

Hence $(M, +, \cdot)$ is a ring.

Since matrix multiplication is not commutative, it is not a commutative ring.

There is a unit element, which is identity matrix $\begin{bmatrix} 1 & 0 \\ 0 & 1 \end{bmatrix}$.

Hence, it possesses unity element.

Zero Divisors: Let $a \neq 0$, $b \neq 0$ be two elements of a ring R. Such that $a.b = 0$, then a and b are called proper divisors of zero.

Note 7.8 In a ring, it is possible to have $ab = 0$ when neither a nor b is zero.

In Example 4. of matrices(M, +, ·), it has proper divisors of zero, for example,

$$\begin{bmatrix} 1 & 0 \\ 0 & 0 \end{bmatrix}\begin{bmatrix} 0 & 0 \\ 1 & 0 \end{bmatrix} = \begin{bmatrix} 0 & 0 \\ 0 & 0 \end{bmatrix}$$

7.11.2 Elementary Properties of a Ring

Let $(R, +, \cdot)$ be a ring, then for every $a, b, c \in R$

1. $a.0 = 0.a = 0$
2. $a.(-b) = -(a.b) = (-a).b$
3. $(-a).(-b) = a.b$
4. $a.(b - c) = a.b - a.c$
5. $(b - c).a = b.a - c.a$

Proof:

1. We know that $\quad 0 + 0 = 0$
   Hence, for every $a \in R$, $\quad a.(0 + 0) = a.0$
   or $\quad a.0 + a.0 = a.0 + 0 \Rightarrow a.0 = 0$
2. $a.0 = a.(b - b)$
   or $0 = a.b - a.(-b) \Rightarrow a.(-b) = -(a.b)$
   Similarly, we can prove that $(-a).b = -(a.b)$
3. $(-a).(-b) = -[a.(-b)]$
   $= -(-a.b) = a.b$
4. $a.(b - c) + a.c = a.b - a.c + a.c$
   $= a.(b - c + c)$
   $= a.(b + 0) = a.b + a.0 = a.b + 0$
   $= a.b$
   $\therefore a.(b - c) = a.b - a.c$

   Similarly, we can prove (v), that is, $(b - c).a = b.a - c.a$

THEOREM 7.37 A ring R is without zero divisors if and only if cancellation laws hold in R, that is, R is without zero divisors $\Leftrightarrow$ cancellation laws hold in R.

Proof: Let R be a ring without zero divisors. Let a, b, and c be any three elements in R such that $a \neq 0$, and $ab = ac$

Now

$$ab = ac \Rightarrow a.(b - c) = 0$$

Since R is without zero divisors,

$$a.(b - c) = 0 \text{ and } a \neq 0 \Rightarrow b - c = 0$$

$$\Rightarrow b = c$$

which shows that left cancellation law holds. Similarly, we can show that right cancellation law holds.

Conversely, suppose that right cancellation law holds in R.

If possible, let $ab = 0, a \neq 0$ and $b \neq 0$ (i.e., R has divisors of zero)

Now $a \neq 0, ab = 0 = a.0 \Rightarrow b = 0$ (Left cancellation law)

This contradicts the assumption that $b \neq 0$. Hence, the ring R is without zero divisors.

7.11.3 Polynomial Ring

Let R be a ring and x an arbitrary symbol not belonging to R which commutes with every element of R. Let x^n denote $x \cdot x \cdot x, \ldots \cdot x$ to n factors, then an expression of the form

$f(x) = a_0x^0 + a_1x + a_2x^2 + \ldots = \sum_n a_nx^n$ $(a_n \in R)$ is called a polynomial over the ring R.

For example,

$f(x) = a_0 + a_1 + x + a_2x^2$, $g(x) = a_1x^2 + a_2x^4 + a_3x^6 + \ldots$ are polynomials over the ring R.

Note 7.9 A polynomial consisting of a single term a_0 of the ring R is called a constant polynomial. The set of all polynomials $f(x)$ over the ring R is denoted by $R[x]$.

Definition 7.9 Addition and multiplication in $R[x]$. Let R be a commutative ring. Let $f(x)$ and $g(x)$ be two polynomials in $R[x]$, where

$$f(x) = a_0x^0 + a_1x + a_2x^2 + \ldots + a_nx^n, (a_n \neq 0)$$

$$g(x) = b_0x^0 + b_1x + b_2x^2 + \ldots + b_mx^m, (b_m \neq 0)$$

where a_i and b_i are elements of R

Then their sum is (where $m < n$)

$$f(x) + g(x) = (a_0 + b)x^0 + (a_1 + b_1)x^1 + (a_2 + b_2)x^2 + \ldots$$
$$+ (a_m + b_m)x^m + a_{m+1}x^{m+1} + a_{m+2}x^{m+2} + \ldots + a_nx^n$$

Their product is

$$f(x).g(x) = a_0b_0 + (a_0b_1 + a_1b_0)x + (a_0b_2 + a_1b_1 + a_2b_0)x^2 + \ldots + a_nb_mx^{m+n}$$

If may be noted that the coefficient of x^k in the product is

$$a_0b_k + a_1b_{k-1} + a_2b_{k-2} + \ldots + a_kb_0$$

It can be shown that the set $R[x]$ of all polynomials over a ring R in symbol x with coefficients in ring R is a ring under polynomial addition and multiplication.

7.11.4 Morphism of Rings

A mapping f from a ring R to a ring S is called a *morphism* if

$$f(ab) = f(a)\,f(b) \qquad \forall a,b \in R$$

A mapping f from a ring R to a ring S *is* called a *homomorphism* if

$$f(a+b) = f(a) + f(b), \text{ and } f(ab) = f(a)\,f(b) \text{ for ever } a,b \in R$$

Monomorphism: A homomorphism $f\colon R \to S$ is called monomorphism if f is one-to-one and epimorphism if f is ONTO.

Isomorphism: A homomorphism $f\colon R \to S$ is called isomorphism if f is both one to one and onto.

Endomorphism: A homomorphism $f\colon R \to R$, that is, ring R into R itself, is called endomorphism.

Automorphism: An isomorphism $f\colon R \to R$ is called automorphism. The mapping f is one-to-one onto and

$$f(a+b) = f(a) + f(b), f(ab) = f(a)\,f(b) \qquad \forall a,b \in R$$

Kernel of a Ring Homomorphism: If f is a homomorphism of a ring R to a ring S, then the set of all elements of R that are mapped to the zero element of S is called Kernel of homomorphism. Thus, Ker $f = \{a \in R : f(a) = 0\}$

7.11.5 Boolean Ring

A ring $(R, +, \cdot)$ is called a Boolean ring if $a^2 = a$ for every $a \in R$.

For example, $R = \{0, 1\}$ is a Boolean ring with respect to operations of addition and multiplication modulo 2. Addition is defined as $0+1=1+0=1, 0+0=0, +1=1$; multiplication is defined as $0.0 = 0$, $0.1 = 0$, $1.1 = 1$.

*R*1: {R, +}is an abelian group.

*R*2: {R, ·} is a semi-group.

*R*3: Multiplication is distributive with respect to addition.

$1.(1+1) = 1.1 + 1.1$

LHS $= 1.(1+1) = 1.0 = 0$

RHS $= 1.1 + 1.1 = 1 + 1 = 0$

Example 1 If $(R, +, \cdot)$ is a Boolean ring, then prove that

(1) $a + a = 0$

(2) $a + b = 0 \Rightarrow a = b$

(3) R is a commutative ring

Solution:

(1) Since $(R, +, \cdot)$ is a Boolean ring, $a^2 = a$, for every $a \in R$. Replacing a by $a + a$, we have $(a + a)^2 = a + a$

or $(a + a)(a + a) = a + a$

or $a(a + a) + a(a + a) = a + a$ (distributive property)

or $aa + aa + aa + aa = a + a$

or $a^2 + a^2 + a^2 + a^2 + = a + a$

or $a + a + a + a = a + a$

or $(a + a) + (a + a) = (a + a) + 0$ (identity o for addition)

$\therefore a + a = 0$ (left cancellation)

(2) Given $a + b = 0$

$\Rightarrow a + b = a + a$

$\Rightarrow b = a$ (left cancellation)

(3) Let a and b be any two elements in R, we have to prove $ab = ba$ for R to be commutative ring. Now $a, b \in R \Rightarrow a + b \in R$

$\therefore (a + b)^2 = a + b$ ($\because R$ is a Boolean ring)

or $(a + b)(a + b) = a + b$

or $a(a + b) + b(a + b) = a + b$

or $aa + ab + ba + b^2 = a + b$

or $a^2 + ab + ba + b^2 = a + b$

or $a + ab + ba + b = a + b$

or $(a + b) + ab + ba = (a + b) + 0$ (addition is commutative and 0 is identity elements)

or $ab + ba = 0$ (left cancellation low)

$\therefore ab = ba$ (from (2))

Hence, R is a commutative ring.

7.12 RING, SUBRING AND IDEALS

Let (R, +, ·) be a ring. A non-empty subset S of the set R is called a subring of R if $(S, +, \cdot)$ itself is a ring. Thus, $(S, +, \cdot)$ is a subring of $(R, +, \cdot)$ if

1. S is closed with respect to operations of addition and multiplication in R
2. $(S, +)$ is a subgroup of $(R, +)$

If R is any ring, then {O} and R itself are always subrings of R. These are known as improper subrings of R. Other subrings of R, if any, are called proper subrings of R.

Example 1 The ring of integers is a subring of the ring of rational numbers.

Example 2 The ring of rational numbers is a subring of the ring of real numbers.

Example 3 The ring of even integers is a subring of the ring of integers.

Example 4 The set of square matrices of order n with its elements as rational numbers is a subring of the ring of all $n \times n$ matrices with elements as real numbers.

THEOREM 7.38 The necessary and sufficient conditions for a non-empty set S of a ring R to be a subring are

(1) $a \in S, b \in S \Rightarrow a - b \in S$
(2) $a \in S, b \in S \Rightarrow ab \in S$

Proof: The conditions are necessary.

Let $(S, +, \cdot)$ be a subring of $(R, +, \cdot)$. Since S is a subring of R, S itself is a ring, and hence $(S, +)$ is an abelian group.

Now $a \in S, b \in S \Rightarrow a \in S, -b \in S$

$$\Rightarrow a + (-b) \in S \Rightarrow a - b \in S$$

Again, since S is a ring, it is closed with respect to multiplication, that is, $a \in S, b \in S \Rightarrow ab \in S$.

The Conditions are Sufficient: Let S be a non-empty subset of R, which satisfies the following conditions:

(1) $a \in S, b \in S \Rightarrow a - b \in S$ and
(2) $a \in S, b \in S \Rightarrow ab \in S$.

From (1) $a \in S, a \in S \Rightarrow a - a \in S \Rightarrow 0 \in S$, that is, zero element $\in S$ and $0 \in S, a \in S \Rightarrow -a \in S$. Hence, every element $a \in S$ has its additive inverse in S. Now, $b \in S \Rightarrow -b \in S$.

Hence, from condition (1) $a \in S, -b \in S \Rightarrow a - (-b) \in S. \Rightarrow a + b \in S$. Therefore, S is closed with respect to addition.

As the elements of S are the elements of R, which satisfy commutative and associative laws of addition, S is commutative and associative with respect to addition. Hence, $(S, +)$ is an abelian group.

From (2), S is closed with respect to multiplication. Since the operation of multiplication is associative and distributive over addition in R; these will also hold for S. Hence, S is a subring of R.

THEOREM 7.39 The intersection of two subrings of a ring R is a subring of R.

Proof: Let $(R_1, +, \cdot)$ and $(R_2, +, \cdot)$ be two subrings of $(R, +, \cdot)$, then $R_1 \cap R_2$ is not empty since $0 \in R_1 \cap R_2$.

To prove that $R_1 \cap R_2$ is a subring, it is sufficient to prove that

(1) $a \in R_1 \cap R_2, b \in R_1 \cap R_2 \Rightarrow a - b \in R_1 \cap R_2$ and
(2) $a \in R_1 \cap R_2, b \in R_1 \cap R_2 \Rightarrow ab \in R_1 \cap R_2$.

We have $a \in R_1 \cap R_2 \Rightarrow a \in R_1$ and $a \in R_2$

$$b \in R_1 \cap R_2 \Rightarrow b \in R_1 \text{ and } b \in R_2$$

Since R_1 and R_2 are both subrings,

$$a \in R_1, b \in R_1 \Rightarrow a - b \in R_1 \text{ and } ab \in R_1$$

and $$a \in R_2, b \in R_2 \Rightarrow a - b \in R_2 \text{ and } ab \in R_2$$

Now, $$a - b \in R_1, a - b \in R_2 \Rightarrow a - b \in R_1 \cap R_2$$

and $$ab \in R_1, ab \in R_2 \Rightarrow ab \in R_1 \cap R_2.$$

$$\therefore\ R_1 \cap R_2 \text{ is a subring of } R.$$

Note 7.10 The sum of two subrings may not be a subring. For example, the multiples of 2 and multiples of 3 are subrings of the ring of integers. The sum will contain 2 and 3, but their sum $2 + 3 = 5$ is not in the sum. Hence, the sum is not a subring.

7.12.1 Ideal

Let S be a non-empty subset of a ring R such that S is an additive subgroup of R, that is, $a - b \in S$, $\forall a, b \in S$, then

1. S is a left ideal of R iff $ra \in S$ for each $r \in R$ and $a \in S$.
2. S is a right ideal of R iff $ar \in S$ for each $r \in R$ and $a \in S$.
3. S is an ideal of R iff it is both a left ideal and a right ideal of R.

If the ring is commutative left and right ideals coincide.

Example 1 The subring of even integers is an ideal of the ring of integers.

Example 2 Let R be a ring of all 2×2 matrices with elements as integers, then the set S of all matrices of the form

$$\begin{bmatrix} a & 0 \\ b & 0 \end{bmatrix}, \quad a, b \in Z \text{ is a left ideal.}$$

Solution: Let $A = \begin{bmatrix} a & 0 \\ b & 0 \end{bmatrix}$ and $B = \begin{bmatrix} c & 0 \\ d & 0 \end{bmatrix}$ be any two elements of S. Then $A - B = \begin{bmatrix} a-c & 0 \\ b-d & 0 \end{bmatrix}$ $\in S$, and hence S is a subgroup of R under addition.

Further, let $\begin{bmatrix} u & v \\ w & x \end{bmatrix}$ be any element in R and $A = \begin{bmatrix} a & 0 \\ b & 0 \end{bmatrix}$ be any element in S, then

$$\begin{bmatrix} u & v \\ w & x \end{bmatrix}\begin{bmatrix} a & 0 \\ b & 0 \end{bmatrix} = \begin{bmatrix} ua + vb & 0 \\ wa + xb & 0 \end{bmatrix} \in S$$

Therefore, S is the left ideal of R.

Note 7.11 S is not a right ideal, since

$$\begin{bmatrix} a & 0 \\ b & 0 \end{bmatrix}\begin{bmatrix} u & v \\ w & x \end{bmatrix} = \begin{bmatrix} au & av \\ bw & bx \end{bmatrix} \notin S$$

Example 3 Let R be a ring of all 2×2 matrices with elements as integers, then the set S of all matrices of the form

$$\begin{bmatrix} a & b \\ 0 & 0 \end{bmatrix}, \quad a, b \in Z \text{ is a right ideal.}$$

Solution:

Let $A = \begin{bmatrix} a & b \\ 0 & 0 \end{bmatrix}$ and $B = \begin{bmatrix} c & d \\ 0 & 0 \end{bmatrix}$ be any two elements of S, then $A - B = \begin{bmatrix} a-c & b-d \\ 0 & 0 \end{bmatrix} \in S$

Hence, S is a subgroup of R under addition. Further, let $\begin{bmatrix} u & v \\ w & x \end{bmatrix}$ be any element in S, then

$$\begin{bmatrix} a & b \\ 0 & 0 \end{bmatrix}\begin{bmatrix} u & v \\ w & x \end{bmatrix} = \begin{bmatrix} au+bw & av+bx \\ 0 & 0 \end{bmatrix} \in S$$

Therefore S is the right ideal of R.

Note 7.12 S is not the left ideal of R.

Example 4 The set $S = \{mx: x \in Z\}$ where m is a fixed integer is an ideal of the ring of all integers R.

Solution: Let mx_1 and mx_2 be any two elements in S, then

$$mx_1 - mx_2 = (x_1 - x_2)m \in S \text{ as } (x_1 - x_2) \text{ is also an integer.}$$

Hence, S is a subgroup under addition. Further, let r be any integer, that is, any element of the ring of integers R, then $r(mx) = (rx)m \in S$ since rx is also an integer. Therefore, S is a left ideal of R. But the ring of integers R is commutative, and hence S is also a right ideal. Therefore, S is an ideal of R.

Example 5 The set of integers Z is only a subring but not an ideal of the ring of rational numbers $(Q, +, \cdot)$.

Solution: The set of integers $(Z, +, \cdot)$ is a ring, and hence is a subring of $(Q, +, \cdot)$. But the product of a rational number and an integer is not necessarily an integer.

For example, $5 \in Z, \frac{2}{3} \in Q,\ 5\frac{2}{3} = \frac{10}{3} \notin Z$

Hence, Z is not an ideal.

Example 6 Prove that the intersection of two ideals of R is also an ideal of R.

Solution: Let S and T be ideals of R, then S and T are subgroups under addition. To show that $S \cap T$ is an ideal of R, we must show that

$$r \in R \text{ and } s \in S \cap T \Rightarrow rs \in S \cap T \text{ and } s\,r \in S \cap T$$

Since S and T are ideals of R,

$$r \in R, s \in S \Rightarrow r\,s \in S \text{ and } sr \in S$$

and $$r \in R, s \in T \Rightarrow r\,s \in T \text{ and } s\,r \in T$$

Now $$r\,s \in S,\ r\,s \in T \Rightarrow r\,s \in S \cap T$$

and $$s\,r \in S,\ s\,r \in T \Rightarrow s\,r \in S \cap T$$

Hence, $S \cap T$ is an ideal.

7.12.2 Quotient Ring or Ring of Residue Classes

THEOREM 7.40 If S is an ideal of a ring R, then the set of $R/S = \{S + a, a \in R\}$ of all residue classes of S in R forms a ring with respect to the operations of addition and multiplication defined as follows:

$$(S + a) + (S + b) = S + (a + b)$$
$$(S + a)\,(S + b) = S + ab$$

Proof: The set R/S is closed with respect to addition and multiplication as $S + (a + b)$ and $S + ab$ are also residue classes. Both these classes are well defined as can be seen from the following:
Let $S + a = S + a'$ and $S + b = S + b'$
Then we have to show that

$$(S + a) + (S + b) = (S + a') + (S + b')$$

and $$(S + a)\,(S + b) = (S + a')\,(S + b')$$

Now $$S + a = S + a' \Rightarrow a' \in S + a$$

and $$S + b = S + b' \Rightarrow b' \in S + b$$

Therefore, there exist $a_1, b_1 \in S$ such that

$$a' = a_1 + a,\ b' = b_1 + b$$

Now, $$a' + b' = (a_1 + a) + (b_1 + b) = (a + b) + (a_1 + b_1)$$

$$\therefore \quad (a' + b') - (a + b) = a_1 + b_1 \in S$$

$$\therefore \quad S + (a' + b') = S + (a + b)$$

$$\Rightarrow (S + a') + (S + b') = (S + a) + (S + b)$$

Thus, addition in R/S is well defined.
Further,

$$a'b' = (a_1 + a)(b_1 + b) = a_1b_1 + ab_1 + a_1b + ab$$
$$= ab + a_1b_1 + a_1b + ab_1$$
$$\therefore a'b' - ab = a_1b_1 + ab_1 + a_1b \in S$$

(Since S is an ideal, $a_1, b_1 \in S$ and $a, b \in R \Rightarrow ab_1, a_1b, a_1b_1 \in S$, and hence $a_1b + ab_1 + a_1b_1 \in S$.)
Since $a'b' - ab \in S$, $S + a'b' = S + ab$

$$\Rightarrow (S + a')(S + b') = (S + a)(S + b)$$

Hence, multiplication in R/S is also well defined.

1. Addition is associative in R/S

$$(S + a) + [(S + b) + (S + c)] = (S + a) + (S + b + c)$$
$$= S + a + (b + c)$$

and $$[(S + a) + (S + b)] + (S + c) = [S + (a + b)] + (S + c)$$
$$= S + (a + b) + c$$

$$= S + a + (b + c) \text{ (since } a, b, c \in R \text{ and addition is associative)}$$

2. Addition is commutative in R/S

$$(S+a)+(S+b) = S+(a+b) = S+(b+a)$$
$$= (S+b)+(S+a)$$

3. Existence of additive identity
We have $S = S + 0 \in R/S$
If $S + a \in R/S$, then $(S+0)+(S+a) = S+(0+a)$
$$= S + a$$

$\therefore S$ is the identity
4. Existence of additive inverse
Let $S + a \in R/S$, then $S + (-a) \in R/S$.
Now, $\quad [S+(-a)]+[S+a] = S+[-a+a] = S+0 = S$
Hence, $S + (-a)$, that is, $S - a$ is the additive inverse of $S + a$.
5. Multiplication is associative

$$[(S+a)(S+b)](S+c) = (S+ab)(S+c) = S+(ab)c$$
$$= S+a(bc) = (S+a)(S+bc) = (S+a)[(S+b)(S+c)]$$

6. Multiplication is distributive with respect to addition

$$(S+a)[(S+b)+(S+c)] = (S+a)[S+(b+c)] = S+a(b+c)$$
$$= S+(ab+ac) = (S+ab)+(S+ac)$$
$$= (S+a)(S+b)+(S+a)(S+c)$$

Hence, R/S is a ring with respect to the operations of addition and multiplication.

7.13 INTEGRAL DOMAIN

Two other algebraic structures that are of significance in which two binary operations are defined are integral domain and field.

Definition of Integral Domain: A commutative ring with unity element containing at least two elements is called an integral domain if it contains no proper divisors of zero.

Thus, the system $(R, +, \cdot)$ is an integral domain if

I_1: $(R, +)$ is an abelian group.
I_2: $(R, \cdot)$ is a monoid.
I_3: Multiplication is commutative, that is, $a.b = b.a, \quad \forall a, b \in R$.
I_4: Multiplication is right and left distributive with respect to addition.
I_5: $a.b = 0 \Rightarrow$either $a = 0$ or $b = 0$

Example 1 The ring of integers is an integral domain.

Example 2 The set of rational numbers Q, the set R of real numbers, and the set C of complex numbers are all integral domains whose zero element is 0 and unity element is 1.

7.14 FIELD

A system $(F, +, \cdot)$ containing at least two elements is called a field if

F_1: The system $(F, +)$ is an abelian group.

F_2: The system $(F^*, \cdot)$ of non-zero elements of F is an abelian group with respect to multiplication.

F_3: The distributive laws are satisfied.

Example 1 The set Q of rational numbers, the set R of real numbers, and the set C of complex numbers are *fields* with respect to ordinary addition and multiplication.

Example 2 The set of numbers of the form $a + b\sqrt{3}$, where a and b are rational numbers, is a field with ordinary addition and multiplication as binary operations. The zero element of the field is $0 + 0\sqrt{3} = 0$ and unity element is $1 + 0, \sqrt{3} = 1$. The multiplicative inverse of $a + b\sqrt{3}$ is $\frac{a - b\sqrt{3}}{a^2 - 3b^2} = \left(\frac{a}{a^2 - 3b^2}\right) - \left(\frac{b}{a^2 - 3b^2}\right)\sqrt{3}$

Example 3 The set of residue classes modulo a prime positive integer p is a *field* with respect to addition and multiplication of residue classes (mod p).

The set of residue classes is

$$\{0\}, \{1\}, \{2\}, \ldots, \{p - 1\}$$

The zero element of the field is $\{0\}$. The unity element is $\{1\}$. The multiplicative inverse of non-zero residue class $\{r\}$ is $\{r_1\}$ such that

$$r.\, r_1 \equiv 1 (\text{mod } p)$$

Example 4 The set $\{0, 1\}$ forms a field with respect to multiplication and addition (mod 2).

(1) The set $\{0, 1\}$ forms an abelian group under addition modulo 2. Identity element is 0 and 1 is its own additive inverse.
(2) The element 1 forms an abelian multiplicative group whose identity element is 1.
(3) The distributive laws hold.

$1.(1 + 1) = 1.1 + 1.1 = 1 + 1 = 0$, L.H.S $= 1.(1 + 1) = 1.0 = 0$
Similarly, $(1 + 1).1 = 0.1 = 0 = 1.1 + 1.1 = 1 + 1 = 0$

THEOREM 7.41 Every field F is an integral domain.

Proof: A field is a commutative ring with unity element. In order to prove that it is an integral domain, we need to show that a field has no zero divisors. Let a, b be any two elements of F such that $ab = 0$, $a \neq 0$. Since a is non-zero, a^{-1} will exist and we have

$$ab = 0 \Rightarrow a^{-1}(ab) = a^{-1}0$$

$$\Rightarrow (a^{-1}a)b = 0 \Rightarrow 1b = 0 \Rightarrow b = 0$$

Similarly, if $ab = 0$, $b \neq 0$, we shall have

$$ab = 0 \Rightarrow (ab)b^{-1} = 0b^{-1}$$

$$\Rightarrow a(bb^{-1}) = 0 \Rightarrow a_1 = 0 \Rightarrow a = 0$$

Thus, in a field $ab = 0 \Rightarrow a = 0$ or $b = 0$. Hence, a field has no zero divisors. Therefore, every field is an integral domain. But the converse is not true, that is, every integral domain is not a field.

For example, the ring of integers is an integral domain but not a field as the non-zero elements do not have multiplication inverse. Only invertible elements are 1 and −1.

EXERCISES

1. Define binary operation over a set. Give two examples of commutative, non-commutative and associative binary operations in each.
2. Which of the following are semigroup:
   (i) The set N of natural numbers with respect to multiplication
   (ii) The set C of complex numbers with respect to addition
   (iii) The set Q of rational numbers with respect to binary operation (*) defined by $a * b = a - b$
   (iv) The set of vectors $V = \{(a_1, a_2, a_3), a_i \in R\}$ with respect to operation of vector product
   (v) The set of matrices of order $m \times n$ with respect to matrix addition
3. Prove that the power set $P(A)$ of a non-empty set A is a semi-group with respect to union of sets.
4. Which of the following are monoids?
   (i) The set N of natural numbers with respect to addition
   (ii) The set of vectors $a = (a_1, a_2, a_3)$, $a_i \in R$ with respect to vector addition
   (iii) The set of square matrices of order n with respect to multiplication
5. Prove that the set Z of integers is a monoid with respect to binary operation (*) defined by,
$$a * b = a + b + 1, a, b \in Z$$
6. Define a group $(G, *)$ giving two examples.
7. Prove that the set N of natural numbers is not a group with respect to addition. It is not a group with respect to multiplication also.
8. Prove that the set R_+ of positive real numbers forms a group with respect to multiplication.
9. Which of the following are groups with respect to indicate binary operations?
   (i) The set $\{1, \omega, \omega^2\}$ with respect to ordinary multiplication where ω is cube root of unity.
   (ii) The set $\{1, -1\}$ with respect to ordinary multiplication.
   (iii) The set of all $m \times n$ matrices with respect to matrix addition.
   (iv) The set of numbers $\{\cos\theta + i \sin\theta, \theta \in R\}$ with respect to multiplication.
10. Prove that the set of rational numbers other than 1 form an abelian group with respect to binary operation (*) defined by $a * b = a + b - ab$.

11. Prove that the set of matrices

$$A = \begin{pmatrix} \cos\alpha & -\sin\alpha \\ \sin\alpha & \cos\alpha \end{pmatrix}$$

where α is a real number forms a group with respect to multiplication.

12. Do the set of positive rational numbers form a group with respect to binary operation * defined by,

$$a * b = \frac{a}{b}$$

13. Prove that the group $\{a, a^2, a^3, a^4, a^5, a^6, a^7, a^8 = e\}$ is cyclic. Also, show that a^3 and a^5 are also its generators.

14. Show that *n*th roots of unity form a cyclic group and $\cos\frac{2\pi}{n} + i\sin\frac{2\pi}{n}$ is a generator.

15. Prove that the multiplicative group of fourth roots of unity, that is, $\{1, -1, i, -i\}$ is cyclic generated by *i*. Is $(-i)$ also its generator?

16. Let *Z* be additive group of integers and $G = \{3^r, r = 0, \pm 1, \pm 2\}$ be a group with respect to ordinary multiplication. Prove that the mapping $f: Z \to G$ defined by,

$$f(r) = 3^r$$

is an isomorphism.

17. Let *R* be additive group of real numbers and R_+ multiplicative group of positive real numbers.
Define $f: R \to R_+$ by $f(x) = e^x$, $x \in R$. Prove that *f* is an isomorphism.

18. Show that the symmetric group on two symbols, that is, $S_2 = \{I, (1, 2)\}$ is isomorphic to multiplication group $G = \{1, -1\}$.

19. Set up isomorphism between the multiplicative group *G* of fourth roots of unity, that is, $G = \{1, -1, i, -i\}$ and the group G' of residue classes modulo 4 with respect to addition.

20. Prove that the set of even integers is a subgroup of the additive group of integers *Z*.

21. Prove that the set of positive rationals Q_+ is a subgroup of the multiplicative group of non-zero rational numbers Q^*.

22. Prove that the set $\{-1, 1\}$ is the subgroup of the multiplication group $\{1, -1, i, -i\}$.

23. Prove that every subgroup of an abelian group is normal.

24. Prove that the subgroup of $n!/2$ even permutations (called alternating group A_n) is a normal subgroup of S_n, the symmetric group on *n* symbols.

25. Define a ring and integral domain. Give an example of a ring, which is not an integral domain.

26. Do the following sets form rings with respect to ordinary addition and multiplication. If so, state if they are integral domains or fields.
   (i) The set N of natural numbers
   (ii) The set of numbers of the form $a\sqrt{3}$ where a is a rational number.
   (iii) The set of numbers of the form $a + b\sqrt{3}$ with a and b as rational numbers

27. Prove that the set of all real numbers of the form $a + b\sqrt{2}$ with a and b as integers is an integral domain with respect to ordinary addition and multiplication. Also show that it is not a field.

   [Hint: the number $10 + 6\sqrt{2}$ is a non-zero element of ring and its inverse is $\frac{1}{10+6\sqrt{2}} = \frac{10-6\sqrt{2}}{(10+6\sqrt{2})(10-6\sqrt{2})} = \frac{10-6\sqrt{2}}{28} = \frac{5}{14} - \frac{3}{14}\sqrt{2}$ is not an element of this ring.]

28. Prove that set of two elements {0, 1} form a field with respect to ordinary addition and multiplication (mod 2) defined as follows:

$$0 + 0 = 0, 0 + 1 = 1 + 0 = 1, 1 + 1 = 0 \text{ and } 0.1 = 1.0 = 0$$

29. Prove that the set $\{\pm 3k, k \in Z\}$ is a subring of the ring of integers Z.
30. Prove that the ring of rational numbers is a subring of the ring of real numbers.
31. Show that the set of integers Z is only a subring of the ring of rational numbers but not an ideal $(Q, +, \cdot)$.
32. Show that the subset S_1 of all 2×2 matrices

$$\begin{pmatrix} 0 & a \\ 0 & b \end{pmatrix}$$

   with a and b integers is a left ideal but not a right ideal for the ring of all 2×2 matrices with elements as integers.

33. Prove that the subset S of all matrices of the form $\begin{pmatrix} a & 0 \\ 0 & b \end{pmatrix}$ with a and b integers is a subring of the ring R of all 2×2 matrices with elements as integers. Further, prove that S is neither a left ideal nor the right ideal of R.

34. Let $R = \left\{\begin{pmatrix} a & b \\ b & a \end{pmatrix}, a, b \in Z\right\}$ be a ring with respect to binary operations of matrix addition and matrix multiplication and $(S, +, \cdot)$ be the ring of integers with ordinary addition and multiplication. Define $f: R \to S$ by

$$f\begin{pmatrix} a & b \\ b & a \end{pmatrix} = a - b$$

   (i) Show that f is a homomorphism.
   (ii) Show that f is not an isomorphism.

8

Ordered Sets and Lattices

LEARNING OBJECTIVES

After reading this chapter, you will be conversant with:

- Introduction
- Partially Ordered Set
- Product of Two Posets
- Hasse Diagram
- Lexicographic Ordering
- Upper and Lower Bounds
- Dual of a Poset
- Isomorphism of Posets
- Well-ordered Set
- Properties of Well-ordered Sets
- Lattices
- Lattice in Terms of Algebraic Structures
- Sublattices
- Bounded Lattices
- Duality
- Complete Lattice
- Isomorphic Lattices
- Complimented Lattice
- Chain and Antichain
- Distributive Lattices
- Modular Lattice
- Boolean Lattice

8.1 INTRODUCTION

The concept of relation discussed in Chapter 2 will now be associated with the idea of ordering the elements of a set. This will lead to the concept of linear ordering and total ordering, which is of great importance in mathematics and computer science. It gives rise to the concept of lattice, which will be discussed in this chapter.

8.2 PARTIALLY ORDERED SET

A relation R on a set A is called a partial order relation if R is reflexive, antisymmetric and transitive. The set A together with partial order relation R is called a partially ordered set or simply a POSET and is denoted by $(A, \leq)$, the symbol $\leq$ is used for partial order relation.

Example 1 Let $P(S)$ be the collection of all subsets of a set S. Then, the relation $\subseteq$ of 'set inclusion' is a partial order relation.

Solution:

(1) The inclusion relation is reflexive as $A \subseteq A, \forall A \in P(S)$
(2) The inclusion relation is antisymmetric as $A \subseteq B$ and $B \subseteq A \Rightarrow A = B$
(3) The inclusion relation is transitive as $A \subseteq B, B \subseteq C \Rightarrow A \subseteq C$

Since all the postulates of partial order relation are satisfied, the given relation of 'set inclusion' is a partial order relation.

Example 2 Show that the relation $\leq$ (less than or equal to) on the set of real numbers R is a partial order relation.

Solution:

(1) The relation is reflexive as $x \leq x, \forall\, x \in R$.
(2) The relation is antisymmetric as $x \leq y$ and $y \leq x \Rightarrow x = y$.
(3) The relation is transitive as $x \leq y, y \leq z \Rightarrow x \leq z, \quad x, y, z \in R$

Since all the postulates of partial order relation are satisfied, the relation (less than or equal to, i.e., $\leq$) is a partial order relation.

Example 3 The relation of divisibility $a\ R\ b$ if a divides b, that is, a/b over the set of natural numbers N is a partial order relation.

Solution:

(1) The relation is reflexive as every natural number is a divisor of itself, that is, $a/a\ \forall a \in N$
(2) The relation is antisymmetric as a/b and $b/a \Rightarrow a = b$
(3) The relation is transitive as

$$
\begin{aligned}
a/b \text{ and } b/c &\Rightarrow b = ak_1 \text{ and } c = bk_2,\ k_1, k_2 \in N \\
&\Rightarrow c = ak_1k_2 \\
&\Rightarrow c = ak \qquad \text{where } k = k_1 k_2 \in N \\
&\Rightarrow a/c
\end{aligned}
$$

Hence, the relation of divisibility over N is a partial order relation.

Example 4 The relation of divisibility over the set of integers Z is not a partial order relation.

Solution: a/b and b/a do not imply $a = b$ and hence the relation is not antisymmetric.
For example, $5/-5$ and $-5/5 \not\Rightarrow 5 = -5$.

Example 5 The relation < (less than) over Z^+ is not a partial order relation.

Solution: The relation 'less than' is not reflexive and hence is not a partial order relation.

Note 8.1 The relation ≥ 'greater than or equal to' is a partial order relation over the set of real numbers. If R is a partial order relation on a set A, then R^{-1}, the inverse relation of R is also a partial order relation on A. The POSET (A, R^{-1}) is called Dual POSET of (A, R). The inverse of symbol ≤ is denoted by $\leq^{-1}$ or ≥. Thus, POSET $(A, \geq)$ is the Dual POSET of $(A, \leq)$, that is, if R is 'less than or equal to', then R^{-1} is 'greater than or equal to'.

8.2.1 Linearly Ordered or Totally Ordered Set

Let $(A, \leq)$ be a poset. If every pair of elements of A is comparable, that is, if for every $a, b \in A$, there exists $a \leq b$ or $b \leq a$, then the set A is called linearly ordered or totally ordered set. It is also called a *chain.*

Example 6 The set of natural numbers N, the set of real numbers R and the set of integers Z with relation 'less than or equal to' are totally ordered.

Example 7 Let $N = \{1, 2, 3,\ldots\}$ be the set of natural numbers ordered by divisibility relation. Find whether the following subsets of N are totally ordered.

(1) $\{2, 4, 8, 16\}$
(2) $\{2, 4, 8, 10\}$
(3) $\{3, 6, 9, 12, 13\}$
(4) $\{1\}$

Solution:

(1) Each pair (2, 4), (2, 8), (2, 16), (4, 8), (4, 16), (8, 16) is comparable with respect to divisibility relation and so the given set is *totally ordered.*
(2) The pairs (4, 10) and (8, 10) are not comparable and so the given set is *not totally ordered.*
(3) The pairs (3, 13), (6, 13), (9, 13), (12, 13) are not comparable and so the given set is *not totally ordered.*
(4) The subset $\{1\}$ is totally ordered, as the set containing a single element is *always totally ordered.*

8.3 PRODUCT OF TWO POSETS

An order relation on the Cartesian product of two sets can be defined. Such a relation is called product order.

THEOREM 8.1 If $(A, \leq)$ and $(B, \leq)$ are posets, then $(A \times B, \leq)$ is a poset with partial order ≤ defined by $(a, b) \leq (a', b')$, if $a \leq a'$ in A and $b \leq b'$ in B. It may be noted that the symbol ≤ being used denotes three distinct partial orders.

Proof: Let $a \leq a'$ in A and $b \leq b'$ in B (given)

(1) If $(a, b) \in A \times B$, then $(a, b) \leq (a, b)$. Since $a \leq a$ in A and $b \leq b$ in B and so satisfies the reflxive property in $A \times B$.

(2) Let us assume that

$$(a, b) \leq (a', b') \text{ and } (a', b') \leq (a, b)$$

where $a, a' \in A$ and $b, b' \in B$.

Then $\quad a \leq a'$ and $a' \leq a$ in A

and $\quad b \leq b'$ and $b' \leq b$ in B

Since A and B are posets, antisymmetry of partial orders in A and B implies $a = a'$ and $b = b'$.

Hence, $\leq$ satisfies antisymmetric property in $A \times B$.

(3) Suppose that $(a, b) \leq (a', b')$ and $(a', b') \leq (a'', b'')$ where $a, a', a'' \in A$ and $b, b', b'' \in B$

Now, $\quad a \leq a'$ and $a' \leq a'' \Rightarrow a \leq a''$

and $\quad b \leq b'$ and $b' \leq b'' \Rightarrow b \leq b''$

and hence $\quad (a, b) \leq (a'', b'')$

showing that $\leq$ satisfies transitive property in $A \times B$. Hence, $(A \times B, \leq)$ is a poset.

8.4 HASSE DIAGRAM

A diagram that presents partial order relation associated with a set is called Hasse diagram named after Helmunt Hasse, professor of mathematics, at Göttingen. This gives a method of representing finite posets $(A, \leq)$, which is as follows:

1. Each element of the poset is represented by a dot.
2. If $a \leq b$, then the dot for b is positioned higher than the dot for a.
3. If $a \leq b$ and there is no intermediate c with $a < c < b$, then a line is drawn from a to b.

Note 8.2

1. Since partial order is reflexive, each element (vertex) is related to itself. In a digraph, the path begins and ends at the same vertex. To simplify, these cycles are deleted in Hasse diagram.
2. Since partial order is transitive, that is, $a \leq b$, $b \leq c \Rightarrow a \leq c$, the edge from a to c is omitted but the edges from a to b and b to c are drawn.

The above consideration results in a simple representation of partial order relation known as Hasse diagram.

Example 1 Let $A = \{1, 2, 3\}$ and define partial order relation $\leq$ by $a \leq b$ if $a \leq b$ (a is less than or equal to b).

The digraph of this relation is

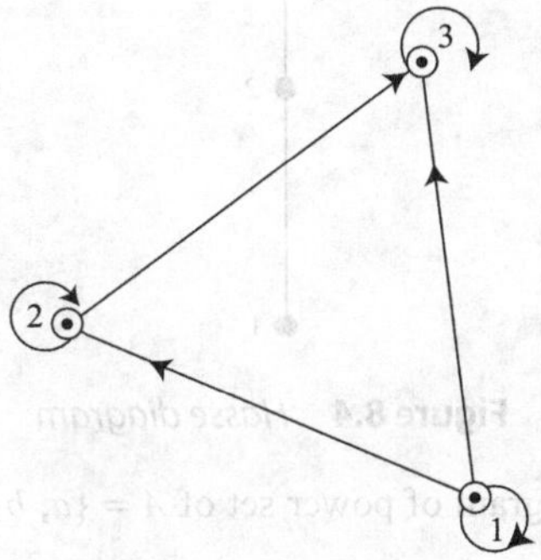

Figure 8.1 *Digraph*

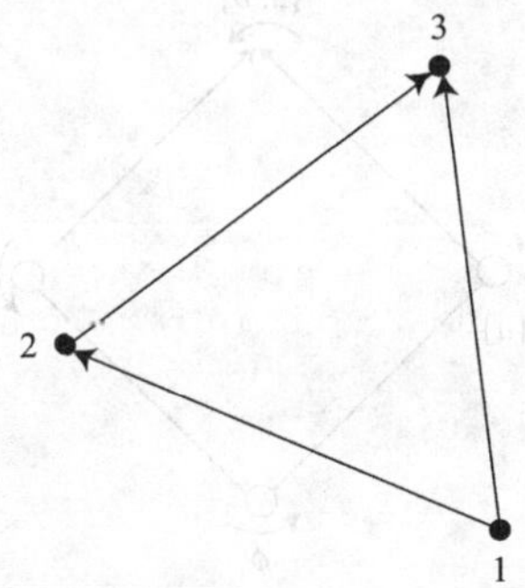

Figure 8.2 *Cycles deleted*

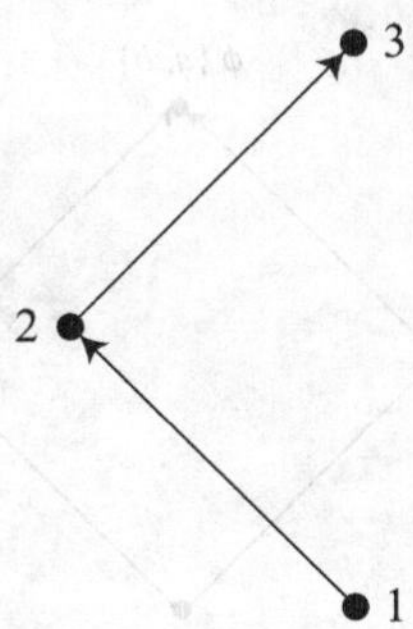

Figure 8.3 *Edge (1, 3) omitted*

Figure 8.4 *Hasse diagram*

Example 2 Draw the Hasse diagram of power set of $A = \{a, b\}$ with set inclusion $\subseteq$ as partial order relation.

Solution: The power set of A, $P(A) = \{\phi, \{a\}, \{b\}, \{a, b\}\}$

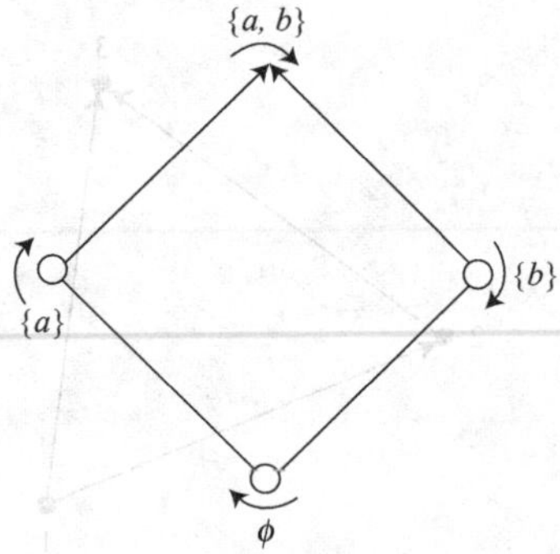

Figure 8.5 *Digraph*

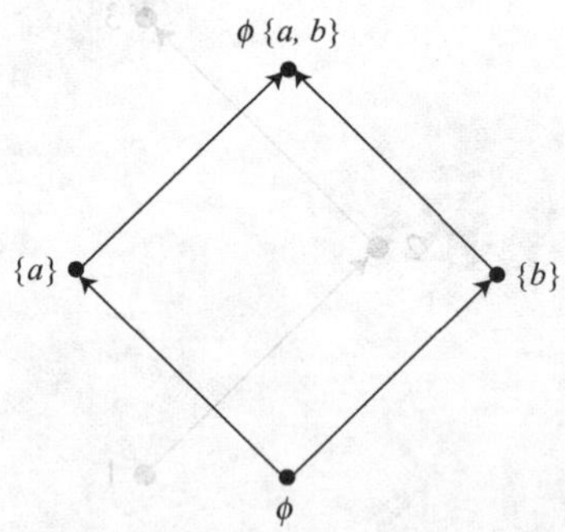

Figure 8.6 *Cycles deleted*

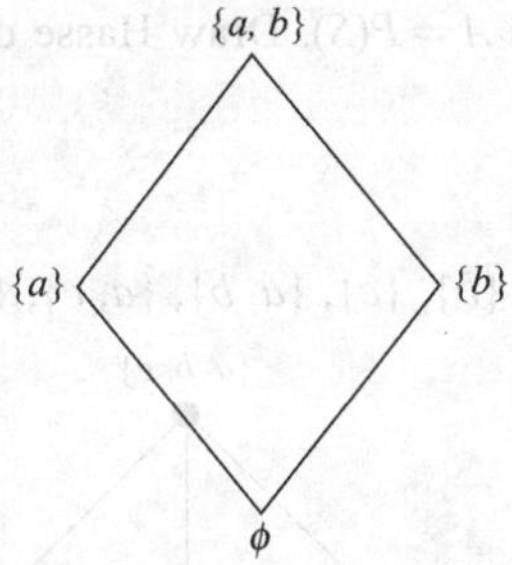

Figure 8.7 *Hasse diagram*

Example 3 Let $A = \{2, 4, 8, 16, 32\}$ and let the partial order relation $\leq$ be defined by $a \leq b$ if a/b. Draw the Hasse diagram for $(A, \leq)$.

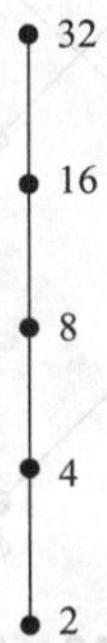

Figure 8.8 *Hasse diagram*

Example 4 Let A be the set of factors of 12 and let the partial order relation $\leq$ be defined by $a \leq b$ if a/b.

Solution:

$$A = \{1, 2, 3, 4, 6, 12\}$$

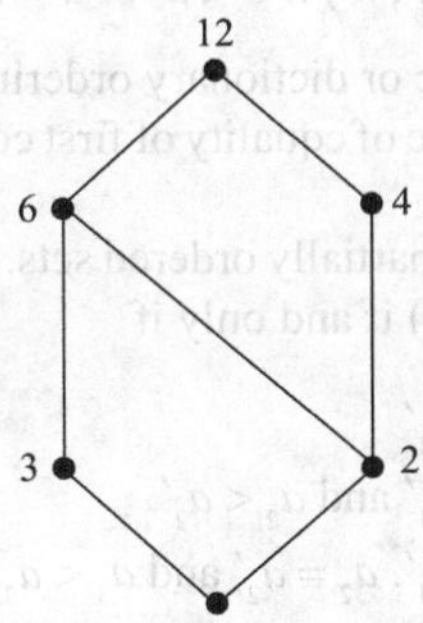

Figure 8.9

Example 5 Let $S = (a, b, c\}$ and $A = P(S)$. Draw Hasse diagram of poset A with partial order relation $\subseteq$ (set inclusion).

Solution:

$$A = \{\phi, \{a\}, \{b\}, \{c\}, \{a, b\}, \{a, c\}, \{b, c\}, \{a, b, c\}\}$$

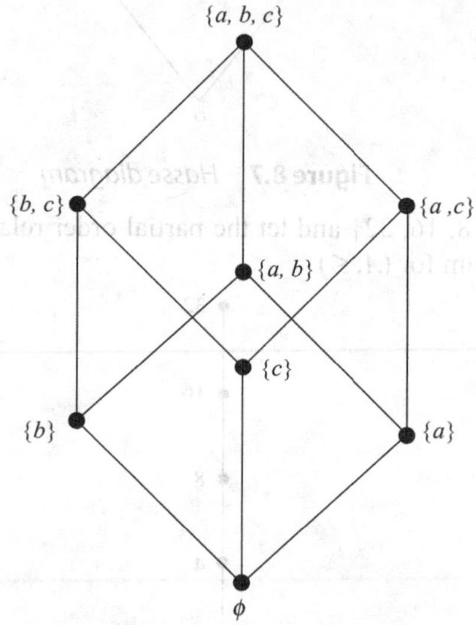

Figure 8.10 *Hasse diagram*

8.5 LEXICOGRAPHIC ORDERING

A partial order relation on $A \times B$ denoted by $<$ where $(A, \leq)$ and $(b, \leq)$ are partially ordered sets is defined by,

$$(a, b) < (a', b') \text{ if } a < a' \text{ or } a = a' \text{ and } b \leq b'$$

This ordering is called lexicographic or dictionary ordering. In this ordering, the first coordinate dominates except for equality. In case of equality of first coordinate, second coordinate dominates and so on.

Let $(A_1, \leq), (A_2, \leq), \ldots, (A_n, \leq)$ be partially ordered sets. Define partial order $<$ on $A_1 \times A_2 \times \ldots \times A_n$ by $(a_1, a_2, \ldots, a_n) < (a_1', a_2', \ldots, a_n')$ if and only if

$$a_1 < a_1'$$

or $$a_1 = a_1' \text{ and } a_2 < a_2'$$

or $$a_1 = a_1', a_2 = a_2' \text{ and } a_3 < a_3''$$

$$\ldots$$

$$\ldots, \ldots$$

$$a_1 = a_1', a_2 = a_2', \ldots, a_{n-1} = a'_{n-1} \text{ and } a_n \leq a'_n$$

Example: Let $A = \{a, b, c, \ldots, z\}$ be English alphabet, which is linearly ordered, that is, $a \le b$, $b \le c$, …, $y \le z$. Then, the set $A^n = A \times A \times \ldots \times A$ (n factors) can be regarded as a set of all words of length n. The lexicographic ordering on A^n has the property that if $v_1 < v_2$, then v_1 would precede v_2 in the dictionary listing. Thus, grape < graph, man < map, can < cap and so on.

8.6 UPPER AND LOWER BOUNDS

Let $(P, \le)$ be a partially order set and let $A \subseteq P$.

Definition 8.1 Upper Bound

An element $x \in P$ is called an upper bound of A, if $a \le x$ for every $a \in A$.

Example 1 Let $B = \{1, 2, 3\}$, then $(P(B), \subseteq)$ with set inclusion relation is a poset.

(1) Let $A = \{\phi, \{1\}, \{2\}, \{3\}, \{1, 2\}\}$, then $\{1, 2, 3\}$ is an upper bound of A because every element of A is contained in $\{1, 2, 3\}$.
(2) Let $A = \{\{3\}, \{1, 3\}\}$, then upper bounds of A are $\{1, 3\}$ and $\{1, 2, 3\}$.
(3) Let $A = \{\phi, \{2\}\}$, then upper bounds of A are $\{2\}$, $\{1, 2\}$, $\{2, 3\}$ and $\{1, 2, 3\}$.

Example 2 Let $A = \{2, 3, 6, 12, 24, 36\}$ be a poset with divisibility relation $\le$ on A.

(1) Consider a subset $B = \{2, 3, 6\}$ of A. The upper bounds are 6, 12, 24, 36.
(2) Let $C = \{12, 24\}$ be a subset of A. The upper bounds of C is 24.
(3) Let $D = \{12, 24, 36\}$; there is no upper bound as 24 does not divide 36 and so 36 cannot be upper bound.

Example 3 Let $N = \{1, 2, 3, \ldots\}$ be the set of natural numbers with the usual relation $\le$ (less than or equal to). Hence, $(N, \le)$ is a poset. All natural numbers ≥ 4 are upper bounds for the set $A = \{1, 2, 3\}$.

Definition 8.2 Lower Bound

Let $(P, \le)$ be a poset and let $A \subseteq P$. An element $x \in P$ is called the lower bound of A, $x \le a$ for every $a \in A$.

Example 4 (1) Let $P = \{1, 2, 3, 4, 5, 6, 7, 8\}$ be a poset whose Hasse diagram is Fig. 8.11(a).

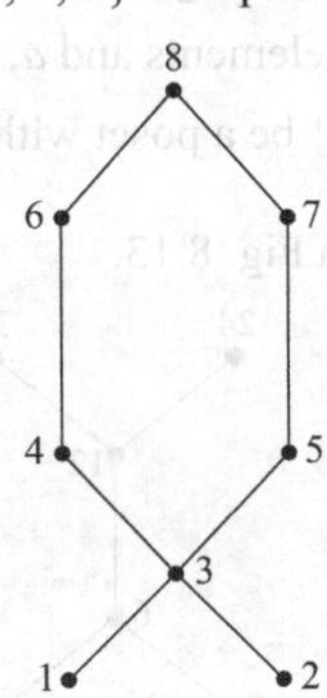

Figure 8.11(a)

Let $A = \{4, 5, 6, 7\}$
Then lower bounds are 1, 2 and 3.

(2) Let $P = \{1, 2, 3, 4, 5\}$ be ordered as shown in Fig. 8.11(b) and let $A = \{3, 4\}$ be its subset. The upper bounds of A are 5, 6 and 7; the lower bounds are 1 and 2.

Figure 8.11(b)

8.6.1 Extremal Elements of Partially Ordered Sets

Let $(A, \leq)$ be a poset with partial order $\leq$.

Maximal Elements: An element $a \in A$ is called a maximal element of A if there is no element $c \in A$ such that $a \leq c$.

Minimal Elements: An element $b \in A$ is called a minimal element of A if there is no element $c \in A$ such that $c \leq b$.

Note 8.3 There can be more than one maximal and minimal element in a poset.

Example 1 Consider the poset A whose Hasse diagram is shown in Fig. 8.12.

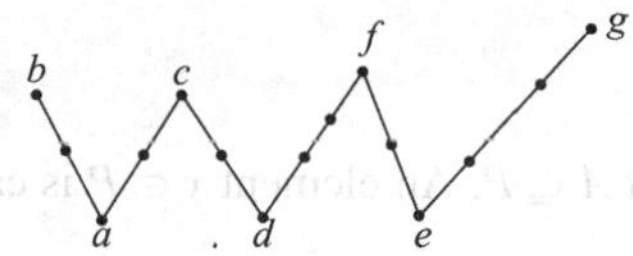

Figure 8.12

The elements b, c, f, g are maximal elements and a, d, e are minimal elements of A.

Example 2 Let $A = \{2, 3, 6, 12, 24, 36\}$ be a poset with partial order of divisibility, that is, $a \leq b$ if a divides b.

The Hasse diagram of A is shown in Fig. 8.13.

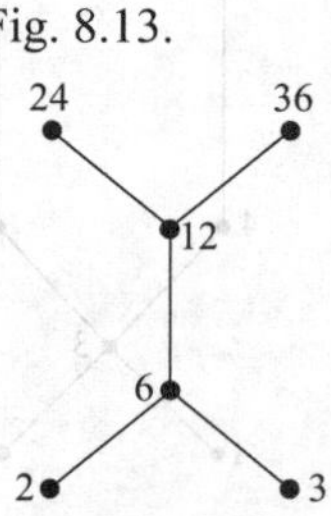

Figure 8.13

The maximal elements are 24 and 36 and the minimal elements are 2 and 3.

Example 3 Let A be the poset of non-negative real numbers with usual partial order less than or equal to ($\le$). Then, 0 (zero) is the minimal element of A. There are no maximal elements.

Example 4 The poset of set of integers Z with usual partial order relation $\le$ has no maximal and no minimal elements.

8.6.2 Least Upper Bound or Supremum

Let $(P, \le)$ be a poset and let $A \subseteq P$. An element x is said to be least upper bound (LUB) of A or supremum of A if x is an upper bound of A and $x \le y$ for all upper bounds of A.

The LUB of a poset, if it exists, is unique.

Example 1 Let $P = \{a, b, c, d, e, f\}$ and let $\le$ be the relation on P represented by Hasse diagram (Fig. 8.14).

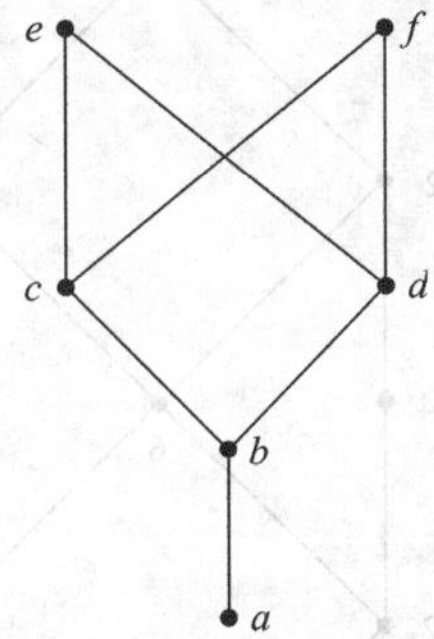

Figure 8.14

Let $A = \{b, c, d\}$

The upper bounds of A are e and f. Since e and f are not comparable, LUB of A does not exist.

Example 2 Let $P = \{1, 2, 3, 4, 5, 6\}$ be ordered and represented by Hasse diagram (Fig. 8.15).

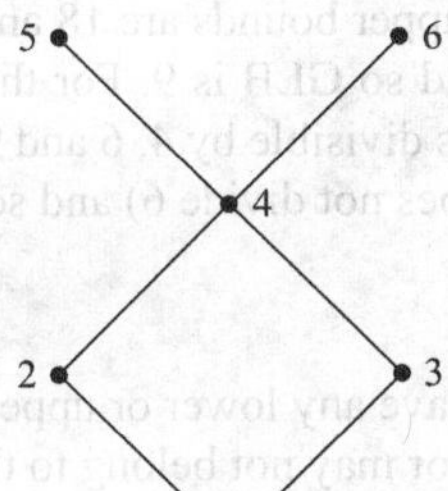

Figure 8.15

If $A = \{1, 2, 3\}$, then upper bounds of A are 4, 5 and 6 and LUB is 4.

8.6.3 Greatest Lower Bound or Infimum

Let $(P, \leq)$ be a poset and $A \subseteq P$. An element $x \in P$ is called the greatest lower bound (GLB) of A or infimum of A if x is the lower bound and $y \leq x$ for all lower bounds y of A.

A GLB is unique if it exists.

Example 1 Let $(N, \leq)$ be the poset of natural numbers. Let $A = \{3, 5, 7, 9, 11\}$, then 1, 2 and 3 are lower bounds of A and GLB or $\inf(A) = 3$.

Example 2 Consider the partially ordered set $P = \{1, 2, 3, 4, 6, 9, 12, 18, 36\}$ with divisibility relation $\leq$. Find GLB and LUB of the sets $\{9, 18\}$ and $\{4, 6, 9\}$.

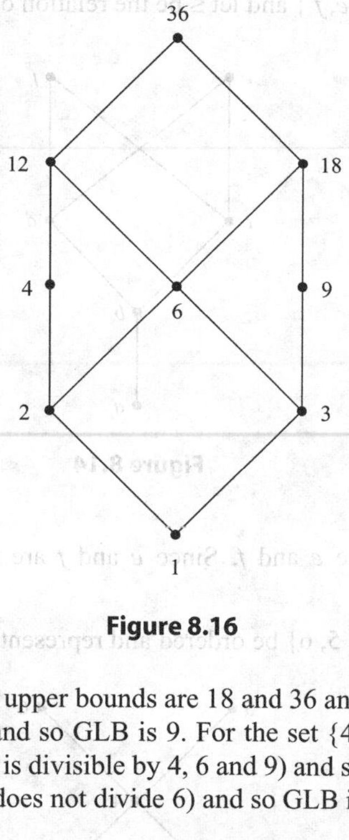

Figure 8.16

Solution: For the set $\{9, 18\}$, the upper bounds are 18 and 36 and so LUB is 18.

The lower bounds are 1, 3, 9 and so GLB is 9. For the set $\{4, 6, 9\}$, the upper bound is 36 (as this is the only element, which is divisible by 4, 6 and 9) and so LUB is 36. The lower bound is 1 (as 4 does not divide 9 and 4 does not divide 6) and so GLB is 1.

Remarks:

1. A subset of a poset may not have any lower or upper bound.
2. A lower or upper bound may or may not belong to the subset of poset.
3. There may exist more than one lower or upper bounds of a set.
4. The greatest element is always the supremum but the converse is not true, that is, $a = \sup(A)$ is the greatest element if and only if $a \in A$.
5. The least element is always the infimum but the converse is not true, that is, $b = \inf(A)$ is the least element if and only if $b \in A$.

8.7 DUAL OF A POSET

Let R be a partial order relation on a set A and let R^{-1} be the inverse relation of R, then R^{-1} is also a partial order relation, defined by a $R^{-1}b$ if and only if $b\ R\ a$, for $a, b \in A$.

THEOREM 8.2 If (A, R) is a poset, then (A, R^{-1}) is also a poset.

Proof:

(1) R^{-1} is reflexive
Since R is reflexive, $a\ R\ a, \forall a \Rightarrow a\ R^{-1}\ a$ by dual property.

(2) R^{-1} is antisymmetric
Let $a\ R^{-1}\ b$ and $b\ R^{-1}\ a$, then we have to prove that $a = b$.
Now,

$$a\ R^{-1}\ b \Rightarrow b\ R\ a \text{ and } b\ R^{-1}\ a \Rightarrow a\ R\ b.$$

But R is antisymmetric, $a\ R\ b, b\ R\ a \Rightarrow a = b$.
Thus, $a\ R^{-1}\ b$ and $b\ R^{-1}\ a \Rightarrow a = b$, and hence R^{-1} is antisymmetric.

(4) R^{-1} is transitive
Let $a\ R^{-1}\ b$ and $b\ R^{-1}\ c$
Now,

$$a\ R^{-1}\ b \Rightarrow b\ R\ a,\ b\ R^{-1}\ c \Rightarrow c\ R\ b$$

Further, $c\ R\ b$ and $b\ R\ a \Rightarrow c\ R\ a \Rightarrow a\ R^{-1}\ c$, and hence, R^{-1} is transitive.

Definition 8.3 Let $(P, \leq)$ be a poset, then $(P, \geq)$ is also a poset. $(P, \geq)$ is called dual of $(P, \leq)$. The least element of $(P, \leq)$ is the greatest element of $(P, \geq)$ and vice versa. For any subset A of P, GLB (A) in $(P, \leq)$ is equal to LUB of $(P, \geq)$.

8.8 ISOMORPHISM OF POSETS

Let $(A, \leq)$ and $(A', \leq')$ be posets. Let $f: A \to A'$ be a one-to-one onto correspondence between A and A'. The function f is called an isomorphism from $(A, \leq)$ to $(A', \leq')$, if for any $a, b \in A$, $a \leq b$ if and only if $f(a) \leq' f(b)$.

If $f: A \to A'$ is an isomorphism, then we say that $(A, \leq)$ and $(A', \leq')$ are isomorphic posets.

Example 1 Let A be the set of positive integers (Z^+) and let $\leq$ be the usual partial order on A. Let A' be the set of positive integers that are multiples of 3 and let $\leq'$ be the usual partial order on A'.

The function $f: A \to A'$ given by $f(a) = 3a$ is an isomorphism from $(A, \leq)$ to $(A', \leq')$.

Solution: f is one-one as $f(a) = f(b) \Rightarrow 3a \Rightarrow 3b \Rightarrow a = b$

Further, if $c \in A'$, then $c = 3a$ for same $a \in Z^+$, which shows that f is onto. Finally, if a and b elements of A, then clearly $a \leq b$ if and only if $3a \leq 3b$. Thus, f is an isomorphism.

Example 2 Let $A = \{1, 3, 4, 12\}$ and let $\leq$ be the divisibility relation. Let $B = \{\phi, \{1\}, \{2\}, \{1, 2\}\}$ and $\leq'$ be the inclusion relation.

Define $f: A \to B$ by

$$f(1) = \phi, f(3) = \{1\}, f(4) = \{2\}, f(12) = \{1, 2\}$$

Then, f is an isomorphism from $(A, \leq)$ to $(B, \leq')$.

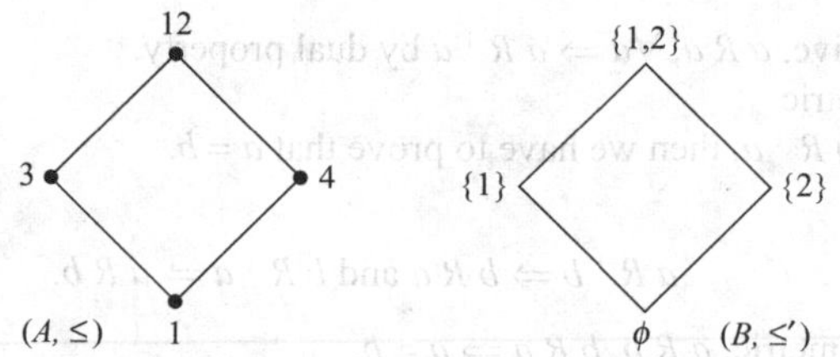

Figure 8.17 *Hasse diagrams of sets A and B*

Note 8.4 Let a poset A be isomorphic to a poset B and $f: A \to B$ be an isomorphism from A onto B. Then

1. An element $a \in A$ is minimal of A iff $f(a)$ is the minimal of B.
2. An element $a \in A$ is the greatest element of A iff $f(a)$ is the greatest element of B.

8.9 WELL-ORDERED SET

A set P with an ordering relation is said to be well-ordered, if every non-empty subset of P has a least element.

8.10 PROPERTIES OF WELL-ORDERED SETS

1. A well-ordered set is a linearly ordered set.
2. Every subset of a well-ordered set is well ordered.
3. If P is a well-ordered set and Q is an isomorphic to P, then Q is well ordered.

Example 1 The set of natural numbers with the usual order less than or equal to ($\leq$) is a well-ordered set.

Example 2 The poset $(Z, \leq)$ is not a well-ordered set, since the subset of negative integers does not contain a minimal element.

8.11 LATTICES

Lattices are used in several applications such as models of information flow and play an important role in Boolean algebra. Lattice structures often appear in computing.

Lattice in Terms of Partially Ordered Sets

A lattice is a partially ordered set $(L, \leq)$ in which every pair of elements $a, b \in L$ has a LUB and a GLB.

The GLB of a subset $\{a, b\} \in L$ is denoted by $a \wedge b$ (meet of a and b) and LUB by $a \vee b$ (join of a and b); since both LUB and GLB produce unique results on the subset $\{a, b\}$, meet and join can be regarded as binary operations. Thus, a lattice is a mathematical structure and can be denoted by $(L, \wedge, \vee)$.

Example 1 Let $(N, \leq)$ be a poset of natural numbers.
Define LUB $= \sup\{a, b\} = \max\{a, b\}$ and GLB $= \text{Inf}\{a, b\} = \min\{a, b\}$
Then, $(N, \leq)$ is a lattice.

Solution: Since $\max\{a, b\}$ and $\min\{a, b\}$ belong to N, $(N, \leq)$ is a lattice.

Example 2 Let $P(A)$ be the power set of a non-empty set A, then $(P(A), \leq)$ is a lattice where the relation $\leq$ is the inclusion $(\subseteq)$ of sets.

Solution: Let A_1 and A_2 be any two elements in $P(A)$.
We have $A_1 \vee A_2 = \sup\{A_1, A_2\} = A_1 \cup A_2 \in P(A)$ and $A_1 \wedge A_2 = \text{Inf.}\{A_1, A_2\} = A_1 \cap A_2 \in P(A)$.

Thus, every two elements of $P(A)$ have their LUB and GLB in $P(A)$. Hence, $[P(A), \subseteq]$ is a lattice.

Example 3 Let Z^+ be the set of positive integers and let $\leq$ define the relation by $a \leq b$ if a divides b, then $(Z^+, \leq)$ is a poset. For any $a, b \in Z^+$, define

$$a \vee b = \text{LCM}(a, b)$$

and

$$a \wedge b = \text{GCD}(a, b)$$

where LCM and GCD, respectively, denote least common multiple and greatest common divisor, then $(Z^+, \wedge, \vee)$ is a lattice.

Example 4 Let $A = \{1, 2, 3, 4, 5, 6, 10, 15, 30\}$ be the set of all divisors of 30. The LCM and GCD of any two elements of A belong to A. Thus, $(A, \leq)$ where $\leq$ denotes divisibility relation, is a lattice.

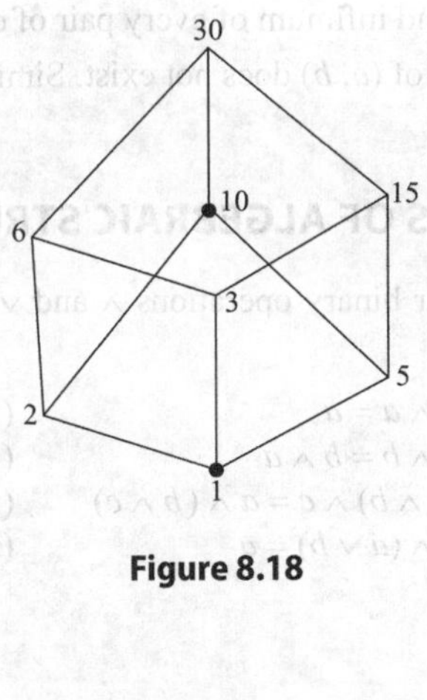

Figure 8.18

Example 5 Let $A = \{2, 3, 4, 6\}$ be a poset with $\leq$ being the divisibility relation. It is not a lattice since $\sup\{4, 6\} = \text{LCM}(4, 6) = 12$ does not belong to A.

Example 6 Find whether the following posets shown in Fig. 8.19b given below are lattices.

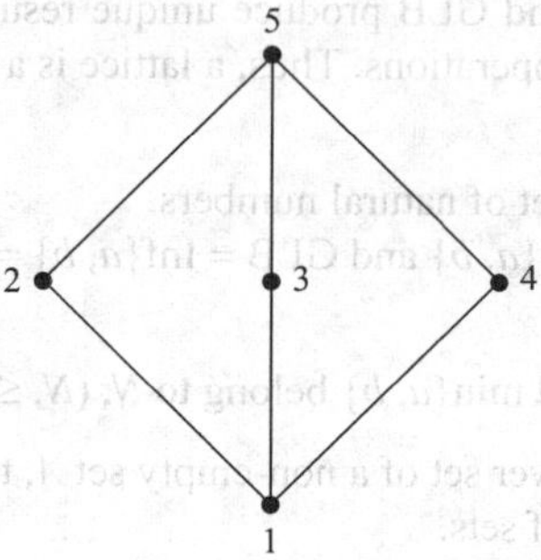

Figure 8.19(a)

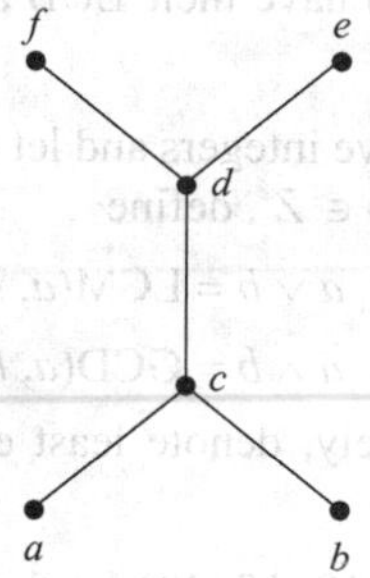

Figure 8.19(b)

Solution:

(a) is a lattice as supremum and infimum of every pair of elements exists.

(b) is not a lattice as infimum of (a, b) does not exist. Similarly, supremum of (e, f) does not exist.

8.12 LATTICE IN TERMS OF ALGEBRAIC STRUCTURES

A non-empty set L closed under binary operations $\wedge$ and $\vee$ is called a lattice $(L, \wedge, \vee)$, if the following axioms are satisfied.

1. Idempotent Law: (i) $a \wedge a = a$ (ii) $a \vee a = a$
2. Commutative Law: (i) $a \wedge b = b \wedge a$ (ii) $a \vee b = b \vee a$
3. Associate Law: (i) $(a \wedge b) \wedge c = a \wedge (b \wedge c)$ (ii) $(a \vee b) \vee c = a \vee (b \vee c)$
4. Absorption Law: (i) $a \wedge (a \vee b) = a$ (ii) $a \vee (a \wedge b) = a$

for all $a, b, c, \in L$.

8.13 SUBLATTICES

Let $(L, \wedge, \vee)$ be a lattice. A non-empty subset S of L is called a sub-lattice of L if S is closed under both operations $\wedge$ and $\vee$.

That is, if $a, b \in S \Rightarrow a \wedge b$ and $a \vee b \in S$ and thus S by itself is a lattice.

Example 1 Let $A=\{1,2,3\}$ and $P(A)=\{\phi,\{1\},\{2\},\{3\},\{1,2\},\{1,3\},\{2,3\},\{1,2,3\}\}$. The subset $S_1=\{\phi,\{2\},\{3\},\{2,3\}\}$ of $P(A)$ is a sub–lattice of $(P(A),\subseteq)$ while $S_2=\{\phi,\{2\},\{3\}\}$ of $P(A)$ is not a sub–lattice of $P(A)$ because sup $\{2, 3\}$ does not exist.

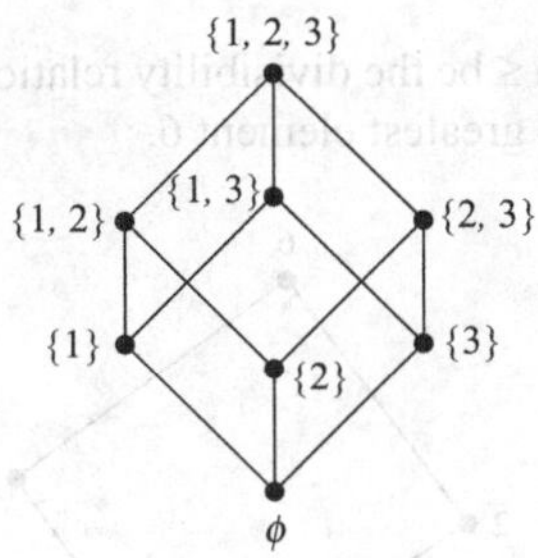

Figure 8.20(a)

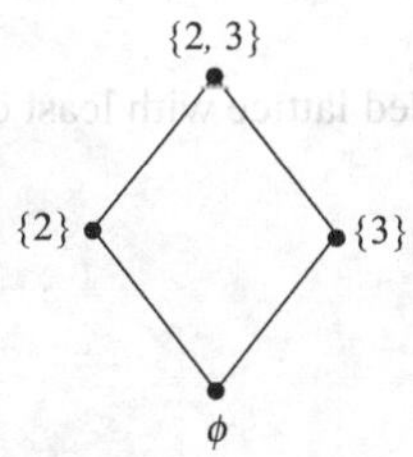

Figure 8.20(b)

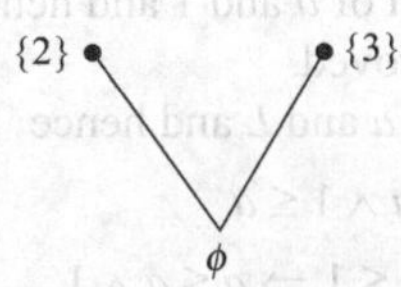

Figure 8.20(c)

Example 2 Let D_n be the set of all positive divisors of a positive integer n. Let Z^+ be the set of positive integers. Prove that D_n is a sub-lattice of Z^+ under the relation of divisibility.

Solution: Since $D_n \subseteq Z^+$ and every pair of elements $a, b, \in D_n$ has a GLB and LUB, D_n is a sublattice of Z^+.

8.14 BOUNDED LATTICES

A lattice L is said to be bounded if it has a least element 0 and a greatest element 1.

The non-negative integers with the usual ordering

$$0 < 1 < 2 < 3\ldots$$

has 0 as a lower bound but has no upper bound. The lattice $P(U)$ of all subsets of a universal set U is a bounded lattice with U as an upper bound and the empty set ϕ as lower bound.

If $L = \{a_1, a_2, \ldots\ldots\ldots, a_n\}$ is a lattice, then $a_1 \vee a_2 \vee \ldots \vee a_n$ and $a_1 \wedge a_2 \wedge \ldots \wedge a_n$ are upper and lower bounds of L.

Example 1 Let $L = \{1, 2, 3, 6\}$ and $\leq$ be the divisibility relation on L. Hence, $(L, \leq)$ is a bounded lattice as it has least element 1 and greatest element 6.

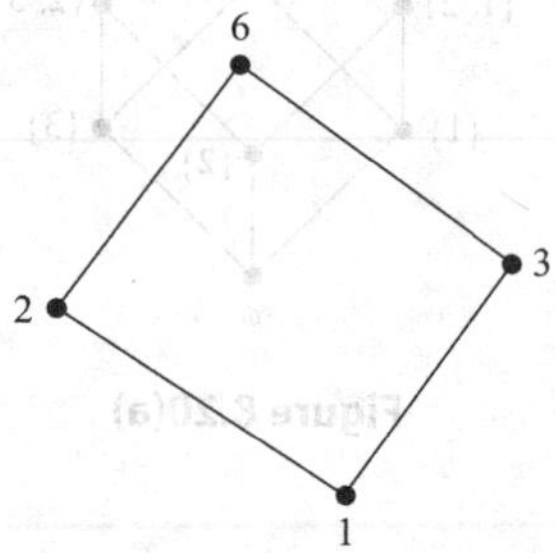

THEOREM 8.3 Let $(L, \leq)$ be a bounded lattice with least element 0 and greatest element 1, then for any element $a \in L$,

(1) $a \vee 1 = 1$ and $a \wedge 1 = a$
(2) $a \vee 0 = a$ and $a \wedge 0 = 0$

Proof:

(1) Let a be any element of L, then
$a \vee 1 \leq 1$ as 1 is the greatest element of L. (i)
However, $a \vee 1$ is the supremum of a and 1 and hence $1 \leq a \vee 1$. (ii)
From (i) and (ii), $a \vee 1 = 1$ Proved
Further, $a \wedge 1$ is the infimum of a and L and hence

$$a \wedge 1 \leq a \quad \text{(iii)}$$

and $a \leq a$ and $a \leq 1 \Rightarrow a \leq a \wedge 1$ (iv)

From (iii) and (iv), we have $a \wedge 1 = a$.

(2) Let $a \in L$ and 0 be the least element of L, then $0 \leq a$. Also, $a \leq a$.
Hence,

$$a \vee 0 \leq a. \quad \text{(v)}$$

As a is the upper bound of a and 0, and $a \vee 0$ is the LUB of $\{a, 0\}$,

$$a \leq a \vee 0 \quad \text{(vi)}$$

From (v) and (vi), we have $a \vee 0 = a$. Hence, proved
Since $a \wedge 0$ is the infimum of a and 0

$$a \wedge 0 \leq 0 \tag{vii}$$

Also,

$$0 \leq a \text{ and } 0 \leq 0$$

Hence,

$$0 \leq a \wedge 0 \tag{viii}$$

From (vii) and (viii), $a \wedge 0 = 0$. Hence, proved

Note 8.5 From the above results, it is clear that in a bounded lattice $(L, \wedge, \vee)$, 0 is identity for the operation $\vee$ and 1 is identity for the operation $\wedge$.

Some Theorems on Lattices

THEOREM 8.4 If $(L, \leq)$ is a lattice, then for every $a, b \in L$,
(1) $a \vee b = b$ if and only if $a \leq b$.
(2) $a \wedge b = a$ if and only if $a \leq b$.
(3) $a \wedge b = a$ if and only if $a \vee b = b$.

Proof:

(1) Let $a \vee b = b$
Since $a \leq a \vee b = b$ and hence $a \leq b$.
Conversely, let $a \leq b$, since $b \leq b$, b is an upper bound of a and b, $a \vee b$ is an upper bound of a and b, $a \vee b \leq b$.
Since $a \vee b$ is an upper bound, $b \leq a \vee b$, and hence, $a \vee b = b$ Proved.

(2) Let $a \wedge b = a$
Since infimum of a and b is a and so $a \leq b$.
Conversely, let $a \leq b$, we have to prove that $a \wedge b = a$.
Since $a \leq a$ and $a \leq b$, $a \leq a \wedge b$.
Further, since $a \wedge b$ is the GLB of a and b,

$$a \wedge b \leq a \qquad \therefore a \wedge b = a$$

(3) Let $a \wedge b = a$, then $a \leq b$ (from (ii) above)
If $a \vee b = b$, then $a \leq b$ (from (i) above)
Hence, $a \wedge b = a$ if and only if $a \vee b = b$.

THEOREM 8.5 Let $(L, \leq)$ be a lattice and $a, b, c \in L$. Then the following results hold:

(1) If $a \leq b$ and $a \leq c$, then $a \leq b \vee c$
(2) If $a \leq b$ and $a \leq c$, then $a < b \wedge c$

Proof:

(1) From the definition of join operation
$b \vee c = \sup\{b, c\}$ and hence $b \leq b \vee c$.
Hence,

$$a \leq b \text{ and } b \leq b \vee c \Rightarrow a \leq b \vee c.$$

(2) Let $a \leq b$ and $a \leq c$, then a is a lower bound of $\{b, c\}$.
Thus,

$$a \leq b \wedge c$$

8.15 DUALITY

The dual lattice of any lattice $(L, \wedge, \vee)$ is obtained by interchanging the operation $\wedge$ and $\vee$ and the relation $\leq$ by $\geq$. Thus, lattices $(L, \leq)$ and $(L, \geq)$ are called duals of each other.

For example, the set of natural numbers N in the partial order relation less than or equal to, that is, $(N, \leq)$, is the dual lattice of the lattice $(N, \geq)$.

8.16 COMPLETE LATTICE

A lattice L is said to be complete if every non-empty subset of L has its supremum and infimum in L.

For example, the poset $N = \{1, 2, 3,\ldots\}$ of natural numbers with usual relation 'less than or equal to' is a complete lattice.

8.17 ISOMORPHIC LATTICES

Two lattices $(L_1, \leq_1)$ and $(L_2, \leq_2)$ are said to be isomorphic if there exists a one-one onto function $f\colon L_1 \rightarrow L_2$, such that

$$f(a \leq_1 b) \Rightarrow f(a) \leq_2 f(b) \text{ for all } a, b \in L_1.$$

Example: Show that the lattice $L_1 = \{1, 2, 3, 6\}$ under divisibility relation and the lattice $L_2 = (P(A), \subseteq)$ where $A = \{a, b\}$ are isomorphic.

Solution:

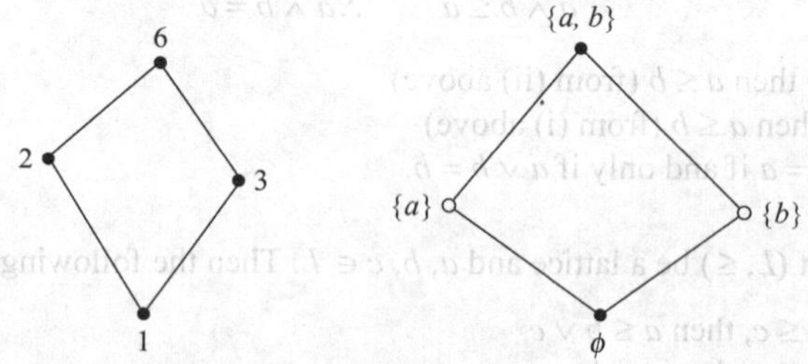

Figure 8.21

Define $f: L_1 \to L_2$ by

$$f(1) = \phi, f(2) = \{a\}, f(3) = \{b\}, f(6) = \{a, b\}$$

The mapping f is one-one and onto.
Also, for all $a, b \in L_1, f(a \vee b) = f(a) \vee f(b)$ and $f(a \wedge b) = f(a) \wedge f(b)$.

THEOREM 8.6 The dual of lattice is a lattice.

Proof: Let (L, R) be a given lattice and let (L, R^{-1}) be its dual where R is defined by

$$a\, R^{-1}\, b \text{ iff } b\, R\, a$$

Let $a, b \in L$, then LUB$\{a, b\}$ exists as R is a lattice.
Then,

$$a \vee b = \text{LUB}\{a, b\} \in L$$

Now,

$$a\, R\, (a \vee b) \text{ and } b\, R\, (a \vee b)$$

$$\Rightarrow (a \vee b)\, R^{-1}\, a \text{ and } (a \vee b)\, R^{-1}\, b$$
$$\Rightarrow a \vee b \text{ is lower bound in } (L, R^{-1})$$

To show that $a \vee b$ is GLB of $\{a, b\}$ in (L, R^{-1}).
Let c be any lower bound of (a, b) in (L, R^{-1}).
Then, $c\, R^{-1}\, a$ and $c\, R^{-1}\, b$

$$\Rightarrow a\, R\, c \text{ and } bRc$$
$$\Rightarrow c \text{ is an upper bound of } \{a, b\} \text{ in } (L, R)$$
$$\Rightarrow (a \vee b)\, R\, c$$

$$\therefore c\, R^{-1}\, (a \vee b) \Rightarrow a \vee b \text{ is GLB } \{a, b\} \text{ in } (L, R^{-1})$$

Similarly, it can be shown that $a \wedge b$ in LUB of (a, b) in (L, R^{-1}).
Hence, (L, R^{-1}) is a lattice.

8.18 COMPLIMENTED LATTICE

Let L be a bounded lattice with lower bound 0 and upper bound 1. An element x in a bounded lattice L is said to be the complement of its another element $y \in L$ provided

1. $x \wedge y = 0$
2. $x \vee y = 1$

Thus, we can say that x is the complement of y and y is the complement of x.

A lattice L is said to be complemented lattice if it is bounded and every element in L has at least one complement.

Note 8.6 An element may have more than one complement and the complementation operation is denoted by (−). $\overline{x}$ denotes the complement of x.

1. The lattice $P(A)$ of all subsets of a non-empty set A is a complemented lattice as each element $A_1 \in P(A)$ has a complement so that $A_1 \vee A_2 = A$ and $A_1 \wedge A_2 = \phi$.

2. $D_{30} = \{1, 2, 3, 5, 6, 10, 15, 30\}$ is a complemented lattice as each element has a complement.
3. The lattice $P(\cup)$ of all subsets of a universal set $\cup$ is complemented lattices as each subset $A \in \cup$ has the unique complement.
4. The lattice whose diagram is given below is a complemented lattice.

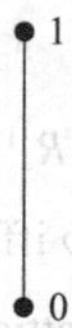

5. The lattice whose diagram is given below

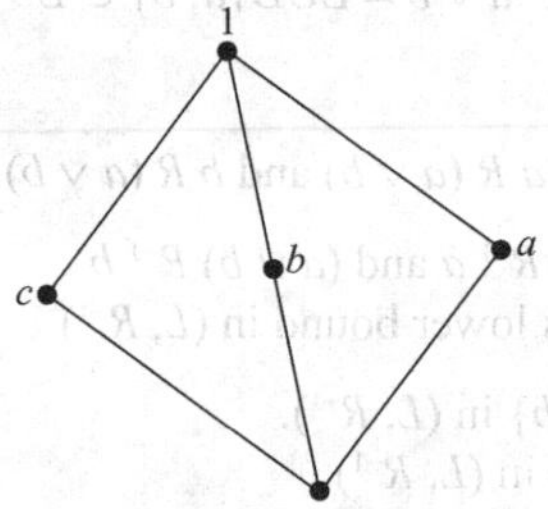

is a complemented lattice.

8.19 CHAIN AND ANTICHAIN

Let $(A, \leq)$ be a poset. A subset of A is called a chain, if every two elements in the subset are related.

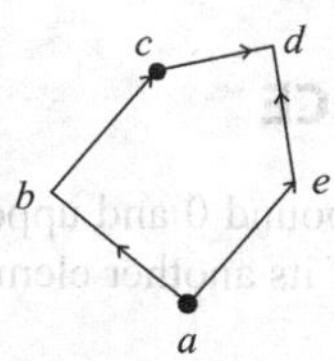

Figure 8.22

In Fig. 8.22 $\{a, b, c, d\}$, $\{a, b, c\}$, $\{a, e, d\}$ and $\{a\}$ are chains.

If A itself is a chain, then poset $(A, \leq)$ is called a totally ordered (or linearly order) set.

Antichain: A subset of A of a poset $(A, \leq)$ is called an antichain, if there exists no relation between two distinct elements in the subset.

1. Let $A = \{a, b\}$, then the subset $\{\{a\}, \{b\}\}$ is antichain. The Hasse diagram for $(A, \subseteq)$ is given in Fig. 8.23(a).

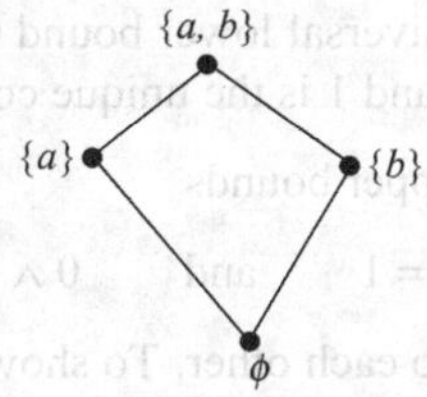

Figure 8.23(a)

The subsets ϕ, $\{a\}$, $\{b\}$, $\{a, b\}$; $\{\phi, \{a\}, \{a, b\}\}$, $\{\phi, \{b\}, \{a, b\}\}$, $\{\phi, \{a\}\}$, $\{\phi, \{b\}\}$ are chain. Here, the length of longest chain is 3 and the number of elements in antichain is 2.

2. The set {1, 2, 4, 8} is a chain under divisibility relation as the least element is 1 and greatest element is 8.

Figure 8.23(b)

3. {2, 3, 5} is not a chain under divisibility relation as there exists no least element in the Hasse diagram [Fig. 8.23(c)].

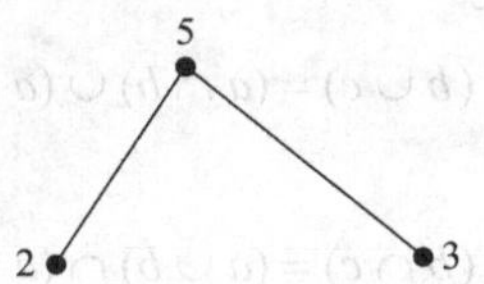

Figure 8.23(c)

4. {1, 2, 5, 10, 30} is a chain under divisibility relation; the least element is 1 and the greatest is 30 in the Hasse diagram [Fig. 8.23(d)].

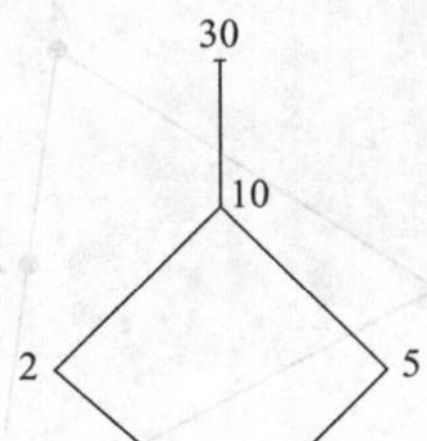

Figure 8.23(d)

THEOREM 8.7 In a lattice L with universal lower bound 0 and universal upper bound 1, show that 0 is the unique complement of 1 and 1 is the unique complement of 0.

Proof: Since 0 and 1 are lower and upper bounds

$$0 \vee 1 = 1 \qquad \text{and} \qquad 0 \wedge 1 = 0$$

and hence 0 and 1 are complements to each other. To show that 1 is a unique complement of 0, let us assume that a is another complement of 0.

Then,

$$0 \vee a = a \qquad 0 \wedge a = 0$$

$$\text{As } 0 \vee 1 = 1 \qquad \therefore a = 1$$

Thus, the complement of 0 is unique.

Similarly, it can be proved that 0 is the unique complement of 1.

8.20 DISTRIBUTIVE LATTICES

A lattice L is said to be distributive if for any elements $a, b, c \in L$

1. $a \wedge (b \vee c) = (a \wedge b) \vee (a \wedge c)$
2. $a \vee (b \wedge c) = (a \vee b) \wedge (a \vee c)$

If the lattice L does not satisfy the above properties, it is called a non-distributive lattice.

Example 1 The lattice of power set $P(S)$ of the finite set S is distributive under the operations of union ($\cup$) and intersection ($\cap$) since

$$a \cap (b \cup c) = (a \cap b) \cup (a \cap c)$$

and also

$$a \cup (b \cap c) = (a \cup b) \cap (a \cup c)$$

for any elements $a, b, c \in P(S)$.

Example 2 The lattice shown in Fig. 8.24 is distributive as it satisfies the distributive properties of all ordered triples 1, 2, 3 and 4.

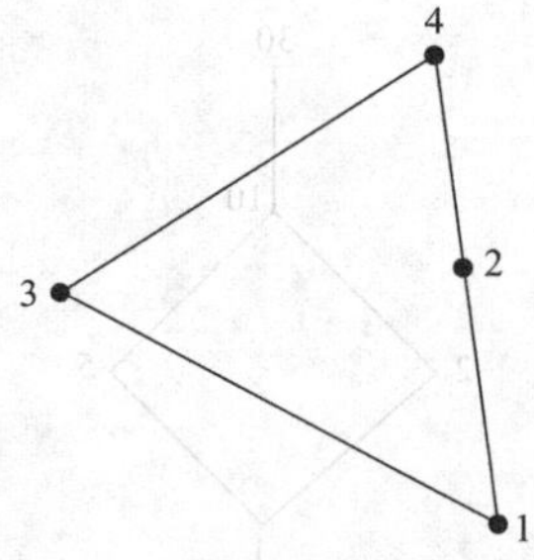

Figure 8.24

Consider, for example,

$$1 \wedge (2 \vee 3) = 1 \wedge 4 = 1$$

and

$$(1 \wedge 2) \vee (1 \wedge 3) = 1 \vee 1 = 1$$

THEOREM 8.8 If $(L, \leq)$ is a distributive lattice, then for any $a, b, c \in L$,

$$a \vee b = a \vee c \text{ and } a \wedge b = a \wedge c \Rightarrow b = c$$

Proof:

$$b = b \vee (b \wedge a) \quad \text{(absorption property)}$$

then

$$\begin{aligned} b &= b \vee (a \wedge b) && \text{(commutative property)} \\ &= b \vee (a \wedge c) && (\because a \wedge b = a \wedge c) \\ &= (b \vee a) \wedge (b \vee c) && \text{(distributive property)} \\ &= (a \vee b) \wedge (c \vee b) && \text{(commutative property)} \\ &= (a \vee c) \wedge (c \vee b) && (\because a \vee b = a \vee c) \\ &= (c \vee a) \wedge (c \vee b) && \text{(commutative property)} \\ &= c \vee (a \wedge b) && \text{(distributive property)} \\ &= c \vee (a \wedge c) && (\because a \wedge b = a \wedge c) \\ &= c \vee (c \wedge a) = c && \text{(absorption property)} \end{aligned}$$

THEOREM 8.9 In a distributive lattice, if an element has a complement, then this complement is unique.

Proof: Suppose that an element a has two complements b and c, then

$$a \vee b = 1 \qquad a \wedge b = 0$$
$$a \vee c = 1 \qquad a \wedge c = 0$$

Now, we have

$$\begin{aligned} b &= b \wedge 1 \\ &= b \wedge (a \vee c) \\ &= (b \wedge a) \vee (b \wedge c) \\ &= 0 \vee (b \wedge c) \\ &= (a \wedge c) \vee (b \wedge c) \\ &= (a \vee b) \wedge c \\ &= 1 \wedge c \\ &= c \end{aligned}$$

THEOREM 8.10 Every chain is a distributive lattice.

Proof: Let $(L, \leq)$ be a chain and let $a, b, c \in L$. Since L is a chain, any two of a, b and c are comparable. Let us consider the following cases.

Case I: Let $a \leq b$ or $a \leq c$

Then,

$$a \wedge (b \vee c) = a \quad \text{(i)}$$

and

$$(a \wedge b) \vee (a \wedge c) = a \vee a = a \quad \text{(ii)}$$

From (i) and (ii), we have $a \wedge (b \vee c) = (a \wedge b) \vee (a \wedge c)$

Case II: $a \geq b$ and $a \geq c$
Then,

$$(a \wedge b) \vee (a \wedge c) = b \vee c$$

Hence,

$$a \wedge (b \vee c) = (a \wedge b) \vee (a \vee c)$$

8.21 MODULAR LATTICE

A lattice $(L, \wedge, \vee)$ is said to be modular lattice if

$$a \vee (b \wedge c) = (a \vee b) \wedge c$$

whenever $a \leq c$.

THEOREM 8.11 In a modular lattice $(L, \wedge, \vee)$, for any $a, b, c \in L$

$$a \leq c \Rightarrow a \vee (b \wedge c) = (a \vee b) \wedge c.$$

Proof: Let $a \leq c$, then $a \vee c = c$
Also, if L is distributive, then

$$a \vee (b \wedge c) = (a \vee b) \wedge (a \vee c)$$
$$= (a \vee b) \wedge c \qquad (\because a \vee c = c)$$

Note 8.7

1. Every distributive lattice is modular.
2. Every modular lattice is not distributive.

Example 1 Show that the lattice in the Fig. 8.25 is a non-distributive modular lattice.

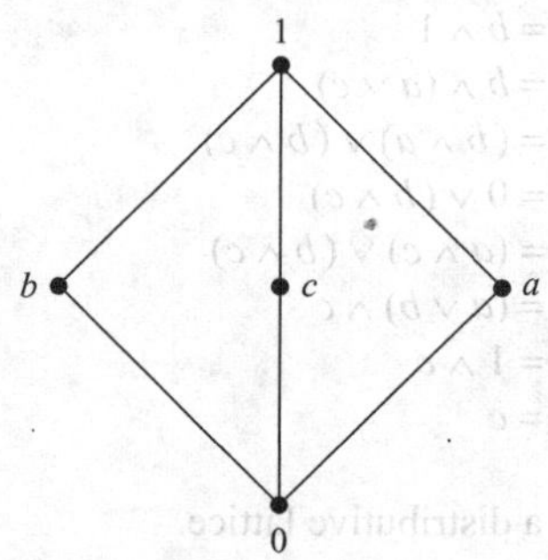

Figure 8.25

Solution: From distributive property of lattice

$$a \wedge (b \vee c) = a \wedge 1 = a$$

and
$$(a \wedge b) \vee (a \wedge c) = 0 \vee 0 = 0$$
$$\therefore a \wedge (b \vee c) \neq (a \wedge b) \vee (a \wedge c)$$

Hence, the result.

Example 2 Show that the pentagonal lattice shown in Fig. 8.26 is not modular.

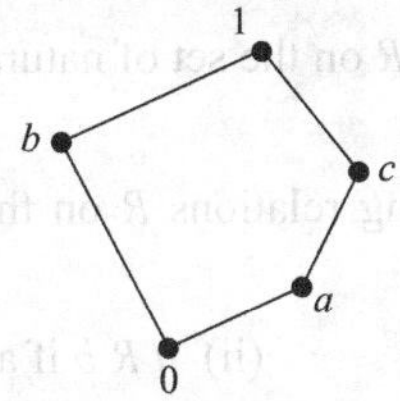

Figure 8.26

Solution: Here $a \leq c$

$$a \vee (b \wedge c) = a$$

and
$$(a \vee b) \wedge c = c$$

Therefore,

$$a \vee (b \wedge c) \neq (a \vee b) \wedge c.$$

and hence the lattice is not modular.

Example 3 Show that every chain is a modular lattice.

Solution: Since we cannot find any triplet a, b, c in a chain such that $a < c$ and b is not comparable with a or c, the lattice is modular.

THEOREM 8.12 Every distributive lattice is modular.

Proof: Let L be a distributive lattice. If $a, b, c \in L$, then

$$a \vee (b \wedge c) = (a \vee b) \wedge (a \vee c) \quad \text{(i)}$$

Also,

$$a \leq c \Rightarrow a \vee c = c \quad \text{(ii)}$$

From (i) and (ii)

$$a \leq c \Rightarrow a \vee (b \wedge c) = (a \vee b) \wedge c$$

Hence, the lattice L is modular.

8.22 BOOLEAN LATTICE

A complemented and distributive lattice is called a Boolean lattice. Let $(L, \leq)$ be a Boolean lattice. Since every element in a Boolean lattice has a unique complement, it defines an algebraic system $(L, \wedge, \vee, -)$, where $\vee$, $\wedge$ and $-$ are join, meet and complementation operations. An algebraic system defined by a Boolean lattice is known as Boolean algebra.

Let A be a finite set, then we know that $(P(A), \subseteq)$ is a lattice. Further, $(P(A), \subseteq)$ is a Boolean lattice in which the supremum (LUB) is A and infimum (GLB) is ϕ and the complement of any set B in $P(A)$ is $A - B$.

EXERCISES

1. Determine whether the relation R on the set of natural number N is a partial order relation where $a\ R\ b$ if and only if $a \le b$.

2. Determine whether the following relations R on the set of integers Z is a partial order relation where

   (i) $a\ R\ b$ if and only if $a = 5b$ (ii) $a\ R\ b$ if and only if $a = b^k$ for some $k \in z^+$

3. If a relation R on a set A is a partial order and an equivalence relation, then what can we say about R.

4. Prove that the relation 'less than or equal to, that is, ≤' on the set of real numbers is a linear order. Is the relation ≥, a linear order.

5. Prove that the set $A = \{3, 6, 12, 14, 18\}$ under the partial order of divisibility relation is linear.

6. Fine whether the post $B = \{5, 10, 30, 60, 120\}$ under divisibility relation is linear.

7. Draw the Hasse diagram of posets under the partial order relation of divisibility

   (i) $A = \{1, 2, 3, 5, 6, 10, 15, 30\}$ (ii) $B = \{6, 12, 36, 72, 144\}$

8. Determine the maximal and minimal elements of the following posets.

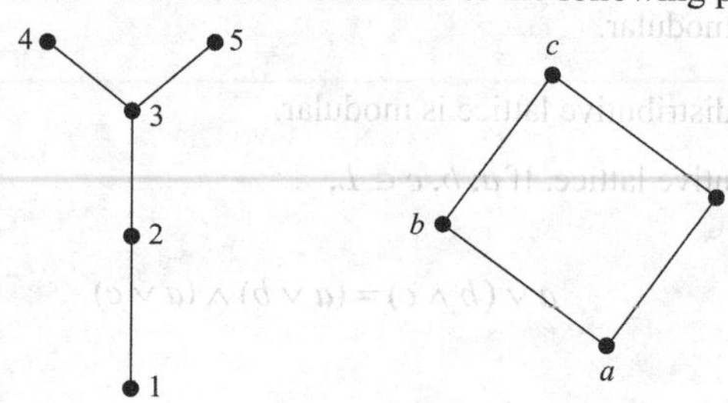

(a) (b)

9. Find the greatest and least elements of the following posets, if they exist.

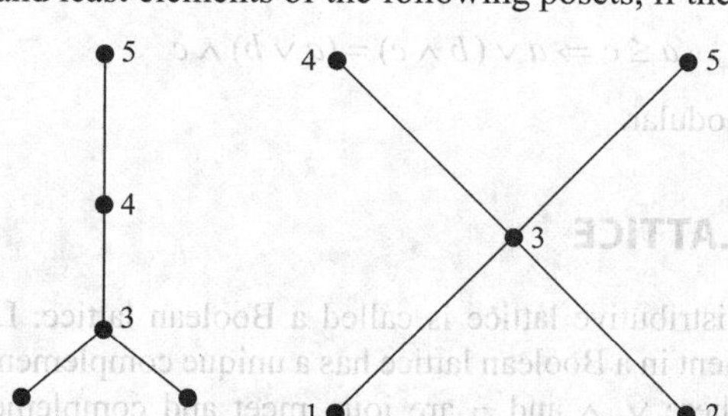

(a) (b)

10. Find the greatest and least elements of the posets A and B

    $A = \{x: \in R \text{ and } 0 < x < 2 \text{ with usual partial order relation} \le\}$

    and $B = \{x: x \in R \text{ and } 0 \le x \le 2 \text{ with usual partial order relation} \le\}$

11. Find the greatest and least elements of poset A with partial order relation of divisibility where

$$A = \{2, 4, 6, 8, 12, 18, 24, 36, 72\}.$$

12. Is the partially ordered set $A = \{2, 3, 6, 12, 24\}$ under the relation of divisibility a lattice?

13. Let $P(\cup)$ be the power set of the universal set $\cup$, then show that the poset $(P(\cup), \leq)$ is a lattice under the relation $\leq$, which is set inclusion $(\subseteq)$ of sets.

14. Let $A = \{1, 2, 4, 6, 8\}$ be a poset under divisibility relation. Find whether every pair of element has GLB and LUB. Draw its Hasse diagram. Is the set A a lattice.

15. Determine whether the following posets are lattice.

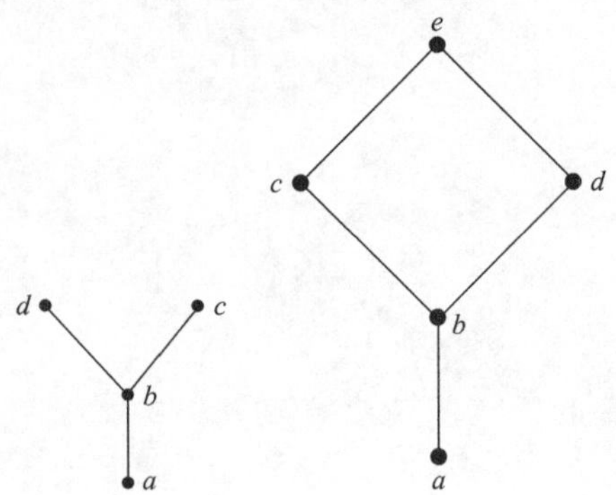

(a) (b)

16. Show that the set $D_{36} = \{1, 2, 3, 4, 6, 9, 12, 18, 36\}$ of positive divisors of 36 is a lattice under the operation $\wedge$ and $\vee$ defined by

$$a \wedge b = \inf(a, b) = \text{GCD}(a, b)$$

and $$a \vee b = \sup(a, b) = \text{LCM}(a, b)$$

17. Let Z^+ be the set of positive integers and D_n the set of all positive divisors of positive integer n, then prove that D_n is a sublattice of Z^+ under the relation of divisibility.

18. Let $A = \{1, 2, 3, 4, 5, 6\}$ be a lattice under the operations $\wedge$ and $\vee$ defined by GLB$\{a, b\}$ $= a \wedge b = \min\{a, b\}$ and LUB$\{a, b\} = a \vee b = \max\{a, b\}$.

    Prove that the following sets are sublattices:
    (i) $B = \{1, 2, 3, 4\}$
    (ii) $C = \{1, 2, 4, 6\}$

19. Let $A = \{a, b, c\}$ and $P(A)$ be the power set of A, then $(P(A), \subseteq)$ is a lattice. Prove that the subset $S_1 = \{\phi, \{b\}, \{c\}, \{b, c\}\}$ is a sublattice of $(P(A), \leq)$, while subset $S_2 = \{\phi, \{b\}, \{c\}\}$ of $P(A)$ is not a sublattice.

20. Show that the lattices $L = \{1, 2, 3, 6\}$ under the divisibility relation and the lattice $(P(A), \subseteq)$ where $A = \{a, b\}$ are isomorphic.

21. Let $A = \{1, 3, 6, 12\}$ be the poset under the relation $\leq$ of divisibility and $B = \{0, 1, 2, 3\}$ be the poset under the usual relation 'less than or equal to', that is, $\leq'$ on integers. Show that $(A, \leq)$ and $(b, \leq')$ are isomorphic.

9

Boolean Algebra

LEARNING OBJECTIVES

After reading this chapter, you will be conversant with:

- Introduction
- Definition: Boolean Algebra
- Disjunctive and Conjunctive Normal Forms (Canonical Forms)
- Switching Network from Boolean Expression
- Karnaugh Map

9.1 INTRODUCTION

Boolean algebra is named after English mathematician George Boole (1815–1864) whose publication in 1854 of the classic 'An investigation of Laws of Thought' led to the development of a mathematical system called Boolean algebra. In 1938, Claude E. Shanon used Boolean algebra to analyse electrical circuits, which opened the door for its applications in design circuitary for personal computers, pocket calculators, compact disc players, and other electronic systems.

Basically, Boolean algebra is a mathematical system consisting of a non-empty set with binary operations of logical addition (OR), logical multiplication (AND) and a unitary operator (Complementation). These operations are represented by symbols (+), (.), and (′), respectively.

9.2 DEFINITION: BOOLEAN ALGEBRA

A non-empty set B consisting of at least two elements 0 and 1 and two binary operations (+) and (.) called addition and multiplication and a unitary operator called complementation (denoted by ′) is called a Boolean algebra provided that the operations satisfy the following axioms for all x, y, and z in B.

1. Commutative Laws $\quad x + y = y + x \quad\quad x.y = y.x$
2. Associative Laws $\quad (x + y) + z = x + (y + z) \quad\quad (x.y).z = x.(y.z)$

3. Distributive Laws $x + (y.z) = (x + y).(x + z)$ $x.(y + z) = (x.y) + (x.z)$
4. Identity Laws $x + 0 = 0 + x = x$ $x.1 = 1.x = x$
5. Complement Laws $x + x' = 1$ $x.x' = 0$

Note 9.1

1. The operators (+) and (.) are symbols and do not stand for addition and multiplication in the sense of number systems.
2. The elements 0 and 1 are symbols, which have no relation with numbers 0 and 1 used in arithmetic. The numbers 1 and 0 are used to represent two possible states, which can be true or false. In electrical systems, they represent on and off states.
3. The elements 0 and 1 are called zero and unit elements, respectively, for addition and multiplication operations.

Example 1 Let $B = \{0, 1\}$ be a given set. Define binary operations of addition (+) and multiplication (.) given in the following tables.

+	0	1
0	0	1
1	1	1

.	0	1
0	0	0
1	0	1

Prove that $\{B, +, ., '\}$ is a Boolean algebra.

Solution: The binary operations (+) and (.) are defined as all the entries in the composition table belong to B.

(1) *Commutative Laws*: $0 + 1 = 1 + 0 = 1;\ 0.1 = 1.0 = 0$

(2) *Associative Laws*: From the table, we have

$(1 + 0) + 1 = 1 + 1 = 1$ $(1.0).1 = 0.1 = 0$
$1 + (0 + 1) = 1 + 1 = 1$ $1.(0.1) = 1.0 = 0$

Hence, associative laws hold for binary operations of addition and multiplication.

(3) *Distributive Laws*

$1 + (0.1) = (1 + 0).(1 + 1)$ $1.(0 + 1) = (1.0) + (1.1)$
$\text{LHS} = 1 + (0.1) = 1 + 0 = 1$ $\text{LHS} = 1.(0 + 1) = 1.1 = 1$
$\text{RHS} = (1 + 0).(1 + 1) = 1.1 = 1$ $\text{RHS} = (1.0) + (1.1) = 0 + 1 = 1$

Hence, the distributive laws are satisfied.

(4) *Identity Laws*: Clearly $x + 0 = x$ and $x.1 = x$ for every $x \in B$. Hence, 0 and 1 are identity elements, for the operations of addition and multiplication, respectively.

(5) *Complementation*: $0 + 0' = 0 + 1 = 1,\ 1 + 1' = 1 + 0 = 1$
$0.0' = 0.1 = 0,\ 1.1' = 1.0 = 0$

Hence, complement laws are satisfied.

Thus, all the axioms are satisfied. Hence, $\{B, +, ., '\}$ is a Boolean algebra.

Example 2 Let B be the set of positive numbers that are divisors of 30. Define binary operations (+), (.), and (′) by,

$$x + y = \text{lcm}\{x, y\},\ \ x.y = \text{gcd}\{x, y\}, \text{ and } x' = \frac{30}{x}$$

Show that $\{B, +, ., '\}$ is a Boolean algebra.

Solution: The given set is $B = \{1, 2, 3, 5, 6, 10, 15, 30\}$.

Since lcm and gcd of any two numbers of B belong to B, the binary operations (+) and (.) are defined.

(1) Commutative Laws

Since $$2 + 5 = \text{lcm}\{2, 5\} = 10 = \text{lcm}\{5, 2\} = 5 + 2$$

and $$2.5 = \text{gcd}\{2, 5\} = 1 = \text{gcd}\{5, 2\} = 5.2$$

Hence, commutative laws are satisfied, that is $a + b = b + a,\ a.b = b.a \quad \forall a, b \in B$

(2) Associative Laws

$$(2 + 3) + 5 = \text{lcm}\{2, 3\} + 5 = 6 + 5 = \text{lcm}\{6, 5\} = 30$$

$$2 + (3 + 5) = 2 + \text{lcm}\{3, 5\} = 2 + 15 = \text{lcm}\{2, 15\} = 30$$

$$\therefore (2 + 3) + 5 = 2 + (3 + 5) \quad \text{(addition is associative)}$$

Also $$(2.3).5 = \text{gcd}\{2, 3\}.5 = 1.5 = \text{gcd}\{1, 5\} = 1$$

and $$2.(3.5) = 2.\text{gcd}\{3, 5\} = 2.1 = \text{gcd}\{2, 1\} = 1$$

$$\therefore (2.3).5 = 2.\text{gcd}(3.5) \quad \text{(multiplication is associative)}$$

Hence, $$(a + b) + c = a + (b + c)$$

$$(a.b).c = a.(b.c) \quad \forall a, b, c \in B$$

(3) Distributive Laws

$$2 + (3.5) = 2 + \text{gcd}\{3, 5\} = 2 + 1\ \text{lcm}\{2, 1\} = 2$$

And $$(2 + 3).(2 + 5) = \text{lcm}\{2, 3\}.\text{lcm}\{2, 5\} = 6.10 = \text{gcd}\{6, 10\} = 2$$

$$\therefore 2 + (3.5) = (2 + 3).(2 + 5)$$

(addition is distributive over multiplication)

Also $$2.(3 + 5) = 2.\text{lcm}\{3, 5\} = 2.15 = \text{gcd}\{2, 15\} = 1$$

And $$(2.3) + (2.5) = \text{gcd}\{2, 3\} + \text{gcd}\{2, 5\} = 1 + 1 = \text{lcm}\{1, 1\} = 1$$

$$\therefore 2.(3 + 5) = (2.3) + (2.5)$$

(multiplication is distributive over addition)

As the above result is true for all the elements of B,

$$a + (b.c) = (a + b).(a + c)$$

and $$a.(b + c) = (a.b) + (a.c) \quad \forall a, b, c \in B$$

(4) Existence of Identity

$$x + 1 = 1 + x = \text{lcm}\{1, x\} = x \quad \forall x \in B$$

and $$x.30 = 30.x = \text{gcd}\{x, 30\} = x$$

Hence, 1 is identity for addition and 30 is identity for multiplication.

Verification

$$2 + 1 = 1 + 2 = \text{lcm}\{1, 2\} = 2$$

and $$2.30 = 30.2 = \text{gcd}\ \{2, 30\} = 2$$

(5) Complement Laws

$$5' = \frac{30}{5} = 6$$

$$5 + 5' = 5 + 6 = \text{lcm}\{5, 6\} = 30 \qquad \text{(Identity for multiplication)}$$

$$5.\ 5' = 5'.5 = \text{gcd}\{5, 6\} = 1 \qquad \text{(Identity for addition)}$$

Thus

$$x + x' = 30,\ x.x' = 1 \qquad \forall x \in B$$

Example 3 Let B be a set $\{0, 1\}$. Define binary operations (+) and (.) on B as follows:

$$x + y = \max\{x, y\},\ x.y = \min\{x, y\}. \text{ Let } 0' = 1 \text{ and } 1' = 0$$

Show that $\{B, +, ., '\}$ is a Boolean algebra.

Solution: Since the maximum and minimum values of the elements of B are in B, the binary operations (+) and (.) are defined.

(1) Commutative Laws

$$1 + 0 = \max\{1, 0\} = 1,\ 0 + 1 = \max\{0, 1\} = 1$$

$$\therefore 1 + 0 = 0 + 1$$

(addition is commutative)

And

$$1.0 = \min\{1, 0\} = 0,\ 0.1 = \min\{0, 1\} = 0$$

$$\therefore 1.0 = 0.1$$

(multiplication is commutative)

(2) Associative Laws

$$(1 + 0) + 1 = \max\{0, 1\} + 1 = 1 + 1 = \max\{1, 1\} = 1$$

$$1 + (0 + 1) = 1 + \max\{0, 1\} = 1 + 1 = \max\{1, 1\} = 1$$

$$\therefore (1 + 0) + 1 = 1 + (0 + 1)$$

(addition is associative)

And

$$(1.0).1 = \min\{1, 0\}.1 = 0.1 = \min\{0, 1\} = 0$$

$$1.\ (0.1) = 1.\min\{1, 0\} = 1.0 = \min\{1, 0\} = 0$$

$$\therefore (1.0).1 = 1.(0.1)$$

(multiplication is associative)

(3) Distributive Laws

$$1 + (1.0) = 1 + \min\{1, 0\} = 1 + 0 = \max\{1, 0\} = 1$$

$$(1 + 1).(1 + 0) = \max\{1, 1\}.\max\{1, 0\} = 1.1 = \min\{1, 1\} = 1$$

$$\therefore 1 + (1.0) = (1 + 1).(1 + 0)$$

(addition is distributive over multiplication)

Similarly, we can prove that multiplication is distributive over addition.

(4) Identity Laws

$$0 + 1 = 1 + 0 = \max\{1, 0\} = 1$$

and

$$0.1 = 1.0 = \min\{1, 0\} = 0$$

Hence, 0 is identity for addition and 1 is identity for multiplication.

(5) Complement Laws

$$\left.\begin{aligned} 1 + 1' &= 1 + 0 = \max\{1, 0\} = 1 \text{ i.e. } x + x' = 1 \\ 1.1' &= 1.0 = \min\{1, 0\} = 0 \text{ i.e. } x.x' = 0 \end{aligned}\right\} \text{ for } x \in B$$

Thus, $\{B, +, ., '\}$ is a *Boolean algebra.*

Note 9.2 As the design and analysis of electronic devices is based on two-element Boolean algebra, we shall confine our discussion to $\{B, +, ., ', 0, 1\}$.

9.2.1 Boolean Variables

Boolean variables are denoted by $x, y, z, \ldots$ or $x_1, x_2, x_3, \ldots, x_n$ and can assume values 0 and 1.

9.2.2 Boolean Expression

Boolean expressions are made up of Boolean Variables and Boolean operators. For example, $x + y, xy, x + y + z, x + (yz)'$ are Boolean expressions; $0, 1, x, y, z, \ldots$ are also Boolean expressions.

Example: Evaluate Boolean expression $y + (xz)'$ for the triplets (1, 0, 0) and (1, 0, 1)

Solution: When $x = 1, y = 0$, and $z = 0$,

$$y + (xz)' = 0 + (1.0)' = 0 + 0' = 0 + 1 = 1$$

When $x = 1, y = 0, z = 1,$

$$y + (xz)' = 0 + (1.1)' = 0 + 1' = 0 + 0 = 0$$

Equality (or Equivalence) of Boolean Expressions

Two Boolean expressions $E_1(x_1, x_2, \ldots, x_n)$ and $E_2(x_1, x_2, \ldots, x_n)$ are equal if they have the same value for all $(x_1, x_2, \ldots, x_n) \in B^n$ and is denoted by $E_1(x_1, x_2, \ldots, x_n) = E_2(x_1, x_2, \ldots, x_n)$ where $B^n = B \times B \times \ldots \times B$ and $B = \{0, 1\}$.

9.2.3 Boolean Function

A Boolean function is an algebraic expression of Boolean variables and logical operators OR, AND, NOT with parentheses, and an equal sign. Thus, a Boolean function is a function $f: B^n \to B$ where $f(x_1, x_2, \ldots, x_n)$ is either 0 or 1.

1. A Boolean function of n variables can be represented by a table having 2^n rows.
2. A Boolean function can be represented by different algebraic expressions.
3. A Boolean function is evaluated in the following order of computation: (1) parentheses, (2) NOT, (3) AND, (4) OR.

4. The complement of the function f is obtained by De Morgan's laws and is denoted by f'.
5. Every Boolean expression in n variables defines a Boolean function.

9.2.4 Minterm and Maxterm

A Boolean variable or its complement is called a literal. A minterm in n variables is a product of n literals in which variable or its complement appears only once. In case of two variables x and y, only four minterms are possible, that is, xy, $x'y$, xy', and $x'y'$. In case of n variables, there are 2^n minterms.

A maxterm of n variables is the sum of n literals in which each variable or its complement appears only once. In case of n variables, there will be 2^n maxterms.

Note 9.3 Each maxterm M_i is the complement of the corresponding minterm (m_i')

Table 9.1 For three variables minterm and Maxterm

Row Number	x	y	z	Minterms	Notation	Maxterms	Notation
0	0	0	0	$x'y'z'$	m_0	$x+y+z$	M_0
1	0	0	1	$x'y'z$	m_1	$x+y+z'$	M_1
2	0	1	0	$x'yz'$	m_2	$x+y'+z$	M_2
3	0	1	1	$x'yz$	m_3	$x+y'+z'$	M_3
4	1	0	0	$xy'z'$	m_4	$x'+y+z$	M_4
5	1	0	1	$xy'z$	m_5	$x'+y+z'$	M_5
6	1	1	0	xyz'	m_6	$x'+y'+z$	M_6
7	1	1	1	xyz	m_7	$x'+y'+z'$	M_7

Example 1 Find a Boolean expression that defines the Boolean function f by below shown table.

x	y	$f(x, y)$
0	0	0
0	1	1
1	0	1
1	1	0

To find the Boolean expression E for a give function f.

Consider rows that have functional value 1. For each such row, construct a Boolean sub-expression, which has the value 1 and has value 0 for all other combinations. The required expression is the sum of all such sub-expressions.

In the table, from the second row, $x = 0, y = 1, f(0, 1) = 1$ and so the sub-expression is $E_1 = x'y$. Similarly, from the third row, $x = 1, y = 0, f(1, 0) = 1$ and so the sub-expression is $E_2 = xy'$. Hence, the required function is $f(x, y) = x'y + xy'$.

Example 2 Construct a logic table for the Boolean function $f(x, y) = x + x'y$

x	y	x'	$x'y$	$f(x, y) = x + x'y$
0	0	1	0	0
0	1	1	1	1
1	0	0	0	1
1	1	0	0	1

Normal Boolean Function: A Boolean function of n variables in which all the n variables are involved is called a normal Boolean function. For example, $x + y, x' + y, x + y', x + xy'$ are normal Boolean functions of two variables x and y.

Some Properties (Theorems) on Boolean Algebra

THEOREM 9.1 For any Boolean algebra $\{B, +, ., '\}$,

(1) Additive identity 0 is unique.
(2) Multiplicative identity 1 is unique.

Proof: (1) Let 0 and 0_1 be two additive identities in B.

Then $0_1 + 0 = 0_1$ (as 0 is the Identity)
and $0 + 0_1 = 0$ (as 0_1 is the Identity)
But $0_1 + 0 = 0 + 0_1$ (Commutative law)
Hence, $0_1 = 0$

Identity element is unique. Similarly (2) can be proved.

THEOREM 9.2 The complement of every element in a Boolean algebra is unique.

Proof: Let x be any element in Boolean algebra B, then by the Complement law $x + x' = 1$ and $xx' = 0, x' \in B$

Let y be a second complement of x, then

$$x + y = 1 \text{ and } xy = 0$$

To show that the complement of an element is unique, we have to prove that $y = x'$

Now $y = y1 = y(x + x') = yx + yx' = xy + x'y$

$= 0 + x'y$

$= x'x + x'y = x'(x + y) = x'1 = x'$

Thus, the complement of every element in a Boolean algebra is unique.

THEOREM 9.3 Boundedness Laws: (1) $x+1=1$ (2) $x\,0=0$

Proof:

(1) $x+1 = x+(x+x')$ (Complement law)

$= (x+x)+x'$ (Associative law)

$= x+x'$ (Idempotent law)

$= 1$ (Complement law)

$\therefore x+1=1$

(2) $x\,0 = x0+0$ (0 being additive Identity)

$= x0+xx'$ (Complement law)

$= x(0+x')$ (Distributive law)

$= xx'$ (Identity law)

$= 0$

$\therefore x0=0$

THEOREM 9.4 Absorption Laws (1) $x+xy=x$ (2) $x\,(x+y)=x$ $\forall x, y \in B$

Proof:

(1) $x+xy = x1+xy$ (1 being multiplicative Identity)

$= x(1+y)$ (Distributive law)

$= x1$ (Boundedness law)

$= x$

(2) $x\,(x+y) = (x+0)(x+y)$ (0 is Identity for addition)

$= x+(0y)$ (Distributive law)

$= x+0$ (0 is Identity for addition)

$= x$

THEOREM 9.5 Idempotent Laws

(1) $x+x=x$ (2) $x.x=x$ $\forall x \in B$

Proof:

(1) $x = x+0 = x+x.x'$

$= (x+x)\cdot(x+x')$

$= (x+x)\cdot 1$

$= x+x$

Hence,

$$x = x+x$$

(2) $x = x.1 = x.(x + x')$
$= x.x + x.x'$
$= x.x + 0$
$= x.x$

Hence,
$x = x.x$

THEOREM 9.6(a) De Morgan's Laws

(1) $(x + y)' = x'y'$ (2) $(xy)' = x' + y'$ $\forall x, y \in B$

Proof: If $x'y'$ is the complement of $x + y$, then we must prove that

(i) $x + y + x'y' = 1$ and (ii) $(x + y)x'y' = 0$

(i) Now $(x + y) + x'y' = (x + y + x')(x + y + y')$ (Distributive law)
$= (y + x + x')(x + y + y')$ (Associative law)
$= (y + 1)(x + 1)$ (Complement law)
$= 11 = 1$ (Boundedness law)

Therefore, $(x + y) + x'y' = 1$ (i)

(ii) $(x + y)x'y' = x(x'y') + y(x'y')$ (Distributive law)
$= (xx')y' + x'(yy')$ (Associative and Commutative laws)
$= 0y' + 0x'$ (Complement laws)
$= 0 + 0 = 0$ (Identity laws)

Therefore, $(x + y)x'y' = 0$ (ii)
From (i) and (ii)

$$(x + y)' = x'y'$$

THEOREM 9.6(b) To prove that $x' + y'$ is the complement of xy, we have to prove

(1) $xy + (x' + y') = 1$ and (2) $(xy)(x' + y') = 0$
Now

$xy + (x' + y') = x' + y' + xy$ (Commutative law)
$= (x' + y' + x)\ (x' + y' + y)$ (Distributive law)
$= (x' + x + y')\ (x' + y' + y)$ (Commutative law)
$= (1 + y')\ (x' + 1)$ (Complement law)
$xy + (x' + y') = 1.1 = 1$ (i) (Boundness law)
And $(xy)(x' + y') = (xy)x' + (xy)y'$ (Distributive law)
$= (yx)x' + (xy)y'$ (Commutative law)
$= y(xx') + x(yy')$ (Associative law)
$= y0 + x0$ (Complement law)
$= 0 + 0 = 0$ (Boundedness law)

Therefore, $(xy)\ (x' + y') = 0$ (ii)
From (i) and (ii), we have $(xy)' = x' + y'$

THEOREM 9.7 Involution law

$$(x')' = x \qquad \forall x \in B$$

Proof: Let x be any element in B, then

$$\begin{aligned}(x')' &= 1.(x')' = (x + x').(x')' \\ &= x.(x')' + x'.(x')' = x.(x')' + 0 \\ &= x.(x')' + x.x' \quad (\because xx' = 0) \\ &= x[(x')' + x'] \\ &= x.1 = x\end{aligned}$$

Hence, $(x')' = x$

Some Illustrative Examples

Example 1 If $a + x = b + x$ and $a + x' = b + x'$, then prove that $a = b$

Solution: $(a + x)(a + x') = (b + x)(b + x')$ (from the given relations)

or $\quad a + (xx') = b + (xx')$

or $\quad a + 0 = b + 0 \quad \therefore a = b$

Example 2 In a Boolean algebra [B, +, ., ′], prove that

(1) $x + b = x + c$ and $x.b = x.c \Rightarrow b = c$

(2) $a.b' = 0 \Leftrightarrow a' + b = 1$

Solution:

(1) Let $x + b = x + c$ and $x.b = x.c$

then $b = b + b.x$ (By Absorption law)

$= b.b + b.x$ (Idempotent law)

$= b.(x + b) = b.(x + c)$ $(\because x + b = x + c)$

$= b.x + b.c = x.c + b.c$ $(\because x.b = x.c)$

$= (x + b).c = (x + c).c = x.c + c.c$

$= x.c + c$

$= c$ (By absorption law)

(2) $(a.b')' = 0' = 1 \quad (\because 0' = 1)$

or $a' + b = 1$ (By De Morgan's law)

Conversely, let $a' + b = 1 \Rightarrow (a' + b')' = 1' = 0$ $\quad \therefore a \,.\, b' = 0$ (By De Morgans law)

Example 3 If $a.x = b.x$ and $a.x' = b.x'$, then prove that $a = b$

Solution: From the given relations

$$a.x + a.x' = b.x + b.x'$$

or $\quad a.(x + x') = b.(x + x')$

or $\quad a.1 = b.1$

$\therefore a = b$

Example 4 Let $[B, +, ., ']$ be a Boolean algebra, then show that

(1) $x+(x+y)=x+y$ (2) $x \cdot (x \cdot y)=x \cdot y$
(3) $x'+x.y=x'+y$ (4) $x'+xy=x'+y$
(5) $x+x'y=x+y$

Solution:

(1) $x+(x+y)=(x+x)+y=x+y$ (By Associative law)
(2) $x \cdot (x \cdot y)=(x \cdot x) \cdot y=x \cdot y$ (By Associative and idempotent law)
(3) $x'+x \cdot y=(x'+x) \cdot (x'+y)$ (By Distributive law)
$= 1 \cdot (x'+y)=x'+y$

(4) $x'+xy=(x'+x) \cdot (x'+y)$ (By Distributive law)
$= 1 \cdot (x'+y)=x'+y$

(5) $x+x'y=(x+x') \cdot (x+y)$ (By Distributive law)
$= 1 \cdot (x+y)=x+y$

Example 5 In a Boolean algebra $[B, +, ., ']$ prove that $(x'.y)'+(x.y')'=1$

Solution: By De Morgan's law, left-hand side is

$$x+y'+x'+y=(x+x')+(y+y')$$
$$=1+1$$
$$=1 \text{ Proved}$$

Example 6 Let $\{B, +, ., '\}$ be a Boolean algebra and $x, y \in B$.
Then show that (1) $x+x'y=x+y$ (2) $x(x'+y)=xy$
(3) $x+(x+y)=x+y$ (4) $x'+xy=x'+y$

Proof:

(1) $x+x'y=x1+x'y$ (Identity law)
$=x(1+y)+x'y$ (Boundedness law)
$=x+xy+x'y$ (Distributive law)
$=x+(x+x')y$ (Distributive law)
$=x+1y$ (Complement law)
$=x+y$ (Identity law)
(2) $x(x'+y)=xx'+xy$ (Distributive law)
$=0+xy$ (Complement law)
$=xy$ (Identity law)
(3) $x+(x+y)=(x+x)+y$ (Associative law)
$=x+y$ (Idempotent law)

(4) $x'+xy=(x'+x)(x'+y)$ (Distributive law)
$=1(x'+y)$ (Complement law)
$=x'+y$ (Identity law)

Example 7 Construct a logic table for the Boolean function

$$f(x, y, z) = (x + y + z)(xyz)'$$

x	y	z	$x + y$	$x + y + z$	xy	xyz	$(xyz)'$	$f(x, y, z) = (x + y + z)(xyz)'$
0	0	0	0	0	0	0	1	0
0	0	1	0	1	0	0	1	1
0	1	0	1	1	0	0	1	1
0	1	1	1	1	0	0	1	1
1	0	0	1	1	0	0	1	1
1	0	1	1	1	0	0	1	1
1	1	0	1	1	1	0	1	1
1	1	1	1	1	1	1	0	0

9.3 DISJUNCTIVE AND CONJUNCTIVE NORMAL FORMS (CANONICAL FORMS)

A Boolean function $f(x_1, x_2, \ldots, x_n)$ of n variables is said to be in disjunctive normal form (DNF) or Canonical form if it is expressed as a sum of products (minterm form) in such a way that each term involves all the variables.

For example,

1. $xyz + x'yz + xy'z$
2. $xy'z' + xy'z + x'y'z'$
3. $xyz' + xy'z + xyz$

These are some DNF in three variables x, y, and z.

A Boolean function $f(x_1, x_2, \ldots, x_n)$ of n variables is said to be in Conjunctive Normal Form (CNF) if it is expressed in a factored form, that is, as a product of sum terms (or maxterms) such that each factor is the sum of all the n variables. This function is known as Dual Canonical form.

For example,

1. $(x + y + z) . (x' + y' + z')$
2. $(x + y + z') . (x' + y + z) . (x' + y' + z)$
3. $(x + y' + z) . (x + y + z') . (x' + y' + z')$

Some Examples on DNF and CNF

Example 1 Find DNF for the Boolean function defined by below table.

x	y	z	f(x, y, z)
0	0	0	0
0	0	1	0
0	1	0	1
0	1	1	0
1	0	0	0
1	0	1	1
1	1	0	1
1	1	1	1

To obtain DNF, we find min term for each row, which has functional value 1. The sum of all such terms is the required DNF.

The minterm corresponding to row 3 is $x'\,y\,z'$.

The minterms corresponding to rows 6, 7, and 8 are $x\,y'\,z$, $x\,y\,z'$, and xyz, respectively.

Hence, the required DNF for the given function is

$$f(x, y, z) = x'\,y\,z' + x\,y'\,z + x\,y\,z' + xyz$$

Example 2 Using laws of Boolean algebra, find the DNF for the Boolean function $f(x, y, z) = y(x + z)$.

Solution:

$f(x, y, z) = y(x + z) = yx + yz$ (Distributive law)

$= yx\,1 + yz\,1$ (Identity law)

$= yx\,(z + z') + yz\,(x + x')$ (Complement law)

$= xyz + xyz' + xyz + x'yz$ (Distributive, Commutative, and associative laws)

$= xyz + x'yz + xyz'$

Example 3 Using laws of Boolean algebra, find DNF of the Boolean function $f(x, y, z) = y + xz$.

Solution:

$$f(x, y, z) = y + xz = y\ 11 + xz\ 1 = y(x + x')(z + z') + xz(y + y')$$
$$= y(xz + xz' + x'z + x'z') + xyz + xy'z$$
$$= xyz + xyz' + x'yz + x'yz' + xyz + xy'z$$
$$= xyz + xyz' + x'yz + x'yz' + xy'z$$

To convert a Boolean function into DNF (disjunctive normal form), multiply those terms that do not contain all the variables by 1 and then replace it by the variables absent in the term and its complements.

Example 4 Obtain CNF of the Boolean function $f(x, y, z) = (x + z)(x + y')$

Solution: In the first factor y is missing and in the second factor z is missing. Adding 0 in each factor and replacing item by, yy′ and zz′. This gives

$$f(x, y, z) = (x + z + 0)(x + 0 + y') \quad \text{(0 is additive identity)}$$
$$= (x + z + yy')(x + y' + zz') \quad \text{(Complement law)}$$
$$= (x + z + y)(x + z + y')(x + y' + z)(x + y' + z') \quad \text{(Distributive law)}$$
$$= (x + y + z)(x + y' + z)(x + y' + z')$$

Example 5 Express the Boolean function $(x + y)(x' + z)$ in CNF form.

Solution: Let $f(x, y, z) = (x + y)(x' + z)$

In the first factor z is missing and in the second factor y is missing. Adding 0 in each factor and replacing these 0s by zz' and yy', we get

$$f(x, y, z) = (x + y + 0)(x' + 0 + z)$$
$$= (x + y + zz')(x' + yy' + z) \quad \text{(Complement laws as } yy' = 0,\ zz' = 0)$$
$$= (x + y + zz')(x' + z + yy')$$
$$= (x + y + z)(x + y + z')(x' + y + z)(x' + y' + z) \quad \text{(Distributive law)}$$

9.3.1 Complete Canonical Form

A canonical form (CNF or DNF) is said to be complete if its maximal or minimal forms include all the variables. A complete canonical form in n variables contains 2^n terms.

Example 1 Prove that the value of complete DNF in two variables is 1 and that of CNF is 0.

Solution: The complete DNF in two variables x and y is given by,

$$f(x, y) = xy + xy' + x'y' + x'y$$

$$= x(y + y') + x'(y' + y) = x1 + x'1 = x + x' = 1$$

The complete CNF in two variables x and y is given by

$$f(x, y) = (x + y)(x + y')(x' + y)(x' + y')$$

$$= [(x + y)(x + y')][(x' + y)(x' + y')] \quad \text{(Associative law)}$$

$$= (xx + yy' + xy' + yx)(x'x' + x'y' + yx' + yy')$$

$$= (x + xy' + xy + 0)\,(x' + x'y' + yx' + 0)$$

$$= [x + x(y' + y)]\,[x' + x'\,(y' + y)]$$

$$= (x + x\,1)\,(x' + x') = xx' = 0$$

Note 9.4 It can be proved that the value of complete DNF in three variables is 1.

Example 2 Find out the disjunctive and conjunctive forms of the Boolean function $f(x, y, z) = [x + (x' + y)'][x + (y'\,z')']$.

Solution:

$$f(x, y, z) = [x + (x' + y)'][x + (y'z')']$$

$$= (x + xy')(x + y + z) \quad \text{(De Morgan's law)}$$

$$= x\,(1 + y')(x + y + z)$$

$$= x.1\,(x + y + z)$$

$$= x(x + y + z) \quad \text{(i)}$$

$$= xx + xy + xz$$

$$= x\,11 + xy\,1 + xz\,1 \quad \text{(Distributive law)}$$

$$= x(y + y')\,(z + z') + xy(z + z') + xz(y + y')$$

$$= x(yz + yz' + y'z + y'z') + xyz + xyz' + xyz + xy'z$$

$$= xyz + xyz' + xy'z + xy'z' + xyz + xyz' + xyz + xy'z$$

$$= xyz + xy'z + xyz' + xy'z'$$

which is the required DNF.

CNF (Conjunctive Normal Form)

$$f(x, y, z) = x(x + y + z) \quad \text{(ii)} \quad \text{(from above relation)}$$

Now
$$x = x + 0 + 0$$
$$= x + yy' + zz'$$
$$= (x + y)(x + y') + zz' \quad \text{(Distributive law in first two terms)}$$
$$= zz' + (x + y)(x + y') \quad \text{(Commutative law)}$$
$$= (zz' + x + y)(zz' + x + y') \quad \text{(Distributive law)}$$
$$= (x + y + zz')(x + y' + zz') \quad \text{(Commutative law)}$$
$$= (x + y + z)(x + y + z')(x + y' + z)(x + y' + z') \quad \text{(Distributive law)}$$

which is the required CNF.

Example 3 Find the minimized Boolean expressions for

(1) $E_1 = y(x + y')$ (2) $E_2 = y + xy'$

Solution:

(1) $E_1 = y(x + y')$
$$= yx + yy' \quad \text{(Distributive law)}$$
$$= xy + 0 \quad \text{(Complement law)}$$
$$= xy$$

(2) $E_2 = y + xy'$
$$= (y + x)(y + y') \quad \text{(Distributive law)}$$
$$= (y + x)\,1 \quad \text{(Complement law)}$$
$$= y + x = x + y \quad \text{(Commutative law)}$$

Example 4 Find the minimized form of the expression

$$E = x'yz + x'yz' + x'y'z'$$
$$= x'y\,(z + z') + x'y'z' \quad \text{(Distributive law)}$$
$$= x'y\,1 + x'y'z' \quad \text{(Complement law)}$$
$$= x'y + x'y'z' \quad \text{(Identity law)}$$
$$= x'(y + y'z') \quad \text{(Distributive law)}$$

$$= x'(y + y')(y + z') \quad \text{(Distributive law)}$$

$$= x' \, 1(y + z') \quad \text{(Complement law)}$$

$$= x'(y + z') = x'y + x'z'$$

Example 5 Simplify the following Boolean function $f(x, y, z)$ to minimum where

$$f(x, y, z) = xyz + xyz' + xy'z + x'yz + xy'z'.$$

Solution:

$$\begin{aligned} f(x, y, z) &= xyz + xyz' + xy'z + xy'z' + x'yz \\ &= xy(z + z') + xy'(z + z') + x'yz \\ &= xy\,1 + xy'\,1 + x'yz \\ &= xy + xy' + x'yz \\ &= x(y + y') + x'yz \\ &= x1 + x'yz \\ &= x + x'yz \\ &= (x + x')(x + yz) \\ &= 1(x + yz) \\ &= x + yz \end{aligned}$$

9.4 SWITCHING NETWORK FROM BOOLEAN EXPRESSION

The Boolean algebra helps in designing electrical circuits from Boolean expressions using logic gates, which are simple devices that accept one or more input signals (i.e. bits 0 or 1) and produce unique output signal 0 or 1.

An *OR gate* receives two bits x and y as input signals and produces a unique output $x + y$ where

$$\begin{aligned} x + y &= 1 \quad \text{if } x = 1 \text{ or } y = 1 \\ &= 0 \text{ otherwise} \end{aligned}$$

An OR gate is represented by the symbol.

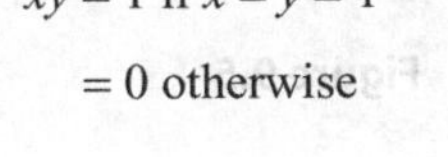

Figure 9.1

An *AND gate* receives two bits x and y as input signals and produces a unique output xy where

$$\begin{aligned} xy &= 1 \text{ if } x = y = 1 \\ &= 0 \text{ otherwise} \end{aligned}$$

An AND gate is represented by the symbol

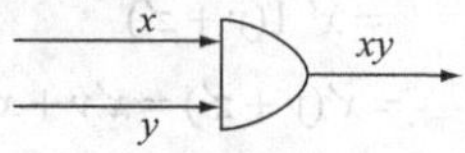

Figure 9.2

A NOT gate receives an input bit x and produces an output x' where

$$x' = 0 \quad \text{if } x = 1$$
$$= 1 \quad \text{otherwise}$$

A NOT gate is also known as an inverter. It produces output opposite to input and is represented by the symbol.

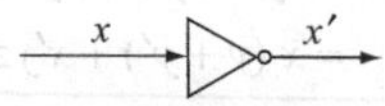

Figure 9.3

A NAND gate receives two input signals x and y and produces an output signal $x \uparrow y$ where

$$x \uparrow y = 0 \quad \text{if } x = y = 1$$
$$= 1 \quad \text{otherwise}$$

A NAND gate is represented by the symbol

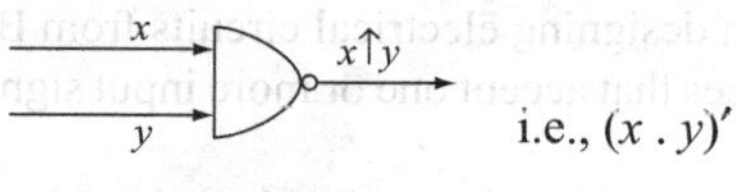

Figure 9.4

A *NOR gate* receives two input signals x and y and produces an output signal denoted by $x \downarrow y$ where

$$x \downarrow y = 1 \quad \text{if } x = y = 0$$
$$= 0 \text{ otherwise}$$

A NOR gate is represented by the symbol

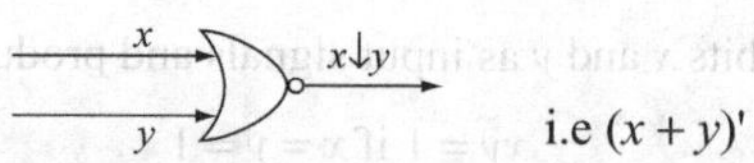

Figure 9.5

The five logic gates discussed above provide necessary basic operations for practical problems. The output is known for given input.

Some Real Life Problems Using Gate Models

1. AND gate
   A machine mower will only move if machine is on and the mower is engaged.

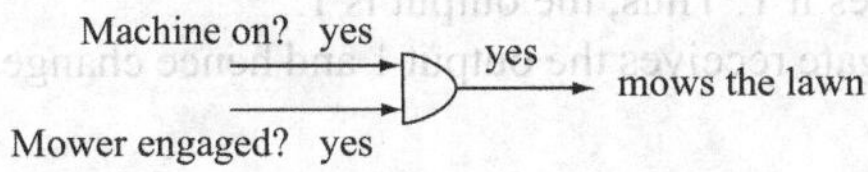

2. NOT gate
   (1) Power plants are generally associated with two generators, one main and one backup. The backup generator works only if the main generator fails.

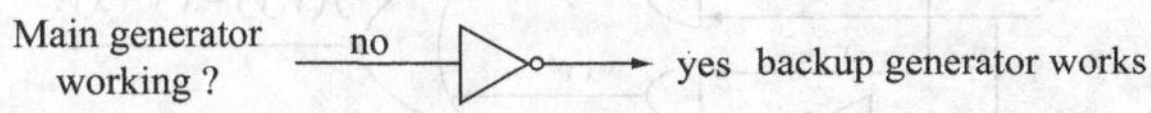

   (2) In our houses, when electricity mains is off from the supply, inverter is on

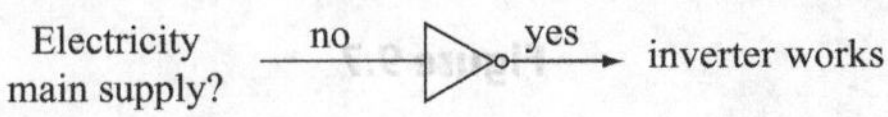

3. NAND gate

   Modern cars are fitted with alarms to warn forgetful drivers. Assume that the car key is in ignition. If the door on the driver's side is not closed or if his shoulder belt is not buckled up, a sound alarm will go automatically.

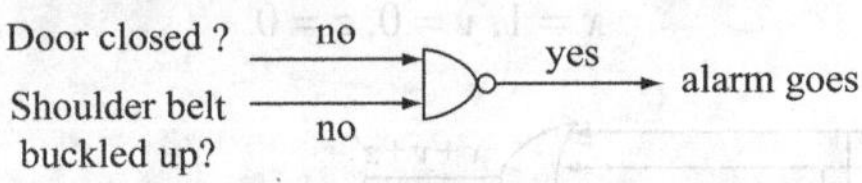

Example 1 Find the output for a given input from the following circuits.
Inputs (1) $x = 1$ and $y = 0$; (2) $x = 0$ and $y = 1$; (3) $x = 1$ and $y = 1$
Circuit (a)

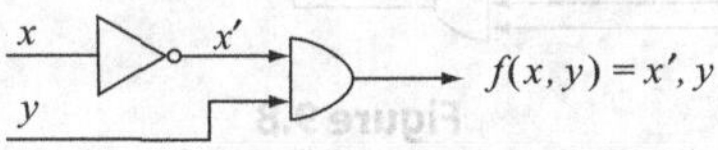

Figure 9.6(a)

Circuit (b)

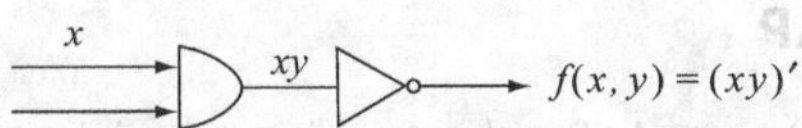

Figure 9.6(b)

Solution: We move from left to right in a given circuit finding the action of each gate for the input values.

(a) (1) The NOT gate changes the input $x = 1$ to 0 and hence both inputs to AND gate are 0. Consequently the output is 0.
(2) The NOT gate changes the input $x = 0$ to 1 and both inputs to AND gate are 1. Consequently, the output is 1.
(3) The NOT gate changes the input $x = 1$ to 0 and hence the AND gate has inputs 0 and 1. Consequently, the product is 0.

(b) (1) The AND gate receives input $x = 1, y = 0$ and so $xy = 0$. The NOT produces the result 1.

(2) The AND gate receives the signal $x = 0$ and $y = 1$ and hence output is 0 and NOT gate changes it 1. Thus, the output is 1.

(3) The NOT gate receives the output 1 and hence changes it to 0. Consequently, the output is 0.

Example 2 Draw the circuit for the function $f(x, y) = x'y + xy'$ and find the output for the input $x = 1$ and $y = 1$.

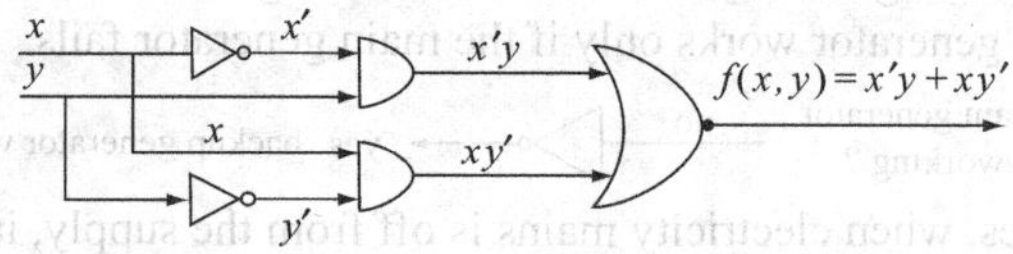

Figure 9.7

Example 3 Construct a circuit that produces the function

$$f(x, y, z) = (x + y + z')(xyz)$$

and find out the output corresponding to input

$$x = 1, y = 0, z = 0$$

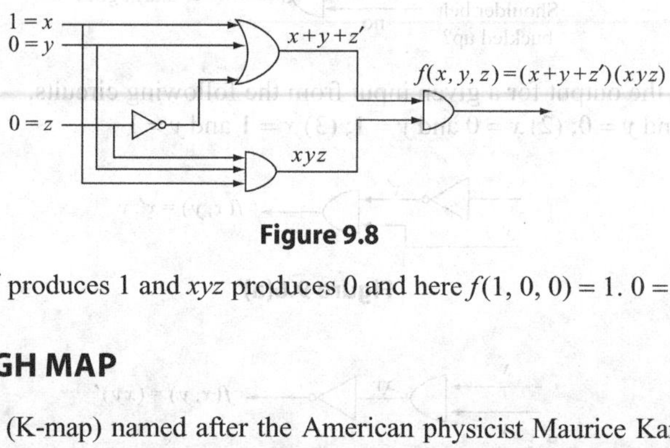

Figure 9.8

Solution: $x + y + z'$ produces 1 and xyz produces 0 and here $f(1, 0, 0) = 1. 0 = 0$.

9.5 KARNAUGH MAP

The Karnaugh Map (K-map) named after the American physicist Maurice Karnaugh is used to simplify Boolean expression. It provides a simple graphical method to simplify the DNF of Boolean expression by grouping minterms, which differ by exactly one literal known as adjacent terms and thus eliminating unwanted variables. The map will be used to simplify expressions involving 2, 3 and 4 variables, although it may be extended to more variables where it becomes different to check the best selection. For a Boolean function of n variables, K-map has rectangular grid of 2^n squares. Each square represents a minterm and 1 is written in corresponding squares of minterms present in the expression and 0 for minterms not present in the expression. The *Canonical Sum of products* can be obtained by OR ring together with those squares that contain 1 always beginning largest block. Groups may be horizontal and vertical but not diagonal. If squares corresponding to min terms are not adjacent, then each forms a block by itself and are included in the sum.

Two-variable K-map: The two variables x and y and their negations can be combined in four ways as shown in Table 9.6(a) and (b) given below.

Table 9.6(a)

	x'	x
y'	$x'y'$	xy'
y	$x'y$	xy

Table 9.6(b)

y \ x	0	1
0	00	10
1	01	11

Table 9.6(b) shows the alternative ways of using binary digits 0 and 1.

Example 1 Consider the Boolean expression $E(x, y) = x'y + xy$

The K-map is

	x'	x
y'	0	0
y	1	1

Two adjacent squares are enclosed by a rectangle. The two terms $x'y$ and xy give a resultant that eliminates the variable x as the variable x occurs in complemented and uncomplimented forms.

$$\therefore \quad E(x, y) = (x' + x)y = 1y = y.$$

Example 2 Simplify the Boolean expression $E(x, y) = xy + x'y + x'y'$ by drawing K-map.

Solution: The Karnaugh is

y \ x	x'	x
y'	1	0
y	1	1

In the map, there are two OR rings as there are two blocks that contain 1. The vertical block corresponds to x' as two adjacent squares correspond to y and y' and are thus eliminated as $y + y' = 1$. Similarly, horizontal block corresponds to y as the two adjacent squares corresponding to x and x' get eliminated. Thus,

$$E(x, y) = x' + y$$

By Boolean algebra we have, $E(x, y) = xy + x'y + x'y'$

$$= (x + x')y + x'y = 1y + x'y'$$

$$= y + x'y' = (y + x')(y + y') \qquad \text{(Distributive law)}$$

$$= y + x' = x' + y$$

= The same as given by K-map

9.5.1 Looping

In a K-map for the Boolean Expression $E(x, y)$ in variables x and y, squares containing binary digits 1 can be represented by two rectangles. The vertical square represents the output x' whereas the horizontal square represents y. Thus, the expression

$$E(x, y) = x' + y$$

This process of combining these digits is called looping.

Note 9.5 While looping to make groups, the following points should be observed:

1. Groups should be as large as possible containing power of two number of squares in each group.
2. Every 1 must be in at least one group.
3. Zeros are not allowed.
4. Diagonal grouping is not allowed.
5. The left most square in a row can be grouped with the right most square and the top square in a column can be grouped with the bottom square.

9.5.2 Three-variable K-Map

A K-map for three variables x, y, z shall have $2 \times 2 \times 2 = 2^3 = 8$ rec grids. Each of the squares will correspond to the possible minterms in x, y, and z.

	$y'z'$	$y'z$	yz	yz'
x'	$x'y'z'$	$x'y'z$	$x'yz$	$x'yz'$
x	$xy'z'$	$xy'z$	xyz	xyz'

	00	01	11	10
0	000	001	011	010
1	100	101	111	110

Example 3 Use K-map to find the minimal form of the following Boolean expressions:

(1) $E_1 = xy'z' + xy'z + xyz + xyz' + x'y'z'$

(2) $E_2 = xy'z' + xy'z + x'y'z' + x'y'z + x'yz + x'yz'$

Solution:

(1) The K-map for the expression E_1 is

x \ yz	$y'z'$	$y'z$	yz	yz'
x'	1			
x	1	1	1	1

The map gives $E_1 = x + y'z'$

Verification

Since 1 occurs in both the loops, the corresponding min term $xy'z'$ are counted twice using idempotent law:

$$
\begin{aligned}
E_1 &= xy'z' + xy'z + xyz + xyz' + (x'y'z' + xy'z') \\
&= x[y'(z' + z) + y(z + z')] + (x' + x)y'z' \\
&= x(y'1 + y1) + 1y'z' \\
&= x(y' + y) + y'z' = x1 + y'z' \\
&= x + y'z'
\end{aligned}
$$

Thus, $E_1 = x + y'z'$

(2) The K-map for the expression E_2 is

yz \ x	y′z′	y′z	yz	yz′
x′	1	1	1	1
x	1	1		

The map gives $E_2 = x' + y'z'$

Example 4 Simplify the Boolean expression

$$E = xy'z' + xy'z + x'y'z' + x'y'z + x'yz + x'yz'$$

by using K-map

Solution:

yz \ x	y′z′	y′z	yz	yz′
x′	1	1	1	1
x	1	1		

The K-map gives

Verification

$$
\begin{aligned}
E &= x' + y' \\
E &= x'y'z' + x'y'z + x'yz + x'yz' + xy'z' + xy'z \\
&= x'\,[y'\,(z' + z) + y(z + z')] + x[y'\,(z' + z)] \\
&= x'\,[y' + y] + xy' = x'1 + xy' = x' + xy'\ (z' + z = 1),\ \text{(Distributive law)} \\
&= (x' + x)(x' + y') = 1(x' + y') = x' + y'
\end{aligned}
$$

9.5.3 Four-Variable K-Map

A K-map for four variables x, y, z, and w can be represented by a grid of $4 \times 4 = 16$ squares which corresponds to 16 possible min terms in x, y, z, and w.

yz \ wx	y′z′	y′z	yz	yz′
w′x′	w′x′y′z′	w′x′y′z	w′x′yz	w′x′yz′
w′x	w′xy′z′	w′xy′z	w′xyz	w′xyz′
wx	wxy′z′	wxy′z	wxyz	wxyz′
wx′	wx′y′z′	wx′y′z	wx′yz	wx′yz′

yz \ wx	00	01	11	10
00	0000	0001	0011	0010
01	0100	0101	0111	0110
11	1100	1101	1111	1110
10	1000	1001	1011	1010

Example 5 Use the K-Map to find the minimal form of the Boolean functions

(1) $f_1(x, y, z) = x'yz + xyz$
(2) $f_2(x, y, z) = x'y'z' + x'yz'$

Solution:

(1) $f_1(x, y, z) = x'yz + xyz$
From the K-map, we have

yz \ x	y′z′	y′z	yz	yz′
x′	0	0	1	0
x	0	0	1	0

$$f_1(x, y, z) = yz$$

(2) $f_2(x, y, z) = x'y'z' + x'yz'$
From the K-map,

yz \ x	y′z′	y′z	yz	yz′
x′	1	0	0	1
x	0	0	0	0

$$f_2(x, y, z) = x'z'$$

Example 6 Use K-map to simplify the Boolean expression

$$E = w'xy'z' + w'xy'z + w'xyz + w'xyz' + wx'y'z' + wx'yz'$$

Solution: The K-map for the function E is the following:

yz \ wx	y′z′	y′z	yz	yz′
w′x′				
w′x	1	1	1	1
wx				
wx′	1			1

The K-map gives

$$E = w'x + wx'z'$$

Verification

$$\begin{aligned} E &= w'xy'z' + w'xy'z + w'xyz + w'xyz' + wx'y'z' + wx'yz' \\ &= w'x[y'(z' + z) + y(z + z')] + wx'z'(y' + y) \\ &= w'x(y' + y) + wx'z' \\ &= w'x + wx'z' \end{aligned}$$

Example 7 Using K-map simplify the Boolean expression

$$E = wxy'z + w'xy'z + w'x'y'z + wx'y'z + wxy'z' + wxyz + wxyz'$$

wx \ yz	y'z'	y'z	yz	yz'
w'x'		1		
w'x		1		
wx	1	1	1	1
wx'		1		

From the K-map $E = wx + y'z$

Example 8 Find the Karnaugh of the Boolean Expression

$$E = A'B'C'D' + AB'C'D' + A'B'CD' + AB'CD' \text{ and simplify it.}$$

Solution: The K-map is as follows:

$$E = A'B'C'D' + AB'C'D' + A'B'CD' + AB'CD'$$

AB \ CD	A'B'	A'B	AB	AB'
C'D'	1			1
C'D				
CD				
CD'	1			1

The top and bottom rows are considered to be adjacent to each other as they are the leftmost and rightmost columns. The variables B' and D' remain unchanged (A and C are in Complemented and uncomplemented forms).

Hence, the required expression is $E = B'D'$.

Note 9.6 These examples suggest that K-maps simplify Boolean expressions very easily in comparison with algebraic laws.

EXERCISES

1. Define Boolean algebra$\{B,+,.,'\}$. Write down the axioms of Boolean algebra.
2. State and prove De Morgan's theorem in both forms.
3. Find the complements of the following expressions by De Morgan's theorem:
   (i) $(A + BC)'$ (ii) $(A' + BC)'$
4. Simplify the following Boolean expressions:
   (i) $y(x + y')$ (ii) $x(x' + y)$
   (iii) $(x + y)(x + y')$ (iv) $xy + xz + y'x$
5. Evaluate the Boolean expression $y(xz' + x'z)$ at ordered triplets (0, 1, 1) and (1, 1, 1).
6. Evaluate the Boolean expression $(x + y + z)(x + y + z')$ at (1, 0, 1) and (1, 1, 0).
7. Construct a logic table for the Boolean expression $(x + y')(x' + y)$.

8. (i) Prove that the set $D_{70} = \{1, 2, 5, 7, 10, 14, 35, 70\}$ of positive factors is a Boolean algebra under the operations $\oplus$, $\odot$, and $'$ defined by,

$$x \oplus y = \text{lcm}(x, y),\ x \odot y = \gcd(x, y), \text{ and } x' = \frac{70}{x}$$

   (ii) Verify (a) $(5\oplus 7)' = 5'\odot 7'$ (b) $5\oplus (2\odot 7) = (5\oplus 2) \odot (5\oplus 7)$
   (iii) Evaluate
      (a) $7\oplus 10$ (b) $5\odot 7$ (c) $(5\oplus 2)'$
      (d) $10\oplus 10$ (e) $5\odot (5\oplus 7)$ (f) $7'\odot 5'$
9. Prove algebraically
   (i) $xx = x$ (ii) $0' = 1$ (iii) $1' = 0$
10. Find DNF of the following Boolean functions in variables x and y:
   (i) $f(x, y) = x + y$ (ii) $f(x, y) = (x + y)xy'$
11. Find DNF of the functions in variables x, y, z using laws of Boolean algebra.
   (i) $f(x, y, z) = x(y + z)$ (ii) $f(x, y, z) = y + zx$
   (iii) $f(x, y, z) = x(y' + z) + z'$ (iv) $f(x, y, z) = (x' + y)' + x'y$
   (v) $f(x, y, z) = (xy' + xz')' + x'$
12. Find DNF of the Boolean function defined by the following table:

x	y	z	f(x, y, z)
0	0	0	0
0	0	1	0
0	1	0	0
0	1	1	1
1	0	0	0
1	0	1	1
1	1	0	1
1	1	1	1

13. Express the following Boolean functions in CNF:
   (i) $f(x, y, z) = (x + y)\,(x + y')\,(x' + z)$
   (ii) $f(x, y, z) = (y + z)\,(y + z')\,(y' + x)$
14. Find the output if the bits 0, 1, and 1 are inputs in each gate:
   (i) AND gate (ii) OR gate
15. (i) Compute the NAND gate output by input of the following pairs of bits:
      (a) 0,0 (b) 1,0
   (ii) Compute the NOR gate output by the input of the pairs of bits
      (a) 0,0 (b) 1,0
16. Simplify the following Boolean expressions by using Karnaugh graph, that is, K-graph:
   (i) $E(x, y) = xy' + x'y$
   (ii) $E(x, y) = xy' + x'y + x'y'$

(iii) $E = ABC + ABC' + A'B'C$
(iv) $E = A'B'CD + A'B'CD' + AB'C'D' + AB'CD'$
(v) $f(x, y, z) = x'yz + xyz' + xyz + x'yz'$

17. When will the combinatorial circuit for each of the Boolean expressions produce 1 as the output:
(i) x' (ii) xy (iii) $x + y$

18. Express the output of inputs x, y, z for logical circuits given in Fig. (a) and (b):

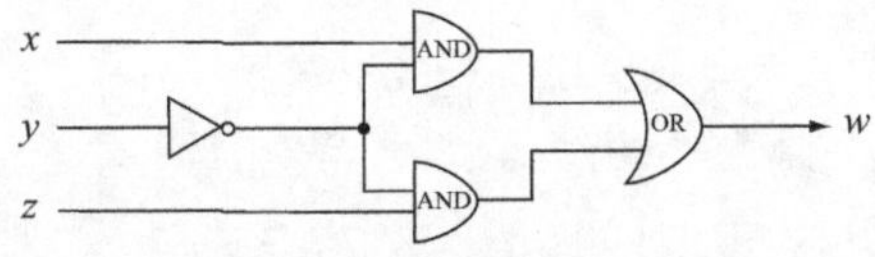

(a)

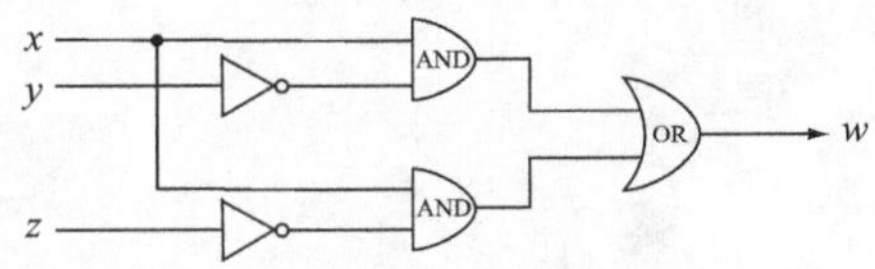

(b)

19. Present a logical circuit for the Boolean expression in inputs A, B, C to output. Y where

$$Y = A'BC + AB'C + AC'$$

20. Construct a combinatorial circuit for the following Boolean expressions:
(i) $(x + y)'z$ (ii) $x + (yz)'$
(iii) $(x + y)z' + yz'$

10

Topics in Graph Theory

LEARNING OBJECTIVES

After reading this chapter, you will be conversant with:

- Introduction
- Graph Definition
- Planar and Non-planar Graphs
- Region
- Operations on Graphs
- Bipartite Graph
- Isomorphism
- Representation of Graphs in Computer Memory
- Representation of Multi Graph
- Walk in a Graph
- Sub-graph
- Connected and Disconnected Graphs
- Graph Colouring
- Chromatic Polynomial
- Shortest Path Problems
- Shortest Path in A Weighted Graph
- Travelling Salesman Problem
- Network Flows
- Matching

This chapter introduces rudiments of graph theory which is used to analyse problems that arise in computer science, communication theory, physical sciences, operations research, and management. Hamiltonian and Eulerian graphs are discussed, which help to solve travelling salesman and other physical problems such as water supply systems and laying out telephone cables in cities. The concept of minimum spanning tree is discussed, which helps in all network design problems. Graph colouring is also introduced due to its widespread applications.

10.1 INTRODUCTION

Graph theory with the passage of time has emerged as a special branch of mathematics which has applications in various branches of engineering such as computer science, electrical network, mechanical engineering, rail roads, highways, bridges, and several other areas of chemistry, genetics, operations research, linguistics, and management science.

A graph can be used to represent almost any physical situation involving discrete objects and relationship between them. Like many other discoveries, graph theory originated from the celebrated königsberg seven bridges problem which was solved in 1736 by Swiss mathematician Leonhard Euler. The problem is as follows: Two islands A and B formed by the Pregel River in Königsberg were connected to each other and to the banks C and D with seven bridges as shown in Fig. 10.1(a)

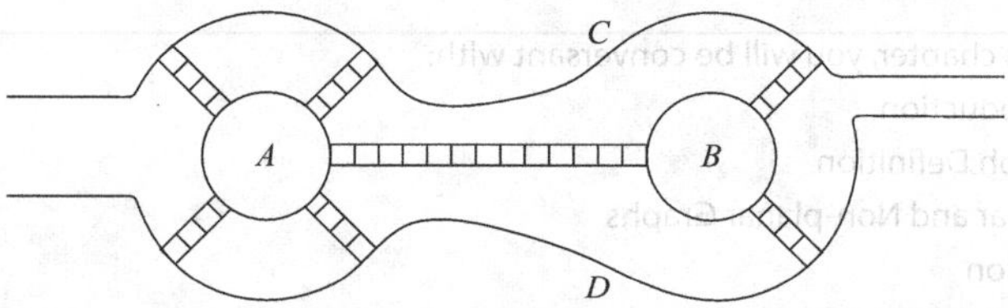

Figure 10.1(a)

The problem was to start at any of the four land areas A, B, C, or D, walk over each of the seven bridges exactly once and return to the starting point (without swimming across the river. Mathematically, the maximum number of possible paths for moving across the seven bridges is $7! = 5040$.

In 1736, Euler constructed a mathematical model (Fig. 10.1(b)) of the problem in which the points A and B represent the islands, and C and D represent the river banks.

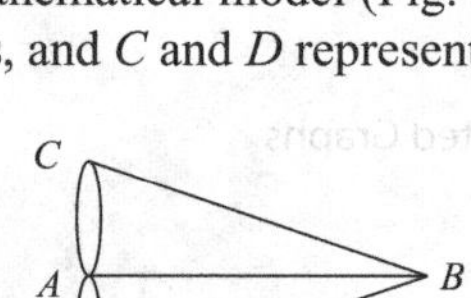

Figure 10.1(b)

The lines/arcs represent seven bridges. Euler concluded that no such walk is possible. Fig. 10.1(b) consisting of points A, B, C, and D and the line segments/arcs joining them is called a graph. The points are called vertices and the lines are called edges.

10.2 GRAPH DEFINITION

A graph $G = (V, E)$ is a system consisting of a set V of vertices $\{v_1, v_2, \ldots\}$ and a set E of edges $\{e_1, e_2, \ldots\}$ such that each edge e_k is identified with an unordered pair of vertices (v_i, v_j). The vertices v_i and v_j are called end vertices of the edge e_k.

Consider the graph (Fig. 10.2) $G = (V, E)$ where

$$V = \{v_1, v_2, v_3, v_4, v_5\} \text{ and } E = \{e_1, e_2, e_3, e_4, e_5, e_6, e_7\}$$

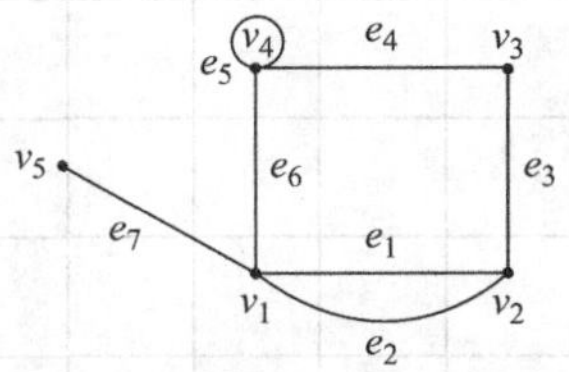

Figure 10.2

Self-Loop: An edge associated with a vertex pair (v_i, v_i), starting and ending at the same vertex is called a self-loop or simply a loop. In Fig. 10.2, edge e_5 is a loop.

Parallel Edges: The edges associated with the same pair of vertices are called parallel edges. In Fig. 10.2, e_1 and e_2 are parallel edges.

Simple Graph: A graph that has neither self-loops nor parallel edges is called a simple graph.

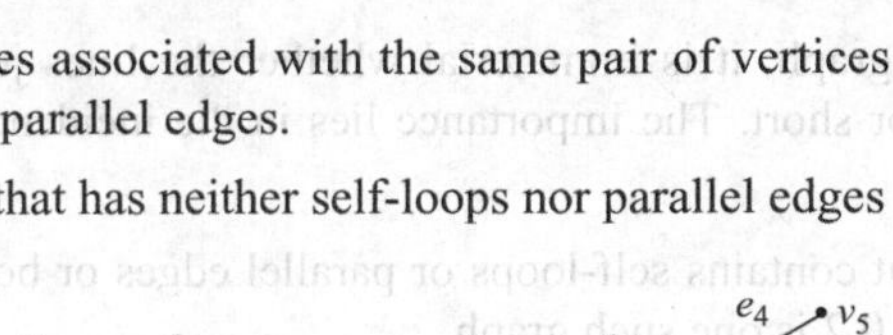

Figure 10.3

The graphs G_1 and G_2 in Figs. 10.3(a) and (b) are simple graphs.

10.2.1 Finite and Infinite Graphs

The definition of graph does not require the vertex set V and the edge set E to be finite; but in almost all the applications, these are finite.

A graph with a finite number of vertices and also a finite number of edges is called a finite graph, else it is called an infinite graph. The study here is confined to finite graphs.

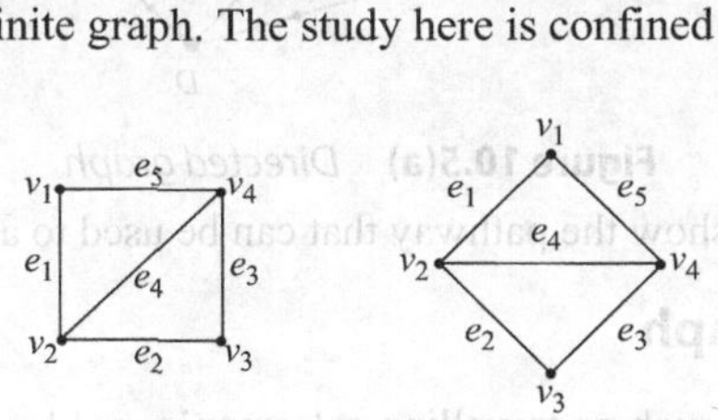

Figure 10.4(a) *Same graph drawn differently*

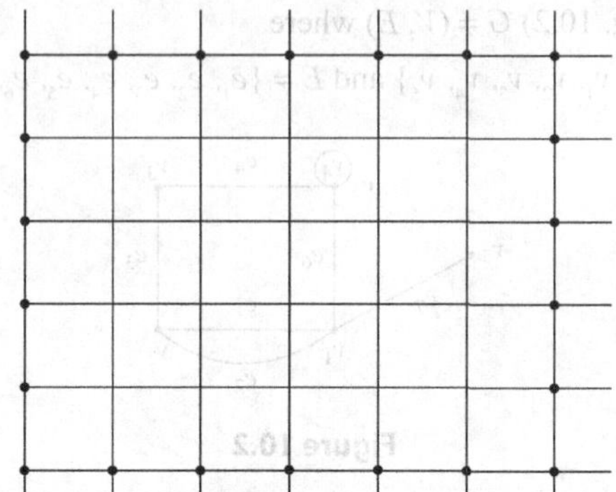

Figure 10.4(b) *Portion of an infinite graph*

Note 10.1 In drawing a graph, it is immaterial whether the lines joining vertices are drawn straight or curved, long or short. The importance lies in the incidence between the edges and vertices [Fig. 10.4(a)].

Multigraph: A graph that contains self-loops or parallel edges or both is called a multigraph. The graph shown is Fig. 10.2 is one such graph.

10.2.2 Directed and Undirected Graph

A graph in which direction is associated with each edge is called a directed graph. A directed graph is a one-way road where traffic flow is allowed only in one direction. A directed edge (v_j, v_k) allows flow from the vertex v_j to v_k. In case of an undirected graph, the edge between the vertices v_j and v_k are considered as an ordered pair and is taken as a two-way road with traffic flows in both directions.

An example of a directed graph is the transportation network as given in Fig. 10.5(a) in which *A* is a factory and *M* is the market connected to *A* through cities *B*, *C*, and *D* via road and the manufactured product will be transported through these roads which are edges in the graph.

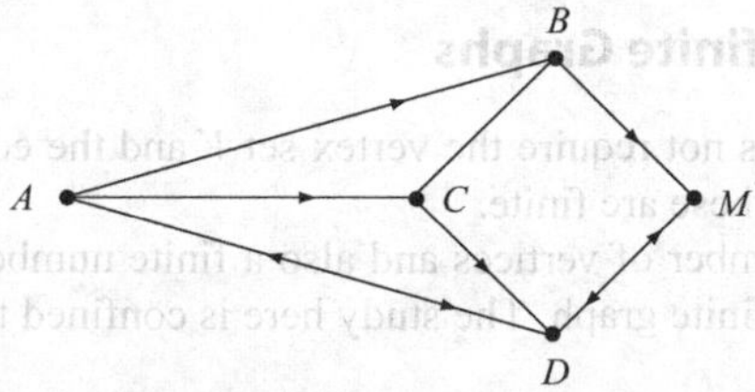

Figure 10.5(a) *Directed graph*

The directions on the graph show the pathway that can be used to approach the market.

10.2.3 Weighted Graph

In many practical problems such as travelling salesman's problem, Chinese postman problem and railway network connecting cities, a non-negative number can be assigned to each edge which is called weight of the edge. Such a graph is called a weighted graph. The weight assigned to an edge e is denoted by $w(e)$. The weights associated with the edges may either be distances or time needed to cover the distances. The problem is to find the minimum path or minimum time or minimum cost to reach the destination.

The graph shown in Fig. 10.5(b) is an example of a weighted graph connecting manufacturing unit P to the market M and the weights represent the distances.

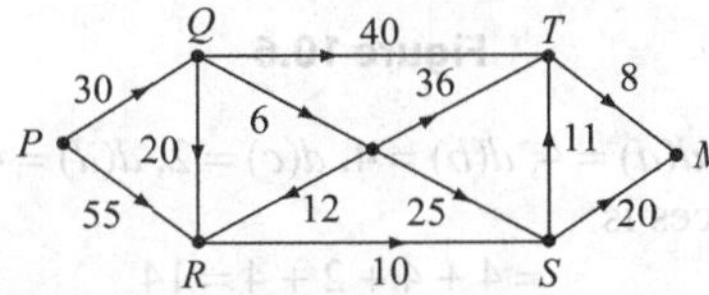

Figure 10.5(b)

The shortest path is

$P \to Q \to U \to R \to S \to T \to M = 77$ km

10.2.4 Incidence and Degree

When a vertex v_k is an end vertex of some edge e_j, then v_k and e_j are said to be incident with (on or to) each other. In Fig. 10.2, the edges e_1, e_2, e_6, and e_7 are incident with vertex v_1.

Two vertices are said to be adjacent if they are the end vertices of the same edge. Two edges are said to be adjacent if they have a common vertex.

Degree of a Vertex: The number of edges incident on a vertex v_k with self-loops counted twice is called the degree of the vertex v_k and is written as $d(v_k)$

In Fig. 10.2, $d(v_1) = 4$, $d(v_2) = 3$, $d(v_3) = 2$, $d(v_4) = 4$, $d(v_5) = 1$

The sum of degrees of all vertices in any graph is twice the number of edges.

Taking Fig. 10.2 as an example, it can be seen that

$$\sum_{k=1}^{5} d(v_k) = d(v_1) + d(v_2) + d(v_3) + d(v_4) + d(v_5)$$

$$= 4 + 3 + 2 + 4 + 1 = 14$$

= twice the number of edges.

THEOREM 10.1 Handshake Theorem

In any graph $G = (V, E)$ with e number of edges and n vertices $v_1, v_2, \ldots, v_n$

$$\sum_{i=1}^{n} d(v_i) = 2e \qquad \text{(i)}$$

Proof: In the case of an undirected graph, the degree of a vertex v is the number of edges meeting at v. Now every edge has exactly two vertices, and hence each edge is counted twice, once at each end. Thus the sum of degrees of all vertices is equal to twice the number of edges.

Note 10.2 The name of the theorem is handshaking theorem because the number of hands involved in handshake of persons will always be even as two hands are involved in every handshake.

Example 1 Consider the graph (Fig. 10.6) with four vertices *a*, *b*, *c*, and *d*.

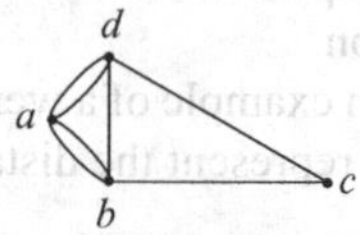

Figure 10.6

$$d(a) = 4,\ d(b) = 4,\ d(c) = 2,\ d(d) = 4$$

The sum of degrees all vertices is

$$= 4 + 4 + 2 + 4 = 14$$

Since the sum is even, there might be a graph with 14/2 = 7 edges.
Fig. 10.6 demonstrates such a graph.

Even and Odd Vertex: A vertex is said to be an even vertex if its degree is even. In case its degree is odd, the vertex is called an odd vertex.

THEOREM 10.2 The number of vertices of odd degree in a graph is always even.

Proof: Considering even and odd vertices separately for a graph, we have

$$\sum_{i=1}^{n} d(v_i) = \sum_{\text{even vertices}} d(v_j) + \sum_{\text{odd vertices}} d(v_k) \qquad \text{(i)}$$

The term on the L.H.S is even. The first term on the R.H.S. is also even being the sum of even terms. Hence, $\sum_{\text{odd vertices}} d(v_k)$ is even which is possible when the number of odd vertices is even.

Indegree and Outdegree: In a directed graph, the number of edges ending at a vertex *v* is called indegree of *v* and is denoted by indegree (*v*). Similarly, the number of edges beginning from the vertex *v* is called its outdegree and is denoted by outdegree (*v*). A self-loop contributes two degrees (one outdegree and one indegree).

10.2.5 Complete Graph

In a graph *G* of *n* vertices $v_1, v_2, \ldots, v_n$ if every vertex is connected to every other vertex, then *G* is called a complete graph and is denoted by K_n.

The complete graphs K_1 to K_5 are the following:

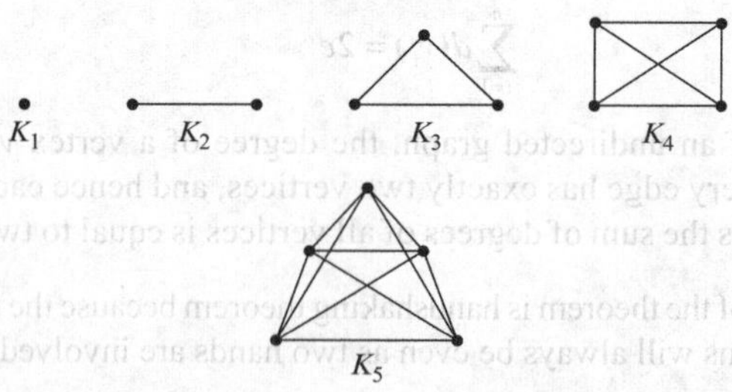

Complete graphs are useful in modelling round-robin tournaments where every team plays every other team exactly once. With n teams participating in the tournament, K_n provides the solution.

THEOREM 10.3 If $G = (V, E)$ is a directed graph, then the sum of outdegree vertices is equal to the sum of indegree vertices which equals to the number of edges.

This is

$$\sum_i in\deg(v_i) = \sum_j out\deg(v_j) = e \text{ (the number of edges).}$$

This is illustrated with the help of an example given in Fig. 10.7.

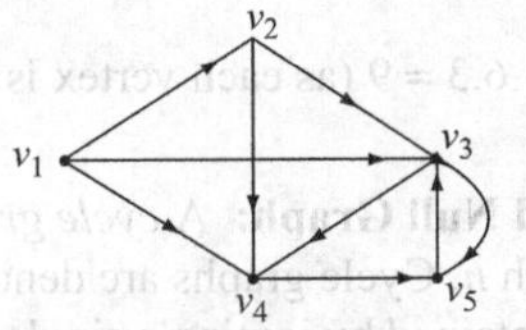

Figure 10.7

Vertex	v_1	v_2	v_3	v_4	v_5
Indegree	0	1	3	3	2
Outdegree	3	2	2	1	1

Sum of indegree vertices = 0 + 1 + 3 + 3 + 2 = 9

Sum of outdegree vertices = 3 + 2 + 2 + 1 + 1 = 9

Number of edges = 9

10.2.6 Regular Graph

A graph in which all vertices are of equal degree is called a regular graph.

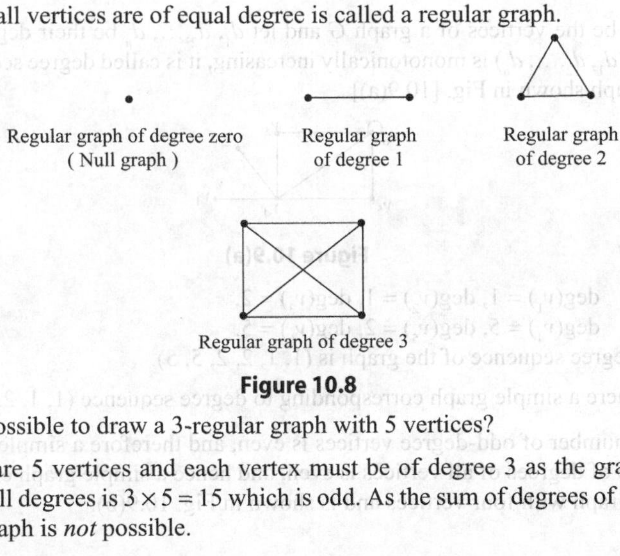

Figure 10.8

Example 1 Is it possible to draw a 3-regular graph with 5 vertices?

Solution: There are 5 vertices and each vertex must be of degree 3 as the graph is 3-regular. Thus, the sum of all degrees is $3 \times 5 = 15$ which is odd. As the sum of degrees of all vertices must be even, such a graph is *not* possible.

Example 2 Draw a 3-regular graph with 6 vertices. Find the number of edges.

Solution:

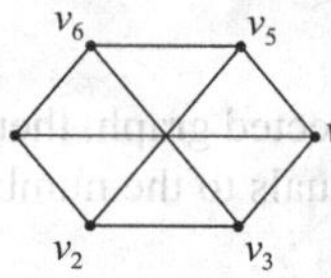

The number of edges

$$= \frac{1}{2} \text{(sum of degrees of all vertices)}$$

$$= \frac{1}{2}.6.3 = 9 \text{ (as each vertex is of degree 3)}$$

Cycle Graph, Wheel Graph, and Null Graph: A *cycle graph* of order n is a connected graph whose edges form a cycle of length n. Cycle graphs are denoted by C_n.

A wheel of order n is a graph obtained by joining a single new vertex (the hub) to each vertex of a cycle graph of order $n - 1$. Wheels of order n are denoted by W_n.

A null graph of order n is a graph with n vertices and no edges. Null graphs of order n are denoted by N_n. Each vertex in a null graph is an isolated vertex.

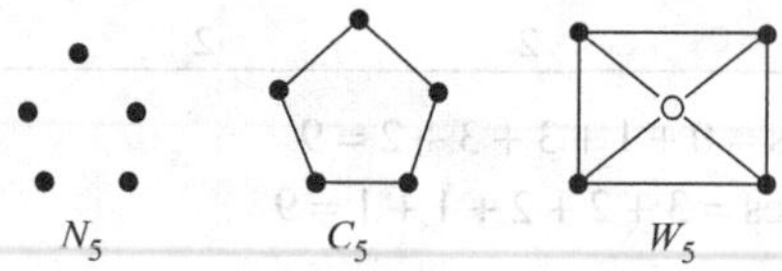

10.2.7 Degree Sequence of Graph

Let $v_1, v_2,\ldots, v_n$ be the vertices of a graph G and let $d_1, d_2,\ldots, d_n$ be their degrees, respectively. If the sequence $(d_1, d_2,\ldots, d_n)$ is monotonically increasing, it is called degree sequence of graph G. Consider the graph shown in Fig. [10.9(a)]

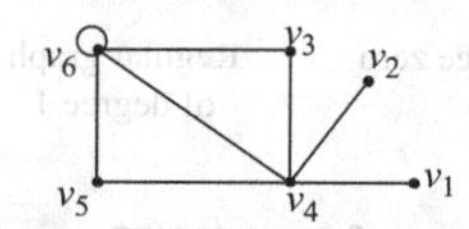

Figure 10.9(a)

Now $\quad \deg(v_1) = 1, \deg(v_2) = 1, \deg(v_3) = 2,$
$\deg(v_4) = 5, \deg(v_5) = 2, \deg(v_6) = 5.$

Hence, the degree sequence of the graph is (1, 1, 2, 2, 5, 5).

Example 1 Is there a simple graph corresponding to degree sequence (1, 1, 2, 2)?

Solution: The number of odd-degree vertices is even, and therefore a simple graph is possible. Further, the sum of degrees of all vertices is even, and hence a simple graph exists. The possible graph can be a graph with four vertices and is shown in Fig. 10.9(b).

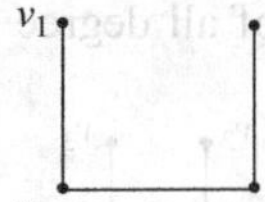

Figure 10.9(b)

Example 2 Is there a simple graph corresponding to degree sequence (1, 1, 4, 4)?

Solution: The number of odd vertices is even and also the sum of degrees of all vertices, that is, $1 + 1 + 4 + 4 = 10$ is even, and hence a simple graph is possible. But in the case of simple graph G of n vertices, the degree of any vertex cannot exceed $(n - 1)$. Here, there are 4 vertices and two of them are of degree 4, which is not possible and hence the graph is not possible.

Example 3 For the graph G, shown in Fig. [10.9(c)], write the degree sequence of G. Also find the number of odd-degree vertices and the number of edges. Verify handshake theorem.

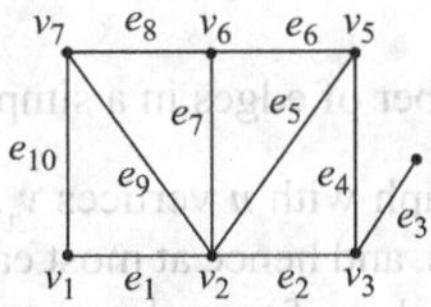

Figure 10.9(c)

Solution:

$\deg(v_1) = 2$, $\deg(v_2) = 5$, $\deg(v_3) = 3$
$\deg(v_4) = 1$, $\deg(v_5) = 3$, $\deg(v_6) = 3$
$\deg(v_7) = 3$

Hence, the degree sequence of G is (1, 2, 3, 3, 3, 3, 5)
The number of odd-degree vertices is 6
Now, the sum of degrees of all vertices is $1 + 2 + 3 + 3 + 3 + 3 + 5 = 20$
The number of edges is 10 (from e_1 to e_{10})
Hence, the sum of degrees of all vertices is equal to twice the number of edges (i.e., handshaking theorem).

Example 4 Draw the graph if it exists for the following degree sequences:

(1) (1, 1, 1) (2) (1, 1, 1, 1) (3) (1, 1, 1, 1, 1, 1)

Solution:

(1) In the given sequence (1, 1, 1), the sum of degrees of all vertices is odd, and hence no simple graph can be drawn.
(2) In the given sequence (1, 1, 1, 1), the sum of degrees of all vertices is even, that is, 4. The number of odd-degree vertices is even, and hence a graph is possible which has 4 vertices of degree 1 each.

The number of edges is half the sum of all degree of vertices, that is, 2. Hence a possible graph is

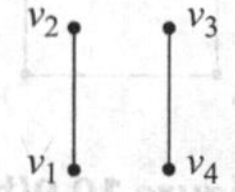

which is a disconnected graph.

(3) In the given sequence (1, 1, 1, 1, 1, 1), the sum of degrees of all vertices is even, that is, 6 and also number of odd vertices is even, that is, 6. Hence, a graph is possible. For the graph, the number of edges will be half the sum of degrees of all vertices, that is, 3. Hence, a possible graph is a disconnected graph.

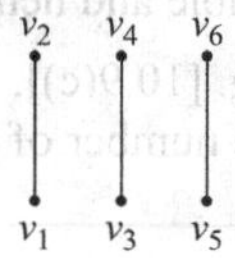

THEOREM 10.4 The maximum number of edges in a simple graph of n vertices is $n(n-1)/2$.

Proof: Let $G = (V, E)$ be a simple graph with n vertices $v_1, v_2, \ldots, v_n$. As the graph is simple, it has neither self-loop nor parallel edges, and hence at most each edge is incident to a two-element subset $\{v_i, v_j\}$ of V. The maximum number of two-element subsets of the set V is ${}^nc_2$.

Hence, the number of edges $\le {}^nc_2$

$$\le n(n-1)/2$$

10.2.8 Isolated and Pendant Vertex

In a graph, a vertex is said to be an isolated vertex if its degree is zero, that is, the vertex has no edge. In Fig. 10.10, vertices v_5 and v_6 are *isolated vertices.*

A vertex of degree 1 is called a pendant vertex. In Fig. 10.10, v_4 is a pendant vertex.

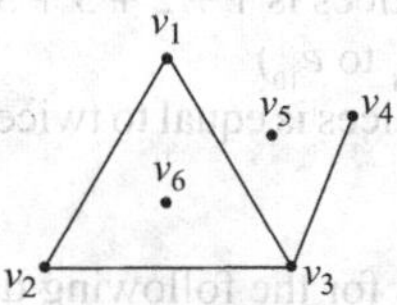

Figure 10.10

10.3 PLANAR AND NON-PLANAR GRAPHS

The concept of planarity finds its application in the design of printed circuit boards in computers.

A graph G is said to be planar if it can be drawn in a plane without crossings, that is, no two edges intersect geometrically except at a vertex to which both are incident. Any such drawing is a plane drawing.

A graph that cannot be drawn without a crossover between its edges is called a non-planar graph.

The following graphs shown in Fig. 10.11 are planar graphs:

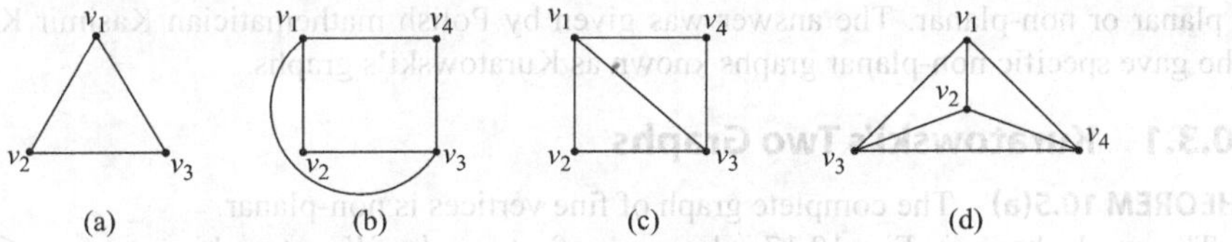

Figure 10.11

Consider a practical problem in which three houses H_1, H_2, and H_3 are to be connected to each of the three utilities water (W), gas (G), and electricity (E) by means of conduits.

The problem can be represented by a graph where conduits are shown as edges and houses and utility centres are vertices.

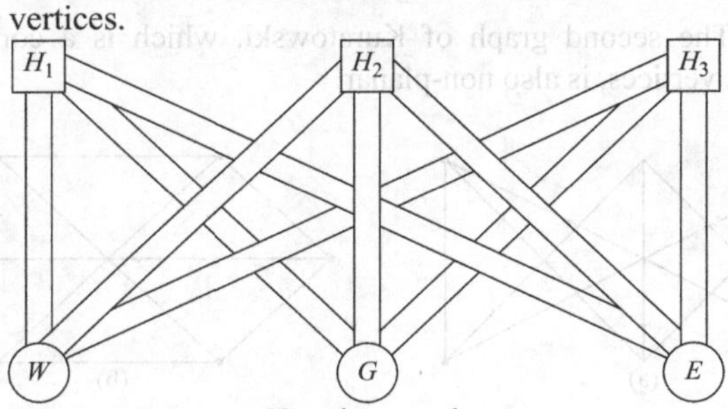

Non-planer graph

Figure 10.12

Figure 10.12 cannot be drawn without edges crossing over each other. Hence, a planar graph for the above problem is not possible. Figure 10.12 shows a non-planar graph as there is a crossover of the edges.

Note 10.3 A graph G with crossover does not mean that it is non-coplanar. There may be another way of drawing the graph in which edges do not cross over. Thus, before coming to a conclusion that a particular graph is non-planer, one must try redrawing it in different ways.

Consider the following graphs given in Figs. 10.13 and 10.14.

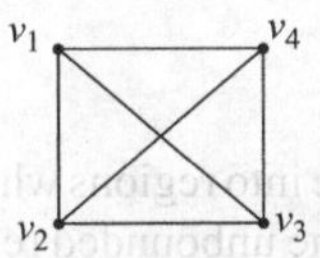

Figure 10.13 *Graph G (intersecting edges, and hence non-planar)*

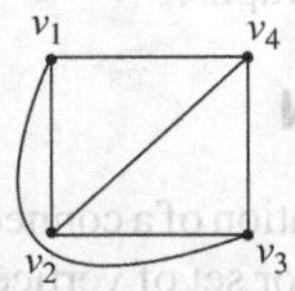

Figure 10.14 *Graph G (redrawn, and hence planar)*

The complete graph G consisting of four vertices v_1, v_2, v_3, and v_4 shown in Fig. 10.15 shows crossover of the edges while redrawn in Fig. 10.16 is planar. Thus, complete graph k_4 is planar.

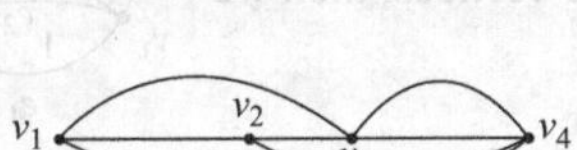

Figure 10.15 *Graph G (intersecting edges)*

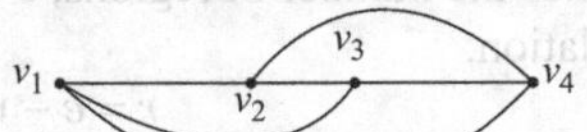

Figure 10.16 *Graph G (planar representation)*

The above discussion gives rise to a question as how to determine whether the graph $G = (V, E)$ is planar or non-planar. The answer was given by Polish mathematician Kasimir Kuratowski who gave specific non-planar graphs known as Kuratowski's graphs.

10.3.1 Kuratowski's Two Graphs

THEOREM 10.5(a) The complete graph of fine vertices is non-planar.

The graph shown in Fig. 10.17 is known as *first graph* of Kuratowski.

A complete graph is a simple graph in which every vertex is joined to every other vertex by an edge. When we join vertices (Fig. 10.17), it can be seen that the edge between v_1 and v_4 cannot be drawn without a crossover and hence the graph is non-planar.

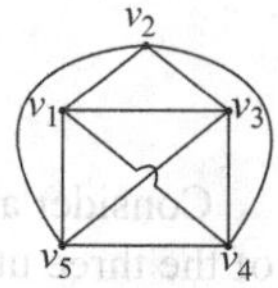

Figure 10.17

THEOREM 10.5(b) The second graph of Kuratowski, which is a complete regular graph with six vertices, is also non-planar.

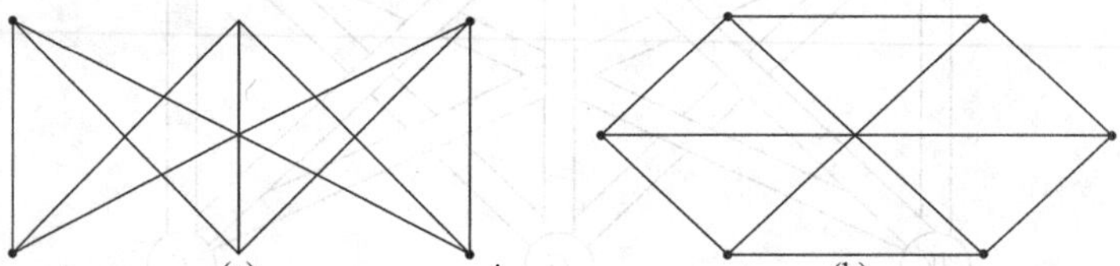

(a) (b)

Figure 10.18

In literature, Kuratowski's first graph is denoted by K_5 and the second graph by $K_{3,3}$, the letter K for honouring Kuratowski.

Note 10.4 Important observations about Kuratowski's two graphs are the following:

(1) Both are regular and non-planar.
(2) Removal of one edge or a vertex makes each a planar graph.
(3) Kuratowski's first graph is the non-planar graph with the smallest number of vertices, and the second graph is non-planer with the smallest number of edges. Hence, both are simplest non-planar graphs.

10.4 REGION

A planar representation of a connected graph divides the plane into regions which are characterized by the set of edges or set of vertices forming its boundary. The unbounded region is also included in the representation.

Consider the planar graph shown in Fig. 10.19

The graph divides the plane in to five regions marked 1–5, the region 5 is infinite region. The graph has 3 vertices and 6 edges.

If r denotes the number of regions, e the edges, and v the vertices, then we have the relation:

$$r = e - v + 2$$

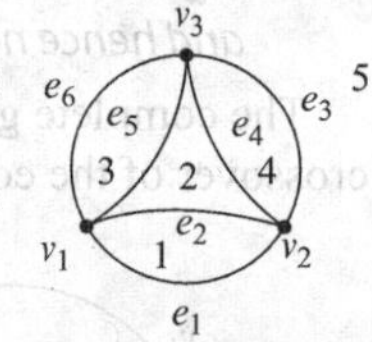

Figure 10.19

This relation is known as Euler's formula which gives the number of regions in a planar graph.

Example 1 Verify Euler's formula for the connected graph shown in Fig. 10.20

In the Fig. 10.20, we have

Number of vertices, $v = 5$

Number of edges, $e = 8$

Number of regions, $r = 5$

Now
$$r = e - v + 2$$
$$= 8 - 5 + 2$$
$$= 3 + 2 = 5$$

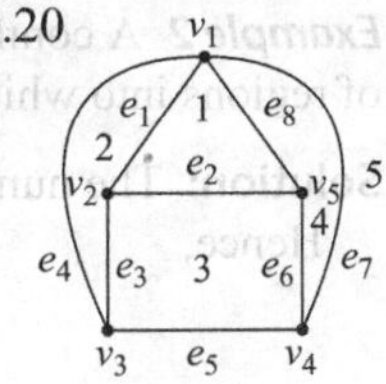

Figure 10.20

Hence, Euler's formula is verified.

THEOREM 10.6 Euler's Formula

In a connected planar graph G with n vertices and e edges, the number of regions r is $e - n + 2$, that is, $r = e - n + 2$.

Proof: The principle of mathematical induction is used for the value of r.

For $r = 1$ and 2, the statement is true as shown in Figs. 10.21(a) and (b).

Figure 10.21(a) **Figure 10.21(b)**

Let the result be true for every connected planar graph G with $r = k + 1$ regions, n vertices, and $e + 1$ edges.

If we select any edge (v_i, v_j) that belongs to regions r_1 and r_2 and delete it from G, then the resulting graph G' will have k regions, n vertices and e edges connected by the relation $r = e - n + 2$.

By re-introducing edge (v_i, v_j) into G', we shall have $k + 1 = (e + 1) - n + 2$ which shows that the result is valid for all values of r and e.

Some Important Results

(1) In a connected simple planar graph G of n (≥ 3) vertices, e edges, and no circuits of length 3, $e \leq 2n - 4$

(2) In a connected simple planar graph G of n (≥ 3) vertices, e (> 2) edges, and r regions,
(i) $3r \leq 2e$ (ii) $e \leq 3n - 6$.

Note 10.5 It may be noted that above conditions (1) and (2) are only necessary conditions for planarity but not sufficient. This means that mere satisfaction of the inequalities does not guarantee the planarity of the graph.

Example 1 Show that the K_5 is non-planar

Solution: The loop-free graph shown in Fig. 10.21(c) has 5 vertices and 10 edges.

For a planar graph
$$e \leq 3n - 6$$
i.e.,
$$3n - 6 \geq e \text{ or } 3n - e \geq 6 \qquad \text{(i)}$$

Here, $n = 5$, $e = 10$ and hence $3n - 6 = 3.5 - 6$
$$= 15 - 6 = 9$$

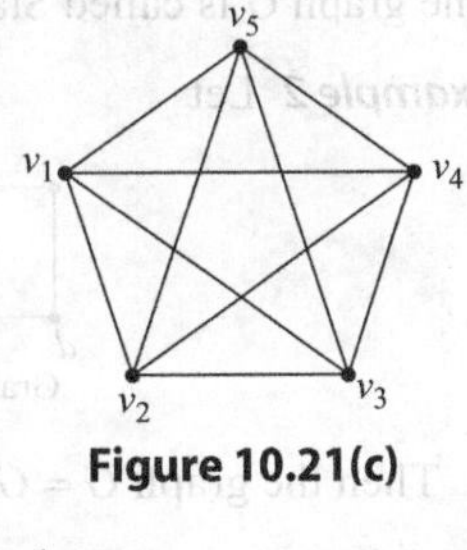

Figure 10.21(c)

which does not satisfy the inequality (i). Hence, the graph K_5 is non-planar.

Example 2 A connected graph has 10 vertices. The degree of each vertex is 3. Find the number of regions into which the planner graph can split.

Solution: The number of vertices $n = 10$

Hence,

$$\sum_{i=1}^{3} d(v_i) = 2 \times \textit{number of edges}$$

$$30 = 2e \qquad \therefore e = 15$$

By Euler's formula, the number of regions $r = e - n + 2$

$$= 15 - 10 + 12$$
$$= 7$$

$\therefore$ The planar graph can split in 7 regions.

10.5 OPERATIONS ON GRAPHS

Let $G_1 = (V_1, E_1)$ and $G_2 = (V_2, E_2)$ be two simple graphs.
The union of G_1 and G_2 is the simple graph $G = (V, E)$ where $V = V_1 \cup V_2$ and $E = E_1 \cup E_2$.

Note 10.6 If there is no common vertex between G_1 and G_2, then the graph $G_1 \cup G_2$ will be disconnected.

Example 1 Let G_1 and G_2 be the two graph

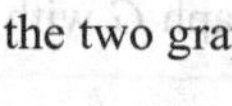

Graph G_1

Graph G_2

Then the graph $G = G_1 \cup G_2$ is

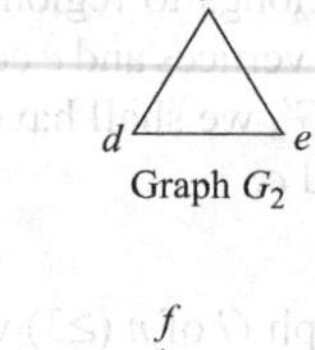

Graph G

The graph *G* is called Star of David

Example 2 Let

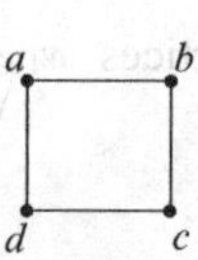

Graph G_1

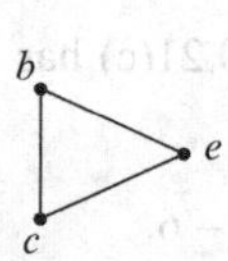

Graph G_2

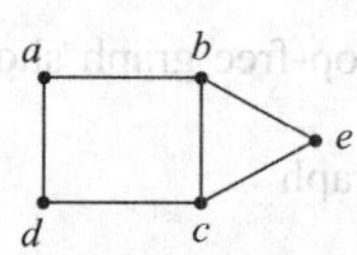

Graph *G*

Then the graph $G = G_1 \cup G_2$

Example 3 Let G_1 and G_2 be the two simple graphs

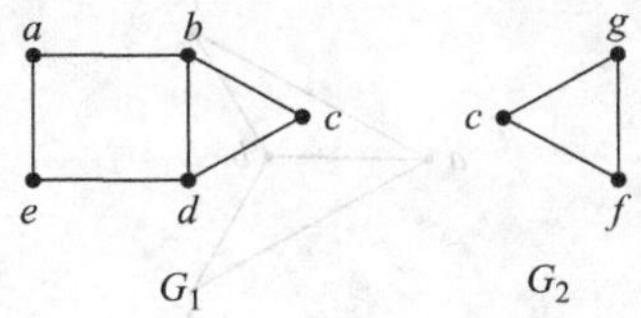

Then the union of the graphs G_1 and G_2, that is, $G = G_1 \cup G_2$ is

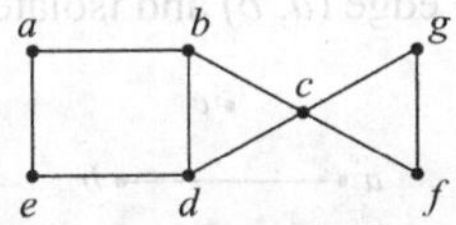

Intersection of Two Graphs: Let $G_1 = (V_1, E_1)$ and $G_2 = (V_2, E_2)$ be two simple graphs. The intersection of G_1 and G_2 written as $G_1 \cap G_2$ is the graph $G = (V, E)$, where $V = V_1 \cap V_2$ and $E = E_1 \cap E_2$

1. The intersection of graphs in Example 1 (10.5) is $G_1 \cap G_2 = \Phi$
2. The intersection of graphs in Example 2 (10.5) is the edge (b, c)

Example 1 Let

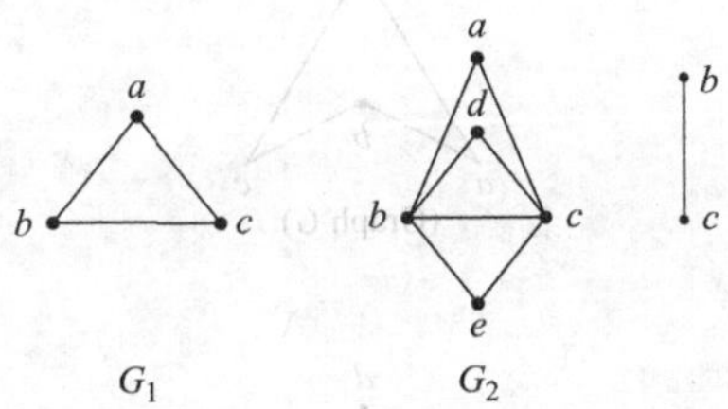

Then $G_1 \cap G_2$ is the graph.

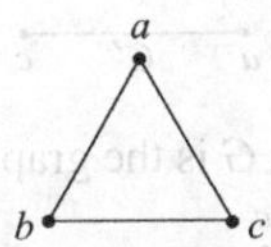

Example 2 Find the union and intersection of the graphs

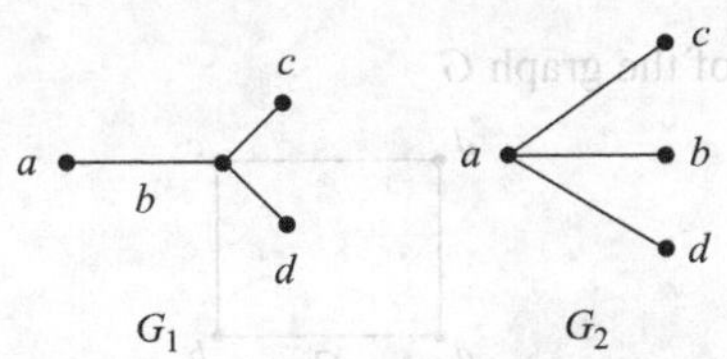

Solution: $G_1 \cup G_2$ is the graph

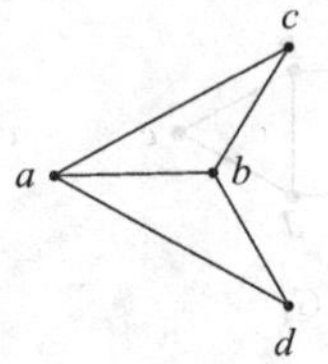

$(G_1 \cup G_2)$

$G_1 \cap G_2$ is the graph consisting of edge (a, b) and isolated vertices c and d

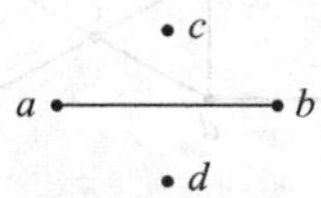

$(G_1 \cap G_2)$

Complement: The complement of a simple graph G is a simple graph G' having the same vertex set of G and where two vertices adjacent in G are not adjacent in G'.

Example 1 Consider the graph G given by

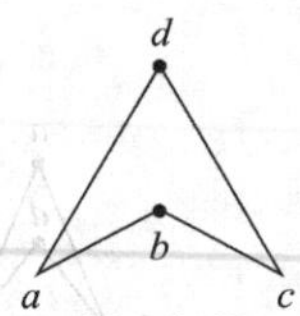

(Graph G)

The complement graph G' is

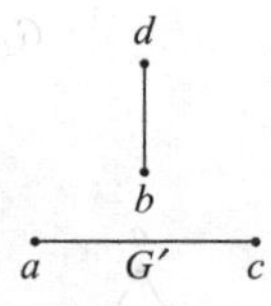

Example 2 The complement of the graph G is the graph G'. It consists of three isolated vertices.

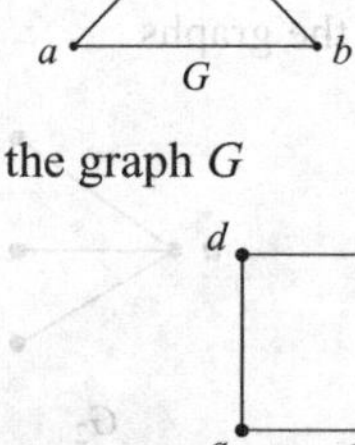

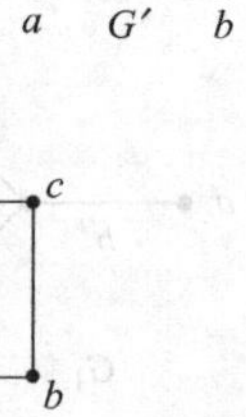

Example 3 The complement of the graph G

is the graph G' given by

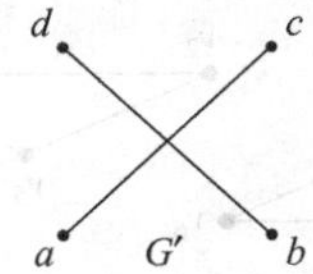

10.5.1 Sum of Two Graphs

The sum of two graphs $G_1 = (V_1, E_1)$ and $G_2 = (V_2, E_2)$ denoted by $G_1 + G_2$ is defined as the graph $G = (V, E)$ if $V_1 \cap V_2 = \Phi$ and the vertex set V is the $V_1 \cup V_2$ and the edge set E consists of edges E_1 and E_2 and the edges joining each vertex of v_1 with each vertex of v_2.

Let

1.
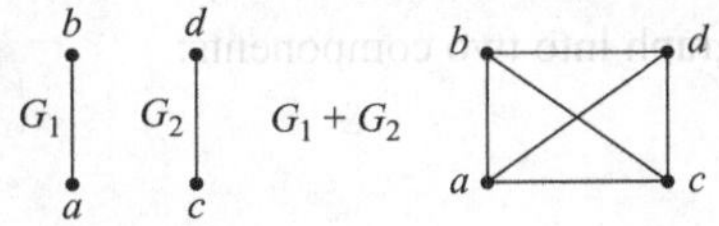

2.
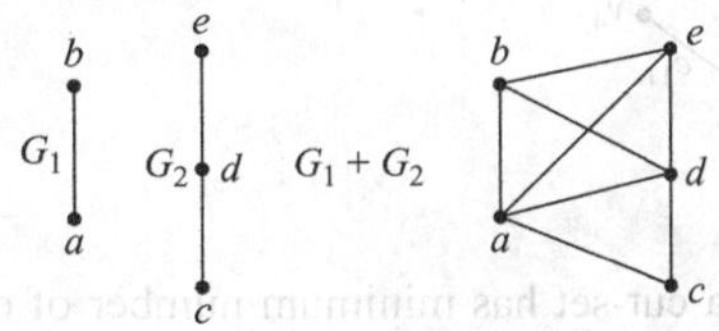

10.5.2 Cut-sets and Cut-vertices

Cut-sets play an important role in the design of communication systems of laying out telephone cables, transportation network of roads, bridges and water supply systems. The cut-sets are used to identify weak spots in a communication network.

Cut-Set: If a graph G is connected and e is an edge such that $G–e$ is disconnected, then e is said to be a bridge or a cut edge.

If a set S of edges is such that its removal from G leaves $G–S$ disconnected, then S is called a *cut-set* provided the removal of no proper subset of S disconnects G.

Consider the graph

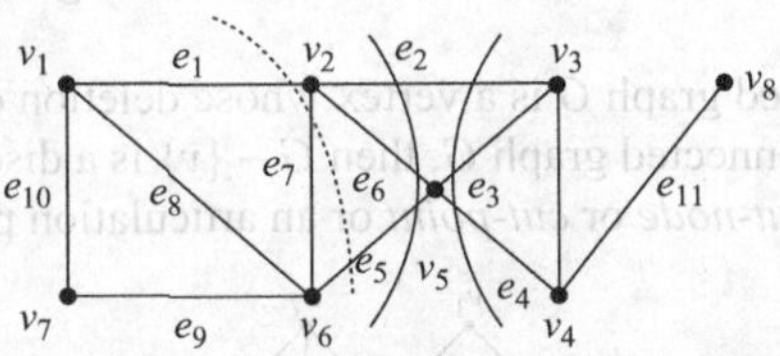

The removal of the set of edges $\{e_2, e_6, e_5\}$ disconnects the graph into two components and hence is a cut-set.

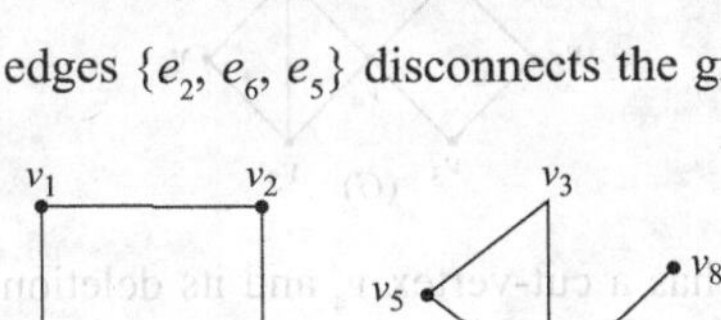

Similarly, the set $\{e_1, e_7, e_5\}$ disconnects the graph into two components:

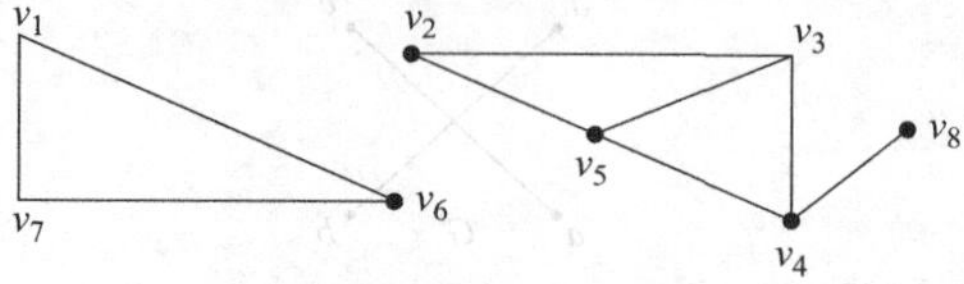

The set $\{e_2, e_3, e_4\}$ disconnects the graph into two components:

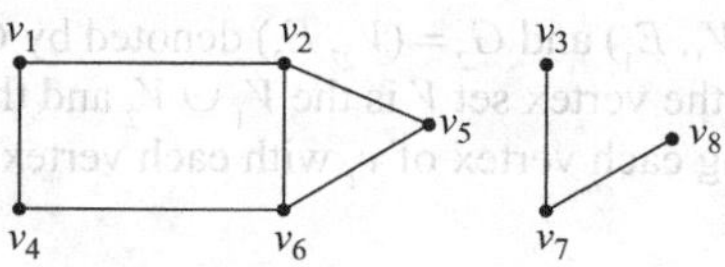

The set $\{e_1, e_8, e_9\}$ is a cut-set as it disconnects the graph into two components:

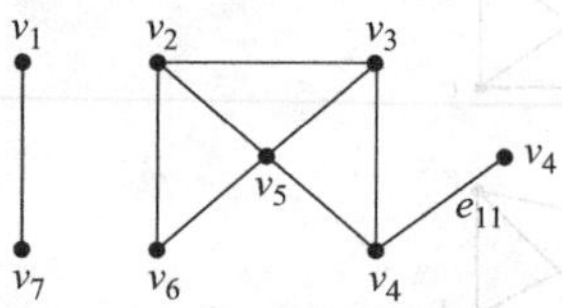

Edge $\{e_{11}\}$ alone is also a cut-set.

As no proper sub-set of a cut set can be a cut-set, a cut-set has minimum number of edges required for disconnecting a connected graph. Hence, a cut-set is also called a *minimal cut-set* or a proper cut-set or a simple cut-set.

Consider the graph G. The cut-set has only one edge *e*. The edge *e* is called a *bridge*.

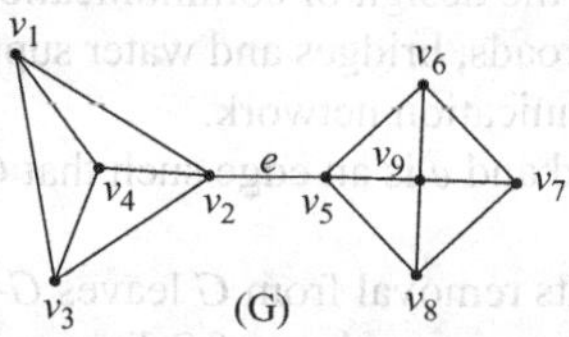

(G)

Cut-Vertex: A concept similar to cut-set (i.e., removal of edges in a connected graph) is for removal of vertices.

A cut-vertex of a connected graph *G* is a vertex whose deletion disconnects the graph.

If *v* is a cut-vertex of a connected graph *G*, then $G - \{v\}$ is a disconnected graph.

A cut-vertex is called a *cut-node* or *cut-point* or an articulation point.

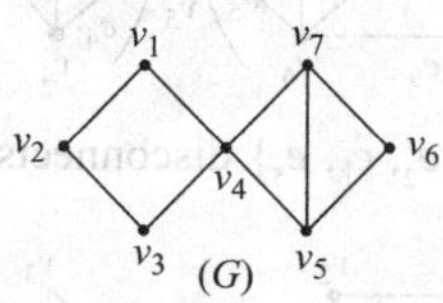

(G)

The graph *G* shown above has a cut-vertex v_4 and its deletion leads to two components G_1 and G_2.

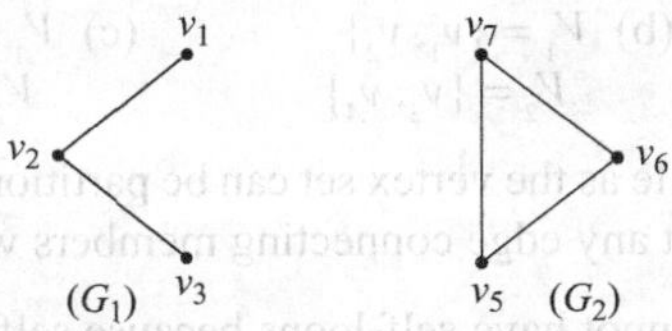

Note 10.7 It should be noted that removal of a vertex implies removal of all edges incident on that vertex. The reason for this is due to the fact that an edge does not exist without both and vertices. Further, every edge of a tree is a cut set since deletion of any edge breaks the tree into two parts.

Separating Set: A *separating set S* in a connected graph *G* is a set of vertices whose deletion disconnects the graph.

Consider the graph

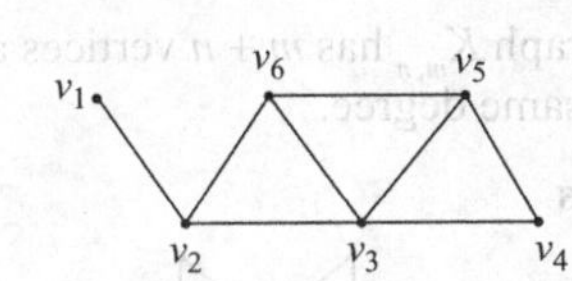

The set of vertices $\{v_6, v_3\}$ separates the graph into two components:

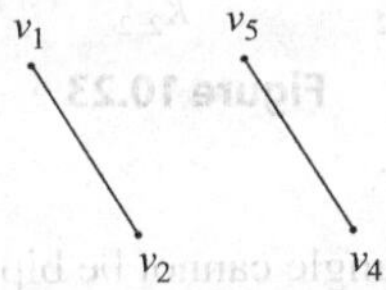

Similarly, the set of vertices $\{v_6, v_3, v_5\}$ separates the graph into two components:

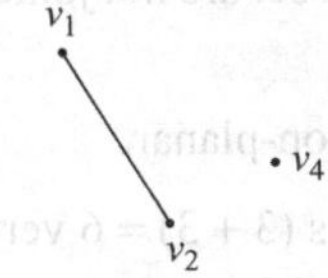

10.6 BIPARTITE GRAPH

A graph $G = (V, E)$ is a bipartite graph if its vertex set V can be partitioned into two disjoint subsets V_1 and V_2 such that every edge in G joins a vertex in V_1 with a vertex in V_2. No edge connects two vertices in V_1 or two vertices in V_2. Consider the following graphs shown in Fig. 10.22:

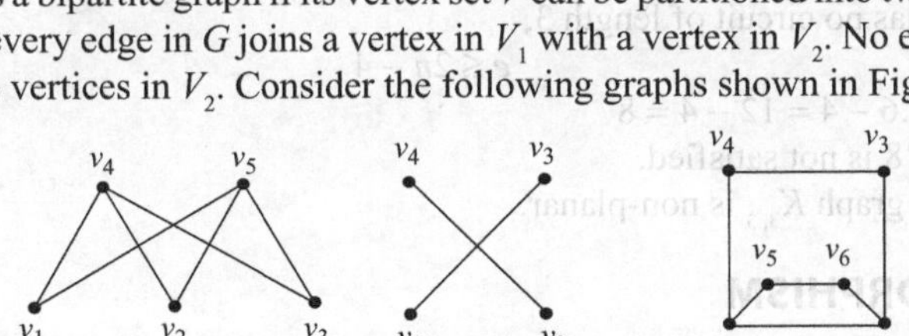

Figure 10.22

(a) $V_1 = \{v_1, v_2, v_3\}$ $V_2 = \{v_4, v_5\}$ (b) $V_1 = \{v_1, v_2\}$ $V_2 = \{v_3, v_4\}$ (c) $V_1 = \{v_1, v_3, v_6\}$ $V_2 = \{v_2, v_4, v_5\}$

Figs. 10.22(a)–(c) are all bipartite as the vertex set can be partitioned into two disjoint sets V_1 and V_2 such that there does not exist any edge connecting members within subsets V_1 or V_2.

Note 10.8 A bipartite graph cannot have self-loops because self-loop connects the same vertex and it is not permitted in bipartite graphs. A bipartite graph may have parallel edges.

10.6.1 Complete Bipartite Graph

A bipartite graph $G = (V, E)$ is said to be a complete bipartite if each vertex of V_1 is connected to each vertex of V_2 where V_1 and V_2 are the partitions of the vertex set V. Complete bipartite graph G is denoted by $K_{m,n}$, where m and n are the number of vertices in vertex sets V_1 and V_2.

Note 10.9 A complete bipartite graph $K_{m,n}$ has $m + n$ vertices and mn edges. If $m = n$, then $K_{m,n}$ is regular as all vertices are of the same degree.

Some Complete Bipartite Graphs

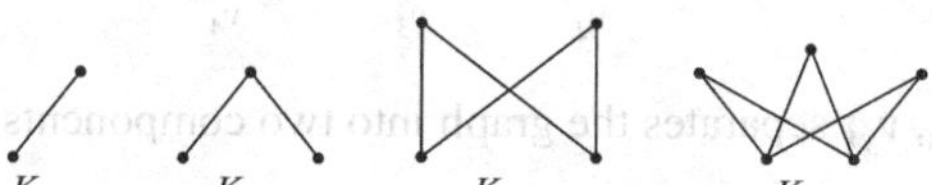

Figure 10.23

The graphs $K_{1,1}$ and $K_{2,2}$ are regular.

Example 1 A graph that contains a triangle cannot be bipartite.

Solution: A triangle consists of 3 vertices, and it is not possible to divide these vertices into two disjoint sets such that the vertices of each set are not joined among themselves. Hence, the graph cannot be bipartite.

Example 2 Show that the graph $K_{3,3}$ is non-planar.

Solution: A complete bipartite graph has $(3 + 3) = 6$ vertices and $3.3 = 9$ edges.

Hence, $$n = 6, e = 9$$

$$\therefore 3n - 6 = 3.6 - 6 = 18 - 6 = 12.$$

Hence, the condition $e \leq 3n - 6$ is satisfied.
As the graph has no circuit of length 3,

$$e \leq 2n - 4$$

But $2n - 4 = 2.6 - 4 = 12 - 4 = 8$
Hence, $e = 9 \nless 8$ is not satisfied.
Therefore, the graph $K_{3,3}$ is non-planar.

10.7 ISOMORPHISM

Two graphs $G = (V, E)$ and $G' = (V', E')$ are said to be isomorphic to each other if there exists a function $f: V \rightarrow V'$ which is one-to-one and onto, that is, there exists one-to-one correspondence between their vertices and also the edges such that incidence relationships are preserved. If the edge e_K is incident on vertices v_i and v_j in G, then the corresponding edge e'_K in G' must be incident on the vertices v_i' and v_j' in G', which corresponds to vertices v_i and v_j. In other words, if $f(v_i) = v_i'$, $f(v_j) = v_j'$ and $f(e_K) = e_K'$, that is, $(v_i, v_j) \in E$ if and only if $[f(v_i), f(v_j)] \in E'$.

Example 1

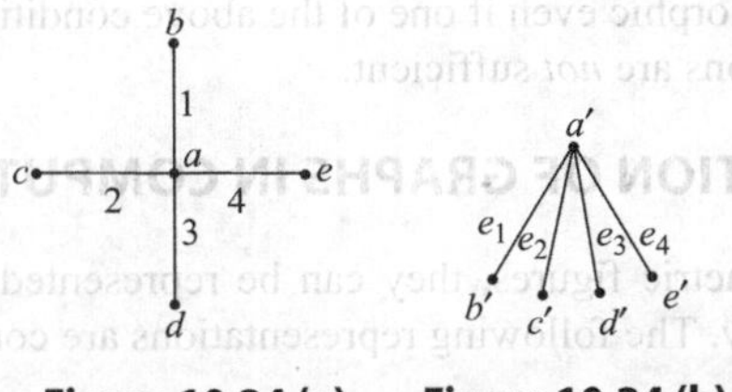

Figure 10.24 (a) **Figure 10.24 (b)**

The graphs (a) and (b) in Fig. 10.24 are isomorphic as the vertices a, b, c, d, and e correspond to the vertices a', b', c', d', and e' and the edges 1, 2, 3, 4 correspond to e_1, e_2, e_3, and e_4.

Example 2

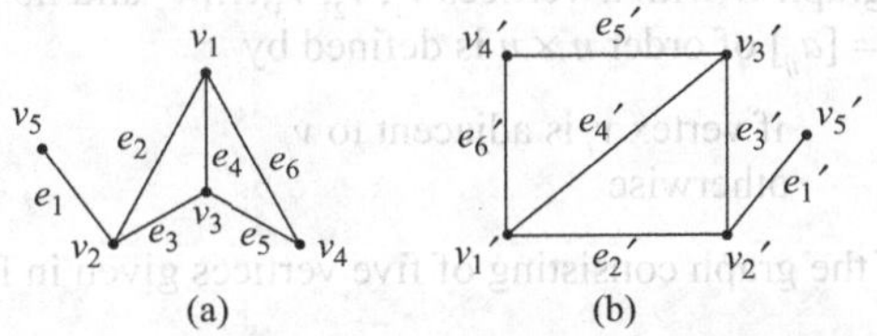

Figure 10.25

The graphs (a) and (b) in Fig. 10.25 are isomorphic. The vertices v_1', v_2', v_3', v_4' and v_5' correspond to v_1, v_2, v_3, v_4, and v_5. The edges e_1', e_2', e_3', e_4', e_5', and e_6' correspond to e_1, e_2, e_3, e_4, e_5, and e_6, respectively, and hence graph (a) is isomorphic to graph (b) given in Fig. 10.25.

From the definition of isomorphism of two graphs, it follows that the graphs must have

(1) The same number of vertices.
(2) The same number of edges.
(3) An equal number of vertices with given degree.

However, these conditions are not sufficient. The graph given in Fig. 10.26(a) and (b) are *not isomorphic*.

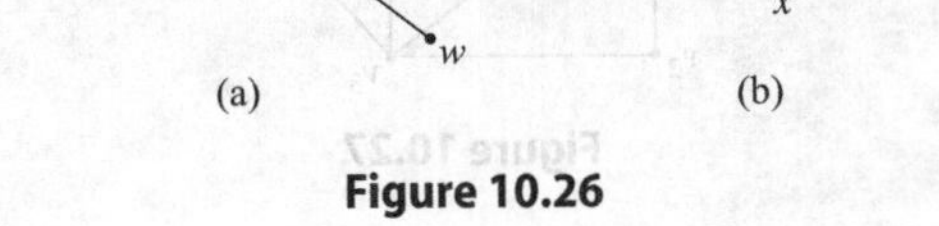

Figure 10.26

If the graphs are isomorphic, vertex u must correspond to the vertex x as there are no other vertex of degree 3. In graph (b), there is only one pendant vertex y adjacent to x, whereas in graph (a), there are two pendant vertices v and w adjacent to u.

Note 10.10 The necessary condition for the graph G and G' to be isomorphic are as follows:

1. Both G and G' have same number of vertices.
2. Both G and G' have same number of edges.
3. Both G and G' have same degree sequences.

The graphs are not isomorphic even if one of the above conditions is not satisfied. However, these conditions are *not* sufficient.

10.8 REPRESENTATION OF GRAPHS IN COMPUTER MEMORY

Although graphs are geometric figures, they can be represented by matrix and hence can be stored in computer memory. The following representations are commonly used:

1. Adjacency matrix (vertex–vertex adjacency matrix)
2. Incidence matrix (vertex–edge incidence matrix)

10.8.1 Adjacency Matrix of Undirected Graph

Consider an undirected graph G with n vertices $v_1, v_2, v_3, \ldots, v_n$ and no parallel edges and loops. The adjacency matrix $A = [a_{ij}]$ of order $n \times n$ is defined by

$$a_{ij} = \begin{cases} 1 & \text{if vertex } v_i \text{ is adjacent to } v_j \\ 0 & \text{otherwise} \end{cases}$$

The adjacency matrix of the graph consisting of five vertices given in Fig. 10.27 is

$$A = \begin{array}{c} \\ v_1 \\ v_2 \\ v_3 \\ v_4 \\ v_5 \end{array} \begin{array}{c} \begin{array}{ccccc} v_1 & v_2 & v_3 & v_4 & v_5 \end{array} \\ \begin{bmatrix} 0 & 1 & 1 & 1 & 0 \\ 1 & 0 & 1 & 0 & 0 \\ 1 & 1 & 0 & 1 & 1 \\ 1 & 0 & 1 & 0 & 0 \\ 0 & 0 & 1 & 0 & 0 \end{bmatrix} \end{array}$$

(adjacency matrix)

Some Important Results of Adjacency Matrix

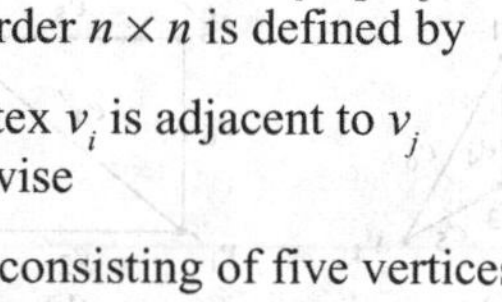

Figure 10.27

1. Adjacency matrix A is symmetric. That is, $a_{ij} = a_{ji}$ for all i and j.
2. All the diagonal elements are zero as there are no loops.
3. Adjacency matrix for a graph with n vertices has n^2 elements.

10.8.2 Incidence Matrix

Let G be a graph with n vertices and m edges. The incidence matrix $M = [m_{ij}]$ of order $n \times m$ is defined by

$$m_{ij} = 1 \quad \text{if vertex } V_i \text{ is incident on edge } e_j$$
$$= 0 \quad \text{otherwise}$$

The matrix has a row for every vertex and a column for every edge.

Consider the graph shown in Fig. 10.28 consisting of 5 vertices and 7 edges which can be represented by the incidence matrix of order 5×7, that is,

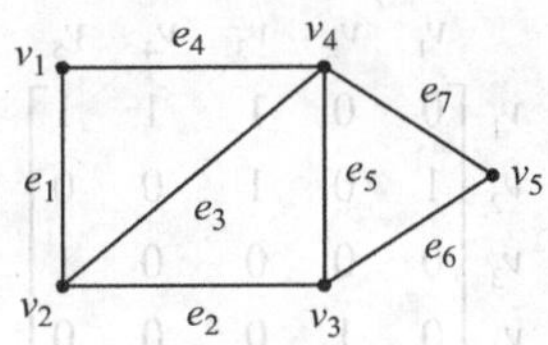

Figure 10.28

$$M = \begin{array}{c} \\ v_1 \\ v_2 \\ v_3 \\ v_4 \\ v_5 \end{array} \begin{array}{c} \begin{array}{ccccccc} e_1 & e_2 & e_3 & e_4 & e_5 & e_6 & e_7 \end{array} \\ \begin{bmatrix} 1 & 0 & 0 & 1 & 0 & 0 & 0 \\ 1 & 1 & 1 & 0 & 0 & 0 & 0 \\ 0 & 1 & 0 & 0 & 1 & 1 & 0 \\ 0 & 0 & 1 & 1 & 1 & 0 & 1 \\ 0 & 0 & 0 & 0 & 0 & 1 & 1 \end{bmatrix} \end{array}$$

(incidence matrix)

Some Important Results for Incidence Matrix

1. Each column contains exactly *two* unit elements as each edge is connected in the two vertices.
2. A row with all its elements zero will correspond to an isolated vertex.
3. A row with single-unit element will correspond to a pendant vertex.
4. The number of elements in row k represents the degree of the vertex v_K.
5. If the graph is connected with n vertices, then the rank of incidence matrix M is $(n - 1)$.

10.8.3 Matrix Representation of Directed Graph: The Adjacency Matrix

The adjacency matrix $A = [a_{ij}]$ for a directed graph G with n vertices is an $n \times n$ matrix defined by

$$a_{ij} = 1 \quad \text{if edge beginning at vertex } v_i \text{ ends at } v_j$$
$$= 0 \quad \text{otherwise}$$

The non-zero elements in the matrix are equal to the number of edges in the graph.

The adjacency matrix of order 5×5 for the directed graph given in Fig. 10.29 is the following:

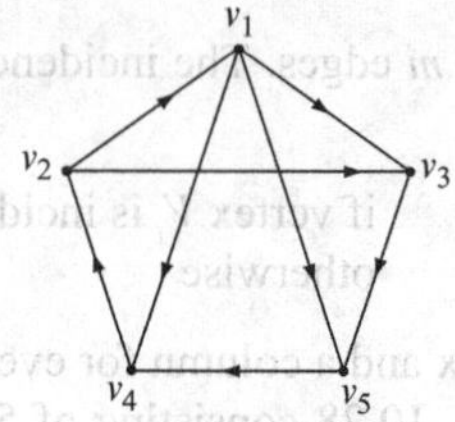

Figure 10.29

$$\begin{array}{c} \\ v_1 \\ v_2 \\ v_3 \\ v_4 \\ v_5 \end{array}\begin{array}{c} \begin{array}{ccccc} v_1 & v_2 & v_3 & v_4 & v_5 \end{array} \\ \begin{bmatrix} 0 & 0 & 1 & 1 & 1 \\ 1 & 0 & 1 & 0 & 0 \\ 0 & 0 & 0 & 0 & 1 \\ 0 & 1 & 0 & 0 & 0 \\ 0 & 0 & 0 & 1 & 0 \end{bmatrix} \end{array}$$

(adjacency matrix)

Incidence Matrix: The incidence matrix $M = [m_{ij}]$ for a directed graph G of n vertices and m edges is of order $n \times m$ and is defined by

$$m_{ij} = \begin{cases} 1 & \text{if } v_i \text{ is the initial vertex of edge } e_j \\ -1 & \text{if } v_i \text{ is the final vertex of edge } e_j \\ 0 & \text{if } v_i \text{ is not incident on edge } e_j \end{cases}$$

The incidence matrix for the graph given in Fig. 10.30 is the following:

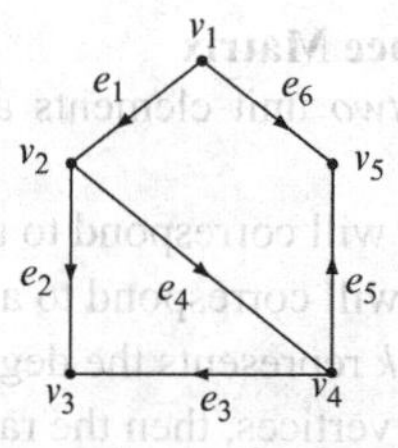

Figure 10.30

$$M = \begin{array}{c} \\ v_1 \\ v_2 \\ v_3 \\ v_4 \\ v_5 \end{array}\begin{array}{c} \begin{array}{cccccc} e_1 & e_2 & e_3 & e_4 & e_5 & e_6 \end{array} \\ \begin{bmatrix} 1 & 0 & 0 & 0 & 0 & 1 \\ -1 & 1 & 0 & 1 & 0 & 0 \\ 0 & -1 & -1 & 0 & 0 & 0 \\ 0 & 0 & 1 & -1 & 1 & 0 \\ 0 & 0 & 0 & 0 & 1 & -1 \end{bmatrix} \end{array}$$

10.9 REPRESENTATION OF MULTI GRAPH

When graph G is a multigraph, then a_{ij} is the number of edges joining the vertex v_i to the vertex v_j.

Let G be a multigraph consisting of n vertices, then the $n \times n$ adjacency matrix $A = [a_{ij}]$ is defined by,

$$a_{ij} = k$$
$$= 0$$

where k is the number of edges joining the vertex v_i to the vertex v_j otherwise.

The adjacency matrix of order 4×4 for the multigraph shown in Fig. 10.31 is the following:

$$A = \begin{array}{c} \\ v_1 \\ v_2 \\ v_3 \\ v_4 \end{array} \begin{array}{c} \begin{array}{cccc} v_1 & v_2 & v_3 & v_4 \end{array} \\ \begin{bmatrix} 0 & 1 & 3 & 2 \\ 1 & 1 & 1 & 0 \\ 3 & 1 & 0 & 0 \\ 2 & 0 & 0 & 1 \end{bmatrix} \end{array}$$

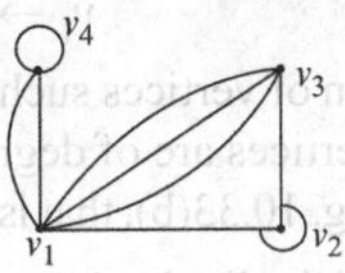

Figure 10.31

Example 1 Find the adjacency matrix and the incident matrix of the multigraph shown in Fig. 10.32

$$A = \begin{array}{c} \\ v_1 \\ v_2 \\ v_3 \\ v_4 \end{array} \begin{array}{c} \begin{array}{cccc} v_1 & v_2 & v_3 & v_4 \end{array} \\ \begin{bmatrix} 0 & 1 & 2 & 1 \\ 1 & 0 & 1 & 0 \\ 2 & 1 & 0 & 1 \\ 1 & 0 & 1 & 1 \end{bmatrix} \end{array}$$

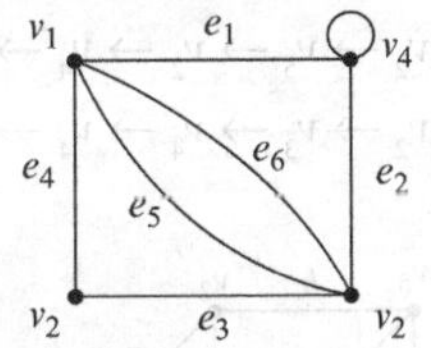

Figure 10.32

The incidence matrix M is the $n \times m$ matrix whose ijth entry is 1 if vertex i is incident to edge e_j and 0 otherwise.

$$M = \begin{array}{c} \\ v_1 \\ v_2 \\ v_3 \\ v_4 \end{array} \begin{array}{c} \begin{array}{cccccc} e_1 & e_2 & e_3 & e_4 & e_5 & e_6 \end{array} \\ \begin{bmatrix} 1 & 0 & 0 & 1 & 1 & 1 \\ 0 & 0 & 1 & 1 & 0 & 0 \\ 0 & 1 & 1 & 0 & 1 & 1 \\ 1 & 1 & 0 & 0 & 0 & 0 \end{bmatrix} \end{array}$$

10.10 WALK IN A GRAPH

A walk is a finite alternating sequence of vertices and edges starting and ending with vertices such that each edge is incident with the vertices preceding and following. In a walk, no edge is traversed more than once although a vertex can appear more than once.

Figs. 10.33(a) and (b) illustrate two walks in the graph:

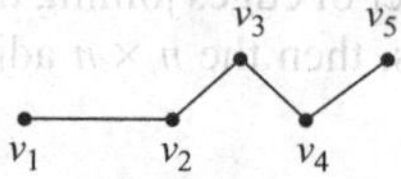

Figure 10.33(a)

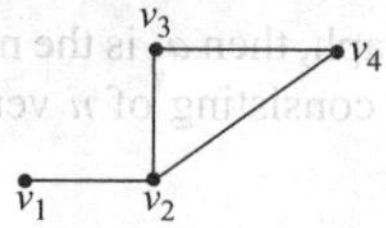

Figure 10.33(b)

1. $v_1 \to v_2 \to v_3 \to v_4 \to v_5$ is a walk
2. $v_1 \to v_2 \to v_3 \to v_4 \to v_2$ is a walk (here vertex v_2 is repeated)

Path: A path is a walk through a sequence of vertices:

$$v_1 \to v_2 \to v_3 \to \ldots v_{n-1} \to v_n$$

without any repetition of vertices such a path is also known as simple path or an elementary path. In a path, terminal vertices are of degree 1, whereas all the intermediate vertices are of degree 2. The walk shown in Fig. 10.33(b), that is, $v_1 \to v_2 \to v_3 \to v_4 \to v_2$ is *not* a path as vertex v_2 is repeated.

Trail: A walk in which all edges are distinct is called a Trail. Thus a trail is a path in which no edge is repeated. The initial and final vertices of a walk are called terminal vertices.

Closed Walk: A walk can begin and end at the same vertex and such a walk is called a closed walk. If terminal vertices are distinct, the walk is called an open walk.

Example 1 $v_1 \to v_2 \to v_5 \to v_2 \to v_4 \to v_4 \to v_5$ is a walk in Fig. 10.34(a)

Example 2 $v_1 \to v_2 \to v_3 \to v_4 \to v_4 \to v_5$ is a walk in Fig. 10.34(a)

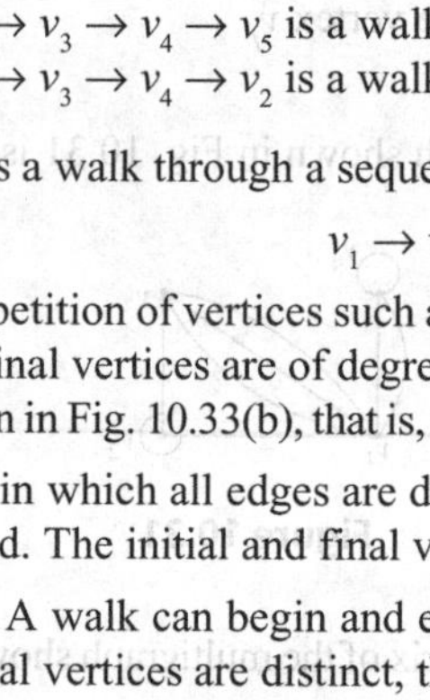

Figure 10.34(a)

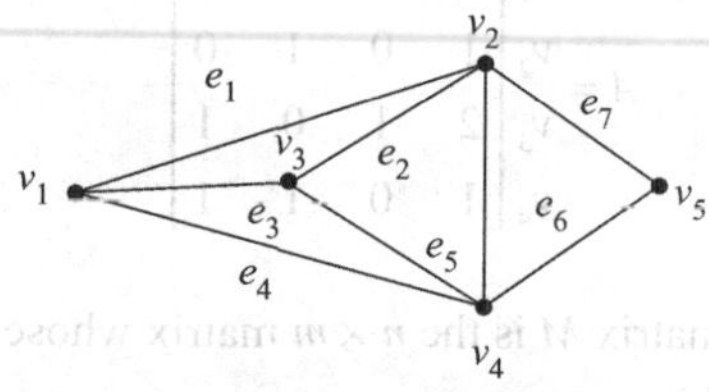

Figure 10.34(b)

Example 3 In Fig. 10.34(a) $v_1 \to v_2 \to v_5 \to v_3$ is a path.

Length of a Path: The number of edges in a path is called the length of the path.

Example 1 In Fig. 10.34(b), $v_1 \xrightarrow{e_1} v_2 \xrightarrow{e_7} v_5 \xrightarrow{e_6} v_4 \xrightarrow{e_5} v_3$ is a path of length 4. This is an open path.

Example 2 In Fig. 10.34(b), the path $v_1 \to v_2 \to v_5 \to v_4 \to v_3 \to v_1$ is a closed path.

Note 10.11 A loop can be included in a walk but not in a path.

10.10.1 Circuit

A closed walk in which no vertex is repeated is called a *circuit*. Thus, a circuit is a *path* that begins and ends at the same vertex.

In other words, it is a closed non-intersecting walk.
In Fig. 10.34(a)

1. $v_1 \xrightarrow{a} v_2 \xrightarrow{b} v_3 \xrightarrow{c} v_1$ is a circuit.

2. $v_1 \xrightarrow{c} v_3 \xrightarrow{e} v_5 \xrightarrow{h} v_2 \xrightarrow{a} v_1$ is a circuit.

Example 1 The following graphs are circuits:

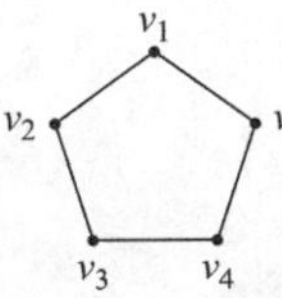

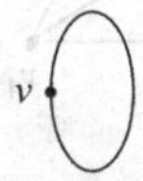

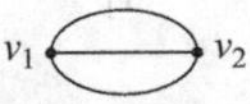

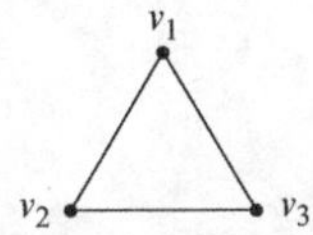

Figure 10.35(a) Figure 10.35(b) Figure 10.35(c) Figure 10.35(d)

Example 2 Find all the circuits of the graph shown in Fig. 10.35(e)

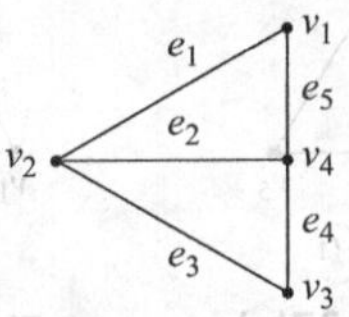

Figure 10.35(e)

(1) $v_1 \to v_2 \to v_4 \to v_1$ is a circuit.
(2) $v_2 \to v_3 \to v_4 \to v_2$ is a circuit.
(3) $v_2 \to v_4 \to v_1 \to v_2$ is a circuit.
(4) $v_3 \to v_4 \to v_1 \to v_2 \to v_3$ is a circuit.

Example 3 Draw a circuit from the graph which includes all the vertices given in Fig. 10.35(f).

Solution:

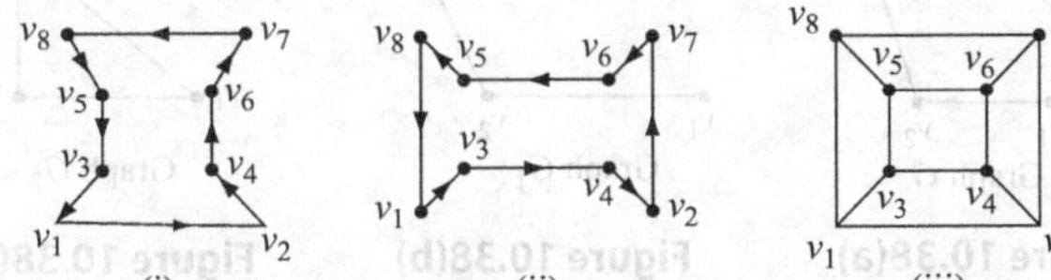

Figure 10.35(f)

The circuits are (i) $v_1\ v_2\ v_4\ v_6\ v_7\ v_8\ v_5\ v_3\ v_1$
and (ii) $v_1\ v_3\ v_4\ v_2\ v_7\ v_6\ v_5\ v_8\ v_1$ (except the initial and final)

10.11 SUB-GRAPH

In some problems, only a part of the entire graph need be considered. Such a graph is called the sub-graph of the given graph. Sub-graphs play an important role in analysing the properties of graphs. The concept of sub-graph is similar to that of a subset, and hence the following results are true:

1. Every graph is a sub-graph of itself.
2. Every vertex in a graph *G* is a sub-graph of *G*.
3. Every edge together with its end vertices is also a sub-graph of *G*.

Definition 10.1 A graph H is said to be a **sub-graph** of a graph G if all its vertices and edges are in G and each edge of H has the same end vertices in H as in G.

In other words, if $G = (V, E)$ and $H = (V_1, E_1)$ are two graphs such that $V_1 \subseteq V$ and $E_1 \subseteq E$, then H is called a **sub-graph** of G.

Example 1 Let G be the original graph shown in Fig. 10.36

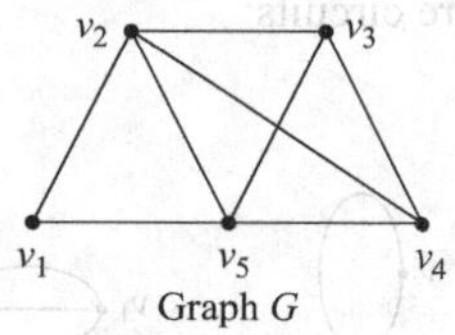

Figure 10.36

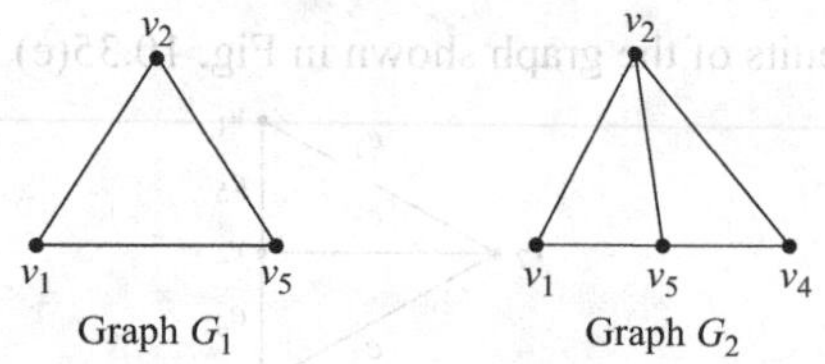

Figure 10.37(a) **Figure 10.37(b)**

The graph G_1 and G_2 shown in Figs. 10.37(a) and (b) are sub-graphs of graph G given in Fig. 10.36.

Example 2

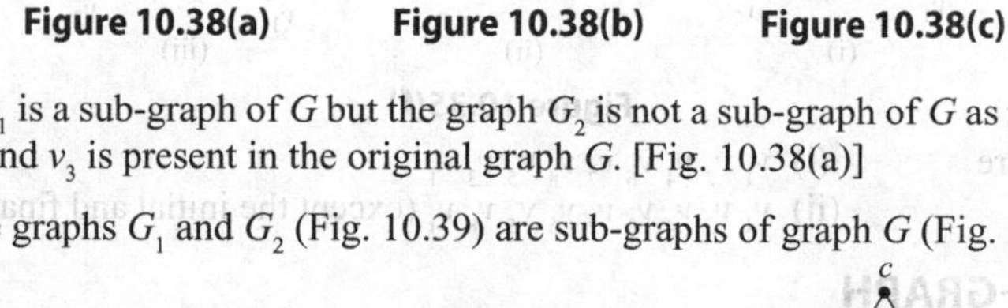

Figure 10.38(a) **Figure 10.38(b)** **Figure 10.38(c)**

The graph G_1 is a sub-graph of G but the graph G_2 is not a sub-graph of G as no edge between the vertices v_1 and v_3 is present in the original graph G. [Fig. 10.38(a)]

Example 3 The graphs G_1 and G_2 (Fig. 10.39) are sub-graphs of graph G (Fig. 10.40).

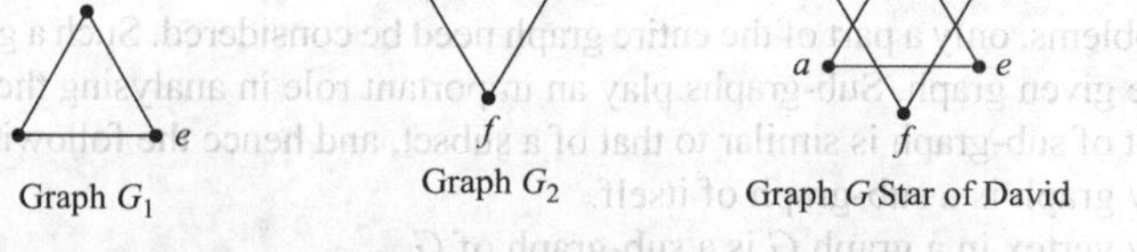

Figure 10.39(a) **Figure 10.39(a)** **Figure 10.40**

10.11.1 Spanning Sub-graph

The graph H is called a spanning sub-graph of G if H contains all vertices of G but not necessarily all edges of G.

Consider the graph G shown in Fig. 10.41.

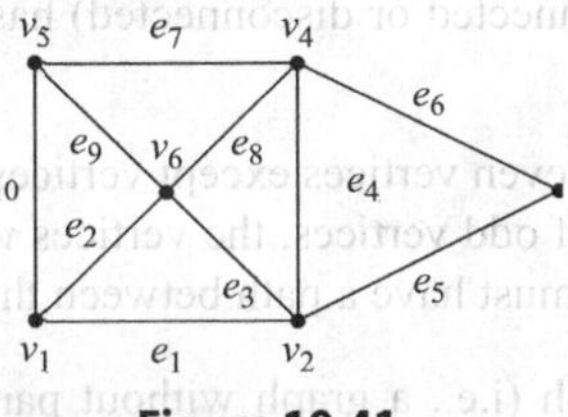

Figure 10.41

The following are spanning sub-graphs G:

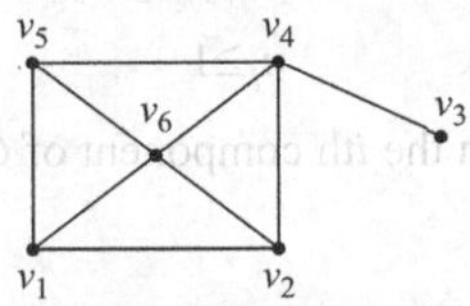

Figure 10.42(a) *Deleting edge e_5*

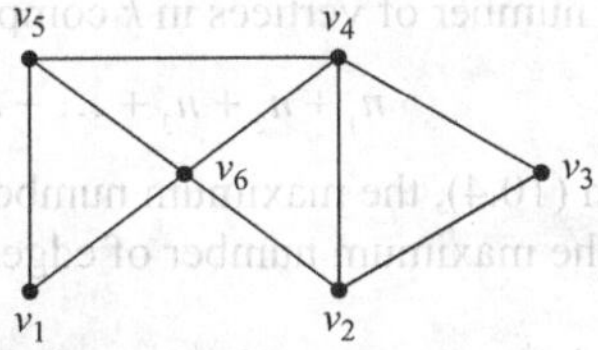

Figure 10.42(b) *Deleting edge e_1*

1. $G\text{–}e_5$
2. $G\text{–}e_1$

10.12 CONNECTED AND DISCONNECTED GRAPHS

A graph G is said to be connected if there exists at least one path between every pair of vertices of G, otherwise it is disconnected.

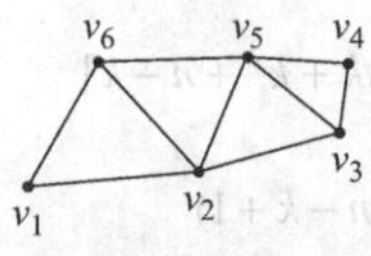

Figure 10.43(a) *connected graph*

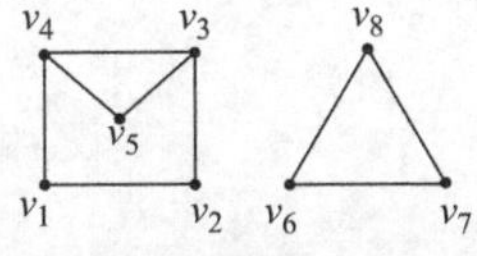

Figure 10.43(b) *Disconnected graph*

The graph in Fig. 10.43(a) is connected, while the graph in Fig. 10.43(b) is disconnected. A null graph of more than one vertex is disconnected. Further, any disconnected graph will have two or more connected sub-graphs which are called components of the original graph.

Let us consider a graph shown in Fig. 10.44 which has three connected components:

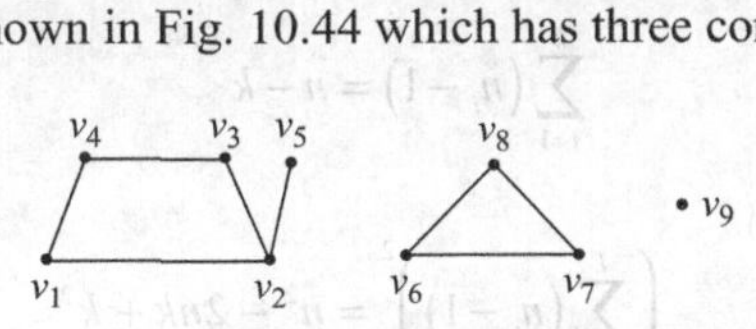

Figure 10.44

The graph has three connected components, the sub-graphs are induced by vertex sets $\{v_1, v_2, v_3, v_4, v_5\}$, $\{v_6, v_7, v_8\}$, and $\{v_9\}$.

The vertex v_9 itself forms a connected component of the graph. There are two useful theorems involving connectedness of the graph.

THEOREM 10.7 If a graph (connected or disconnected) has two vertices of odd degree, there must be a path joining them.

Proof: Let G be a graph with all even vertices except vertices v_1 and v_2, which are odd. Since no graph can have an odd number of odd vertices, the vertices v_1 and v_2 in graph G must belong to the same component, and hence must have a path between them.

THEOREM 10.8 A simple graph (i.e., a graph without parallel edges and self-loops) with n vertices and k components can have at most $(n-k)(n-k+1)/2$ edges.

Proof: Let the number of vertices in k components of a graph G be $n_1, n_2, n_3, \ldots, n_k$. Then

$$n_1 + n_2 + n_3 + \ldots + n_k = n \qquad n_i \geq 1$$

By Theorem (10.4), the maximum number of edges in the ith component of G is $n_i(n_i - 1)/2$. Therefore, the maximum number of edges in G is

$$\begin{aligned}
\frac{1}{2}\sum_{i=1}^{k} n_i(n_i - 1) = \frac{1}{2}\sum_{i=1}^{k} n_i\,(n_i^2 - n_i) &= \frac{1}{2}\sum_{i=1}^{k} n_i^2 - \frac{1}{2}\sum_{i=1}^{k} n_i \\
&= \frac{1}{2}\sum_{i=1}^{k} n_i^2 - \frac{1}{2}(n_1 + n_2 + \ldots + n_k) \\
&= \frac{1}{2}\sum_{i=1}^{k} n_i^2 - \frac{n}{2} \\
&\leq \frac{1}{2}[n^2 - (k-1)(2n-k)] - \frac{n}{2} \\
&\leq \frac{1}{2}[n^2 - 2nk + k^2 + n - k] \\
&= \frac{1}{2}(n-k)(n-k+1)
\end{aligned}$$

by Algebraic inequality $\sum_{i=1}^{k} n_i^2 \leq n^2 - (k-1)(2n-k)$

Proof:

$$\sum_{i=1}^{k}(n_i - 1) = n - k$$

Squaring both sides,

$$\left(\sum_{i=1}^{k}(n_i - 1)\right)^2 = n^2 - 2nk + k^2$$

or $$\left[(n_1-1)+(n_2-1)+\ldots+(n_i-1)+\ldots+(n_k-1)\right]^2 = n^2 - 2nk + k^2$$

or $$\sum_{i=1}^{k}(n_i-1)^2 + \text{Non-negative crossterms} = n^2 + k^2 - 2nk$$

or $$\sum_{i=1}^{k}\left(n_i^2-2n_i\right)+k+\text{ Non-negative crossterms } = n^2 + k^2 - 2nk$$

Hence $$\sum_{i=1}^{k} n_i^2 \le n^2 + k^2 - 2nk - k + 2n = n^2 + (k-1)(2n-1) \text{ proved}$$

10.12.1 Eulerian and Hamiltonian Graphs

We now consider two broad categories of problems where graph theory is used. In the first type of problems, we are interested to traverse a path using each edge of the graph exactly once. Such a problem arises in the delivery of goods through a network of roads where minimum distance is to be covered up in view of economy and time saving. The resulting graph is known as Euler Graph.

In the second type of problems, one has to visit each vertex exactly once. Such a path is useful to service engineers who must service machines on a regular basis. Each machine can be represented by a vertex. The resulting graph is known as Hamiltonian graph.

Euler Path: A path in a graph G is called an Euler path if it includes every edge exactly once. To find Euler path, vertices may be repeated.

Euler Circuit: If the initial and terminal vertices are same on an Euler path, then that path is called an Euler circuit. This is also called Euler line. Cleary, a graph may contain Euler path but not necessarily an Euler circuit.

Definition: Euler Graph

A graph G is called an Euler graph if it has at least one Euler circuit.

Example 1 Consider the graph shown in Fig. 10.45

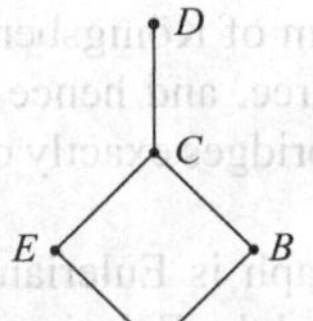

Figure 10.45

The path *DCEABC* contains all the edges exactly once, and hence is an *Euler path*.

However, it is *not* an Euler circuit as starting and ending at the same vertex is not possible without repeating an edge. Thus, the graph of Fig. 10.45 is not an Euler graph.

Example 2 Consider the graphs shown in Figs. 10.46(a) and (b)

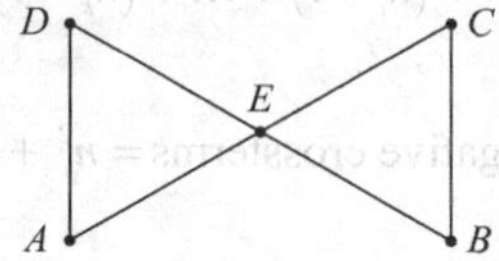

Figure 10.46(a)

The path *AECBEDA* is an Eulerian circuit, and hence the graph is an Euler graph.

Example 3 Consider the graph is Fig. 10.46(b)

The path *A B C D E F C G A* is an Eulerian circuit. Similarly, the path *A G C F E D C B A* is an Eulerian circuit, and hence the graph is an Eulerian graph. It can be observed that each vertex is of even degree.

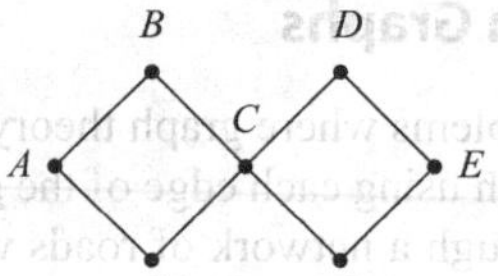

Figure 10.46(b)

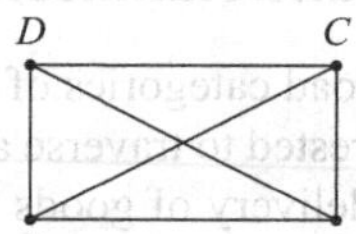

Figure 10.46(c)

A little experimentation will show that no Euler path or circuit is possible for the graph in Fig. 10.46(c). Hence, the graph is not an Euler graph. By its definition, path is always connected and hence Euler graph is always connected.

THEOREM 10.9 A connected graph *G* is an Euler graph if and only if each vertex of *G* is of even degree.

Proof: Let *G* be an Euler graph, and hence there will exist an Euler circuit, that is, a closed walk. In tracing this walk, every time the walk meets the vertex v_k, it will go through two new edges incident on v_k, the one we enter and the other to exist. This is true not only for all intermediate vertices of the walk but also for the starting vertex. Thus if *G* is an Euler graph, all vertices are of even degree. Conversely, if all vertices of *G* are of even degree, then *G* is an Euler graph can be proved.

10.12.2 Königsberg Bridge Problem

From the graph representing the problem of Königsberg Bridge shown in Fig. 10.1(b), we find that all its vertices are not of even degree, and hence it is not an Euler graph. Thus, it is not possible to walk over each of the seven bridges exactly once and return to the starting point. Thus this problem has no solution.

According to Euler, a connected graph is Eulerian if each vertex is of even degree. This theorem does not give any method to find the Eulerian circuit

We can use the following algorithm to find the Eulerian circuit.

10.12.3 Fleury's Algorithm

Let $G = (V, E)$ be a connected graph with each vertex of even degree.

Step 1: Select any starting vertex $v \in V$.

Step 2: Traverse any edge, choosing an edge that will disconnect the graph only if there is no other choice.
Step 3: Repeat Step 2 until no edge remains in E.
END

Example: Use Fleury's algorithm on the graph to find an Eulerian circuit.

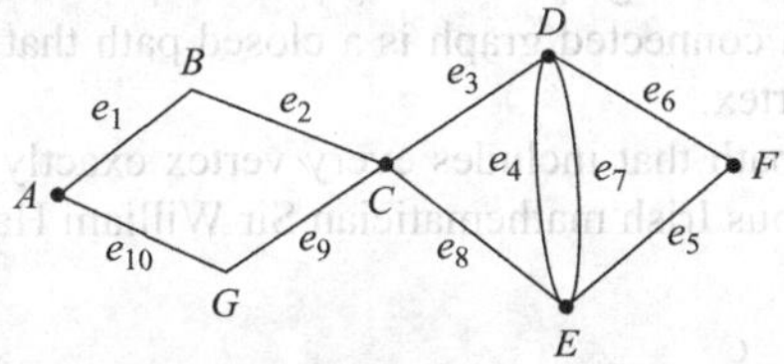

Solution: Each vertex is of even degree, and hence an Euler circuit is possible.

Step 1: We can start from any vertex. Let us start at the vertex A. We can traverse either edge AB (i.e., e_1) or AG (i.e., e_{10})

We choose AB. At B, there is only one edge BC (i.e., e_2).

At C, we can choose either CD (i.e., e_3) or CE. Let us choose CD. Proceeding step by step, we obtain

$$A \xrightarrow{e_1} B \xrightarrow{e_2} C \xrightarrow{e_3} D \xrightarrow{e_4} E \xrightarrow{e_5} F \xrightarrow{e_6} D \xrightarrow{e_7} E \xrightarrow{e_8} C \xrightarrow{e_9} G \xrightarrow{e_{10}} A$$

10.12.4 Unicursal Graph

Consider the graph shown in Fig. 10.46(d)

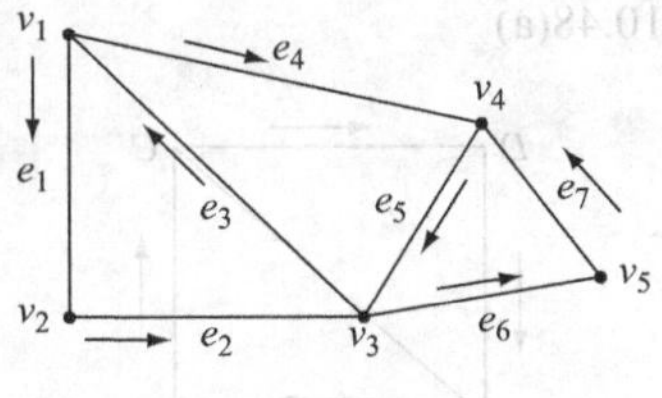

Figure 10.46(d)

The walk $v_1 \to v_2 \to v_3 \to v_1 \to v_4 \to v_3 \to v_5 \to v_4$ includes all edges without retracing any edge is not *closed*. The initial vertex is v_1 and the terminal vertex is v_4. Such an open walk is called an open Euler line or unicursal line. A graph having unicursal line is called *unicursal graph*.

Some Useful Results on Euler Graph

1. If every vertex of a connected graph G is of even degree, then the graph is an Euler graph.
2. If a connected graph has more than two vertices of odd degree, then there can be no Euler path. But if there are exactly two vertices of odd degree, then there exists an Euler path. Any Euler path in G must begin at one vertex of odd degree and end at the other.
3. If a graph G is connected and each vertex has even degree, then there is an Euler circuit in G.
4. If a graph G is connected and has one vertex of odd degree, then there can be no Euler circuit in G.

We now consider the second category of graph problems in which each vertex is to be visited exactly once. Such a path is useful to persons serving a set of vending machines on a regular basis. Each vending machine can be regarded as a vertex.

10.12.5 Hamiltonian Paths and Circuits

A *Hamiltonian path* in a connected graph G with $|V| \geq 3$ is a path that uses each vertex exactly once. A *Hamiltonian circuit* in a connected graph is a closed path that traverses each vertex exactly once except the starting vertex.

A graph with a closed path that includes every vertex exactly once is called a Hamiltonian graph named after the famous Irish mathematician Sir William Hamilton.

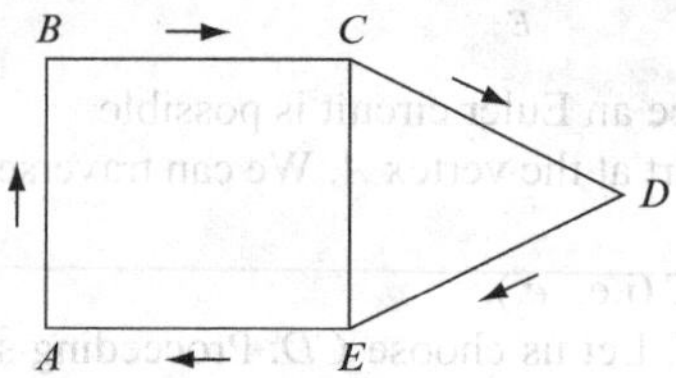

Graph G_1

Figure 10.47(a)

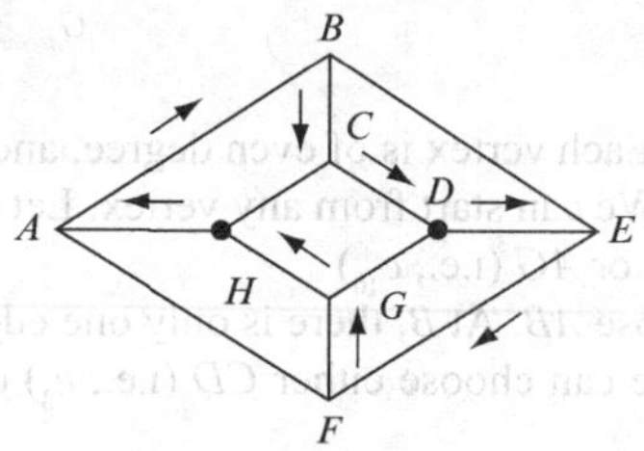

Graph G_2

Figure 10.47(b)

The graph G_1 shown in Fig. 10.47(a) is Hamiltonian as it contains a Hamiltonian circuit (i.e., closed path) $A \to B \to C \to D \to E \to A$

The graph G_2 shown is Fig. 10.47(b) $A \to B \to C \to D \to E \to F \to G \to H \to A$ is Hamiltonian

The graph G_3 shown in Fig. 10.48(a)

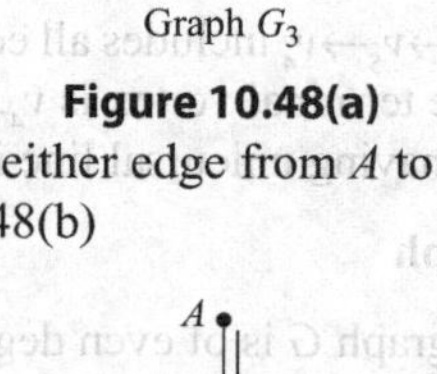

Graph G_3

Figure 10.48(a)

$A \to B \to C \to D \to A$ (choosing either edge from A to B) is Hamiltonian.

In the graph G_4 shown in Fig. 10.48(b)

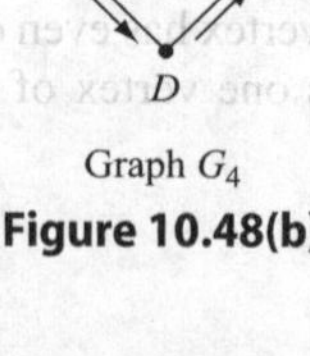

Graph G_4

Figure 10.48(b)

$A \to B \to C \to D \to E$ is a Hamiltonian path because it contains each vertex only once. It is easy to see that there is no Hamiltonian circuit for this graph. Hence, the graph G is not Hamiltonian.

The graph G_5 shown in Fig. 10.48(c) has no Hamiltonian path. It has no Hamiltonian circuit, and hence the graph G_5 is not Hamiltonian.

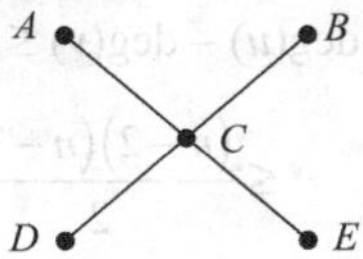

Graph G_5

Figure 10.48(c)

The graph G_6 is not Hamiltonian because we have to pass through vertex D to return the home vertex choosing any vertex A, B, C to start with.

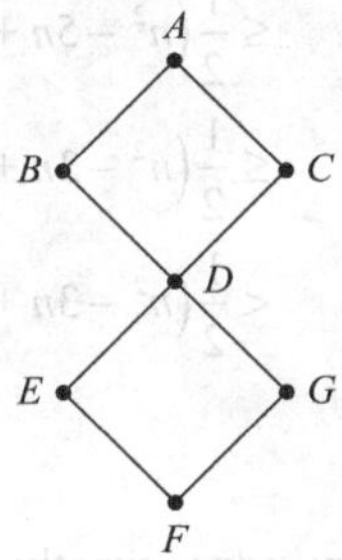

Graph G_6

Figure 10.48(d)

The following two theorems provide sufficient conditions for a simple connected graph to be Hamiltonian.

ORE'S THEOREM 10.10 Let G be a simple connected graph with $n \geq 3$ vertices, then G is Hamiltonian (i.e., G has a Hamiltonian circuit) if

$$\deg(u) + \deg(v) \geq n$$

for every pair of non-adjacent vertices u and v.

DIRAC'S THEOREM 10.11 A simple connected graph G with $n \geq 3$ is Hamiltonian is if $\deg(v) \geq \frac{n}{2}$ for every v in G.

Proof: The proof follows from Ore's theorem, that is, $\deg(u) + \deg(v) \geq n$ for every pair of non-adjacent vertices.

Corollary: The connected graph G with n vertices has a Hamiltonian circuit provided the number of edges in G, $e \geq \frac{1}{2}\left(n^2 - 3n + 6\right)$ *for* $n \geq 3$

Proof: Let the graph G be non-Hamiltonian. Then by Dirac's theorem, there will exist a pair of non-adjacent vertices u and v such that

$$\deg(u) + \deg(v) \leq n - 1$$

Let H be the sub-graph of G obtained by deleting the vertices u and v from G. The graph H will have $(n-2)$ vertices and e-deg(u) – deg(v) edges. Now the maximum number of edges of H can be ${}^{n-2}c_2$.

Hence,

$$e\text{-deg}(u) - \deg(v) \leq {}^{n-2}c_2$$
$$\leq \frac{(n-2)(n-3)}{2}$$
$$\leq \frac{1}{2}\left(n^2 - 5n + 6\right)$$
$$\therefore e \leq \frac{1}{2}\left(n^2 - 5n + 6\right) + (n-1)$$
$$\leq \frac{1}{2}\left(n^2 - 5n + 6 + 2n - 2\right)$$
$$\leq \frac{1}{2}\left(n^2 - 3n + 4\right)$$
$$< \frac{1}{2}\left(n^2 - 3n + 6\right)$$

which is a contradiction.

Note 10.12 Eulerian graph uses every edge exactly once but may repeat vertices while Hamiltonian graph uses each vertex exactly once (except for the first and last) but may skip edges.

The graphs given below illustrate comparison between Eulerian and Hamiltonian graphs. Eulerian as each vertex is of even degree and Hamiltonian as each vertex is of degree $\geq 4/2 = 2$

Figure 10.49(a)

Eulerian but non-Hamiltonian (Each vertex is of even degree).

$$a \to b \to c \to d \to e \to f \to d \to a$$

Since vertex d is repeated, and hence non-Hamiltonian.

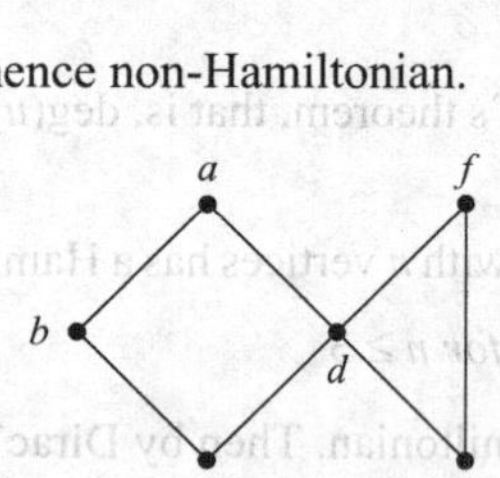

Figure 10.49(b)

Hamiltonian ($a \rightarrow b \rightarrow c \rightarrow d \rightarrow a$) but non-Eulerian (Four vertices of odd degree)

Figure 10.49(c)

Neither Eulerian (more than two vertices of odd degree) nor Hamiltonian (No circuit through all vertices is possible without repetition of vertices)

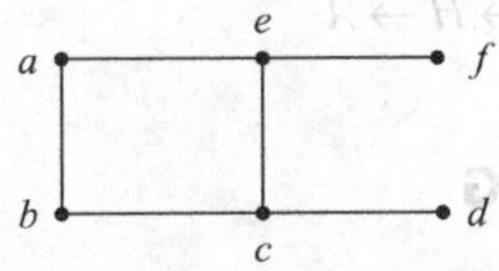

Figure 10.49(d)

Note 10.13

1. The graph in Fig. 10.49(a) is Hamiltonian as each vertex is of degree $\geq \frac{4}{2} = 2$ (there are four vertices).
2. The graph in Fig. 10.49(b) is non-Hamiltonian as there are 6 vertices and there is only one vertex of degree $\geq \frac{n}{2}$, that is, ≥ 3, and other are of degree < 3.
3. The graph in Fig. 10.49(c) is Hamiltonian. The numbers of vertices $n = 4$. Each vertex is of degree $\geq \frac{4}{2}$, that is, 2.
4. The graph in Fig. 10.49(d) is non-Hamiltonian. The number of vertices $n = 6$. There are only two vertices of degree $\geq \frac{n}{2}$, that is $\frac{6}{2} = 3$ and others are of degree <3.

Some Observations on Hamiltonian Circuit for the Graph $G = (V, E)$

1. If G has Hamiltonian Circuit of n vertices, it must consist of exactly n edges. If we remove any edge from the circuit, we shall have Hamiltonian path.
2. Hamiltonian path is a sub-graph of Hamiltonian circuit.
3. The length of a Hamiltonian path in a connected graph of n vertices is $(n - 1)$.
4. If G has Hamiltonian circuit, then each vertex of G must be of degree ≥ 2.

THEOREM 10.12 If a graph G has m edges and n vertices, then G has a Hamiltonian circuit if

$$m \geq \frac{1}{2}\left(n^2 - 3n + 6\right)$$

The converse of the theorem is not true, that is, the given condition is sufficient but not necessary as can be seen from the graph consisting of 8 vertices and 8 edges given in Fig. 10.49(e).

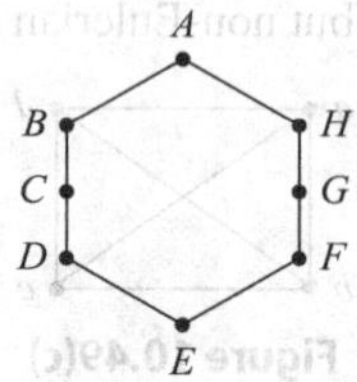

Figure 10.49(e)

The given condition is not satisfied, but there is Hamiltonian circuit for the graph given be $A \rightarrow B \rightarrow C \rightarrow D \rightarrow E \rightarrow F \rightarrow G \rightarrow H \rightarrow A$

10.13 GRAPH COLOURING

Graph colouring problems have gained importance due to their wide practical applications in which 'colour' has different meaning. For example, if the graph represents a connected grid of cities, each city can be marked with the name of airline having most flights to and from the city. In this case, vertices are cities and colours are airline names.

The subject of graph colouring can be viewed to have its origin from the map colouring problem which was conjectured that four distinct colours are required to colour any map drawn on a plane such that no two regions (countries) sharing a common border have the same colour. The problem could not be solved for over a century. In 1976, two American mathematicians Kenneth Appel and Wolfan Haken found the solution with the aid of computer computations performed on almost 2000 configurations of graph that a minimum of four district colours are needed to colour the map. Graph colouring finds its application in scheduling conflict-free examinations for various courses minimizing time slots and also in the design of traffic light patterns at intersections in big cities.

Definition 10.2 Let $G = (V, E)$ be an undirected graph with no multiple edges and $C = \{c_1, c_2, ..., c_n\}$ a set of n colours.

A function f: $V \rightarrow C$ is called colouring of graph G if $f(v_i) \neq f(v_j)$, v_i and v_j being adjacent vertices of V.

Thus colouring of a graph G means colouring of vertices with different colours.

A colouring is called proper colouring if any two adjacent vertices have different colours. The minimum number of colours needed for a proper colouring of graph G is called the chromatic number of G.

A graph that requires K different colours for its proper colouring is called K-chromatic. The number K is called chromatic number of G.

A graph with n vertices can be coloured by n colours by assigning a colour to each vertex. The proper colouring which is of interest in practical applications is the requirement of minimum number of colours.

Let us consider the graph with 5 vertices shown in Fig. 10.50(a).

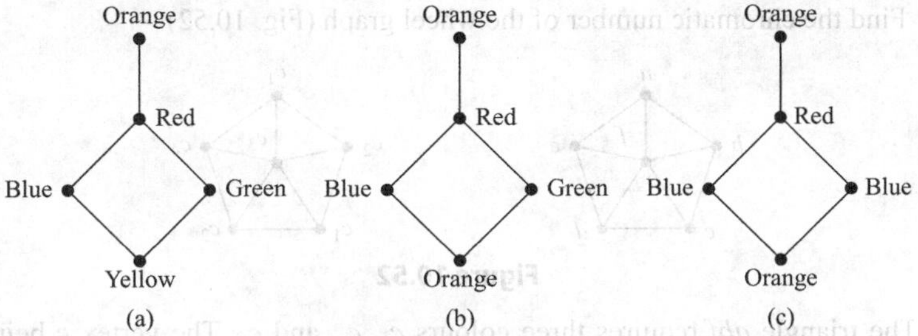

Five colours are required Four colours are required Three colours are required

Figure 10.50

Fig. 10.50(c) shows that a minimum of three colours are required for proper colouring, and hence its chromatic number is 3.

The definition of chromatic number leads to the following obvious results:

1. The chromatic number of a null graph is 1.
3. The chromatic number of a complete graph with n vertices is n.
3. The chromatic number of any graph with two or more vertices is ≥ 2.
4. The chromatic number of a graph having one or more edges is at least 2.
5. Every graph having a triangle is at least 3-chromatic.
6. A graph consisting of simply one circuit with the number of vertices $n \geq 3$ is 2-chromatic if n is even and 3-chromatic if n is odd. (If n is odd, nth and the first vertex will be adjacent and will have same colour and thus requiring a third colour.
7. Chromatic number of a bipartite graph $k_{m,\, n}$ is 2.

Note 10.14

1. The maximum number of colours required for colouring a connected graph G with n vertices is n.
2. If $d_{\max}$ is the maximum degree of vertices in a graph G, the chromatic number of $G \leq d_{\max} + 1$.

Example 1 Show that the chromatic number of the graph (Fig. 10.51) is 4.

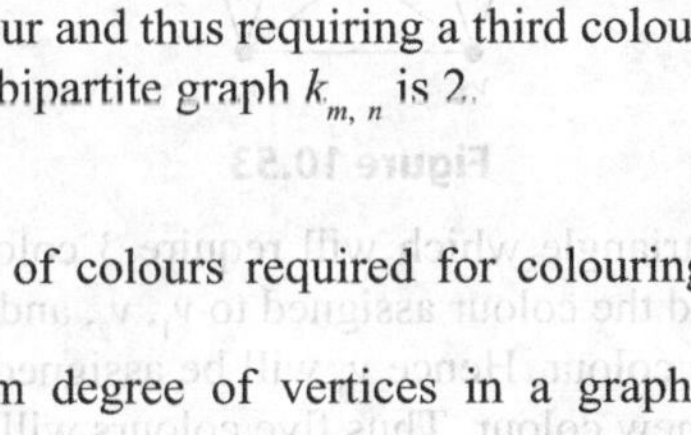

Figure 10.51

Solution: The triangle of vertices v_1, v_2, v_3 require three different colours say c_1, c_2, and c_3. The vertex v_4 is adjacent to v_1 and v_3 and hence must be assigned a colour different from that of v_1 and v_3. Hence, the colour of v_4 is c_2. Now v_5 must be assigned a colour different from c_1, c_2, and c_3 as v_5 is adjacent to v_1, v_4, and v_3. Let c_4 colour be assigned v_5. Hence chromatic number of the graph is 4.

Example 2 Find the chromatic number of the wheel graph (Fig. 10.52)

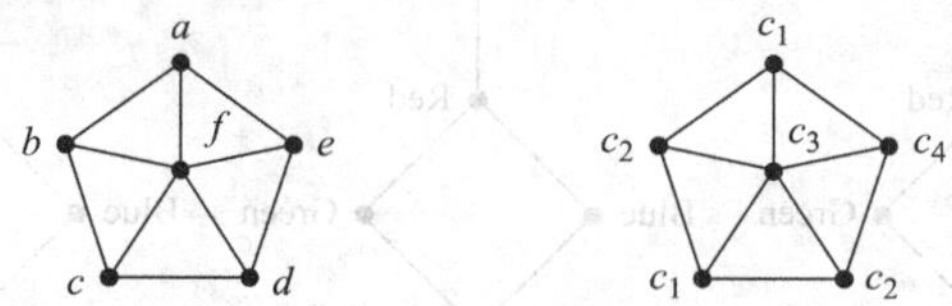

Figure 10.52

Solution: The triangle *abf* requires three colours c_1, c_2, and c_3. The vertex *c* being adjacent to vertices *b* and *f* cannot be assigned colours c_2 and c_3 and assign it c_1. The vertex *d* can be assigned c_2. The remaining vertex *e* cannot be assigned c_1, c_2, or c_3 as adjacent vertices *a*, *d*, and *f* have these colours. Hence, a colour c_4 will be assigned to vertex *e*. Hence, the wheel graph w_6 is 4 chromatic.

Example 3 Find the chromatic number of the graph K_5, that is, the number of colours necessary for proper colouring of the graph shown in Fig. 10.53.

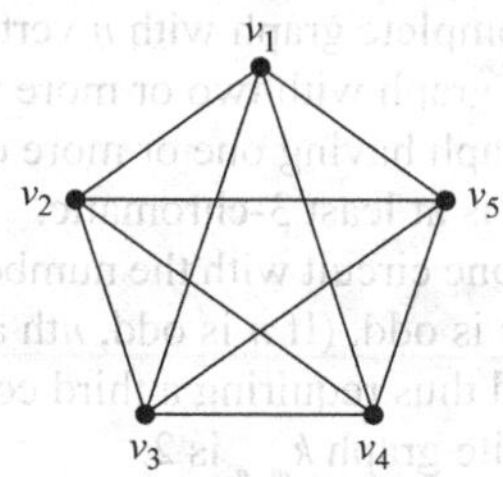

Figure 10.53

Solution: The graph contains triangle which will require 3 colours say for vertices v_1, v_2, v_3. The vertex v_4 cannot be assigned the colour assigned to v_1, v_2, and v_3 as this will amount to have adjacent vertices with the same colour. Hence v_4 will be assigned a fourth colour. Similarly, v_5 will also have to be assigned a new colour. Thus five colours will be needed. That is, chromatic number of K_5 is 5.

Example 4 Find the number of colours for proper colouring of $K_{2,3}$ (Fig. 10.54).

Solution: The chromatic number of any bipartite graph $K_{m,n}$ is 2. Since $K_{3,2}$ is a bipartite, two colours are required for its proper colouring.

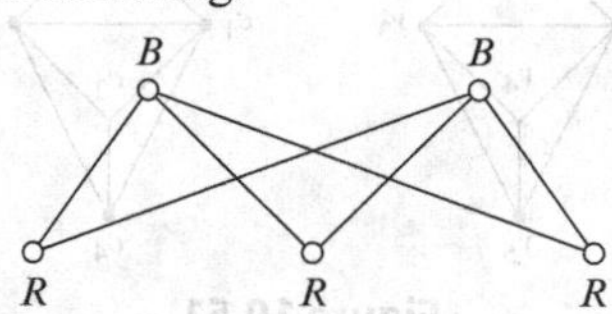

Figure 10.54

The vertex set $K_{2,3}$ can be partitioned into two disjoint subsets V_1 and V_2 such that every vertex in V_1 is adjacent to every vertex in V_2 and vice versa. In addition, no vertices in V_1 and in V_2 are adjacent. Let every vertex in V_1 be assigned one colour (black) and in V_2 another colour (red). Thus, only two colours are required for the graph $K_{2,3}$.

Example 5 Find the chromatic number of the cycle graph C_n.

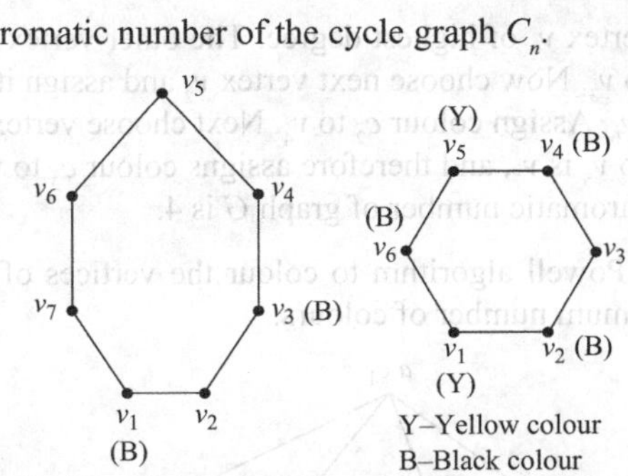

Solution: Let n be even and vertices be $v_1, v_2, v_3, v_4, \ldots, v_{2n}$. Then odd vertices $v_1, v_3, v_5, \ldots, v_{2n-1}$ can be assigned yellow colour and the even numbered vertices $v_2, v_4, \ldots, v_{2n}$ black colour. If n is odd, let the vertices be $v_1, v_2, \ldots, v_{2n+1}$. If we assign black colour to odd vertices, then v_1 and v_{2n+1} (here v_1, v_7) are adjacent vertices which receive the same colour. Since this not acceptable, a third colour (say green) shall be assigned to v_{2n+1} (here v_7). Therefore, exactly three colours are needed. Hence, the chromatic number of cycle graph C_n is 2 if n is even and 3 otherwise.

10.13.1 Welch and Powell Algorithm

An algorithm developed by Welch and Powel can be used to find the number of colours for proper colouring of a graph.

Step 1: Arrange vertices in a sequence in the descending order of their degrees.

Step 2: Assign colour c_1 to the first vertex and then assign c_1 to each other vertex not adjacent to the previous vertex.

Step 3: Assign colour c_2 to the next non-coloured vertex in the sequence and repeat Step 2.

Step 4: Repeat the process until all the vertices have been painted.

It may be noted that the process may not always yield minimum number of colours.

Example 1 Use Welch and Powell algorithm to colour the vertices of graph G (Fig. 10.55) with minimum number of colours.

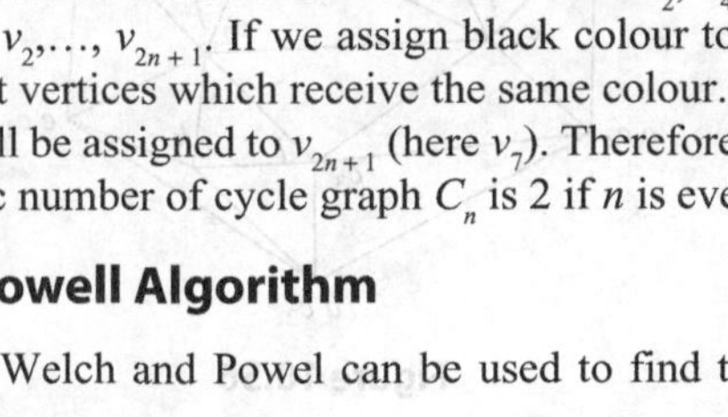

Graph G

Figure 10.55

Solution: The vertices of graph G are arranged in sequential order (descending order) of their degrees.

Vertex	v_1	v_2	v_4	v_5	v_3	v_6	v_7
Degree	5	4	4	4	3	3	3
Colour	c_1	c_2	c_2	c_3	c_1	c_4	c_3

Assign colour c_1 to the vertex v_1 of highest degree. The other vertex not adjacent to v_1 is v_3, and hence assign colour c_1 to v_3. Now choose next vertex v_2 and assign it colour c_2. The vertices not adjacent to v_2 are v_4 and v_6. Assign colour c_2 to v_4. Next choose vertex v_5 and assign it colour c_3. The vertex not adjacent to v_5 is v_7, and therefore assigns colour c_3 to v_7. The vertex v_6 can be assigned colour c_4. Hence, chromatic number of graph G is 4.

Example 2 Use Welch and Powell algorithm to colour the vertices of the graph (Fig. 10.56) having 12 vertices with minimum number of colours.

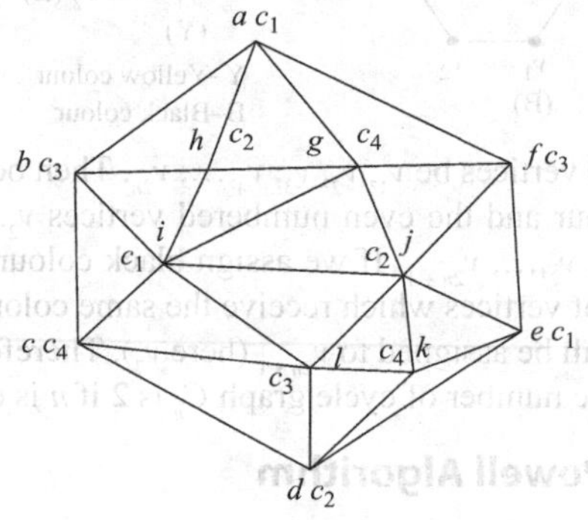

Figure 10.56

Solution:

Vertices	i	j	g	l	k	h	a	b	c	d	e	f
Degree	6	6	5	5	4	4	4	4	4	4	4	4
Colour	c_1	c_2	c_4	c_3	c_4	c_2	c_1	c_3	c_4	c_2	c_1	c_3

We can write the vertices in descending order of their degrees. Assign colour c_1 to the vertex of highest degree i or j. Let us assign to the vertex i and assign c_1 to vertices not adjacent to i, that is, to vertex a and e (f cannot be assigned c_1 otherwise a and c will have same colour). Now assign colour c_2 to vertex j(which also has highest degree) and assign colour c_2 to vertices d and h not adjacent to j. Assign colour c_3 to the next lower degree vertex l and assign colour c_3 to vertices non-adjacent vertices b and f. Now assign colour c_4 to the vertex g and assign c_4 to non-adjacent vertices c and k. Thus four colours are required for proper colouring of vertices.

10.14 CHROMATIC POLYNOMIAL

There are many different ways in which a graph G of n vertices can be properly coloured by using sufficiently large number of colours. This property of the graph can be expressed by a polynomial called chromatic polynomial of graph G.

Let $P_n(\lambda)$ denote the chromatic polynomial of a graph G with n vertices and its value given the number of ways of properly colouring using λ or lesser number of colours.

Let c_k be the different ways of properly colouring G using exactly k different colours. Now total number of ways of selecting k colours out of λ colours is

$$\binom{\lambda}{k} \text{ or } \frac{\lambda(\lambda-1)(\lambda-2)\ldots(\lambda-k+1)}{k!}$$

Hence, there are $c_k\left|\begin{matrix}\lambda\\ k\end{matrix}\right|$ different ways of properly colouring G using exactly k colours. Since k can assume n positive integral values (at most n colours can be used for n vertices), the chromatic polynomial is the sum of these terms. Thus

$$P_n(\lambda) = \sum_{k=1}^{n}\binom{\lambda}{k}c_k$$

$$= c_1\lambda + c_2\frac{\lambda(\lambda-1)}{2!} + c_3\frac{\lambda(\lambda-1)(\lambda-2)}{3!} + \ldots + c_n\frac{\lambda(\lambda-1)(\lambda-2)\ldots(\lambda-n+1)}{n!}$$

For each graph, c_k will be evaluated. For example, a graph of one edge requires at least two colours for its proper colouring and so $c_1 = 0$.

A graph with n vertices and n different colours can be properly coloured in $n!$ ways. Hence, $c_n = n!$

Example: Find the chromatic polynomial of the graph shown in Fig. 10.57.

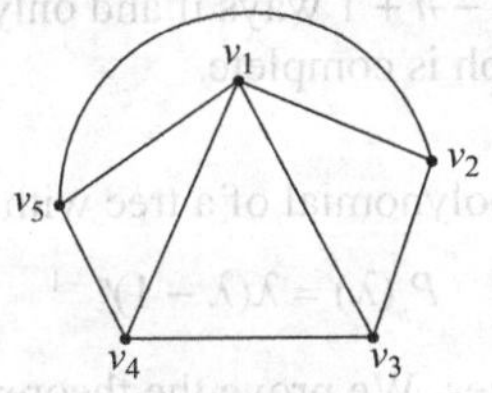

Figure 10.57

Since the graph consists of 5 vertices, the chromatic polynomial is denoted by

$$P_5(\lambda) = c_1\lambda + c_2\frac{\lambda(\lambda-1)}{2!} + c_3\frac{\lambda(\lambda-1)(\lambda-2)}{3!} + c_4\frac{\lambda(\lambda-1)(\lambda-2)(\lambda-3)}{3!}$$
$$+ c_5\frac{\lambda(\lambda-1)(\lambda-2)(\lambda-3)(\lambda-4)}{5!}$$

Since the graph has a triangle, it will require at least three different colours for proper colouring. Therefore

$$c_1 = 0,\ c_2 = 0, \text{ and } c_5 = 5!$$

To evaluate c_3, let three colours a, b, and c be assigned properly to vertices v_1, v_2, and v_3. This can be done in $3! = 6$ different ways. Now vertex v_5 must have the same colour as v_3 and v_4 must have the same colour as v_2. Hence

$$c_3 = 6$$

Similarly, with four colours v_1, v_2, and v_3 can be properly coloured in 4.3.2 = 24 different ways. The fourth colour can be assigned to v_4 or v_5 and thus provide two choices. Therefore,

$$c_4 = 2.24 = 48$$

The fifth vertex provides no additional choice. Substituting these coefficients in $P_5(\lambda)$, we get

$$P_5(\lambda) = 0 + 0 + \lambda(\lambda - 1)(\lambda - 2) + 2\lambda(\lambda - 1)(\lambda - 2)(\lambda - 3) + \lambda(\lambda - 1)(\lambda - 2)(\lambda - 3)(\lambda - 4)$$

or

$$\begin{aligned} P_5(\lambda) &= \lambda(\lambda - 1)(\lambda - 2)[1 + 2\lambda - 6] + \lambda(\lambda - 1)(\lambda - 2)(\lambda - 3)(\lambda - 4) \\ &= \lambda(\lambda - 1)(\lambda - 2)[2\lambda - 5 + \lambda^2 - 7\lambda + 12] \\ &= \lambda(\lambda - 1)(\lambda - 2)(\lambda^2 - 5\lambda + 7) \end{aligned}$$

The presence of $(\lambda - 1)$ and $(\lambda - 2)$ indicates that G is at least 3-chromatic.

THEOREM 10.13 A graph of n vertices is a complete graph if and only if its chromatic polynomial is

$$P_n(\lambda) = \lambda(\lambda-1)(\lambda-2)...(\lambda-n+1)$$

Proof: Let G be a graph of n vertices. With λ colours, any selected graph can be coloured in λ ways. A second vertex can be coloured properly in $\lambda - 1$ ways, the third in $\lambda - 2$ ways, and so on. The nth vertex can be coloured in $\lambda - n + 1$ ways if and only if every vertex is adjacent to every other. That is, if and only if the graph is complete.

THEOREM 10.14 The chromatic polynomial of a tree with n vertices is

$$P_n(\lambda) = \lambda(\lambda - 1)^{n-1}$$

Proof: Let G be a tree with n vertices. We prove the theorem by induction. If $n = 1$, then G has one vertex that can be coloured in λ distinct ways and so $P_n(\lambda) = \lambda$.

If $n = 2$, then G has one edge (two vertices) and hence two colours are required for proper colouring of the graph. Therefore,

$$c_1 = 0, c_2 = 2$$

Then
$$P_n(\lambda) = 0 + 2\frac{\lambda(\lambda-1)}{2!} = \lambda(\lambda-1)$$

Let us assume that the result is true for all trees having vertices less than n. Now consider a tree with n vertices. Let v be a pendent vertex of G and $G' = G - v$. Now G' is a tree with $n - 1$ vertices. Hence, by hypothesis

$$P_n(\lambda) = \lambda(\lambda - 1)^{n-2}$$

Now G' can be properly coloured if vertex v is painted with the colour other than that of the vertex adjacent to v (one and only one vertex is adjacent to v as it is a pendant vertex).

Thus, v can be coloured with $(\lambda - 1)$ colours for each proper colouring of G'. Therefore, the number of ways of proper colouring of G is

$$P_n(\lambda) = \lambda(\lambda - 1)^{n-2}.(\lambda - 1) = \lambda(\lambda - 1)^{n-1}$$

10.15 SHORTEST PATH PROBLEMS

The problem of finding shortest distance connecting cities by road/rail network is of great importance in view of reduction in cost as well as time. We shall discuss two algorithms to determine the shortest distance between two designated vertices: one in a graph without weights and other in weighted graphs.

10.15.1 Shortest Path in a Graph without Weights

In a loop-free graph $G(V, E)$ without weights, the shortest path between two vertices u and v is the one that uses the least number of edges. Two algorithms known as breadth-first search (BFS) and back-tracking search (BTS) are discussed to find the shortest path in an unweighted graph.

The basic idea in BFS is that if the starting vertex is u, we process all the neighbours of u and then neighbours of all the neighbours of u. In the process, no vertex is processed twice. BFS can be applied to both directed and undirected graphs. In case of unweighted graph, it leads to the shortest path from its starting vertex.

10.15.2 Algorithm for BFS

/* $G = (V, E)$ is an unweighted graph */

Step 1: Set count $j = 0$ and assign label 0 to the starting vertex p.

Step 2: Find all unlabelled vertices in G adjacent to p labelled j.
If there is no such vertex/vertices, then p is an isolated vertex otherwise label them $j + 1$.

Step 3: If q is labelled, go to Step 4.
If not, set the count $j = j + 1$ and go to Step 2.

Step 4: The length of the path from p to q is $j + 1$.
STOP.

After finding the shortest path from the above algorithm, we use BTS to find the actual shortest path, which uses label $\lambda(v)$ generated in BFS algorithm.

10.15.3 BTS Algorithm

Step 1: Step $j = \lambda(q)$ and assign $v_j = q$.

Step 2: Find a vertex u adjacent to v_j and assign $v_{j-1} = u$ where $\lambda(u) = j - 1$.

Step 3: If $j = 1$, STOP otherwise assign $j = j - 1$ and go to Step 2.

Note 10.15 There can be several shortest paths from p to q and this algorithm finds one of them.

Example 1 Find the shortest path from the vertex v_1 to v_8 and its length from the graph given below (Fig. 10.58).

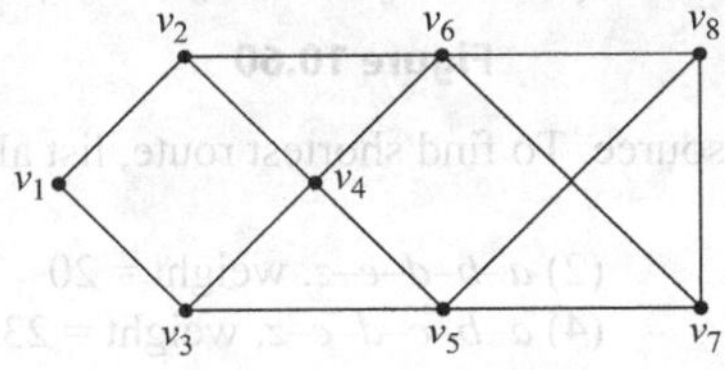

Figure 10.58

Solution: Applying BFS, the vertex v_1 is labelled as 0 for $j = 0$. Then the vertices v_2 and v_3 (adjacent vertices of v_1) are labelled $j = j + 1 = 0 + 1 = 1$. The vertices v_4, v_5 and v_6 are labelled $j = j + 1 = 1 + 1 = 2$. The vertices v_7 and v_8 are labelled $2 + 1 = 3$. Since v_8 is labelled 3, the length of a shortest path from v_1 to v_8 is 3.

Now using back-tracking algorithm, since $\lambda(v_8) = 3$, we start with $j = 3$ and $v_j = v_8$. The vertices adjacent to v_8 are v_5 and v_6. Hence, we choose v_5 or v_6 with $\lambda(v_5) = 2$ and assign $v_5 = 2$.

Next we choose v_3 adjacent to v_5 with $\lambda(v_3) = 1$ and assign $v_3 = 1$. Finally, we take v_1 adjacent to v_3 with $\lambda(v_1) = 0$ and assign $v_1 = 0$. Hence, the shortest path from v_1 to v_8 is $v_1\ v_3\ v_5\ v_8$ and path length is 3.

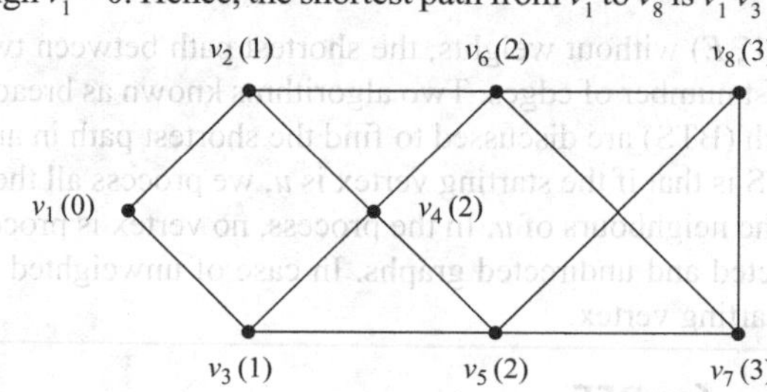

Figure 10.59

Note 10.16 There can be other shortest paths from v_1 to v_8. The other shortest path is $v_1\ v_2\ v_6\ v_8$.

10.16 SHORTEST PATH IN A WEIGHTED GRAPH

There are several problems of considerable importance that arise in industry, which can be modelled by graphs with weights assigned to their edges. The primary goal in these problems is to find shortest paths from a specified vertex to all the destinations. An efficient method was developed in 1959 by the Dutch mathematician Edsger Wybe Dijkstra known as Dijkstra Algorithm, which can find shortest path from any vertex to every vertex in the digraph, if it exists.

The following example finds the shortest route for the truck driver working in a company.

A computer company has a manufacturing plant at location *a* and a warehouse at location *z*. The possible streets that a truck driver can take from *a* to *z* are given by the digraph, arrows indicating one-way streets and the weights, the distances in kilometres (Fig. 10. 60).

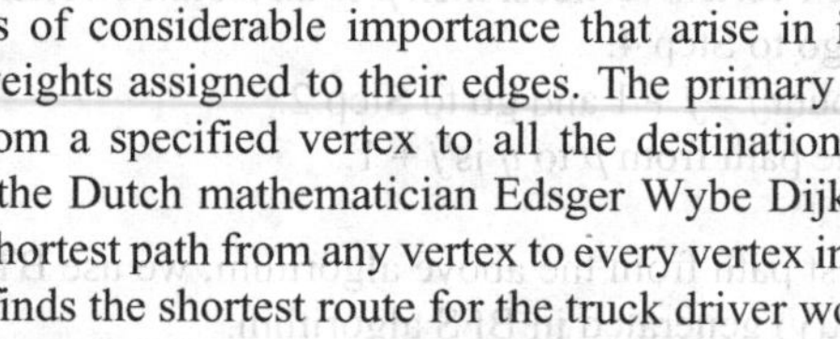

Figure 10.60

Solution: The vertex *a* is the source. To find shortest route, list all simple paths from *a* to *z* and compute their weights.

(1) *a–b–d–z*, weight = 16 (2) *a–b–d–e–z*, weight = 20
(3) *a–b–c–d–z*, weight = 19 (4) *a–b–c–d–e–z*, weight = 23
(5) *a–c–d–z*, weight = 11 (6) *a–c–d–e–z*, weight = 15

Clearly the path with minimum weight is *a–c–d–z* and the minimum distance is 11 km.

The above method of finding the shortest path involves finding the distance along all possible paths from source to destination is quite laborious, time consuming, and is certainly not the best with a diagraph with many vertices and edges. Dijkstra's algorithm is not only efficient but also provides the advantage of finding the shortest path from the source vertex to every vertex, which is discussed here.

10.16.1 Single-source Shortest Path Problem

In a weighted connected graph, we are interested in finding the shortest path from the source vertex to all other vertices. Several algorithms, such as Moore's, Belman–Ford's and Dijkstra's exist for the solution. Dijkstra's algorithm is discussed here due to its greater efficiency.

10.16.2 Dijkstra's Algorithm

Let $G = (V, E)$ be a connected graph with n vertices $v_1, v_2, \ldots, v_n \in V$. Let v_1 be the starting vertex and v_n the destination vertex. Let edges $(v_1, v_2), (v_2, v_3), \ldots, (v_i, v_j), \ldots, (v_{n-1}, v_n)$ provide paths in G from v_1 to v_n. Let $w(i, j)$ be the weight of the edge (v_i, v_j). If no edge exists between vertices v_i and v_j, we define $w(i, j) = +\infty$ to indicate large value.

Let S be the set of all vertices in the graph whose lengths of shortest paths from the starting vertex v_1 to u are known. If $L(u)$ denote the length of shortest path from v_1 to u, then

$$L(u) = \begin{cases} 0 & v_1 = u \\ w(v_1, u) & \text{if edge}(v_1, u)\text{exists} \\ \infty & \text{if edge}(v_1, u)\text{does not exist} \end{cases}$$

The algorithm provides an iterative process for the solution.

Step 0:

$$\text{Initially,} \quad V = \{v_1, v_2, v_3, \ldots, v_n\}, \qquad S = \{v_1\}$$

$$V - s = \{v_2, v_3, \ldots, v_n\}$$

Vertices	v_1	v_2	v_3	$\ldots v_i$	$\ldots v_n$
$L(v)$	0	$w(1, 2)$	$w(1, 3)$	∞	∞

If vertices v_2 and v_3 are connected to v_1 by edge and $L(v) = \infty$ for all vertices not connected to v_1 by an edge.

Step 1: Select a vertex v in $V - S$ for which $L(v)$ is minimum: $v = v_k$ (say).

Step 2: Add v_k to S and update $V - S$.

$$S = \{v_1, v_k\} \text{ and } V - S = \{v_2, v_3, \ldots, v_{k-1}, v_{k+1}, \ldots, v_n\}$$

Step 3: Update the array L of distances of vertices x in $V - S$ using v_k as a vertex along the path from v_1 to x.

Step 4: Update $L(x)$ using

$$L(x) = \text{Min}\ \{L(x), L(v_k) + w(v_k, x)\}$$

Step 5: Repeat Steps 1–4 until $|V - S| = 1$

Step 7: STOP

Example 1 Using Dijkstra's algorithm, find the shortest path from v_1 to v_5 in the weighted graph (Fig. 10.61).

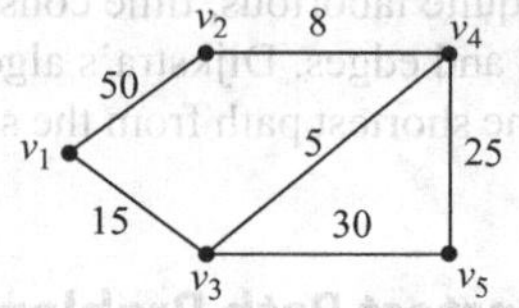

Figure 10.61

Solution: Let V be the set of all vertices. Let S be the set of all vertices whose shortest distances from the starting vertex are known.

Let $S = \{v_1\}$ where v_1 is the starting vertex. Let $L(u)$ denote the length of the shortest path from v_1 to u passing through vertices in S.

Define initially

$$L(u) = \begin{cases} 0 & \text{if } u = v_1 \\ w(v_1, u) & \text{if edge } (v_1, u) \text{ exists} \\ \infty & \text{if edge } (v_1, u) \text{ does not exist} \end{cases}$$

Thus, initially $V = \{v_1, v_2, v_3, v_4, v_5\}$, $S = \{v_1\}$
$V - S = \{v_2, v_3, v_4, v_5\}$

Vertex	v_1	v_2	v_3	v_4	v_5
$L(v)$	0	50	15	∞	∞

Step 1: Select a vertex v in $V - S$ with $L(v)$ at minimum.
Step 2: Add v to S and updated $V - S$ and S.
Step 3: Update distance L of vertices x in $V - S$ using v as a vertex along the path from v_1 to x.

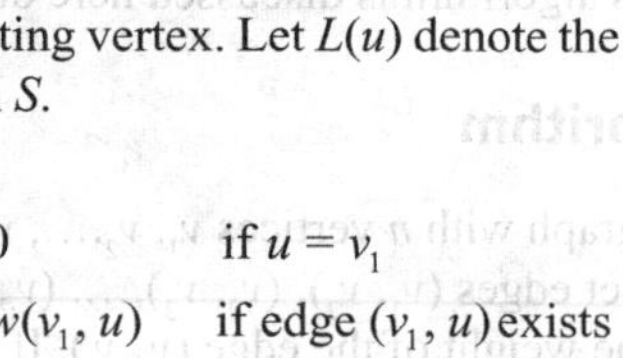

$$L(x) = L(v) + w(v, x)$$

Now $L(V_2) = \text{Min}\{L(v_2), L(v_1) + w(v_1, v_2)\}$
$= \text{Min}\{50, 0 + 50\} = 50$
$L(V_3) = \text{Min}\{L(v_3), L(v_1) + w(v_1, v_3)\} = \text{Min}\{15, 0 + 15\} = 15$
$L(V_4) = \text{Min}\{L(v_4), L(v_1) + w(v_1, v_4)\} = \text{Min}\{\infty, 0 + \infty\} = \infty$
$L(V_5) = \text{Min}\{L(v_5), L(v_1) + w(v_1, v_5)\} = \text{Min}\{\infty, 0 + \infty\} = \infty$

Hence, we have

Vertex	v_1	v_2	v_3	v_4	v_5
$L(v)$	0	50	15	∞	∞

Step 4: Select a vertex in $V - S$ with $L(v)$ minimum:

$$v = v_3$$

Step 5: Add v to s and update $V - S$:

$$S = \{v_1, v_3\},\ V - S = \{v_2, v_4, v_5\}$$

Step 6: Update length of paths of vertices x in $V - S$ using.

$$L(v_2) = \text{Min}\{L(v_2), L(v_3) + w(v_3, v_2)\} = \text{Min}\{50, 15 + \infty\} = 50$$

$$L(v_4) = \text{Min}\{L(v_4), L(v_3) + w(v_3, v_4)\} = \text{Min}\{\infty, 15 + 5\} = 20$$
$$L(v_5) = \text{Min}\{L(v_5), L(v_3) + w(v_3, v_5)\} = \text{Min}\{\infty, 15 + 30\} = 45$$

Consequently

Vertex	v_1	v_2	v_3	v_4	v_5
$L(v)$	0	50	15	20	45

Step 7: Select a vertex v in $V - S$ with $L(v)$ minimum:

$$v = v_4$$

Step 8: Add v_4 to S and update $V - S$:

$$S = \{v_1, v_3, v_4\},\ V - S = \{v_2, v_5\}$$

Step 9: Update the length of paths of vertices x in $V - S$.

$$L(v_2) = \text{Min}\{L(v_2), L(v_4) + w(v_2, v_4)\} = \text{Min}\{50, 20 + 8\} = 28$$

$$L(v_5) = \text{Min}\{L(v_5), L(v_4) + w(v_4, v_5)\} = \text{Min}\{45, 20 + 25\} = 45$$

Consequently

Vertices	v_1	v_2	v_3	v_4	v_5
$L(v)$	0	28	15	20	45

Step 10: Select a vertex v in $V - S$ with $L(v)$ minimum:

$$v = v_2$$

Step 11: Add v_2 to S and update $V - S$;

$$S = \{v_1, v_3, v_4, v_2\},\ V - S = v_5$$

Since $|V - S| = 1$, STOP.
Hence, the minimum distance of v_5 from v_1 is 45 and the path is v_1–v_3–v_4–v_5.

Example 2 Apply Dijkstra's algorithm to find the shortest path from v_1 to v_6 in the weighted graph (Fig. 10.62).

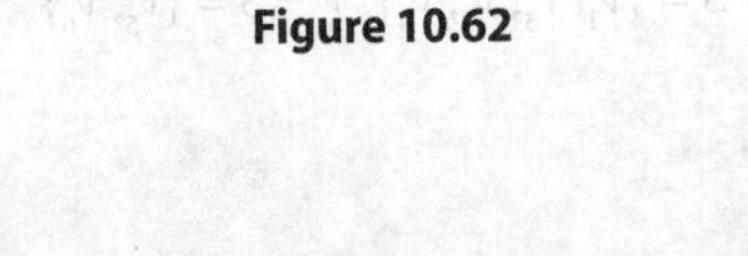

Figure 10.62

1. Initially $S = \{v_1\}$, $V - S = \{v_2, v_3, v_4, v_5, v_6\}$

Vertex	V_1	V_2	V_3	V_4	V_5	V_6
$L(v)$	0	10	2	14	∞	∞

2. Select a vertex in $V - S$ with $L(v)$ minimum: $v = v_3$
3. Add v_3 to S and upgrade $V - S$:

$$S = \{v_1, v_3\},\ V - S = \{v_2, v_4, v_5, v_6\}$$

4 Update the array L of distance of vertices x in $V - S$ using v_3 as a vertex along the path from v_1 to x.

$$L(v_2) = \text{Min}\{L(v_2), L(v_3) + w(v_3, v_2)\} = \text{Min}\{10, 2 + 3\} = 5$$

$$L(v_4) = \text{Min}\{L(v_4), L(v_3) + w(v_3, v_4)\} = \text{Min}\{14, 2 + 5\} = 7$$

$$L(v_5) = \text{Min}\{L(v_5), L(v_3) + w(v_3, v_5)\} = \text{Min}\{\infty, 2 + \infty\} = \infty$$

$$L(v_6) = \text{Min}\{L(v_6), L(v_3) + w(v_3, v_6)\} = \text{Min}\{\infty, 2 + \infty\} = \infty$$

Consequently

Vertex	V_1	V_2	V_3	V_4	V_5	V_6
$L(v)$	0	5	2	7	∞	∞

Now above steps are repeated until $|V - S| = 1$.

1. Select the vertex with minimum $L(v)$: $v = v_2$
2. Add v_2 to S and upgrade $V - S$: $S = \{v_1, v_3, v_2\}$, $V - S = \{v_4, v_5, v_6\}$
3. Update array L of distances from v_1 to each vertex x in $V - S$ using v_2 as a vertex along the path from v_2 to x

$$L(v_4) = \text{Min}\{L(v_4), L(v_2) + w(v_2, v_4)\} = \text{Min}\{7, 5 + \infty\} = 7$$

$$L(v_5) = \text{Min}\{L(v_5), L(v_2) + w(v_2, v_5)\} = \text{Min}\{\infty, 5 + 4\} = 9$$

$$L(v_6) = \text{Min}\{L(v_6), L(v_2) + w(v_2, v_6)\} = \text{Min}\{\infty, 5 + 15\} = 20$$

Consequently

Vertex	V_1	V_2	V_3	V_4	V_5	V_6
$L(v)$	0	5	2	7	9	20

1. Select a vertex in $V - S$ with minimum $L(v)$: $v = v_4$
2. Add V_4 to S and upgrade $V - S$:

$$S = \{v_1, v_3, v_2, v_4\},\ V - S = \{v_5, v_6\}$$

3. Update array $L(v)$ of distances of vertices x in $V - S$ using v_4 as a vertex along the path from v_1 to x

$$L(v_5) = \text{Min}\ \{L(v_5), L(v_4) + w(v_4, v_5)\} = \text{Min}\ \{9, 7 + 4\} = 9$$

$$L(v_6) = \text{Min}\ \{L(v_6), L(v_4) + w(v_4, v_6)\} = \text{Min}\ \{20, 7 + 5\} = 12$$

Consequently

Vertex	v_1	v_2	v_3	v_4	v_5	v_6
$L(v)$	0	5	2	7	9	12

1. Select a vertex in $V - S$ with minimum $L(v)$: $v = v_5$.
2. Add v_5 to S and update $V - S$:

$$S = \{v_1, v_3, v_2, v_4, v_5\},\ V - S = \{v_6\}$$

3. Update $L(v)$

$$L(v_6) = \text{Min}\ \{L(v_6), L(v_5) + w(v_5, v_6)\}$$

$$= \text{Min}\ \{12, 9 + 6\} = 12$$

Since $V - S = \{V_6\}$, that is, $|V - S| = 1$ the process is over. The length of shortest distance from v_1 to any vertex v is $L(v)$.

Hence, the shortest distance of v_6 from v_1 is $L(v_6) = 12$.

Note 10.17 The method provides the shortest distances from the starting vertex v_1 to every other vertex in the graph; it did not find the shortest paths.

10.17 TRAVELLING SALESMAN PROBLEM

An interesting real-life problem named as travelling salesman problem arises when a salesman assigned to a city A would like to visit every other city exactly once and return to his home city so that the total distance travelled is minimum. Such a problem is faced by a service personnel who is serving a set of vending machines or computer systems on a regular basis. In this problem, the cities can be considered as vertices of a graph G, the edges, the paths of transportation, and the weight of an edge as the cost of travelling along that edge. The goal is to visit all the cities only once, which starts and ends at the same city. This essentially means to find a Hamiltonian circuit in the connected graph G so that the sum of the weights along the cycle is minimum. The problem of finding a minimum Hamiltonian cycle in a weighted (Hamiltonian) graph is called the Travelling Salesman Problem. Its solution lies in calculating all Hamiltonian cycles and then selecting the most economical which is quite laborious. It is an open question as to whether there exits an efficient (polynomial time) algorithm for its solution. However, a procedure known as Nearest Neighbourhood Method, which gives a reasonably good result, is discussed here but it does not ensure to be minimum is illustrated by the following example.

Method of Nearest-Neighbourhood

Step 1: Start with the specified vertex (home vertex) of the graph G.

Step 2: Find the vertex closest to it with minimum weight to form the initial path. This path will be augmented from vertex by vertex process as given in Step 3.

Step 3: Let v denote the latest vertex. Find the vertex closest to v with minimum weight and add to the path the edge connecting v.

Step 4: Repeat Step 3 until all vertices in G are included in the path.

Step 5: Form a circuit by adding the edge joining the starting (home) vertex and the last vertex.

The sum of all the weights of edges forming the circuit is the required solution.

Consider the weighted Hamiltonian graph G with starting (home) vertex A. (Fig. 10.63)

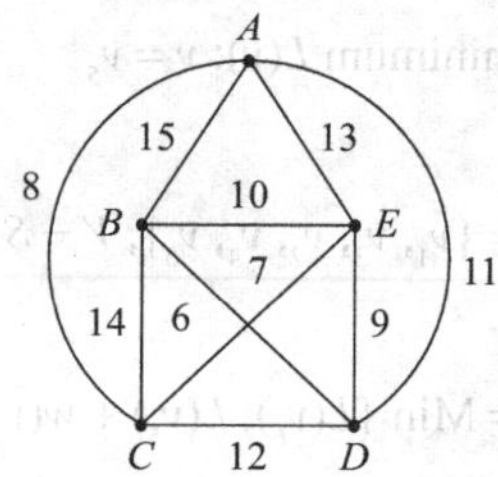

(garaph G)

Figure 10.63

If we start from vertex A, a vertex by vertex construction of Hamiltonian circuit by using nearest neighbour, we obtain the following: $A\ C\ E\ D\ B\ A$ circuit (Fig. 10.64).

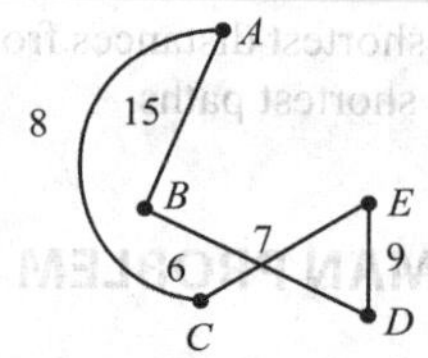

Figure 10.64

Total distance = 8 + 7 + 9 + 6 + 15
= 45

The Hamiltonian circuit giving minimum distance is ($A\ C\ E\ B\ D\ A$ circuit) (Fig. 10.65)

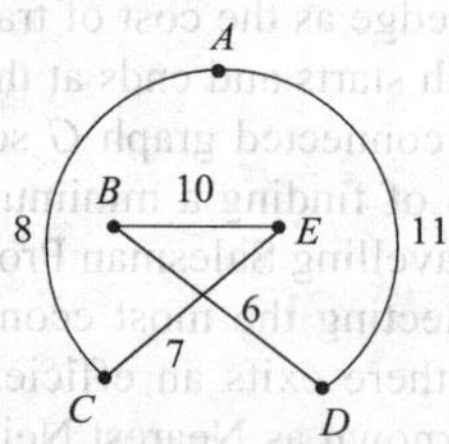

Figure 10.65

The distance is = 8 + 7 + 10 + 6 + 11
= 42

10.18 NETWORK FLOWS

Network systems arise in transportation, transmission, communication and flow of commodities such as oil, gas, water, etc. The other commodity may be automobiles, manufactured goods, kilowatts of electricity, and something intangible such as information and messages. The network of roads, railways, telephone lines, interconnection of factories, warehouses, and retail outlets involve flow of commodities. All these physical systems can be visualized as a connected weighted digraph. We can think of the edges of digraph representing roads, railways tracks, power lines and pipe-lines, that is, channels through which commodity flows. The vertices of digraphs can represent railway stations, cities (connected by roads), oil reservoirs, power relay stations, and telephone outlets. The weights on edges can be viewed as time or cost required to complete certain activities. Here, we think of edges (arcs) of digraph as pipes and the weight on an arc as its capacity, that is, the maximum amount of commodity that can flow through it per unit of time.

The general problem in all the situations is to find the maximum flow of commodity per unit time.

Definition 10.3 A network $N(V, E)$ is a simple connected weighted graph, which satisfies the following conditions:

1. there exists a distinguished (unique) vertex $S \in V$ called *source* whose in degree is zero.
2. there exists a distinguished (unique) vertex $t \in V$ called *sink* whose out degree is zero.
3. every directed edge $e = (u, v) \in E$ is assigned a non-negative real number called *Capacity* of e denoted by $c(e)$. It represents the maximum rate at which commodity can be transported along the edge e.

The vertices distinct from the source S and the sink t are called *intermediate* vertices.

Consider the network Fig. 10.66, which has four vertices a, b, c, d other than the source S and sink t; the capacity of each is shown in it.

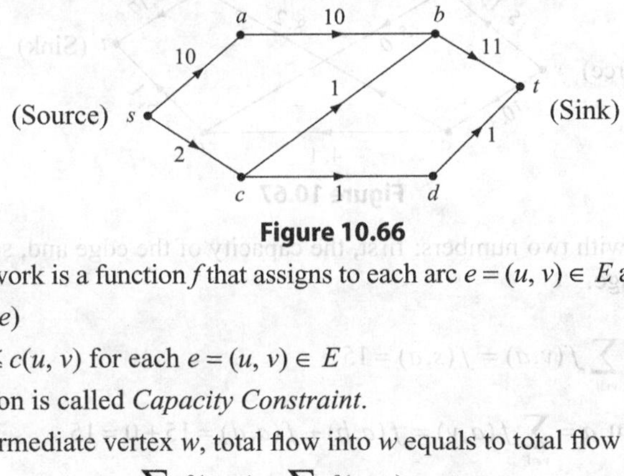

Figure 10.66

A *flow* in a network is a function f that assigns to each arc $e = (u, v) \in E$ a real number such that

1. $0 \le f(e) \le c(e)$

   i.e., $f(u, v) \le c(u, v)$ for each $e = (u, v) \in E$

   This condition is called *Capacity Constraint.*

2. For any intermediate vertex w, total flow into w equals to total flow out of w, that is,

$$\sum_{v \in V} f(v, w) = \sum_{v \in V} f(w, v) \quad \text{for all } w \in V$$

   This condition ensures that there is no accumulation or generation of commodity at any intermediate vertex along the way and is known as *Flow Conservation.*

3. $f(e) = 0$ for every edge e incident to source S or incident from the sink t. This ensures that the flow moves from source to sink.

Saturated and Unsaturated Flow:

A flow along an edge $e = (u, v)$ is said to be

Saturated if $f(e) = c(e)$

i.e., $f(u, v) = c(u, v)$

If $f(e) < c(e)$, i.e., $f(u, v) < c(u, v)$, it is called *unsaturated.*

Slack (or Residual Capacity): If an edge e is unsaturated then

$$S(e) = c(e) - f(e)$$

is called the Slack or Residual capacity of edge e.

The Value of Flow: In a directed network with source s and sink t, the net amount of flow per unit time leaving the source or equivalently the net amount of flow per unit time entering the sink is called the value of flow. Mathematically,

Value (f) = $\sum_{v \in V} f(s,v)$, where s is the source and the vertices v are incident from the source s.

= $\sum_{v \in V} f(v,t)$, where t is the sink and the vertices v are incident to the sink t.

Thus, the total outgoing flow from the source is equal to the total incoming flow at the sink.

Maximal Flow: A maximal flow in a network is a flow that achieves the largest possible value. Thus, a flow f in a network N is maximal if value (f) > value (f') for every flow (f') in N.

The general problem is to find the maximum value of a flow through a directed network without exceeding the capacity of any edge.

Example 1 Consider the network flow shown in Fig. 10.67 with source s and sink t.

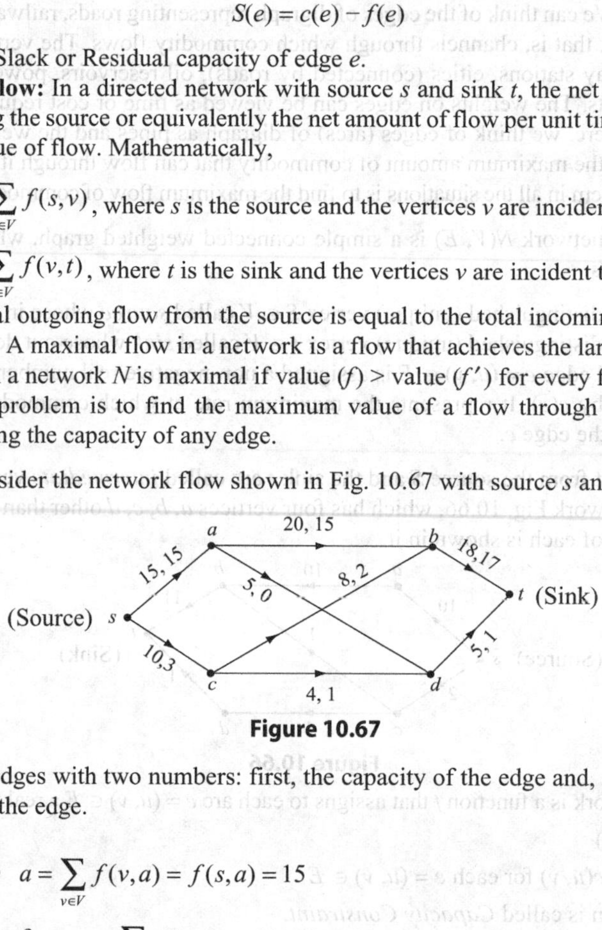

Figure 10.67

Let us label edges with two numbers: first, the capacity of the edge and, second, the amount of flow through the edge.

The flow into $a = \sum_{v \in V} f(v,a) = f(s,a) = 15$

The flow away from $a = \sum_{v \in V} f(a,v) = f(a,b) + f(a,d) = 15 + 0 = 15$

The value of flow value (f) = $\sum_{v \in V} f(s,v) = f(s,a) + f(s,c) = 15 + 3 = 18$

Or the value of flow value (f) = $\sum_{v \in V} f(v,t) = f(b,t) + f(d,t) = 17 + 1 = 18$

The edge (s, a) is saturated as the capacity and flow are equal.

The edges (a, b), (b, t), (s, c), (c, d), (c, b), (a, d), and (d, t) are unsaturated. The slack for edge (a, b) is $c(a, b) - f(a, b) = 20 - 15 = 5$.

Similarly, the slack for other unsaturated edges can be calculated. At every intermediate vertex, flow conservation law is satisfied.

10.18.1 Cut-Set and Capacity

In the transport network $N(V, E)$, let us consider a cut-set with respect to vertices s and t ignoring the directions of edges. A cut-set with respect to s and t is a cut that separates source s and sink t. Let (S, T) denote a cut that partitions the set V of vertices into two subsets S and T where S contains the source s and T contains the sink t. The capacity of cut $C(S, T)$ is the sum of capacities of all those edges that are directed from S to T, that is,

$$C(S, T) = \sum_{\substack{u \in S \\ v \in T}} c(u, v)$$

With reference to the network in Fig. 10.67, the sets S and T from $(s - t)$ cut.

$S = \{s\}$, $T = \{a, b, c, d, t\}$, cut-set $(S, T) = \{sa, sc\}$

Capacity = 15 + 10 = 25

$S = \{s, a\}$, $T = \{b, c, d, t\}$, $(S, T) = \{sc, ab, ad\}$

Capacity = 10 + 20 + 5 = 35

$S = \{s, b\}$, $T = \{a, c, d, t\}$, $(S, T) = \{sa, sc, bt\}$

Capacity = 15 + 10 + 18 = 43

$S = \{s, c\}$, $T = \{a, b, d, t\}$, $(S, T) = \{sa, cb, cd\}$

Capacity = 15 + 4 + 8 = 27

$S = \{s, a, b\}$, $T = \{c, d, t\}$, $(S, T) = \{sc, ad, bt\}$

Capacity = 10 + 5 + 18 = 33

And so on.

There are four vertices other than source and sink in the network and so there are $2^4 = 16$ cut-sets. The calculation for all the cut-sets will show that value of flow (18) (Fig. 10.67) does not exceed the capacity of all the possible cut-sets. This does not happen by chance but leads to the following theorem.

THEOREM 10.15 In a directed network $N(V, G)$, the value of an (s, t) flow does not exceed the capacity of any (s, t) cut, that is, cut separating s from t.

Proof: Let (S, T) be an arbitrary cut such that s is in vertex set S and t is in vertex set T.

$$\text{Val}(f) = \sum_{v\in V} f(s, v) - \sum_{v\in V} f(v, s)$$

$$= \sum_{u\in S}\left(\sum_{v\in V} f(u, v) - \sum_{v\in V} f(v, u)\right)$$

$$= \sum_{u\in S, v\in V} f(u, v) - \sum_{\substack{u\in S\\ v\in V}} f(v, u)$$

Since (S, T) is a partition, we have

$$\text{Val}(f) = \sum_{\substack{u\in S\\ u\in S}} f(u, v) + \sum_{\substack{u\in S\\ v\in T}} f(u, v) - \sum_{\substack{u\in S\\ v\in S}} f(v, u) - \sum_{\substack{u\in S\\ v\in T}} f(v, u)$$

$$= \left(\sum_{\substack{u\in S\\ v\in S}} f(u, v) - \sum_{\substack{u\in S\\ v\in S}} f(v, u)\right) + \left(\sum_{\substack{u\in S\\ v\in T}} (f(u, v) - f(v, u))\right)$$

$$\text{Val}(f) = \sum_{\substack{u\in S\\ v\in T}} [f(u, v) - f(v, u)] \quad \text{(the first two terms being same cancel)}$$

Since $f(u, v) \le c(u, v)$ and $f(v, u) \ge 0$, $f(u, v) - f(v, u) \le c(u, v)$ for all u and v.

Thus,
$$\text{Val}(f) \le \sum_{\substack{u\in s\\ v\in T}} c(u, v) = \text{Cap}(S, T)$$

which was desired.

THEOREM 10.16 (Max-Flow Min Cut Theorem)

In any transport network $N(V, E)$, the maximum value of a flow from s to t is equal to the minimum value of the capacities of all the cuts in G that separates s from t.

Proof: Let s and t be the source and sink of a transport network $N(V, E)$ with vertices $v_1, v_2, \ldots, v_k$.

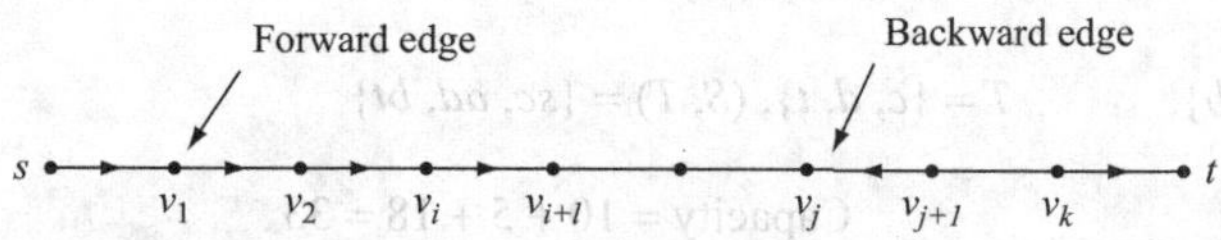

Figure 10.68

Let there be some flow pattern in G such that the flow from s to t assumes its maximum possible value w_0. Define a vertex set S in G recursively as follows:

(1) $s \in S$

(2) If vertex $v_i \in S$ and either $f(v_i, v_j) < C(v_i, v_j)$ or $f(v_j, v_i) > 0$, then $v_j \in S$.

Any vertex not in S belongs to T.

Since the vertex t cannot be in S, there will exist a path P from s to t, say s, $v_1, v_2, \ldots, v_i, v_{i+1}, \ldots, v_j, v_{j+1}, \ldots, v_k$, t for which in every edge either flow is $f(v_j, v_{j+1}) < c(v_j, v_{j+1})$ or $f(v_{j+1}, v_j) > 0$. In the path P, an edge (v_j, v_{j+1}) directed from v_j to v_{j+1} is called forward edge and an edge (v_{j+1}, v_j) directed from v_{j+1} to v_j is called a backward edge (Fig. 10.68).

Let δ_1 be the minimum of all differences $c(v_j, v_{j+1}) - f(v_j, v_{j+1})$ in forwarded edges and δ_2 be the minimum of all flows in backward edges in the path P. Let $\delta = \min(\delta_1, \delta_2)$. Then the flow in the network can be increased by increasing the flow in each forward edge and decreasing the flow in each backward edge by an amount δ satisfying conservation law at each intermediate vertex and also the capacity constraint for all edges. This contradicts the assumption that w_0 was the maximum flow.

Thus, the cut (S, T) separates s from t and according to condition (2), for each vertex $v_i \in S$ and $v_j \in T$, we have

$$f(v_i, v_j) = c(v_i, v_j)$$

and
$$f(v_j, v_i) = 0$$

Therefore, the value of the flow

$$w_o = \sum_{\substack{v_i \in S \\ v_j \in T}} f(v_i, v_j) - \sum_{\substack{v_i \in S \\ v_j \in T}} f(v_j, v_i)$$

$$= \sum_{\substack{v_i \in S \\ v_j \in T}} c(v_i, v_j) = C(S,T)$$

which proves the theorem.

Flow Augmenting Path: A flow f along a path (Fig. 10.68) from s to t is called flow augmenting path if

1. every forward edge of the path has excess capacity (i.e., some unsaturated edge).
2. every backward edge on the path has non-zero flow. Mathematically, if

$$f(e) < c(e) \quad \text{for each forward edge } e \in E$$

and
$$f(e) > 0 \quad \text{for each backward edge } e \in E$$

In such a path, we can increase flow in each forward arc and decrease the flow by the same amount in backward arcs and thus conservation law will be satisfied at each vertex.

Example 1 Find the maximum flow for the following network:

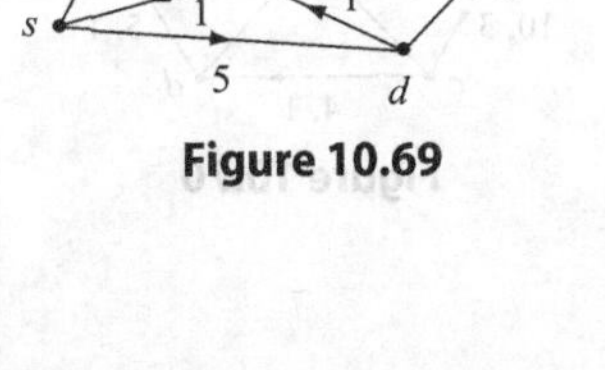

Figure 10.69

A transport network is shown in Fig. 10.69 where the numbers written along the edge are edge capacities. Since the network has 3 vertices, it has eight ($2^3 = 8$) cuts, which separate s from t.

	Vertex set S	Cut set	C(S,T)
1	{s}	{sb, sc, sd} 5 + 0 + 5	10
2	{s, b}	{sd, bt} 5 + 3	8
3	{s, c}	{sb, sd, cb, ct} 5 + 5 + 6 + 4	20
4	{s, d}	{sb, dc, dt} 5 + 1 + 2	8
5	{s, b, c}	{sd, bt, ct} 5 + 3 + 4	12
6	{s, b, d}	{bt, dc, dt} 3 + 1 + 2	6
7	{s, c, d}	{sb, cb, ct, dt} 5 + 6 + 4 + 2	17
8	{s, b, c, d}	{bt, ct, dt} 3 + 4 + 2	9

The cut with minimum capacity is the one in which $s = \{s, b, d\}$ and T is $\{c, t\}$. The maximum possible flow in the network is 6.

In Summary: it has been established that the value of a flow f can be computed in the following different ways:

(1) Value (f) = The total flow out of the source s.
(2) Value (f) = The total flow into the sink t.
(3) Value (f) = The flow across any S–T cut.

The methods (1) and (2) are in fact special cases of (3) as in (1) flow out of source s equals (S, T) cut where $S = \{s\}$. In case (2) the total flow into the sink t corresponds to the cut where $T = \{t\}$.

Maximal Flows: The solution of many practical problems requires finding maximal flow for a network. Hence, our goal here is to present a procedure for maximum flow in a directed network.

Consider the network given in Fig. 10.70.

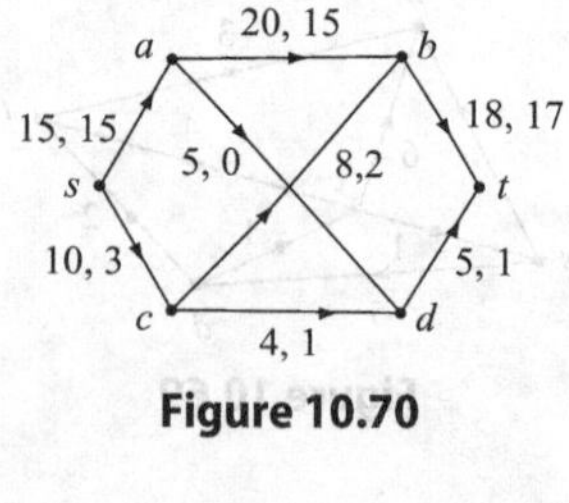

Figure 10.70

The flow out of the source s equals $15 + 3 = 18$ and the flow into the sink t equals $17 + 1 = 18$, and so, physically this equality makes sense in view of conservation law, which requires that no flow is lost at any vertex. Hence, whatever flow leaves the source must reach the sink. The value of flow $(f) = 18$.

Now we want to find (s, t) flow with the largest possible value, that is, maximal flow for which flow augmenting is considered for unsaturated edges.

Consider the path P: *scbadt* from s to t in Fig. 10.70 as the path *sabt* cannot be flow augmented due to the edge (s, a) being a saturated edge.

The path P is flow augmenting because, checking forward arcs,

$3 = f(s, c) < c(s, c) = 10,\quad$ slack $= 10 - 3 = 7$

$2 = f(c, b) < c(c, b) = 8,\quad$ slack $= 8 - 2 = 6$

$0 = f(a, d) < c(a, d) = 5\quad$ slack $= 5 - 0 = 5$

$1 = f(d, t) < c(d, t) = 5\quad$ slack $= 5 - 1 = 4$

and for backward arc $15 = f(a, b) > 0$

Minimum of slacks $(7, 6, 5, 4) = 4$

Increasing by 4, the flow in forward arcs of path p, and decreasing by 4, the flow in the backward arcs, we obtain the network and flow shown in Fig. 10.71.

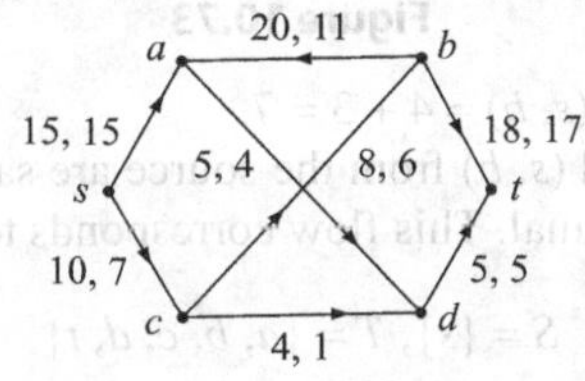

Figure 10.71

The value of maximum flow

$$= f(s, a) + f(s, c)$$
$$= 15 + 7 = 22$$

The above procedure shows that for obtaining maximal flow, the unsaturated edges are saturated so that maximum capacity of the channels is used. Thus, a flow is maximal if there is no augmenting path from source to sink.

Example 1 Consider the network shown in Fig. 10.72.

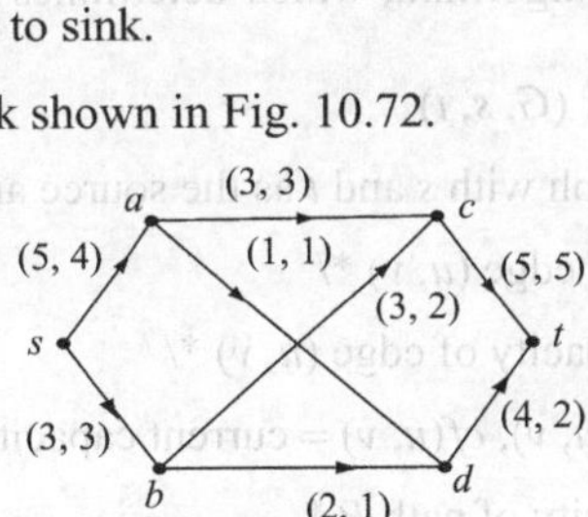

Figure 10.72

The value of flow = $f(s, a) + f(s, b) = 4 + 3 = 7$

This is the maximal flow as the edges (a, c), (c, t), and (s, b) are saturated. For the edge (s, a) slack is 1. If we consider the path sact, flow augmentation at vertex a is not possible since (a, c) is saturated. This flow corresponds to the (S, T) cut where

$$S = \{s, a\},\ T = \{b, c, d, t\}$$

$$\text{The capacity of the cut is} = \sum_{u\in s, v\in T} c(u, v) = c(s,b) + c(a,c) + c(a,d)$$

$$= 3 + 3 + 1 = 7$$

Example 2 Consider the network shown in Fig. 10.73.

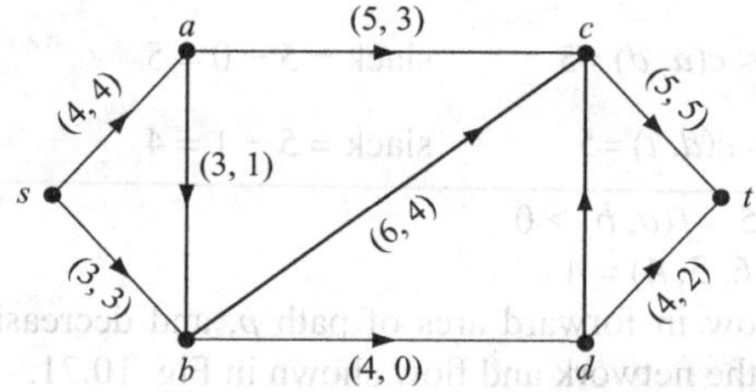

Figure 10.73

The value of flow is $f(s, a) + f(s, b) = 4 + 3 = 7$.

Since both the edges (s, a) and (s, b) from the source are saturated, no flow augmentation is possible. Hence, this flow is maximal. This flow corresponds to (S, T) cut where

$$S = \{s\},\ T = \{a, b, c, d, t\}$$

$$\text{The capacity of the cut} = \sum_{u\in s, v\in T} c(u, v) = c(s,a) + c(s,b)$$

$$= 4 + 3 = 7$$

10.18.2 Ford–Fulkerson Algorithm

The high-level procedure of the algorithm, which determines maximum flow of a network, is given below.

Ford–Fulkerson Algorithm for (G, s, t)

/* G is a connected weighted graph with s and t as the source and sink vertices. */

/* $f(u, v)$ represents flow through edge (u, v) */

/* $cf(u, v)$ represents residual capacity of edge (u, v) */

/* Residual capacity of an edge (u, v), $cf(u, v)$ = current capacity − current flow */

/* $cf(P)$ represents residual capacity of path P */

/* Initialize all edges $(u, v) \in G$*/

begin

for each edge $(u, v) \in G$ do

$f(u, v) \leftarrow 0$

while (there is some path P from s to t in the residual network Gf)

begin

$cf(P){:}\leftarrow \min\{cf(u, v) | (u, v) \text{ is in } P\}$

for each edge $(u, v) \in P$ do

begin

$f(u, v) \leftarrow f(u, v) + cf(P)$

$cf(u, v) \leftarrow cf(u, v) - cf(P)$

end

end

end

10.18.3 Ford–Fulkerson Algorithm for Maximum Flow

In this algorithm, which is iterative in nature, one starts with $f(u, v) = 0$ for each edge and thus initial flow value is zero. At each iteration, an augmenting path is considered and the flow is increased by the residual capacity of the path, that is, $\min[c(u, v) - f(u, v)]$ for each edge (u, v) of the path. The process is repeated until no augmenting path can be found.

Example 1 Find the maximum flow for the following (s, t) networks given in Fig. 10.74.

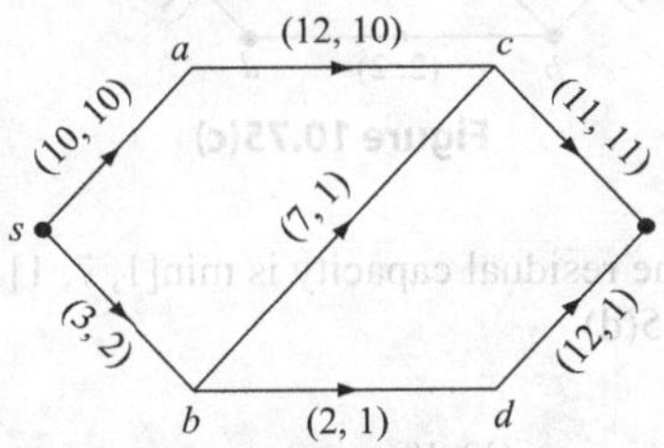

Figure 10.74

Solution: We start with zero flow in every edge, as shown in Fig. 10.75(a).

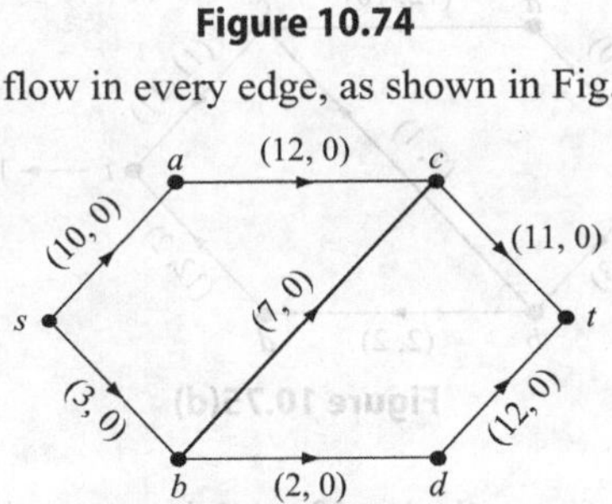

Figure 10.75a

There is capacity available along the paths (s, b, d, t), (s, b, c, t), and (s, a, c, t).

Consider the path P_1: (s, b, d, t).

The residual capacity of the path P_1 is min[3, 2, 12] = 2 and so we increase the flow along P_1 by 2, which leads to the network (Fig. 10.75b).

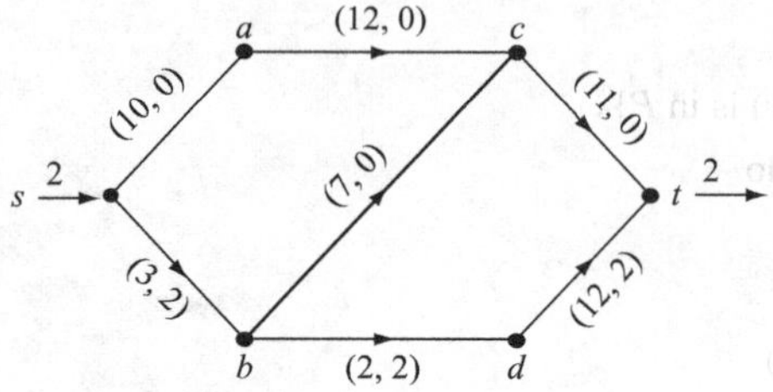

Figure 10.75(b)

For the path P_2: (s, a, c, t), the residual capacity is min[10, 12, 11] = 10 and so the flow along P_2 is increased by 10 (Fig. 10.75c).

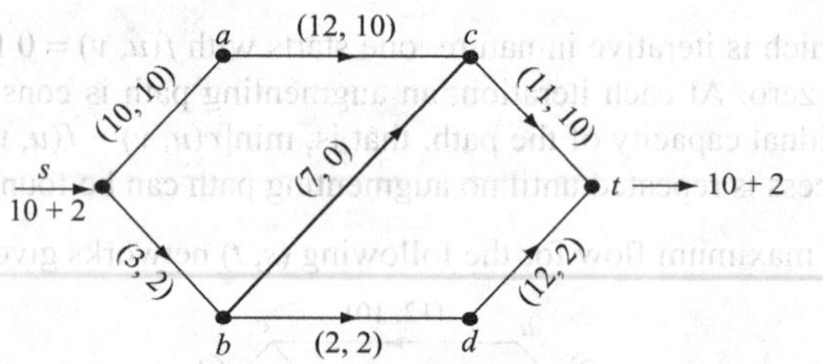

Figure 10.75(c)

For the path p_3: (s, b, c, t), the residual capacity is min[1, 7, 1] = 1 and so the flow along this path is increased by 1 Fig. 10.75(d).

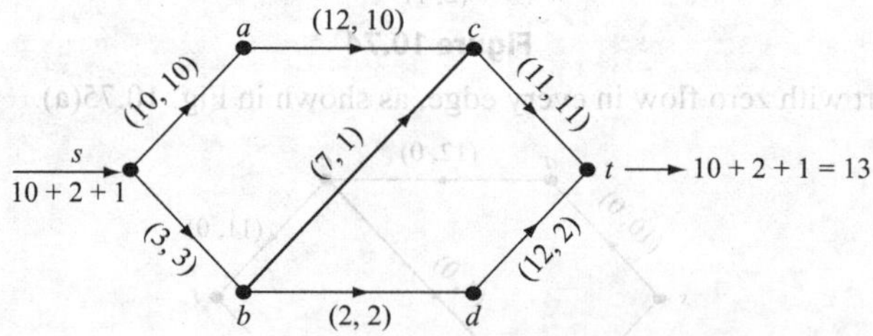

Figure 10.75(d)

Hence, the value of the maximum flow is 13 as no increment is possible due to saturation of edges (s, a), (s, b), (b, d), and (c, t).

10.19 MATCHINGS

Matchings in graphs play an important role in providing solution to problems of assigning jobs to job applicants, marriage problems and various other situations.

Suppose that three applicants A_1, A_2, and A_3 apply for five vacant positions P_1, P_2, P_3, P_4, and P_5. Applicant A_1 is suitable for position P_1 or P_2 or P_5. Applicant A_2 is suitable for P_2, P_3, P_5 and A_3 for P_4 or P_5. This situation can be represented by a graph shown in Fig. 10.76. The applicants and the vacant positions can be regarded as vertices. The edges represent the qualifications for different positions.

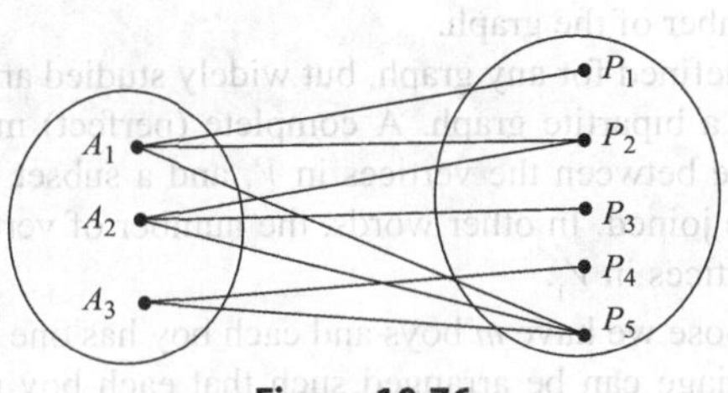

Figure 10.76

The graph is obviously bipartite, the vertices being partitioned into two sets $V_1 = \{A_1, A_2, A_3\}$ and $V_2 = \{P_1, P_2, P_3, P_4, P_5\}$. Now the question arises: Is it possible to assign a position to every applicant for which he is qualified. If the answer is no, what is the maximum number of positions that can be filled. This is a problem of matching from one set of vertices V_1 to another set V_2.

Definition 10.4 A *matching* in a graph is a set of edges such that no vertex is incident with more than one edge in the set, that is, matching is a subset of edges in which no two edges are adjacent. A vertex which is incident with an edge is said to be saturated. A matching is perfect (complete) if and only if each edge is saturated.

Example 1 Consider the graph in Fig. 10.77.

(1) The single edge *bc* is a matching as it saturates *b* and *c*.
(2) The set $\{bc, ab\}$ is not a matching because vertex *b* is incident with two edges.
(3) The set $\{ac, bd\}$ is a perfect matching.

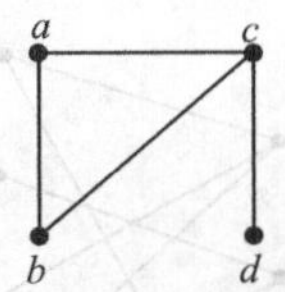

Figure 10.77

It can be observed that if a matching is perfect, the vertices in a graph can be partitioned into two sets of equal size and the edges of the matching provide a one-to-one correspondence between these sets.

Maximal Matching: A maximal matching is a matching to which no edge can be added. For example, in a complete graph of three vertices (i.e., a triangle), any single edge is a maximal matching. Similarly, in the graph of Fig. 10.78.

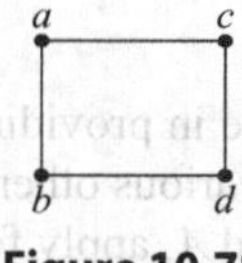

Figure 10.78

$\{ac, bd\}$ represents maximal matching. Also, $\{ab, cd\}$ represents maximal matching. Thus, a graph may have different maximal matchings. The maximal matchings with largest number of edges are called largest maximal matchings. The number of edges in a largest maximal matching is called the matching number of the graph.

Although matching is defined for any graph, but widely studied are bipartite graphs.

Let $G = G(V_1, V_2)$ be a bipartite graph. A complete (perfect) matching in $G(V_1, V_2)$ is a one-to-one correspondence between the vertices in V_1 and a subset of vertices in V_2 such that corresponding vertices are joined. In other words, the number of vertices in V_2 must be at least equal to the number of vertices in V_1.

Marriage Problem: Suppose we have m boys and each boy has one or more girlfriends. Under what conditions, the marriage can be arranged such that each boy marries his girlfriend. We construct a bipartite graph $G = (V_1, V_2)$ where $V_1 = \{b_1, b_2, \ldots, b_m\}$ represents the set of m boys and $V_2 = \{g_1, g_2, \ldots, g_n\}$ represents their n $(n \geq m)$ girlfriends). An edge joins the vertex b_i to vertex g_j if and only if g_j is a girlfriend of b_i.

For example, if there are five boys $\{b_1, b_2, b_3, b_4, b_5\}$ and six girls $\{g_1, g_2, g_3, g_4, g_5, g_6\}$ and the friendships are given by

b_1	g_1, g_3, g_4
b_2	g_2, g_5, g_6
b_3	g_3, g_4, g_5
b_4	g_1, g_6
b_5	g_6

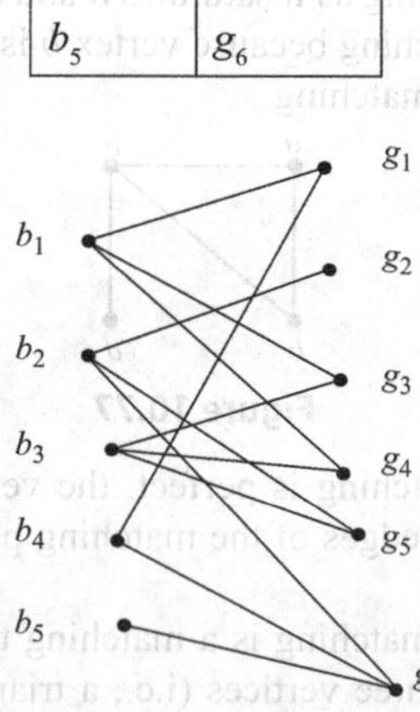

A possible solution is for b_1 to marry g_3, b_2 to marry g_5, b_3 to marry g_4, b_4 to marry g_1, and b_5 to marry g_6.

In terms of matrimonial terminology, we can observe that, for the solution of the marriage problem, every set of k boys must have at least k girlfriends for all integers k where $1 \le k \le m$.

This condition is known as marriage condition. It is a necessary condition, which if not satisfied boys would not be able to marry the girls in that set. This marriage condition also turns out to be sufficient.

THEOREM 10.17 (Hall's Marriage Theorem)
A necessary and sufficient condition for the solution of marriage problem is that each set of k girls collectively knows at least k boys for $1 \le k \le m$.

This result can be presented in a general form as follows:

A complete (perfect) matching of V_1 into V_2 in a bipartite graph $G = G(V_1, V_2)$ exists if and only if every subset of k vertices in V_1 is collectively adjacent to k or more vertices in V_2 for all values of k.

10.19.1 Assignment Problem

Any assignment of workers to jobs is a matching problem since no worker can be given two jobs and each job is filled by at most one worker. Hence, the graph is bipartite in which vertices correspond to workers and jobs and where edges join workers to jobs for which they have required skill (or qualifications).

Suppose that $G = (V_1, V_2)$ is a bipartite graph in which V_1 represent 'workers' and V_2 the 'jobs'.

In case there are more workers than jobs, that is, $|V_1| > |V_2|$ then it is not possible to provide job to every worker. Thus, the condition $|V_1| \le |V_2|$ is the necessary condition for the assignment of job to every worker, that is, the set of vertices V_1 will be saturated. In other words, if X is any subset of V_1 and $A(X)$ is the set of vertices in V_2 that correspond to a vertex in X,

$$A(X) = \{v \in V_2 | (v, x) \text{ is an edge for some } x \in V_2\}$$

Then, we require $|X| \le |A(X)$.

The above necessary condition

$$|X| \le |A(X)| \text{ for all } X \sqsubseteq V_1$$

is also the sufficient condition for the existence of matching, which saturates V_1.

This was discovered in 1935 by English mathematician Philip Hall and is also known as Hall's marriage theorem given in Theorem 10.17.

Example 1 Consider the problem of job assignment in which three workers W_1, W_2, and W_3 apply for jobs J_1, J_2, J_3, and J_4.

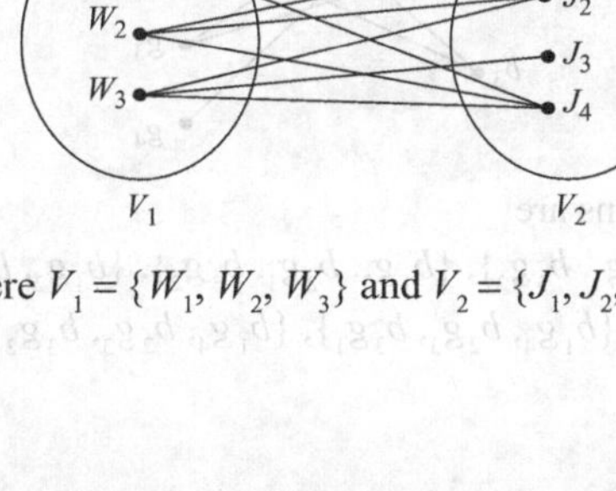

The graph is bipartite where $V_1 = \{W_1, W_2, W_3\}$ and $V_2 = \{J_1, J_2, J_3, J_4\}$ and the information can be arranged in tabular form.

Workers	Jobs
W_1	J_1, J_4
W_2	J_1, J_2, J_4
W_3	J_2, J_3, J_4

A particular solution is the match $\{W_1J_1, W_2J_2, W_3J_4\}$.

Thus, each worker has been assigned a job for which he or she is qualified. Applying assignment condition $|X| \le A(X), X \subseteq V_1$, we have

X:	W_1	W_2	W_3	W_1W_2	W_1W_3	W_2W_3	$W_1W_2W_3$
|X|:	1	1	1	2	2	2	3
|A(X)|:	2	3	3	3	4	4	4
	(J_1, J_4)	(J_1, J_2, J_4)	(J_2, J_3, J_4)	(J_1, J_2, J_4)	(J_1, J_2, J_3, J_4)	(J_1, J_2, J_3, J_4)	(J_1, J_2, J_3, J_4)

Thus, $|X| \le |A(X)|$ for each subset X of $V_1 = \{W_1, W_2, W_3\}$ and hence matching is perfect, that is, each worker will get a job.

Example 2 Suppose three boys b_1, b_2, and b_3 know four girls g_1, g_2, g_3, and g_4 as given below:

Boy	Girls known to boys
b_1	g_4, g_2, g_3
b_2	g_1, g_3
b_3	g_1, g_2

(1) Draw the bipartite graph corresponding to the table of relationship.
(2) Find five different solutions of the above marriage problem.
(3) Check the marriage condition for this problem.

Solution:

The bipartite graph corresponding to the given relationship is the following:

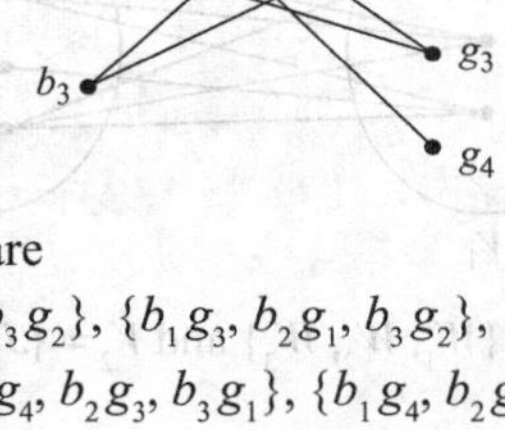

(1) The five different solutions are

$$\{b_1g_4, b_2g_1, b_3g_2\}, \{b_1g_3, b_2g_1, b_3g_2\}, \{b_1g_2, b_2g_3, b_3g_1\}$$
$$\{b_1g_4, b_2g_3, b_3g_1\}, \{b_1g_4, b_2g_3, b_3g_2\}$$

(2) The condition $|X| \leq |A(X)|$, for every $X \subseteq V_1$

X:	b_1	b_2	b_3	b_1b_2	b_1b_3	b_2b_3	$b_1b_2b_3$
|X|:	1	1	1	2	2	2	3
|A(X)|:	3	2	2	4	4	3	4

(g_2, g_3, g_4) (g_1, g_3) (g_1, g_2) (g_1, g_2, g_3, g_4) (g_1, g_2, g_3, g_4) (g_1, g_2, g_3) (g_1, g_2, g_3, g_4)

Thus, $|X| \leq |A(X)|$ for each subset X of $\{b_1, b_2, b_3\}$, which shows that matching is perfect, that is, every boy can marry a girl known to him.

Example 3 Show that a complete graph K_n has a perfect matching if n is even.

Solution: Let n be even, say $2m$, then

$V(K_n) = \{u_1, u_2, u_3, \ldots, u_{2m}\}$. Since every pair of vertices $(u_1, u_2), (u_3, u_4), \ldots, (u_{2m-1}, u_{2m})$ is joined by an edge, it will give perfect matching.

Example 4 Show that the graph $K_{m,n}$ has perfect matching if and only if $m = n$.

Solution: Let $m = n$ and the vertex sets be $V_1 = \{u_1, u_2, \ldots, u_m\}$ and $V_2 = \{v_1, v_2, \ldots, v_m\}$.

Then $\{(u_1, v_1), (u_2, v_2), \ldots, (u_m, v_m)\}$ is a perfect matching, conversely, let $K_{m,n}$ be perfect matching and $m < n$. Since each edge in V_1 in perfect matching must join to vertex of V_2 and so there will not be m edges. If $m < n$, some vertex in v_2 will not be a part of any edge in the matching, which contradicts the assumption of perfect matching. Hence, $m = n$.

EXERCISE – A

1. In the graphs $G = (V, E)$, give V, the set of vertices, and E, the set of edges.

(a)

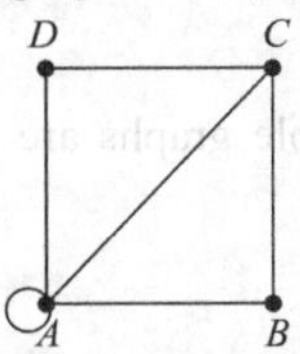

(b)

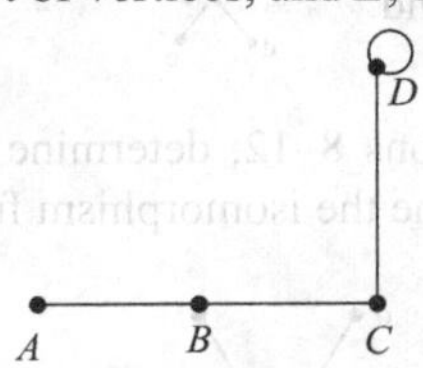

(c)

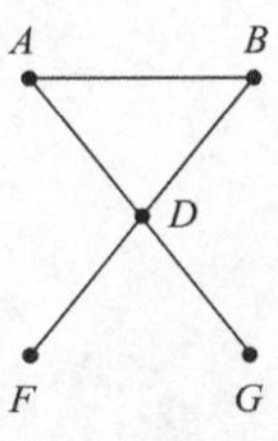

(d)

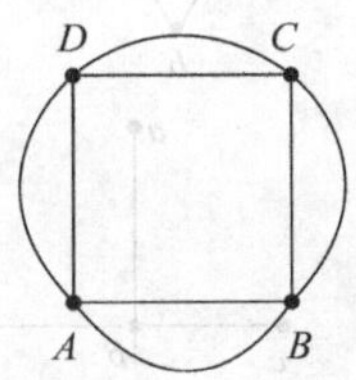

(e)

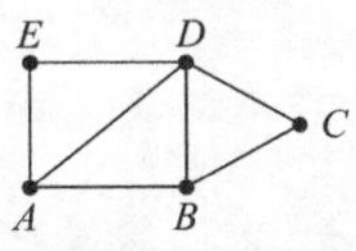

(f)

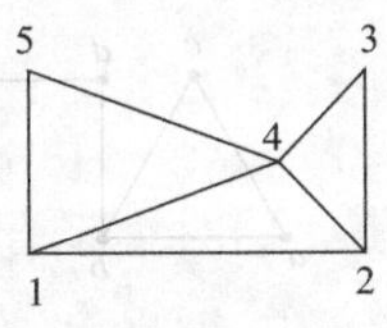

2. Give the degree of each vertex of graphs in question 1(a), 1(e), and 1(f).
3. Draw null graph (N_3) with three vertices.
4. Draw complete graphs K_3, K_4, and K_5.
5. (i) Draw regular graphs of degree 2 and 3.
   (ii) Is complete graph K_n a regular graph of degree $(n - 1)$.
6. (i) Find the isolated and pendant vertices of the graph

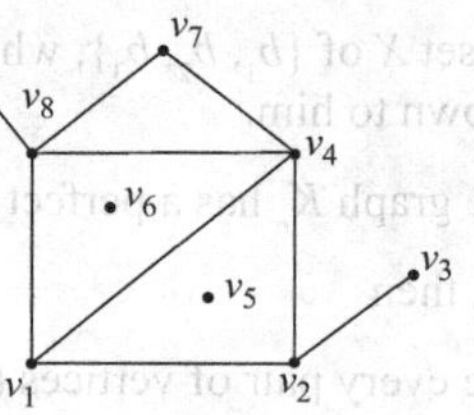

   (ii) A graph G has five vertices. Find the maximum number of edges in G if
      (a) G is a graph.
      (b) G is a multigraph.
   (iii) Can a simple graph of 7 vertices have 25 edges?
7. By suitably labelling the vertices, show that graphs pictured as letters are isomorphic.

   (a) 

 and 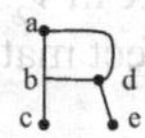

 are isomorphic

   (b) 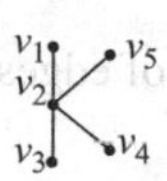

 and 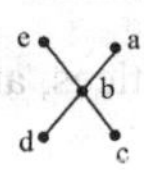

 are isomorphic

In the following questions 8–12, determine whether the simple graphs are isomorphic. When they are, determine the isomorphism function f.

8. 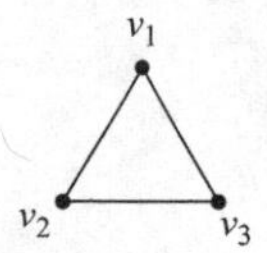

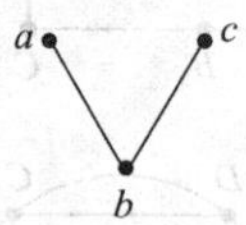

9. 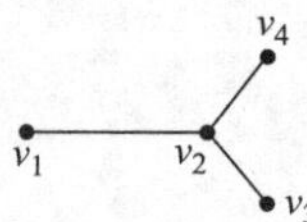

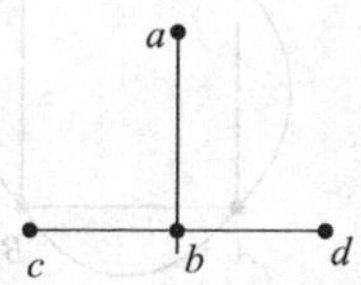

10. 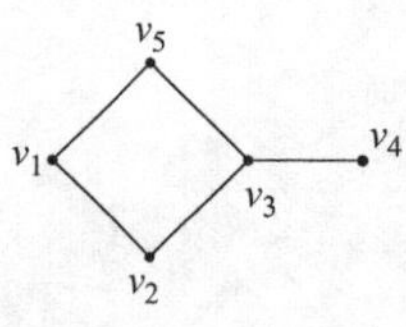

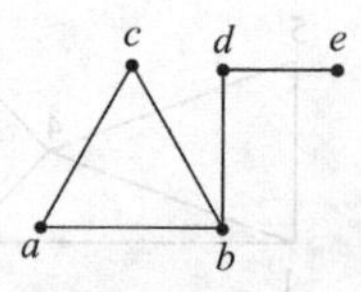

11. 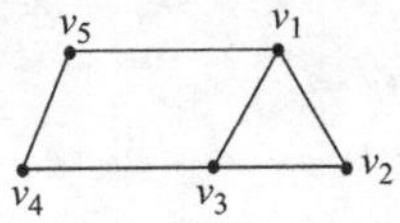

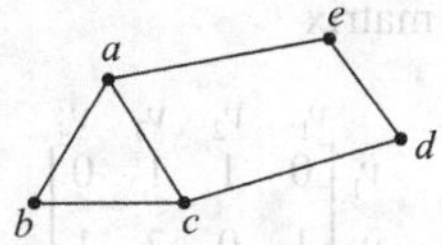

12. 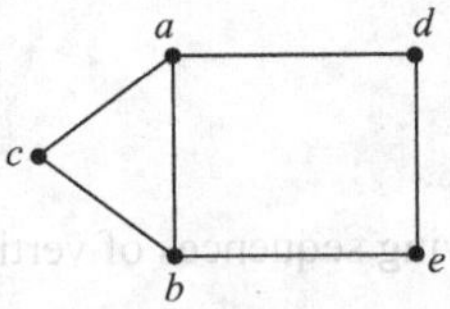

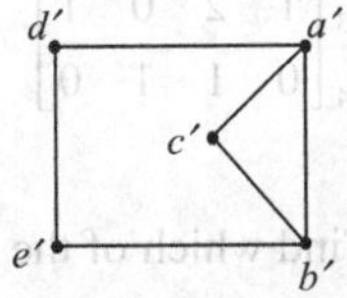

13. Show that the graphs shown in Figs. (a) and (b) are not isomorphic.

(a) 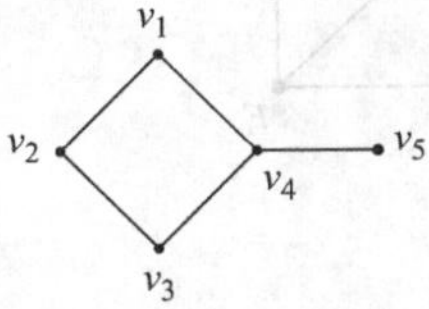

(b) 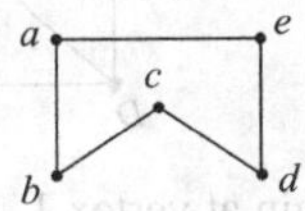

14. Show that the following graphs are isomorphic:

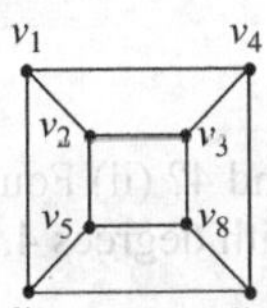

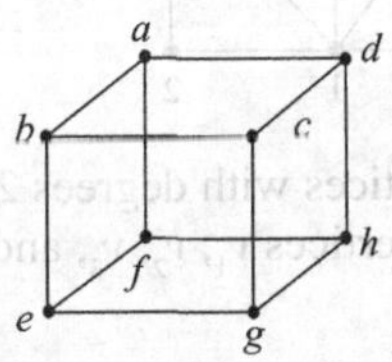

15. Find the adjacency matrix A = $[a_{ij}]$ for the graphs given in Figs. (a) and (b).

(a) 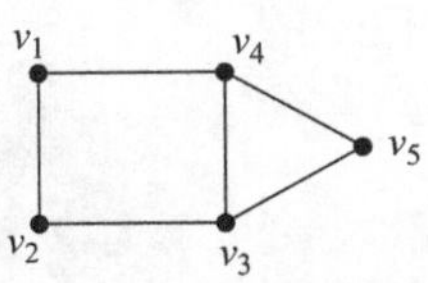

(b) 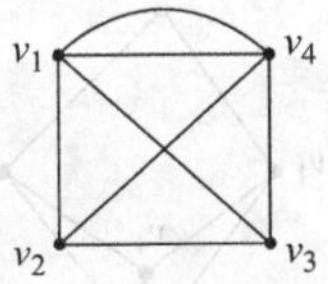

16. Find the adjacency and incidence matrices of the graph

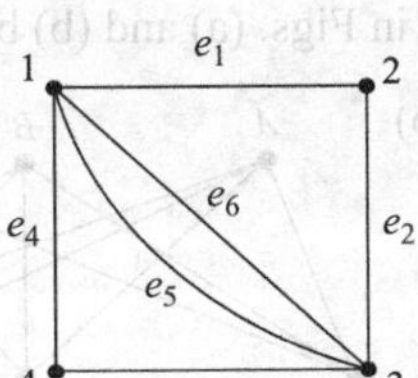

17. Draw the graph of the adjacency matrix

$$A = \begin{array}{c c} & \begin{array}{cccc} v_1 & v_2 & v_3 & v_4 \end{array} \\ \begin{array}{c} v_1 \\ v_2 \\ v_3 \\ v_4 \end{array} & \begin{bmatrix} 0 & 1 & 1 & 0 \\ 1 & 0 & 2 & 1 \\ 1 & 2 & 0 & 1 \\ 0 & 1 & 1 & 0 \end{bmatrix} \end{array}$$

18. Consider the graph G given below. Find which of the following sequences of vertices are simple paths, walks, and circuits.
    (i) (A, B, C, F)
    (ii) (A, B, C, F, E, D, A)
    (iii) (A, B, F, E, D, B, A)
    (iv) (A, B, F, E, D)
    (v) (D, E, F, C, B, D)

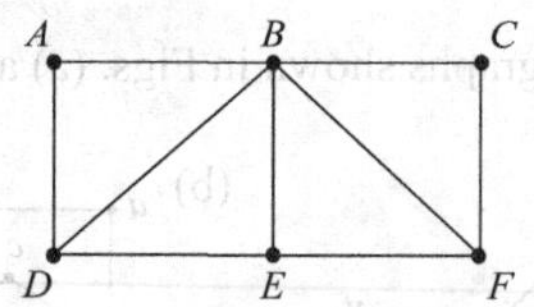

19. List all circuits in the graph that begin at vertex 1.

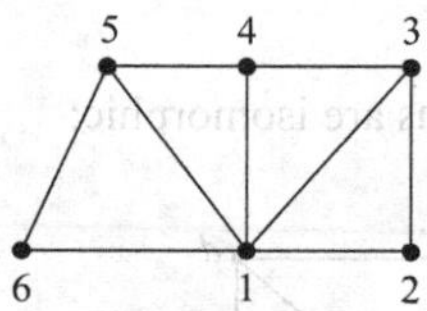

20. Can there be a graph that has (i) three vertices with degrees 2, 3, and 4? (ii) Four vertices of degrees 2, 2, 2, 2. (iii) There are four vertices v_1, v_2, v_3, and v_4 with degrees 4, 5, 2, and 5, respectively. Draw a graph if possible.

21. Is the graph bipartite? If so, identify the vertex sets V_1 and V_2.

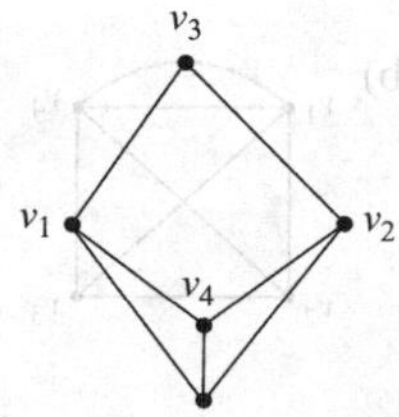

22. Are the following graphs given in Figs. (a) and (b) bipartite?

(a)

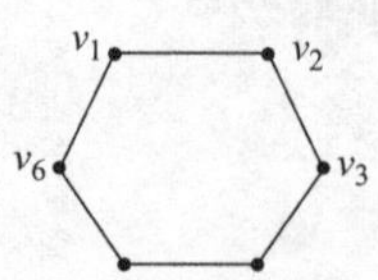

(b)

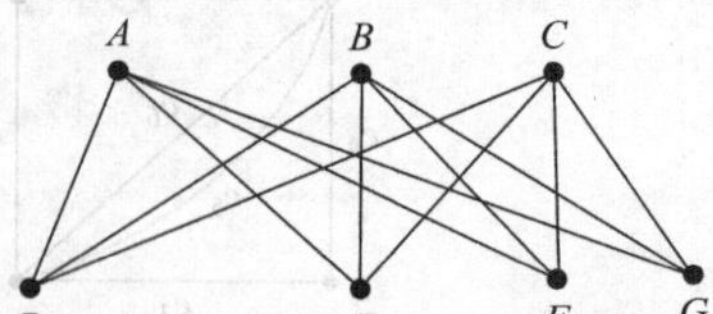

23. Determine which of the following graphs (a) to (g) have an Eulerian circuit, an Eulerian path but not an Eulerian circuit or neither.

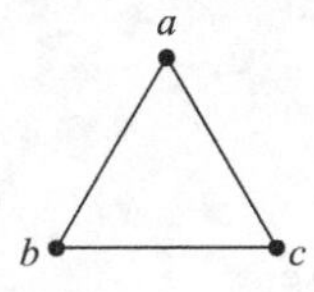

(a)

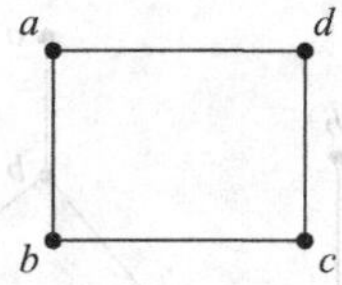

(b)

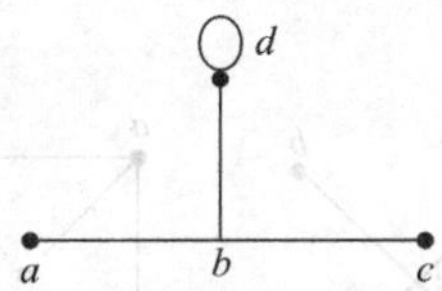

(c)

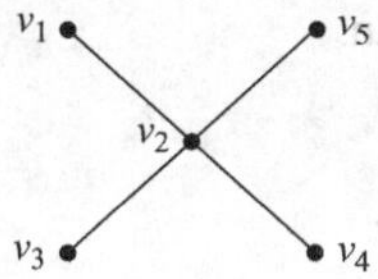

(d)

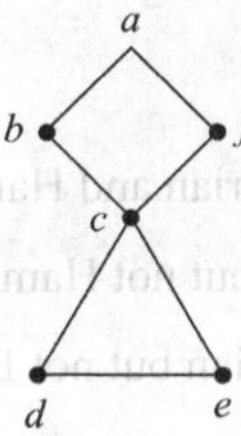

(e)

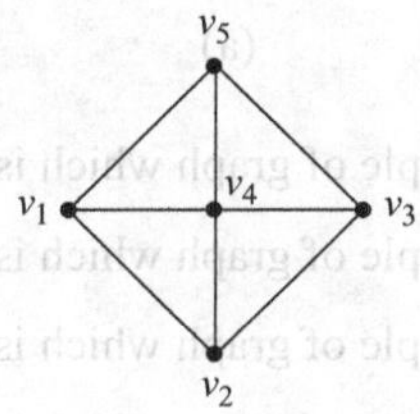

(f)

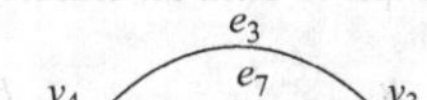

(g)

24. (i) Show that the graph

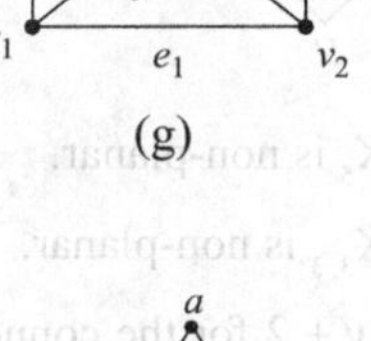

is an Eulerian graph.

(ii) Is the graph traversable (has an Eulerian path)? Is it an Euler graph that has an Euler circuit?

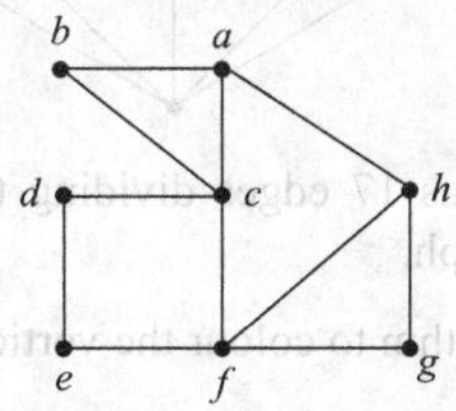

25. Determine whether the graphs shown have a Hamiltonian circuit, Hamiltonian path but no Hamiltonian circuit.

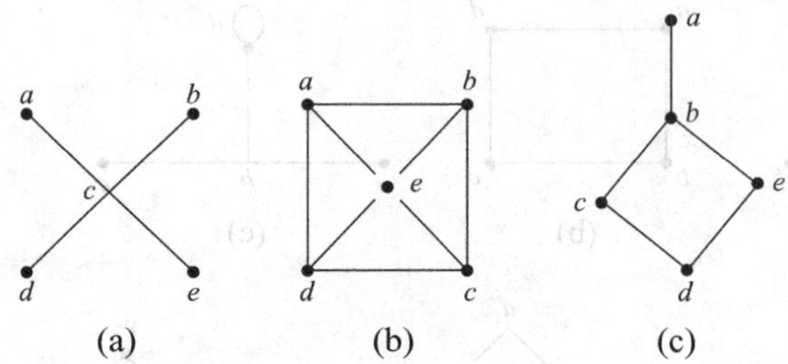

26. Give an example of graph which is both Eulerian and Hamiltonian.
27. Give an example of graph which is Eulerian but not Hamiltonian
28. Give an example of graph which is Hamiltonian but not Eulerian.
29. Give an example of graph which is neither Eulerian nor Hamiltonian.
30. Use Fluery's algorithm on the graph to find an Eulerian circuit starting from the vertex A.

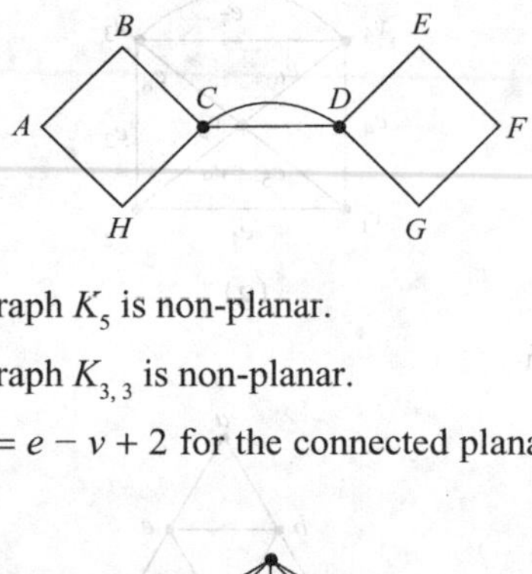

31. Prove that the complete graph K_5 is non-planar.
32. Prove that the complete graph $K_{3,3}$ is non-planar.
33. Verify Euler's formula $r = e - v + 2$ for the connected planar graph where symbols have their usual meaning.

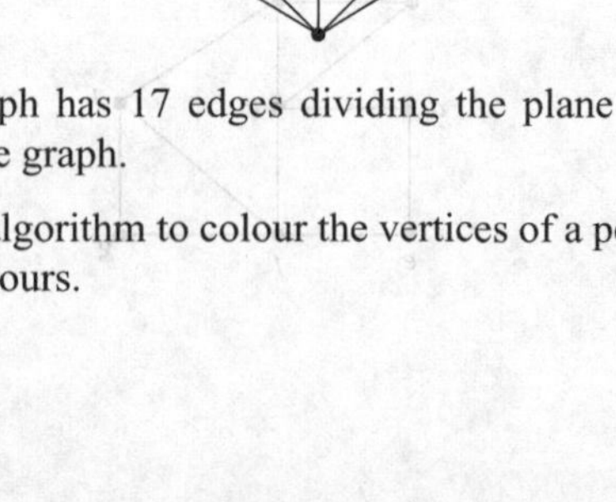

34. A connected planar graph has 17 edges dividing the plane into nine regions. Find the number of vertices of the graph.
35. Use Welch and Powell algorithm to colour the vertices of a polygon of eight vertices with minimum number of colours.

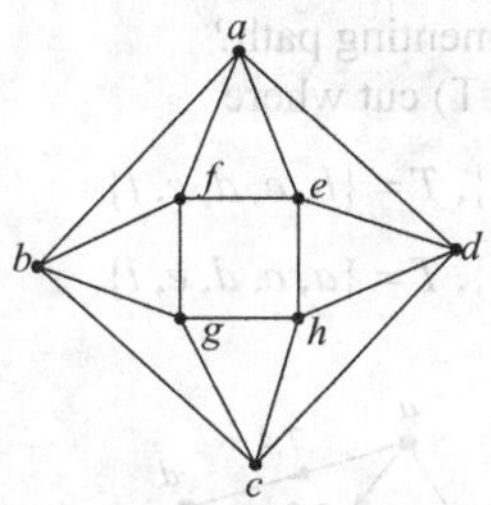

36. Find the chromatic number of the complete graph K_n and hence find the chromatic number of K_4.

37. Find the chromatic number of $K_{m,n}$.

38. Find the chromatic number of the following graphs so that adjacent portions have different colours.

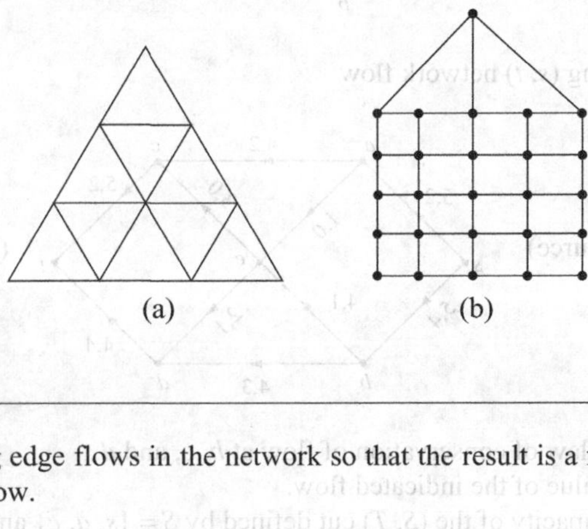

(a) (b)

EXERCISE – B

1. Fill in the missing edge flows in the network so that the result is a flow. Also, determine the value of the flow.

(i)

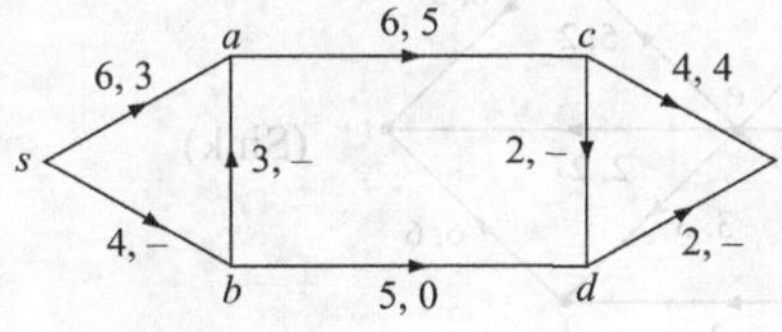

(ii)

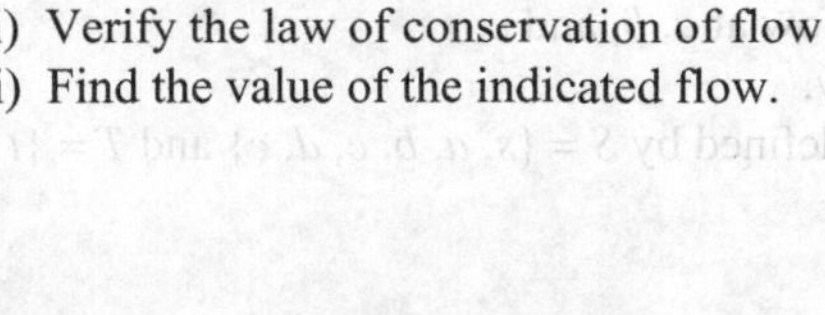

2. For the network shown below
   (i) Verify the law of conservation of flow at vertices at a and c.
   (ii) Find the value of the indicated flow.

(iii) Is there some flow augmenting path?

(iv) Find the capacity of (S, T) cut where

$$S = \{s, a\},\ T = \{b, c, d, e, t\}$$

and $$S = \{s, b\},\ T = \{a, c, d, e, t\}$$

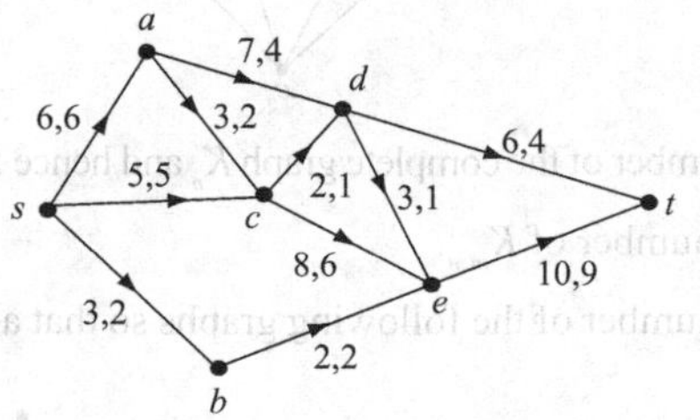

3. For the following (s, t) network flow

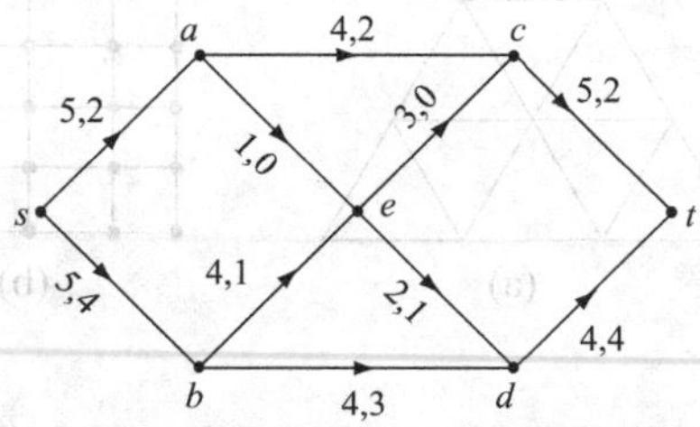

(i) Verify the law of conservation of flow at *b*, *c*, and *d*.

(ii) Find the value of the indicated flow.

(iii) Find the capacity of the (S, T) cut defined by $S = \{s, a, c\}$ and $T = \{b, d, e, t\}$.

(iv) Can the flow be increased along the path *sbedt*?

(v) Is the given flow maximum?

(vi) Can the flow along the path *sact* be increased?

4. For the following (s, t) network flow:

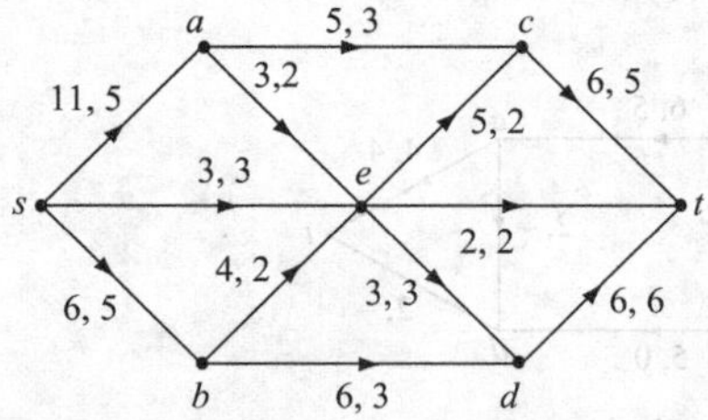

(i) Verify the law of conservation of flow at *a*, *b*, and *c*.

(ii) Find the value of the indicated flow.

(iii) Find the capacity of the (S, T) cut defined by $S = \{s, a, b, c, d, e\}$ and $T = \{t\}$.

(iv) Name all the saturated arcs.
(v) Is the given flow saturated?

5. For the network flow with source s and sink t,

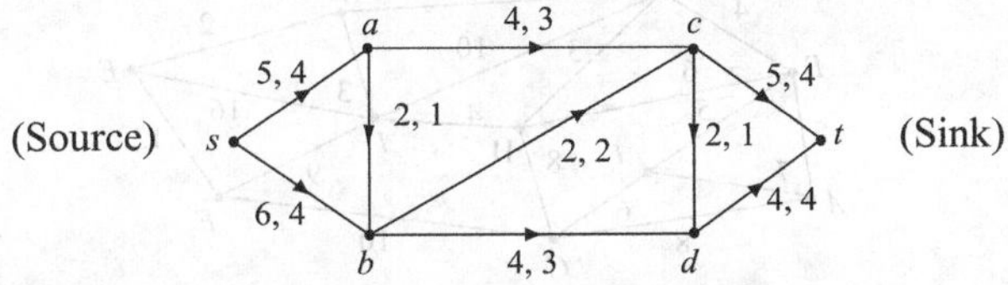

(i) Verify the law of conservation of flow at a, b, and d.
(ii) Find the value of the indicated flow.
(iii) Is the given flow maximum?
(iv) Can the flow along the path $s\ b\ c\ t$ be increased?
(v) Find the capacity of (S, T) cut where

$$S = \{s, a, b\} \text{ and } T = \{c, d, t\}$$

6. Find the maximum flow for the network with edges labelled with their capacities shown in the figure by listing all (S, T) cuts.

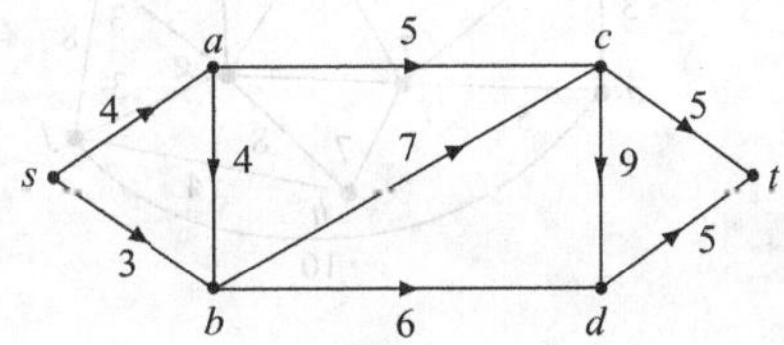

7. For the (s, t) network flow
(i) Find the value of flow.
(ii) Find the maximum flow.
(iii) Find the cut for maximum flow.

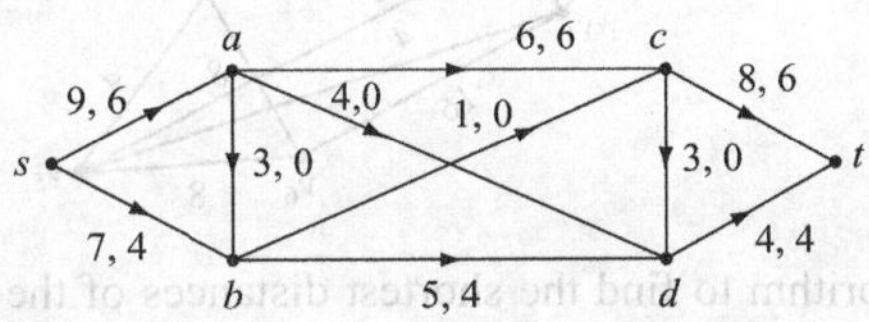

8. Find the length of the shortest path from the vertex a to h in the weighted graph by using Nearest Neighbourhood method.

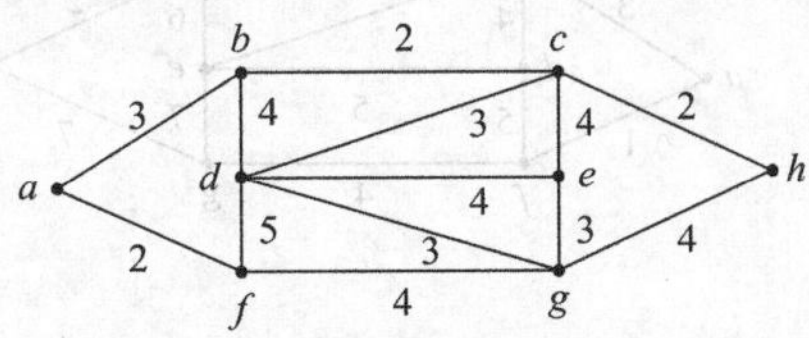

9. Find the length of the shortest path from vertex A to the vertex E in the following weighted graph using Nearest Neighbourhood method.

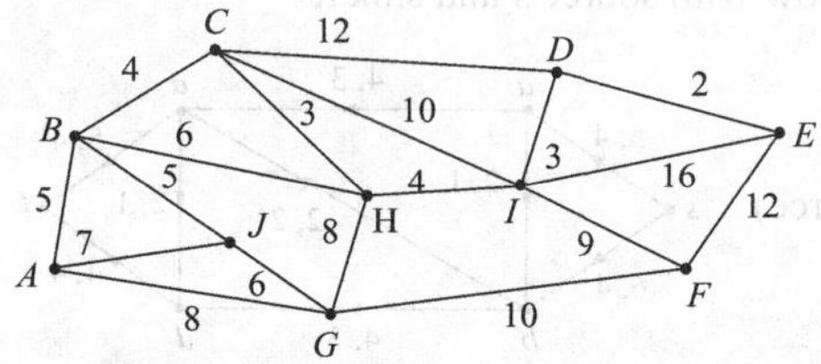

10. A service engineer has to visit 9 institutions for maintenance of computer systems whose distances from his office in kilometres are known. Find the path of minimum distance such that he visits every institution and is back to his office.

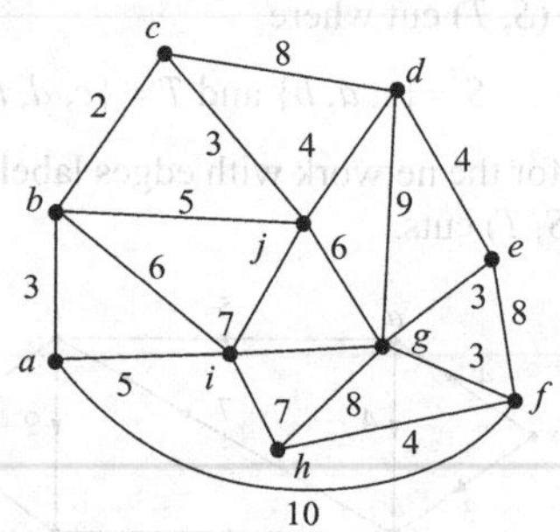

11. Use Dijkstra's algorithm to find the shortest distance between the indicated vertices in the weighted graph.

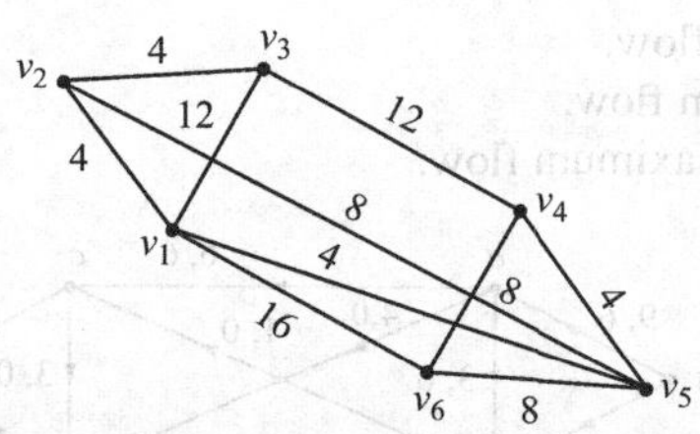

12. Use Dijkstra's algorithm to find the shortest distances of the indicated vertices from the vertex a in the weighted graph.

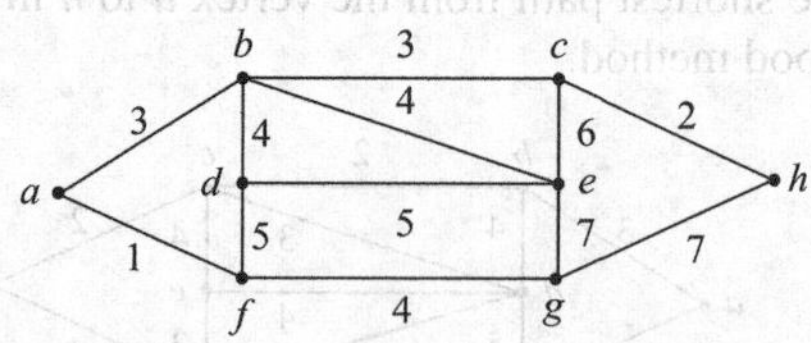

13. Suppose three girls a, b, c know four boys x, y, z, w as given below:

Girls	Boys known to girls
a	$y\ z\ w$
b	$x\ z$
c	$x\ y$

(i) Draw the bipartite graph to this table of relationship.
(ii) Find five different solutions of the marriage problem.
(iii) Check the marriage condition for this problem.

14. Suppose that three applicants a_1, a_2, and a_3 apply for five vacant positions p_1, p_2, p_3, p_4, and p_5 in a company. Each person is qualified for one or more jobs. The following information is available from the applications:

Applicants	Positions for their qualifications
a_1	p_2, p_4, p_5
a_2	p_1, p_2, p_3
a_3	p_1, p_2, p_3, p_4

(i) Draw the bipartite graph for the relationship.
(ii) Give two solutions of the assignment problem.

11

Trees

LEARNING OBJECTIVES

After reading this chapter, you will be conversant with:

- Introduction
- Rooted Structure
- Binary Tree of An Algebraic Expression
- Tree Searching or Tree Traversal
- Binary Search Tree
- Spanning Trees
- Minimal Spanning Trees

11.1 INTRODUCTION

One of the most important class of graphs are trees that are extensively used in computer science, switching network, corporate management, business, and educational organizations.

A corporate management structure can be presented as follows:

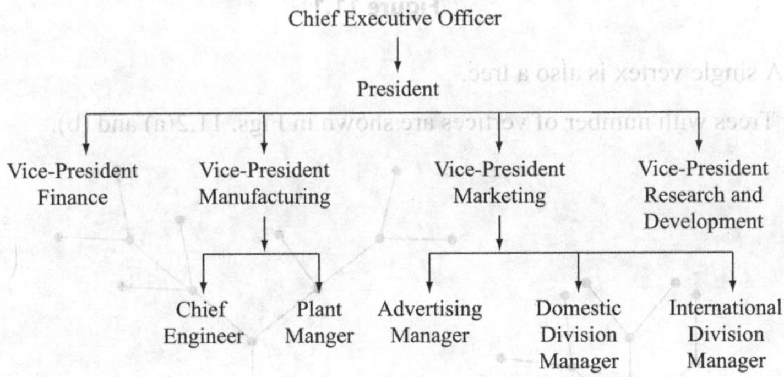

Thus, a *tree* is a simple graph having no loops and no parallel edges.

11.1.1 Definition

A graph is acyclic if it has no cycles. A *tree* is a connected acyclic graph.

11.1.2 Properties of Tree

1. There is one and only one path between every pair of vertices of a tree.
2. If in a graph G, there is one and only one path between every pair of vertices, then G is a tree.
3. A tree with n vertices has $(n - 1)$ edges.
4. Any connected graph with n vertices and $(n - 1)$ edges is a tree.
5. A graph is a tree if and only if it is minimally connected.
6. In any tree with two or more vertices, there are at least two pendant vertices.

Examples

1. The genealogy of a family is represented by a tree. The term *tree* comes from family tree.
2. A river with its tributaries and sub-tributaries can be represented by a tree.
3. The spies in a spy ring can be represented by a tree.
4. A corporation's management system can be represented by a tree.

Example 1 Draw trees of one, two, three, and four vertices.

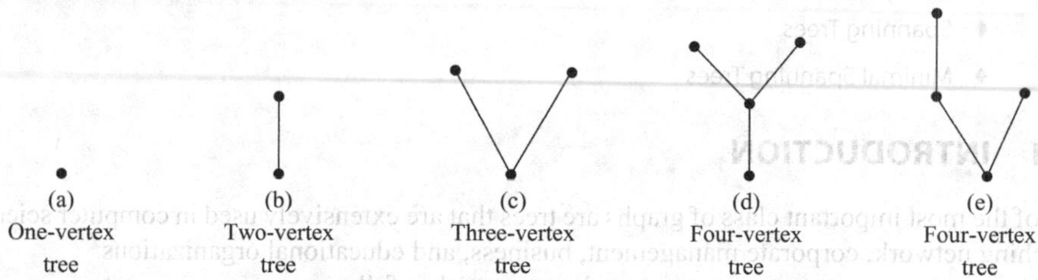

Figure 11.1

Note 11.1 A single vertex is also a tree.

Example 2 Trees with number of vertices are shown in Figs. 11.2(a) and (b).

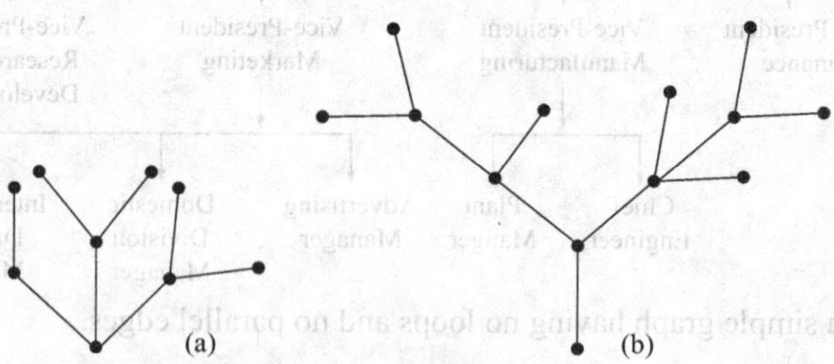

Figure 11.2

11.1.3 Directed and Undirected Trees

Mathematically, a *treeT* can be defined as $T = (V, E)$ where $V = \{v_0, v_1, v_2, \ldots, v_n\}$ is a set of vertices (or notes) and $E = \{e_1, e_2, e_3, \ldots, e_k\}$ is a set of edges if there is a vertex v_0 in V with the property that there exists a path from v_0 to every other vertex in V. The vertex v_0 is called the root of the tree. The tree T is called rooted tree and can be denoted by (T, v_0).

The vertex v_0 has in-degree zero, whereas every other vertex of V has in-degree one. A tree may be directed or undirected. The graph in Fig. 11.2 are undirected trees.

Directed Tree: A directed tree is an acyclic directed graph having one vertex with in-degree zero and other having in-degree one.

Note 11.2 In every tree directed or undirected, a unique path must exist.

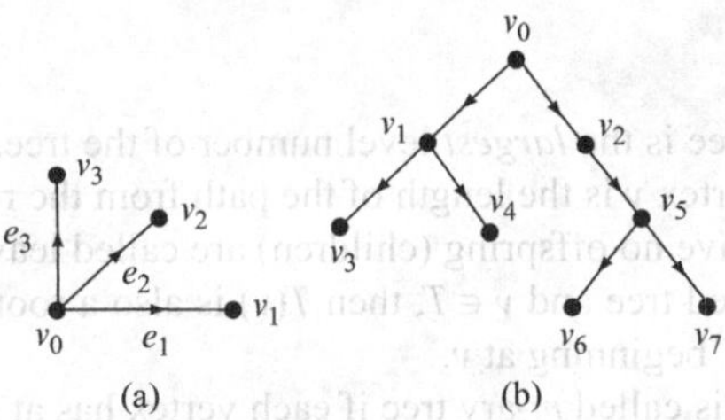

Figure 11.3

11.2 ROOTED STRUCTURE

There are many occasions when we come across rooted structures that can be represented as rooted trees.

Let us consider the family tree of a person who has three sons, eldest having no children, the second having two and the youngest having three children. This can be represented by the following presentation of a tree.

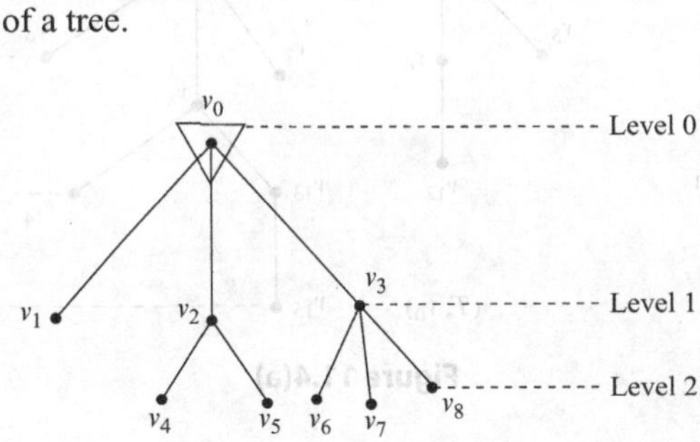

Figure 11.4

Figure 11.4 presents a tree with its roots v_0. The vertex v_0 is such that no edge enters v_0 but may leave, and hence the edges are drawn outwards, and thus the in-degree of the vertex v_0 is zero. This example leads to the following definition of a *rooted tree*.

Definition 11.1 A rooted tree is a directed tree graph having exactly one vertex with incoming degree zero and all other vertices having incoming degree one.

A vertex whose outgoing degree is zero is called a leaf or terminal node and a vertex whose outgoing degree is non-zero is called an internal node or branch node.

The terminal vertices of the edge beginning at v_0 are called level-1 vertices. The root vertex v_0 of the tree has level 0. The vertex v_0 is called the parent of the vertices at level 1. The vertices at level 1 are called children (offspring) of v_0. The edges leaving level-1 vertices are drawn downwards and terminate at various vertices, which are called level 2 and so on. The vertices v_4 and v_5 are called children of v_2, and v_6, v_7, v_8 are called children of v_3. The vertices having the same parent are called siblings to each other, and therefore v_4 and v_5 are siblings.

A vertex c is said to be descendent of the vertex a if there is a directed path from a to c. Also a is called ancestor of c. Thus, in Fig. 11.4, v_4 is descendent of v_0 and v_3 is ancestor of vertices v_6, v_7 and v_8.

Some Definitions

Height: The height of the tree is the *largest* level number of the tree.

Depth: The depth of any vertex v is the length of the path from the root v_0 to v.

Leaves: The vertices that have no offspring (children) are called leaves of the tree.

Sub-tree: If (T, v_0) is a rooted tree and $v \in T$, then $T(v)$ is also a rooted tree with root v. We say that $T(v)$ is the sub-tree of T beginning at v.

m-ary tree: A rooted tree T is called m-ary tree if each vertex has at most m children.

Example 1 Consider the rooted tree (T, v_0) shown in Fig. 11.4(a)

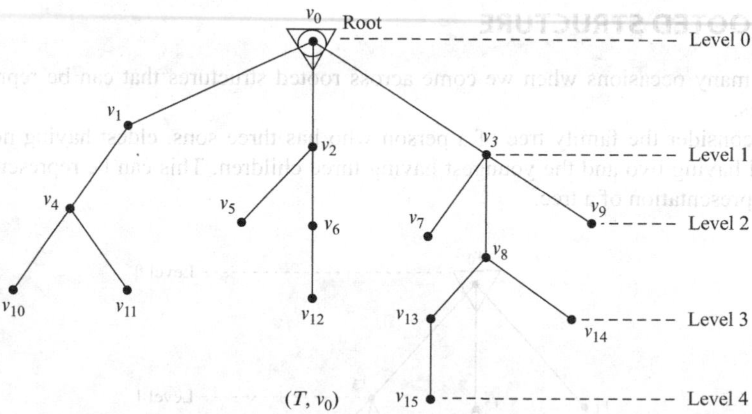

Figure 11.4(a)

(1) List all level-3 vertices.
(2) List all leaves.
(3) What is the height of $T(v_3)$?
(4) Compute sub-tree $T(v_2)$.
(5) Compute sub-tree $T(v_3)$.

Solution:

(1) Level-3 vertices are v_{10}, v_{11}, v_{12}, v_{13}, and v_{14}.
(2) Leaves are v_{10}, v_{11}, v_5, v_{12}, v_7, v_{15}, v_{14}, and v_9.
(3) The height of $T(v_3)$ is 3.
(4) $T(v_2)$ is

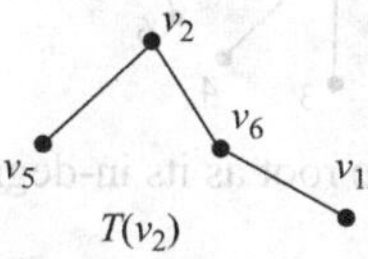

$T(v_2)$

(5) $T(v_3)$ is

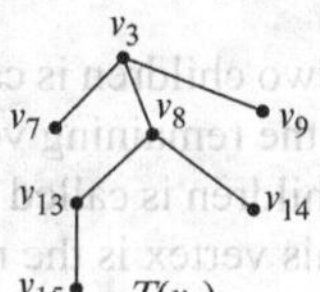

$T(v_3)$

Example 2 Consider the rooted tree shown in Fig. 11.4(b).

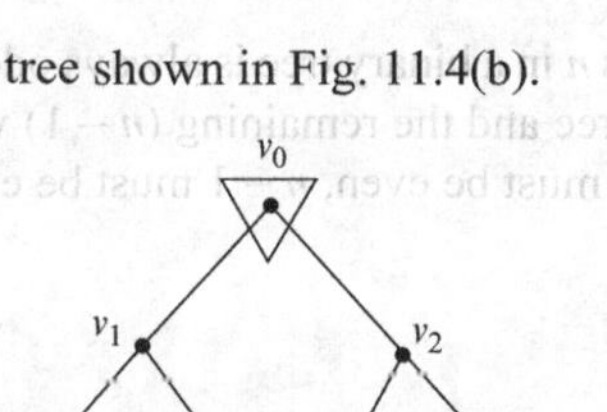

Figure 11.4(b)

Identify the path length from its root v_0 to each of the vertices v_1, v_4, and v_8.

Solution: The path length in a rooted tree to any vertex is the number of edges from the root vertex to the vertex under consideration.

From v_0 to v_1: (v_0, v_1), path length is one.
From v_0 to v_4: (v_0, v_1, v_4), path length is two.
From v_0 to v_8: (v_0, v_2, v_5, v_8), path length is three.

Example 3 Let $V = \{a, b, c, d, e\}$. If $T = \{(a, b), (f, a), (f, d), (d, e), (f, c)\}$ defines a tree, find the root.

Solution: The graph shows that f is the root as its in degree is zero, and there is a path from f to every vertex.

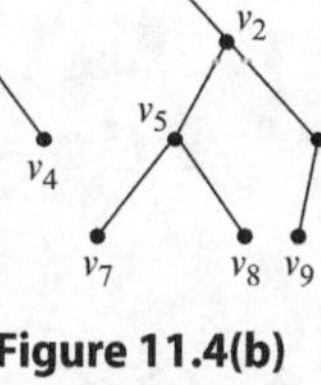

Example 4 Let $V = \{1, 2, 3, 4, 5, 6, 7\}$ If $T = \{(1, 2), (2, 3), (2, 4), (1, 5), (5, 6), (5, 7)\}$ defines a tree, find the root.

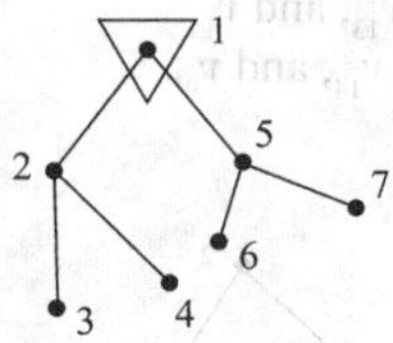

Solution: The graph shows that 1 is the root as its in-degree is zero and there is a path from 1 to every vertex.

11.2.1 Binary and Ternary Trees

A tree in which each vertex has at most two children is called a binary tree. In a *binary tree*, there is exactly one vertex of degree two and the remaining vertices are of degree one or three. A tree in which each vertex has at most three children is called a ternary tree. Since the vertex of degree two is distinct from all other vertices, this vertex is the root.

The following properties follow directly from the definition:

Example: The number of vertices n in a binary tree is always odd. This follows from the fact that exactly one vertex is of even degree and the remaining $(n - 1)$ vertices are of odd degree. Since the sum of degrees of all vertices must be even, $n - 1$ must be even. Hence, n is odd.

Properties of a Binary Tree

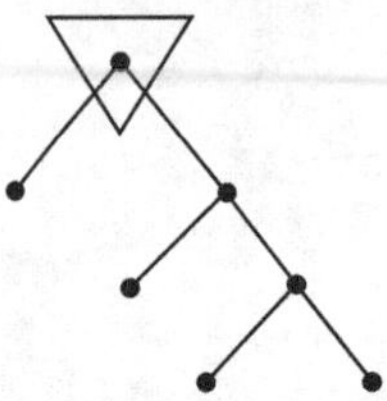

Seven vertex
Binary tree

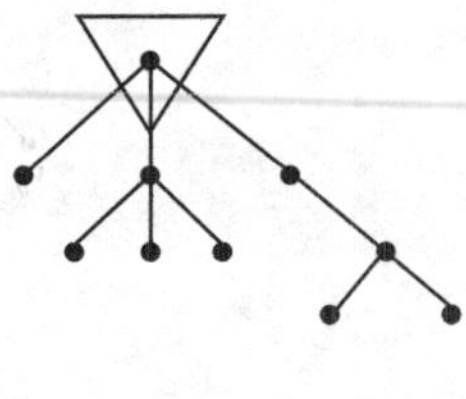

Ten vertex
Ternary tree

1. A binary tree is a rooted tree.
2. The number of vertices n in a binary tree is always *odd*.
3. Let p be the number of pendent vertices in a binary tree of n vertices, then $(n - p - 1)$ is the number of vertices of degree 3.

   Hence

$$
\begin{aligned}
& 2\times \text{ number of edges} = \text{Total degree of vertices} \\
\text{or}\quad & 2(n-1) = [\text{one vertex of degree } 2 + p \text{ vertices of degree } 1 + (n-p-1) \text{ vertices of degree } 3] \\
\text{or}\quad & 2n-2 = 2 + p + 3\,(n-p-1) \\
\text{or}\quad & 2n-2 = 2 + p + 3n - 3p - 3 \\
\text{or}\quad & 2p = n + 1 \text{ or } p = \frac{n+1}{2}
\end{aligned}
$$

4. The number of internal (i.e. non-pendent) vertices is always one less than the number of pendent vertices.

11.2.2 Complete Binary Tree

If every internal vertex (node) of a rooted tree has exactly m offsprings (children), then the tree is called a complete m-ary tree. In particular, if $m = 2$, then the tree is called complete binary tree. Figure 11.5 shows a three level binary tree with 15 vertices.

Level	vertices
1	2
2	4
3	8

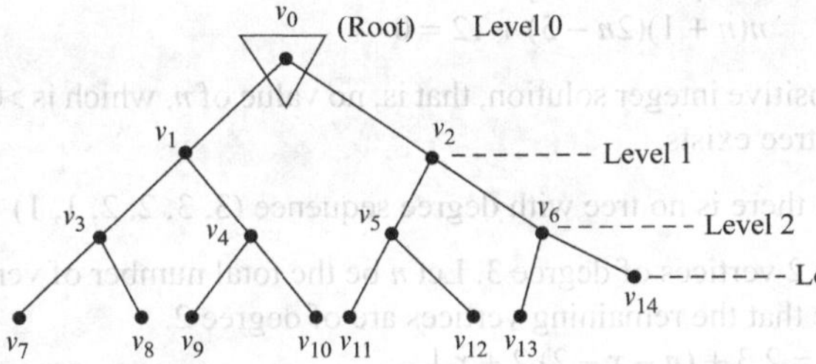

Figure 11.5 *Complete binary tree of height 3*

The number of vertices in levels 1, 2, and 3 are 2, 4, and 8, respectively, i.e., 2, 2^2, 2^3. This shows that the maximum number of vertices at level k will be 2^k.

Example 1 A tree has three vertices of degree 2, 2 vertices of degree 3, and 2 vertices of degree 4. How many vertices of degree 1 the tree has if it has no vertex of degree > 4.

Solution: Let λ be the number of vertices of degree 1, then the total number of vertices in the tree are $3 + 2 + 2 + \lambda = 7 + \lambda$.

The total number of degrees of all vertices is $6 + 6 + 8 + \lambda = 20 + \lambda$.

A tree having $7 + \lambda$ vertices has $(7 + \lambda - 1)$, i.e., $6 + \lambda$ edges.

Since the sum of degree of vertices = 2 number of edges.

Hence,
$$20 + \lambda = 2\,(6 + \lambda)$$
or
$$20 + \lambda = 12 + 2\lambda \Rightarrow \lambda = 8$$

Therefore, the total number of vertices of degree 1 is 8.

Example 2 A tree has 1 vertex of degree 1, 2 vertices of degree 2, 3 vertices of degree 3, and n vertices of degree n. Prove that no such tree exists.

Solution: Total number of vertices $= 1+ 2 +3 + \ldots + n = \dfrac{n(n+1)}{2}$

Hence, the number of edges $= \dfrac{n(n+1)}{2} - 1 = \dfrac{n^2+n-2}{2}$

The sum of degree of all vertices

$$= 1^2 + 2^2 + \ldots + n^2 = \frac{n\,(n+1)\,(2n+1)}{6}$$

Since the sum of degrees of all vertices = 2 × number of edges, we have

$$\frac{n\,(n+1)\,(2n+1)}{6} = 2\left[\frac{n(n+1)}{2} - 1\right]$$

or $\qquad n(n+1)(2n+1) = 6[n(n+1) - 2]$

or $\qquad n(n+1)(2n+1) - 6n(n+1) + 12 = 0$

or $\qquad n(n+1)[2n+1-6] + 12 = 0$

or $\qquad n(n+1)(2n-5) + 12 = 0$

This equation has no positive integer solution, that is, no value of *n,* which is >0 and is an integer, exists. Hence, no such tree exists.

Example 3 Shows that there is no tree with degree sequence (3, 3, 2, 2, 1, 1)

Solution: The tree has 2 vertices of degree 3. Let *n* be the total number of vertices and *x* be leaf vertices. Let us assume that the remaining vertices are of degree 2.

The sum of degrees $= 2.3 + (n - x - 2).2 + x.1$

$= 6 + 2n - 2x - 4 + x$

$= 2n - x + 2 = 2(n - 1)$ as a tree of *n* vertices has $(n - 1)$ edges

$\therefore 2n - x + 2 = 2n - 2 \;\; \therefore x = 4$

That is, there must be four leaves. But according to the problem, there are only two leaves. Hence, such a degree sequence tree is not possible.

11.2.3 Ordered Rooted Tree

A rooted tree in which the vertices at each level are ordered as first, second, third, and so on is called an ordered rooted tree. Vertices at the same level are arranged along the same horizontal line from left to right. The vertices *b*, *c*, and *d* can also be labelled as 1, 2, and 3, respectively. The vertices following 1 can be assigned labels 1.1, 1.2, 1.3, etc. Similarly, vertices following 2 can be assigned 2.1, 2.2. The vertices follow 3.1 can be labelled as 3.1.1, 3.1.2, and so on.

Here

$$b < c < d.$$

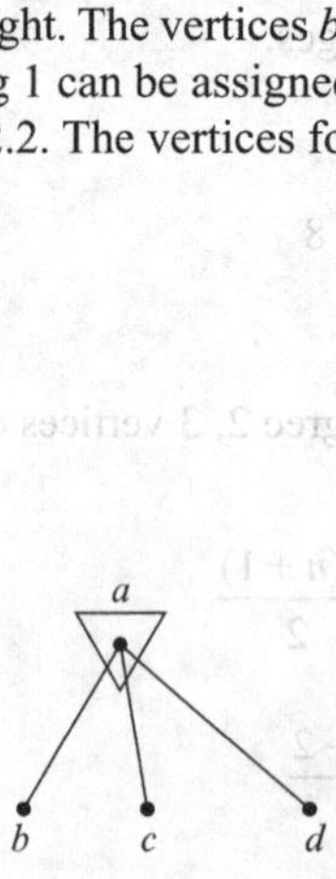

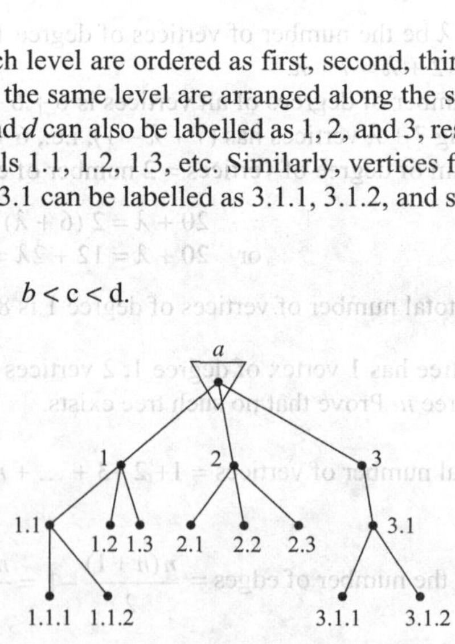

11.3 BINARY TREE OF AN ALGEBRAIC EXPRESSION

Binary trees are fine tools for representing and evaluating algebraic expressions involving binary operators. An algebraic expression containing operands and binary operators can be represented by a binary expression tree.

Expression tree is constructed by storing the operator at the nodes (vertices) of the tree. The first operand is represented on the left sub-tree and the second operand on the right sub-tree. A vertex representing an operand is a leaf and a vertex representing an operator is a non-leaf. Operators are placed following their precedence rule, that is, operators with higher precedence are placed in the lowermost sub-trees of the tree so that they are evaluated first.

Example 1 Consider the expression

$$E = a * (b - c)/d + e$$

Solution: The tree is build up from the bottom. Construct a binary tree for each of the expressions $b - c$, $a * (b - c)$, $a * (b - c)/d$, and $a * (b - c)/d + e$ successively in the Figs. 11.6–11.9.

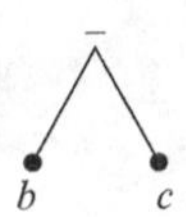

Figure 11.6

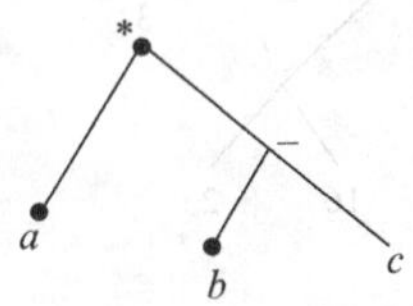

Figure 11.7

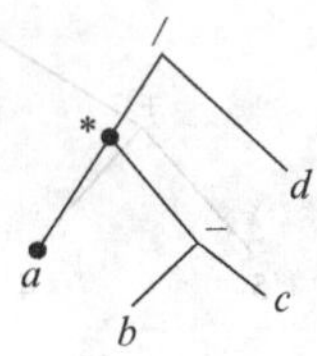

Figure 11.8

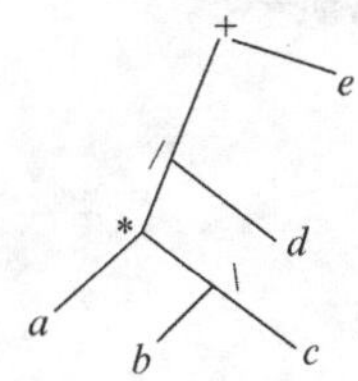

Figure 11.9

Example 2 Draw the tree corresponding to the algebraic expression

$$E = (3y + x)^4 \{a + (b/c)\}$$

Solution: Figure 11.10 represents the algebraic expression.

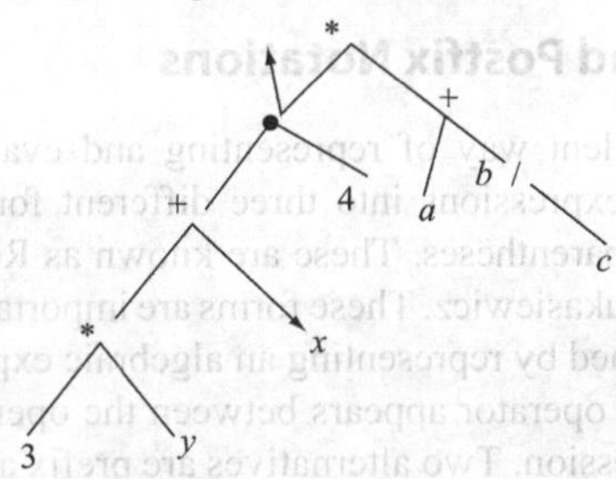

Figure 11.10

Example 3 Represent the following expressions by binary tree:

(1) a/b

(2) $(a * b)/c$

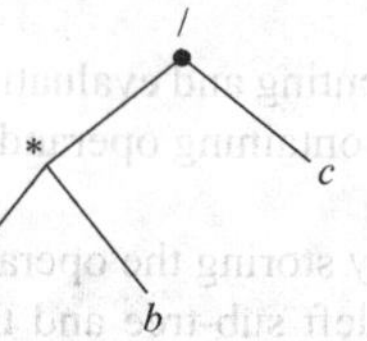

(3) $(a + \text{b}) * (c/d)$

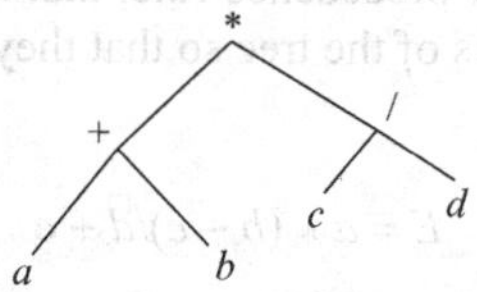

(4)

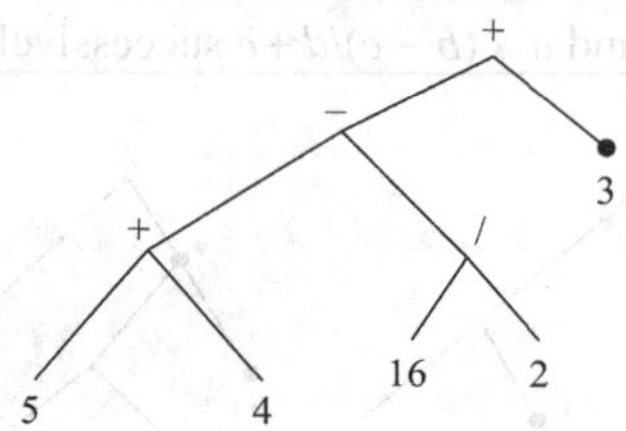

Figure 11.11

The expression represented by the binary tree is

$$\begin{aligned} &(5+4)-(16/2)+3 \\ &=9-8+2 \\ &=3 \end{aligned}$$

11.3.1 Infix, Prefix, and Postfix Notations

Binary trees provide an excellent way of representing and evaluating algebraic expressions. Computer converts algebraic expressions into three different forms, that is, infix, prefix, and postfix, which do not involve parentheses. These are known as Reversed Polish Notation in the honour of Polish logician Jan Lukasiewicz. These forms are important for compilation of computer programmers and can be obtained by representing an algebraic expression by rooted tree.

The familiar form in which operator appears between the operands is called infix form. For example, $x + y$ is an infix expression. Two alternatives are prefix and postfix forms. In the prefix notation, binary operator precedes the two operands and in postfix notation binary operator succeeds operands.

The three forms can be summarized as follows:

Notation	General Form	Example
Infix	operand operator operand	$x+y, x*y$
Prefix	operator operand operand	$+xy, *xy$
Postfix	operand operand operator	$xy+, xy*$

The main advantage of Polish notations are that algebraic expressions can be written without parentheses and having no ambiguity.

To convert an Infix expression into Prefix or Postfix, we move from left to right until we find a string of the form *Fxy* where *F* is the symbol for binary operation, that is, +, –, *, ÷, etc. and *x* and *y* are numbers. We evaluate *xFy* and substitute the answer for the string *Fxy*. The above form of algebraic expression is named as prefix as operator symbol precedes the operands. Postfix form of algebraic expression is the same as prefix except that the operator symbol comes after operands.

Infix	Prefix	Postfix
$a*(b-c)/d+e$	$+/*a-bcde$	$abc-*d/e+$
$(b+c-a)/(d*e)$	$*/-a+bcde$	$bc+-a/de*$
$(a-b)*(c+d)$	$*-ab+cd$	$ab-cd+*$
$(a+b)^3*(a-b)$	$*+ab\uparrow 3-ab$	$ab+\uparrow 3*ab-$

Example 1 Find the prefix, infix, and postfix form of the expression tree.

Solution: Prefix: $*+ab-cd$
Infix: $(a+b)*(c-d)$
Postfix: $ab+cd-*$

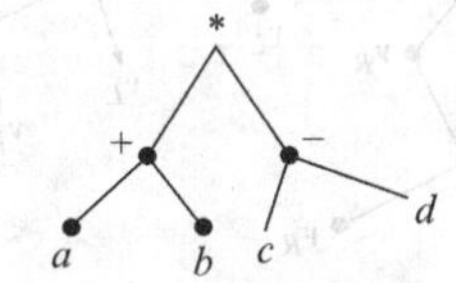

Figure 11.12

Example 2 Write the expression *E* given below in prefix and postfix forms where $E = a*(b+c)/d-e$.

Solution:

Prefix form

$a*(b+c)/d-e$

$+bc$

$*a+bc$

$/*a+bcd$

$-/*a+bcde$

Postfix form

$a*(b+c)/d-e$

$bc+$

$abc+*$

$abc+*d/$

$abc+*d/e-$

Example 3 Evaluate the prefix expression $-\uparrow/* 4 + 87529$.

Solution: $-\uparrow / * 4 + 87529$

15	$8 + 7 = 15$
60	$15 * 4 = 60$
12	$60 / 5 = 12$
144	$(12)^2 = 144$
136	$144 - 9 = 136$

11.4 TREE SEARCHING OR TREE TRAVERSAL

There are situations where one must pass through the tree considering each vertex exactly once in some specified order. For example, at each vertex of a tree that represents an algebraic expression, we perform the computation indicated by the operator at that vertex. The process of visiting each vertex in a systematic way and performing the required computation is called searching or performing **tree search**. The process is also called walking or traversing the tree.

Let us consider a binary tree. Each vertex has exactly two offsprings that can be specified as left offspring and right offspring denoted by v_L and v_R. Such a labelled tree is called positional tree (see Fig. 11.13).

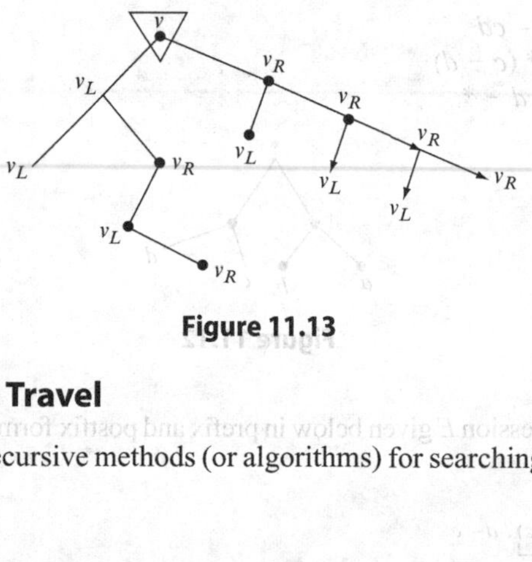

Figure 11.13

11.4.1 Binary Tree Travel

There are *three* standard recursive methods (or algorithms) for searching (traversing) a positional binary tree *T* with root *V*.

1. Preorder
2. Inorder
3. Postorder

Preorder Search: The steps of preorder search are the following:

1. Visit the root.
2. Traverse the left sub-tree if v_L exists.
3. Traverse the right sub-tree if v_R exists.

Example 1 Traverse the positional tree in Fig. 11.14(a) in preorder.

Preorder is *A B C D E F*

Explanation

1. Visit the root A and output A.
2. Traverse the left sub-tree rooted at B and so output B (visiting & root in preorder) now traverse left sub-tree rooted at C and the output C.
3. C has no left and right sub-tree the hence back up to B and traverse the right sub-tree rooted at D. Now, D has no left or right sub-tree and so output D.
4. Back track to A. Traverse the right sub-tree and output E.
5. E has no left sub-tree. Traverse right sub-tree rooted at F. As F has no left and right sub-tree, output F.

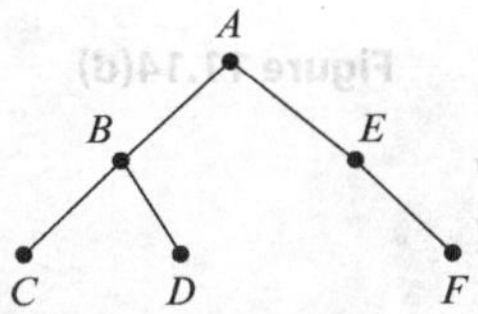

Figure 11.14(a)

Example 2 Traverse the positioned tree in Fig. 11.14(b) in preorder.

Preorder is

A B D E C F H I G J

Explanation will be as in Example 1.

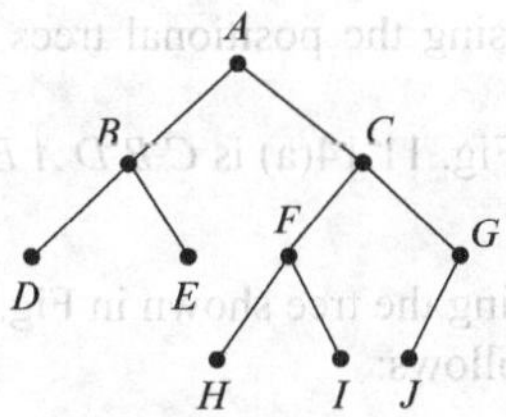

Figure 11.14(b)

Example 3 Consider the binary tree given in Fig. 11.14(c) and determine the data sequence for preorder, inorder, and postorder traversal.

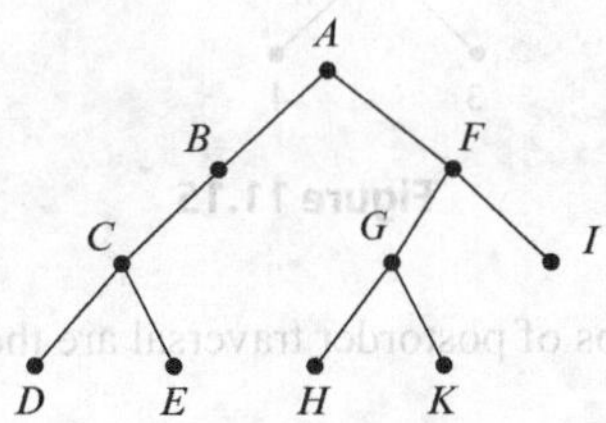

Figure 11.14(c)

Solution:

Preorder:	*A B C D E F G H K I*
Inorder:	*D C E B A H G K F I*
Postorder:	*D E C B H K G I F A*

Example 4 Determine preorder, in order and postorder traversal for the binary tree given in Fig. 11.14(d).

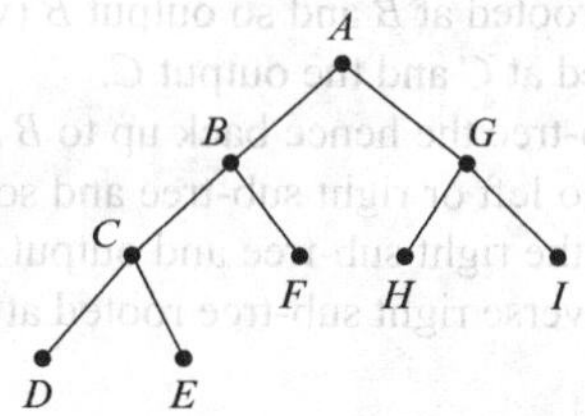

Figure 11.14(d)

Preorder: *A B C D E F G H I*
Inorder: *D C E B F A H G I*
Postorder: *D E C F B H I G A*

INORDER Traversal: The steps of inorder traversal are the following:

1. Traverse the left sub-tree.
2. Visit the root.
3. Traverse the right sub-tree.

Example 1 Give the output traversing the positional trees shown in Fig. 11.14(a) and (b) in INORDER.

Inorder output for tree given in Fig. 11.14(a) is *C B D A E F* and for Fig. 11.14(b) is *D B E A C H F I J G*.

Example 2 Give the output traversing the tree shown in Fig. 11.15 inorder.

Inorder traversing output is as follows:
3–2–4–1–5

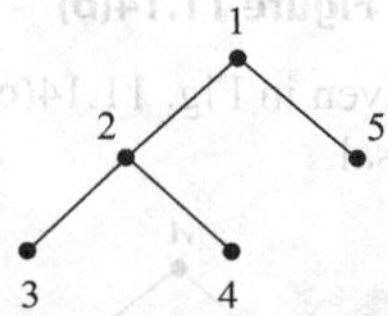

Figure 11.15

POSTorder Traversal: The steps of postorder traversal are the following:

1. Traverse left sub-tree.
2. Traverse right sub-tree.
3. Visit the vertex.

Postorder traversal output for positional tree shown in Fig. 11.15 is

3–4–2–5–1

Example 1 Find preorder, inorder and postorder traversal for the positional tree shown in Fig. 11.16.

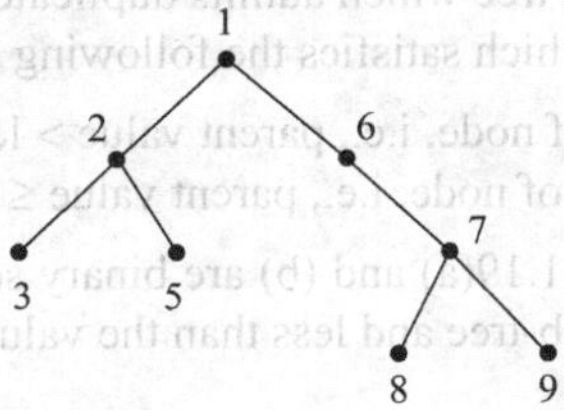

Figure 11.16

Solution:

(1) Preorder traversal output is 1–2–3–5–6–7–8–9

(2) Inorder traversal output is 3–2–5–1–6–8–7–9

(3) Postorder traversal output is 3–5–2–8–9–7–6–1

Example 2 Find preorder, inorder and postorder traversal outputs for the binary tree shown in Fig. 11.17.

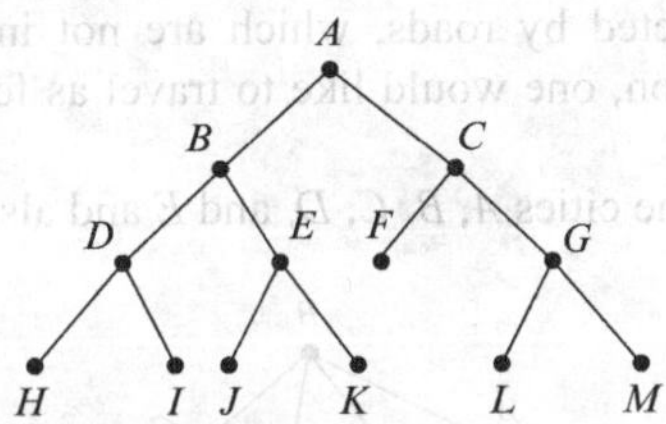

Figure 11.17

Preorder:	*A B D H I E J K C F G L M*
Inorder:	*H D I J E K B A F C L G M*
Postorder:	*H I D J K E B F L M G C A*

11.5 BINARY SEARCH TREE

A binary tree *T* is called a binary search tree if the value at each vertex (NODE) is greater than the value at left sub-tree and less than the value at right sub-tree.

The binary search tree helps search for and finds an element with average running time $f(n) = \log_2 n$ where n is the number of data items and also to insert and delete an information.

A simple binary search tree is shown in Fig. 11.18.

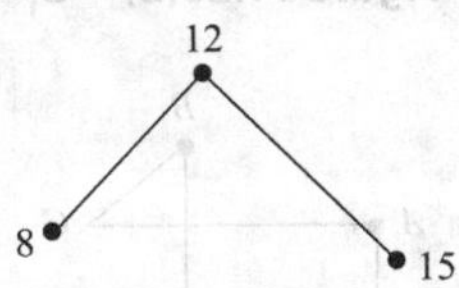

Figure 11.18

Note 11.3 In the above definition, it has been assumed that all the NODE values are distinct.
The definition of binary search tree which admits duplicates is as follows:
A binary search tree is a tree which satisfies the following condition at each vertex (NODE).

1. Node value > left sub-tree of node, i.e., parent value > left child
2. Node value ≤ right sub-tree of node, i.e., parent value ≤ right child

The binary trees shown in Figs. 11.19(a) and (b) are binary search tree because every vertex is greater than the value in its left sub-tree and less than the values in its right sub-tree.

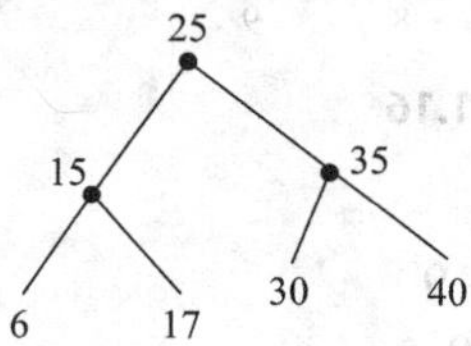

Figure 11.19(a)

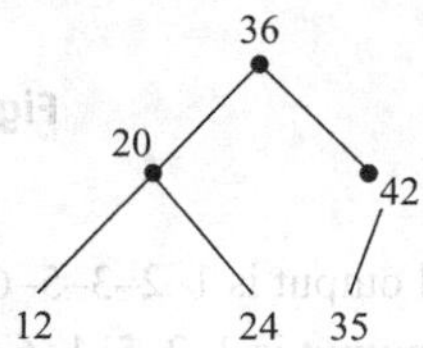

Figure 11.19(b)

11.6 SPANNING TREES

A state has five towns connected by roads, which are not in good conditions due to poor maintenance. In such a condition, one would like to travel as few roads as possible to visit all the towns.

Let the Fig. 11.20(a) show the cities A, B, C, D, and E and also the roads connecting them by the graph G.

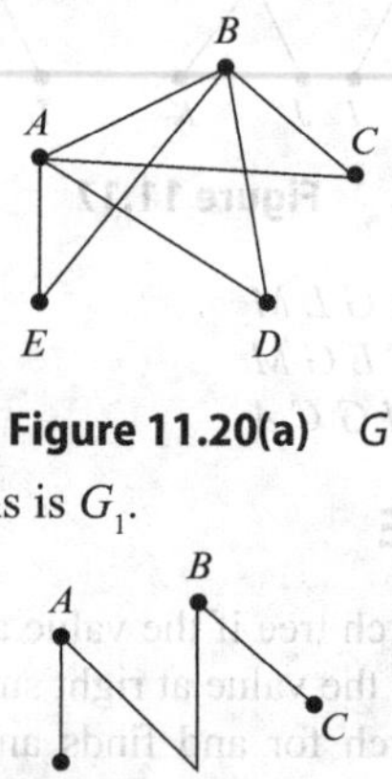

Figure 11.20(a) G

One possible way to visit all the towns is G_1.

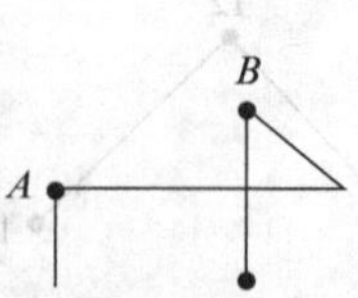

Figure 11.20(b) G_1

The next possible way can be G_2.

Figure 11.20(c) G_2

Another possible way can be G_3 [Fig. 11.20(d)].

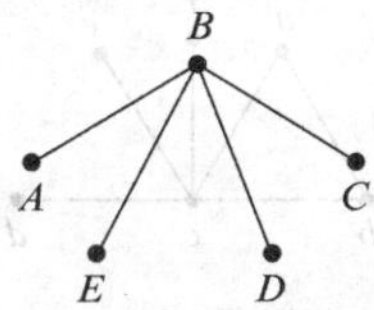

Figure 11.20(d) G_3

The graphs G_1, G_2, and G_3 shown in Figs. 11.20(b)–(d) are subgraphs of the graph G [Fig. 11.20(a)], which consists of all the vertices of G but not necessarily the edges.

These are called spanning trees of G iff it is a tree. Spanning trees find their application to obtain independent set of circuit equation for electrical network and to connect different destination by road, rail, or air to minimize distance (or time).

11.6.1 Definition

Let $G = (V, E)$ be an undirected connected graph. A subgraph $S = (V, E')$ of G is called a spanning tree of G if it is a tree.

Obviously, a spanning tree consists of all vertices of G, and hence can be obtained by deleting one or more edges of G.

Consider the graph G shown in Fig. 11.21.

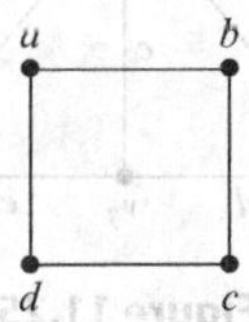

Figure 11.21

The possible spanning trees are given in Fig. 11.22.

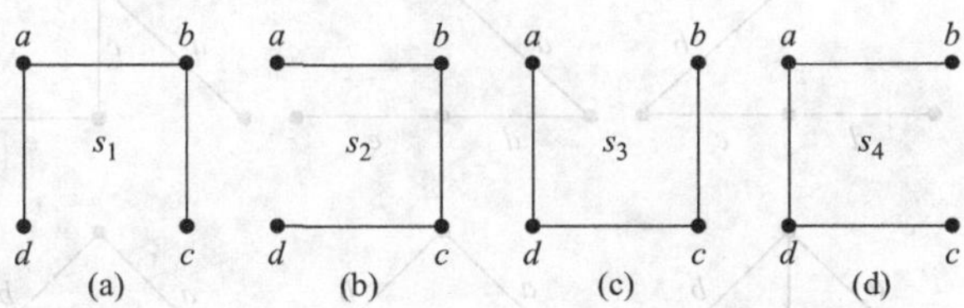

Figure 11.22

This suggests that many spanning trees may exist for an undirected connected graph.

Finding a spanning tree of a connected graph G is simple. If there is no circuit, then G itself is a spanning tree. In case G has a circuit, deleting an edge will lead to a spanning tree as the graph remains connected and also has all its vertices.

Example 1 Consider the graph shown in Fig. 11.23.

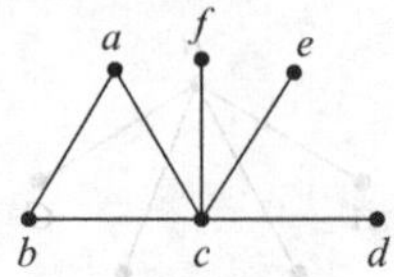

Figure 11.23

Solution: The graph contains only one circuit (cycle) formed by vertices *a*, *b*, and *c*. As the circuit contains three edges, spanning trees can be obtained by deleting one edge shown in Fig. 11.24(a)–(c).

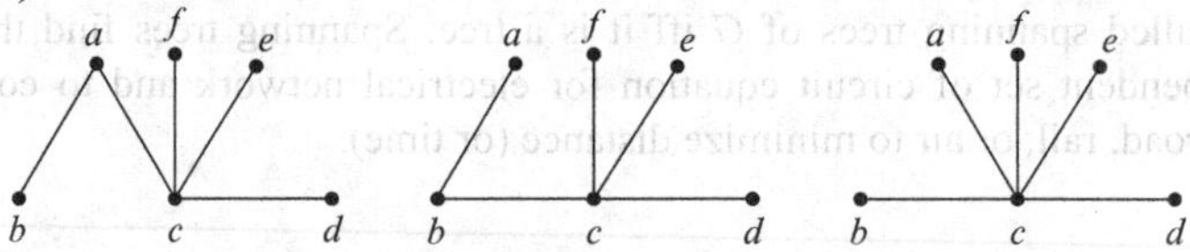

(a) (b) (c)

Figure 11.24

Example 2 Find all the spanning trees of the graph *G* shown in Fig. 11.25.

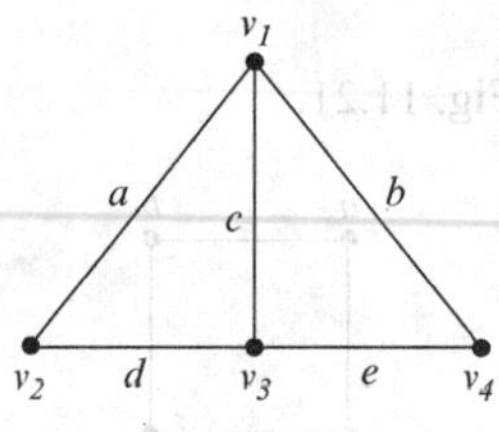

Figure 11.25

Solution: The graph *G* has 4 vertices, and hence spanning tree must have 4 – 1 = 3 edges. Out of five edges *a*, *b*, *c*, *d*, and *e*, two edges can be deleted in $5C_2 = 10$ ways. Of these, two will lead to disconnected graphs. Hence, there are eight spanning trees, which are shown below in Fig. 11.26.

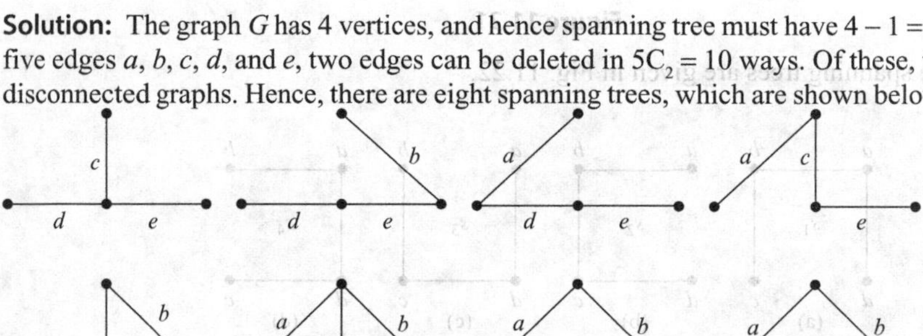

Figure 11.26

The above example of connected graphs leads to the following theorem.

THEOREM 11.1 Every connected graph G has a spanning tree.

Proof: Let G be connected graph. If G is a tree, then it is a spanning tree.

If G is not a tree, it must contain a cycle. If we remove an edge from the cycle, the resulting graph is still connected. If this graph is acyclic, then it is a tree and hence a spanning tree.

If G contains another cycle, remove an edge from the cycle. The process is continued until we have a subgraph H which is acyclic. Since H is connected and acyclic, it is a tree. As H contains every vertex of G, it is a spanning tree of G.

THEOREM 11.2 A graph (non-directed) G is connected if and only if G contains a spanning tree.

Proof: If G is a spanning tree, then there is a path between every pair of vertices in G along the tree T. Hence, G is connected. Conversely, let the connected graph G have k cycles. If $k = 0$, then G is connected with no cycles and hence G is a tree.

Let us assume that all connected graphs with lesser than k cycles have a spanning tree. Now suppose that G is a connected graph with k cycles. If we remove an edge from one of the cycles, then G-e is still connected and has a spanning tree by hypothesis because G-e has lesser cycles than k. But G-e has all vertices of G. The spanning tree for G-e is also the spanning tree for G. Hence, the theorem by mathematic induction.

11.6.2 Branch and Chord of a Spanning Tree

An edge of a spanning tree S is called a branch of S. An edge of G, which is not in a spanning tree, is called a chord. In electrical engineering, a chord is sometimes referred to as a link.

Consider the graph G shown in Fig. 11.27.

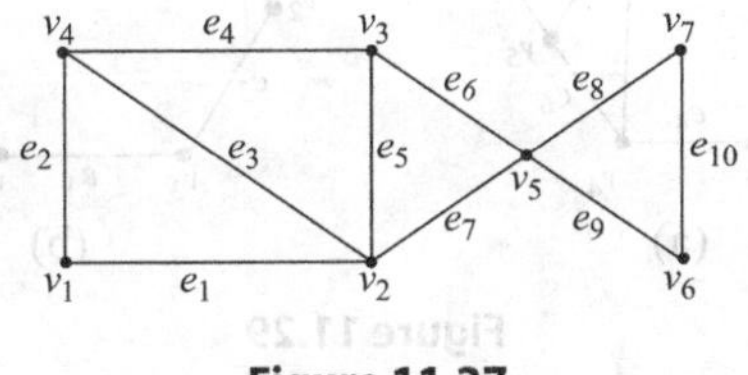

Figure 11.27

A spanning tree of the graph is $v_1\ e_2\ v_4 e_4\ v_3 e_5 v_2\ e_7\ v_5\ e_8\ v_7\ e_{10}\ v_6$. The branches are e_2, e_4, e_5, e_7, e_8, and e_{10} and the chords are e_1, e_3, e_6, e_9.

Note 11.4 The branches and edges are defined with respect to a given spanning tree.

Example: Find the branches and chords of the spanning tree S shown in Fig. 11.28(b) of the graph G in Fig. 11.28(a).

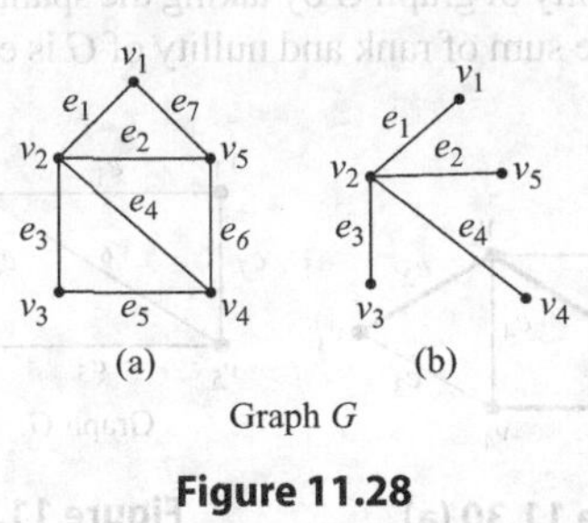

Figure 11.28

The branches are e_1, e_2, e_3, and e_4. The chords are e_5, e_6, and e_7.

Here, the number of vertices $n = 5$ and the number of edges $e = 7$.

The spanning tree has $n - 1$, that is, $5 - 1 = 4$ branches and $e - n + 1 = 7 - 5 + 1 = 3$ chords.

Any spanning tree will also give the same number. This suggests the following theorem (proof not given).

THEOREM 11.3 In any connected graph of n vertices and e edges, every spanning tree has $(n - 1)$ tree branches and $e - n + 1$ chords.

Rank and Nullity: Let G be a connected graph of n vertices and e edges, then a natural question arises as to what can be the number of branches and chords for any spanning tree of G.

The number of branches in any spanning tree S of G

$$= \text{Rank } (r) \text{ of } G = n - 1$$

The number of chords with respect to any tree S of G

$$= \text{Nullity } (\mu) \text{ of } G = e - n + 1$$

And Rank of G + Nullity of $G = e$ = The total number of edges in G.

For example, consider the graph G shown in Fig. 11.29(a) and a spanning tree S of G in Fig. 11.29(b).

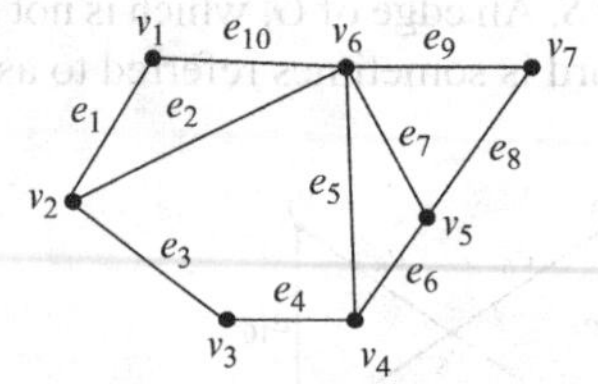

(a)

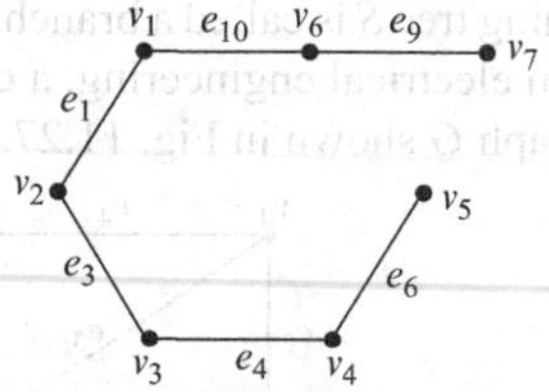

(b)

Figure 11.29

The number of vertices $n = 7$

The number of edges $e = 10$

$\therefore$ Rank $= 7 - 1 = 6$

Nullity $= 10 - 7 + 1 = 4$

Rank + Nullity = Total number of edges

The number of branches in $S = 6$

The number of chords $= 4$

Example 1 Find the rank and nullity of graph G by taking the spanning tree S shown in Fig. 11.30 (a) by dark lines. Also, verify that the sum of rank and nullity of G is equal to the number of edges in G.

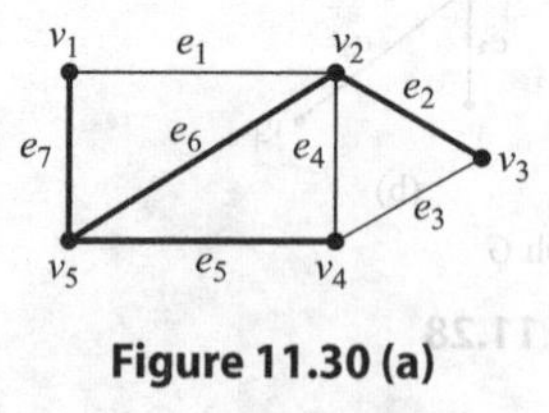

Figure 11.30 (a)

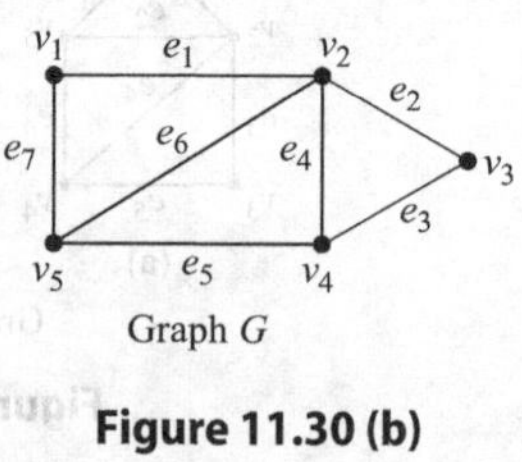

Graph G

Figure 11.30 (b)

The number of vertices $n = 5$
The number of edges $e = 7$
The number of branches of spanning tree $S = 4$ (dark line) $= 5 - 1 = n - 1$
The number of chords of spanning tree $S = 3$, that is, edges e_1, e_4, and e_3
$\therefore$ Rank of $G = n - 1 = 5 - 1 = 4$
Nullity of $G = e - n + 1 = 7 - 5 + 1 = 3$
Further Rank of G + Nullity of $G = 4 + 3 = 7$ = Number of edges in G

Example 2 A farmhouse consists of five walled plots of land as shown in Fig. 11.31(a) and these are full of water. Find the number of walls that will have to be broken so that the water is drained out completely.

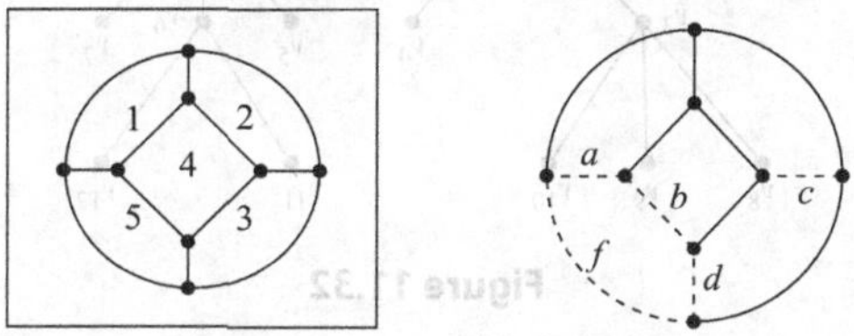

Figure 11.31(a) **Figure 11.31(b)**

Solution:

The number of vertices $n = 8$
The number of edges $e = 12$
The number of chords $= e - n + 1 = 12 - 8 + 1 = 5$

Hence, to form a spanning tree which does not have any circuit, we have to remove all the chords and so removal of 5 chords will drain out the water completely, that is, the removal of edges a, b, c, d, and f is needed for draining out the plots.

11.6.3 Methods for Finding Spanning Trees

There are two efficient search methods for finding a spanning tree of a connected graph. These are known as breadth-first search (BFS) and depth-first search (DFS) methods. In both these methods, all the vertices are visited but in different order. Both methods have their advantages depending upon the nature of the problem. For example, BFS is used in finding the shortest path, whereas DFS is used in finding network flows.

Breadh-first Search Algorithm: In the BFS method, the vertices are visited sequentially on a given level until all are visited. This means that we visit all vertices adjacent to the current vertex before proceeding to another vertex at next level.

An outline of BFS algorithm is as follows:

1. Start at vertex v_0.
2. Mark it visited.
3. Visit all unvisited vertices adjacent to v_0.
4. Visit all unvisited vertices adjacent to each of them.
5. Repeat step 4 until all vertices are visited.

To illustrate the procedure for BFS, let us consider the tree shown in Fig. 11.32. In order to perform, BFS method, we start at vertex v_0 and visit vertices v_1 and v_2 that are adjacent to v_0. We then visit vertices v_3 and v_4 adjacent to v_1 and then vertices v_5, v_6, and v_7 which are adjacent to v_2. Finally, we visit vertices v_8, v_9, and v_{10} which are adjacent to v_3 and then the vertices v_{11} and v_{12} adjacent to v_6.

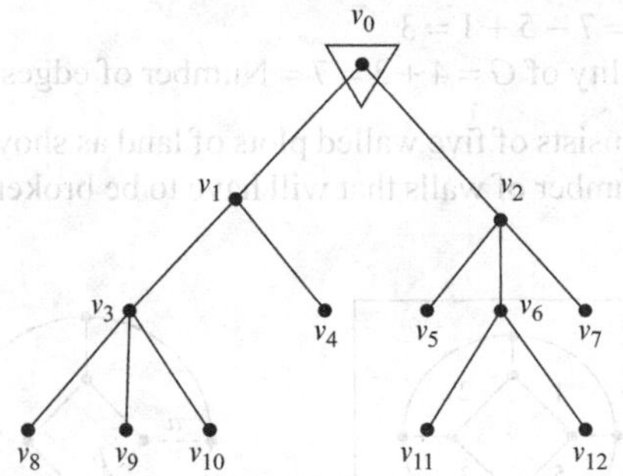

Figure 11.32

11.6.4 Depth-first Search Algorithm

In the DFS method, we proceed to all successive levels possible in a tree before fanning out to other vertices.

An outline of DFS algorithm is as follows:

Step 1: Start at a vertex v_0.
Step 2: Mark it as visited.
Step 3: Pick a vertex v_i adjacent to v_0 and not yet visited and traverse the graph as for as possible marking each vertex from which no more path is available.
Step 4: Backtrack to the vertex v_i and repeat Step 3.
Step 5: If a vertex v_j is reached such that all its adjacent vertices have been visited, backtrack to the last visited vertex v_k and repeat Step 3.
Step 6: Stop when all the vertices have been visited.

To illustrate the procedure, we apply DFS to the tree shown in Fig. 11.32.

We start at the vertex v_0 and move down to vertices v_1, v_3, and v_8. At v_8, no further path is available, and therefore we backtrack to v_3 and go down to visit the vertex v_9. We backtrack to v_3 and then go down to visit v_{10}. We backtrack to v_1 via v_3 and then visit v_4. Backtracking to v_0, then takes us to v_2 and v_5. Again backtrack to v_2 and then visit v_6 and v_{11}. We again backtrack to v_6 and visit v_{12}. Finally, backtracking via v_6 to v_2, we visit v_7. Thus all the vertices are visited. Due to backtracking involved in the procedure, DFS method is also called backtracking.

11.7 MINIMAL SPANNING TREES

A weighted connected graph may have several spanning trees, but the tree with minimum weight is of special importance. The finding of minimal spanning tree in a connected weighted graph has several useful applications. Let us consider the following practical problem:

A company is required to lay natural gas pipelines between five towns *a*, *b*, *c*, *d*, and *e*. The weights (in millions of rupees) of the edges represent the cost of building the various pipelines. The pipelines must be laid in such a way that the gas can be sent from any town to any other town holding the construction cost to a minimum.

This problem leads to finding a spanning tree for the weighted graph with the sum of the weights of the trees edge is minimum. Such a tree is the minimal spanning tree.

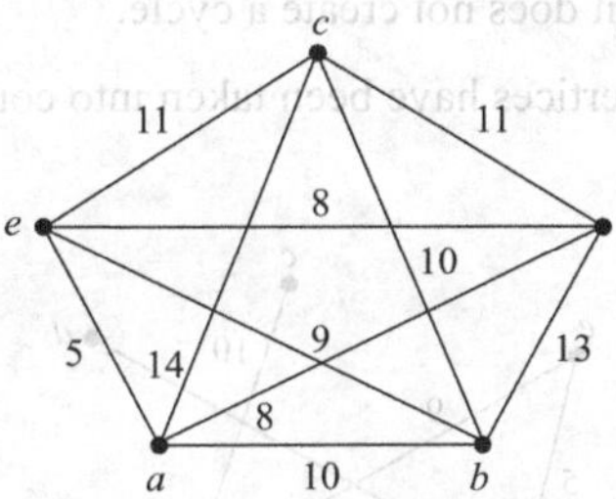

Figure 11.33

Minimal Spanning Tree: Let *G* be a connected weighted graph. The weight of a spanning tree of *G* is the sum of the weights of its edges. A minimal spanning tree of *G* is the one whose weight is minimum. Several algorithms can find minimal spanning tree *T* of a connected weighted graph. Two algorithms due to Joseph Kruskal and Robert Prim are generally used to construct an optimal spanning tree. These algorithms are known as greedy algorithms because these involve choosing edges of minimum weight in such a way that no cycle is created.

11.7.1 Kruskal's Algorithm

The steps of constructing minimal spanning tree *T* from a connected weighted graph *G* is as follows:

Step 1: Arrange the edges in *G* in an increasing order of their weights.

Step 2: Choose an edge e_j in *G* with minimum weight and add this in *T*. If more than one edge has the same minimum weight, choose arbitrarily any one of them.

Step 3: Choose another edge e_k with next higher weight and add this to the edge already selected if it does not form a cycle.

Step 4: Repeat step 3 until all the $(n - 1)$ edges are added to *T* and do not form a cycle where *n* denotes the number of vertices in *G*.

Example 1 Use Kruskal's algorithm to find a minimal spanning tree for the graph shown in Fig. 11.33.

Solution:

Step 1: Arrangement of edges in increasing order of their weights is as follows:

Weight	5	8	8	9	10	10	11	11	13	14
Edge	(*a*, *e*)	(*a*, *d*)	(*e*, *d*)	(*b*, *e*)	(*a*, *b*)	(*b*, *c*)	(*c*, *d*)	(*c*, *e*)	(*b*, *d*)	(*a*, *c*)

Step 2: Choose the edge (a, e) since it has smallest weight and include in T.

Step 3: An edge with next smallest weight is (a, d) and (e, d). Let us choose since it (a, d) does not form any cycle with the edge in T, include it in T.

Step 4: Edge (e, d) forms a cycle with the edges in T and so it is discarded.

Step 5: Next smallest edge (b, e) does not produce a cycle and hence include in T.

Step 6: Edge (a, b) creates a cycle and so rejected.

Step 7: Edge (b, c) is selected as it does not create a cycle.

Having selected four edges, all vertices have been taken into consideration. Figure 11.34 shows the resulting spanning tree.

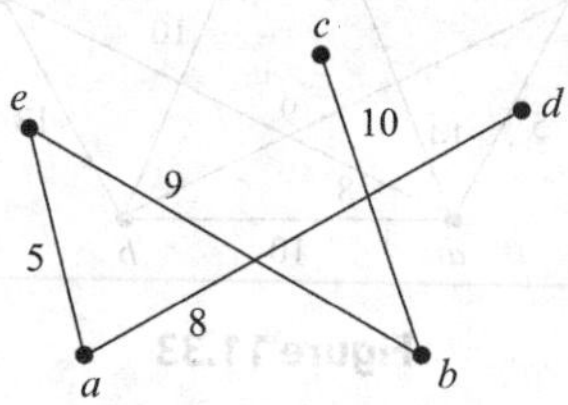

Figure 11.34

The following table summarizes the steps indicating that minimal spanning tree has 32 as its weight and so the minimum cost of building the pipelines is 32 million rupees.

Edge	Weight	Decision	Spanning Tree	Number of Edges	Weight of T	Done
{a, e}	5	Select		1	5	No
{a, d}	8	Select	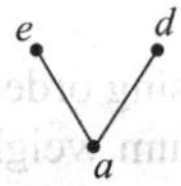	2	13	No
{e, d}	8	Reject	–	2	13	No
{b, e}	9	Select	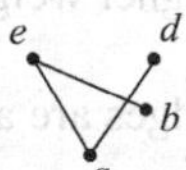	3	22	No
{a, b}	10	Reject	–	3	22	No
{b, c}	10	Select	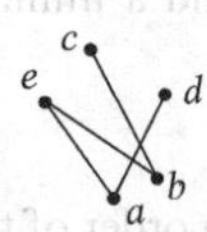	4	32	Yes

Example 2 Use Kruskal's algorithm to determine the road network of minimal cost for eight cities in Fig. 11.35 in thousands of rupees.

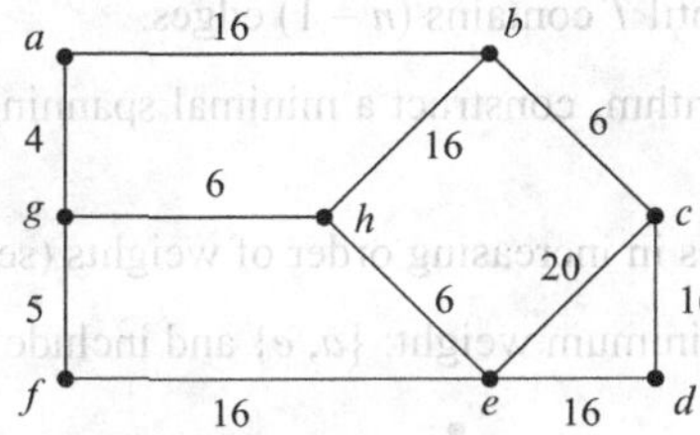

Figure 11.35

Solution:

Step 1: The arrangement of edges in increasing order of their weights is as follows:

Weight	4	5	6	6	6	16	16	16	16	16	20
Edge	(a, g)	(g, f)	(h, g)	(b, c)	(e, h)	(a, b)	(b, h)	(c, d)	(e, d)	(f, e)	(e, c)

Step 2: Choose edge (a, g) since it has smallest weight and include in T.

Step 3: An edge with next the smallest weight is (g, f) since it does not form a cycle with the edge in T and hence include it in T.

Step 4: An edge with next the smallest weight is (g, h), which does not form a cycle, and hence include in T.

Step 5: An edge (h, e) is included in T as it is next the smallest weight not forming a cycle.

Step 6: Add edges (a, b) and (b, c) to T as these do not form a cycle.

Step 7: Add edge (c, d) to T as it does not form a cycle.

Having selected seven edges, all the vertices have been taken into account. Fig. 11.36 shows the resulting spanning tree.

The minimal cost $= 5 + 4 + 16 + 6 + 16 + 6 + 6 = 59$ thousand rupees.

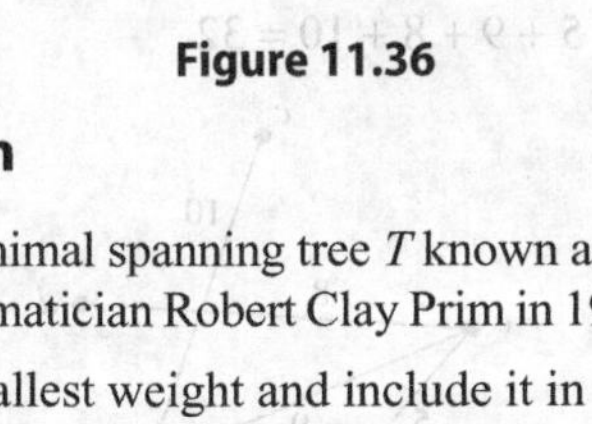

Figure 11.36

11.7.2 Prim's Algorithm

Another method for finding a minimal spanning tree T known as Prim's algorithm was discovered by American engineer and mathematician Robert Clay Prim in 1957. It requires the following steps:

Step 1: Choose an edge with smallest weight and include it in T.

Step 2: Select an edge of least weight which is incident with a vertex of an edge in T.

Step 3: If the edge selected in Step 2 does not create a cycle with the edges in *T*, include it in *T*, otherwise discard it.

Step 4: Repeat Steps 2 and 3 until *T* contains $(n - 1)$ edges.

Example 1 Using Prim's algorithm, construct a minimal spanning tree for the graph shown in Fig. 11.33.

Solution: Arrange all the edges in increasing order of weights (see table appearing on Pg. 23).

Step 1: Choose an edge with minimum weight: $\{a, e\}$ and include in *T*.

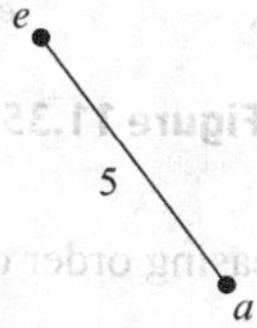

Step 2: Select edge with minimum weight incident with *a* or *e*. Edges $\{e, d\}$ and $\{a, d\}$ are two candidates. Both of these do not form a cycle. Let us choose one of them, say $\{e, d\}$, and so add it to *T*.

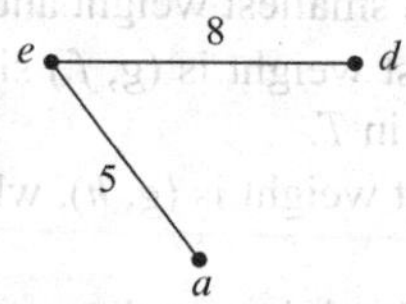

Step 3: Now look for the edge incident with a vertex in *T*: $\{a, d\}$.

Step 4: Edge $\{a, d\}$ creates a cycle and hence reject it.

Step 5: Edge $\{e, b\}$ is the next edge with minimum weight and hence add it to *T*.

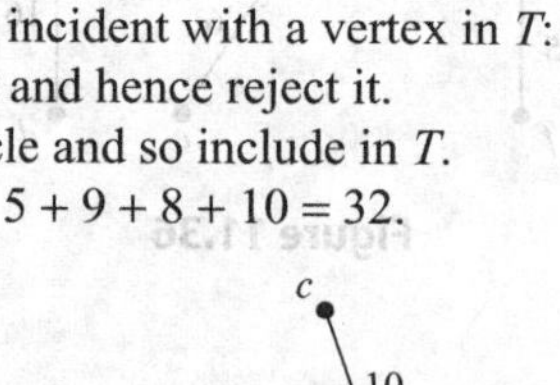

Step 6: Look for the lightest edge incident with a vertex in *T*: $\{a, b\}$ and $\{b, c\}$.

Step 7: Edge $\{a, b\}$ forms a cycle and hence reject it.

Step 8: $\{b, c\}$ does not form a cycle and so include in *T*.

The weight of spanning tree is $5 + 9 + 8 + 10 = 32$.

Table below summarizes the steps required in the formation of minimal spanning tree by Prim's algorithm. Arrangement of edges with increasing weights is given.

Edge	Weight	Decision	Spanning Tree	Weight of Tree	Number of Edges	Done
{e, a}	5	Select		5	1	No
{e, d}	8	Select	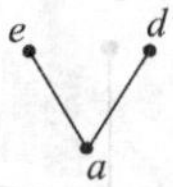	13	2	No
{a, d}	10	Reject	–	13	2	No
{e, b}	9	Select	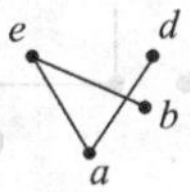	22	3	No
{a, b}	10	Reject	–	22	3	No
{b, c}	10	Select	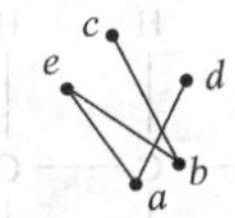	32	4	Yes

Example 2 Apply Prim's algorithm to find minimal cost for network as shown in Fig. 11.35.

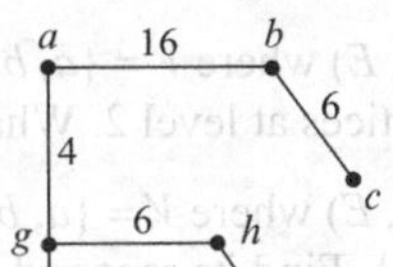

Figure 11.37

Step 1: The edges in increasing order of weights are given in table appearing on Pg. 11.25.

Step 2: An edge with minimum weight is $\{a, g\}$ and so include in T.

Step 3: Edge $\{g, f\}$ with next minimum weight is incident at g and does not form a cycle and so include in T.

Step 4: Edge $\{g, h\}$ is another edge with next minimum weight and does not form a cycle, and hence include in T.

Step 5: The next edge incident of h is $\{h, e\}$ and incident at e is $\{e, d\}$. These do not form a cycle and so included in T.

Step 6: The edge $\{f, e\}$ incident at f forms a cycle and hence rejected.

Step 7: The edge $\{a, b\}$ is incident at a and edge $\{b, c\}$ is incident at b. These are included in T as they do not form a cycle.

Hence, Fig. 11.37 is the required minimal spanning tree as all vertices have been taken into account. The minimal cost = 6 + 16 + 4 + 5 + 6 + 6 + 16 = 59 thousand rupees.

Note 11.5 Kruskal's and Prim's algorithm present fine paradigms of a greedy algorithm. Although greedy algorithms may not always yield optimal solution, but they do in these cases.

EXERCISES

1. Determine if each of the following graphs is a tree.

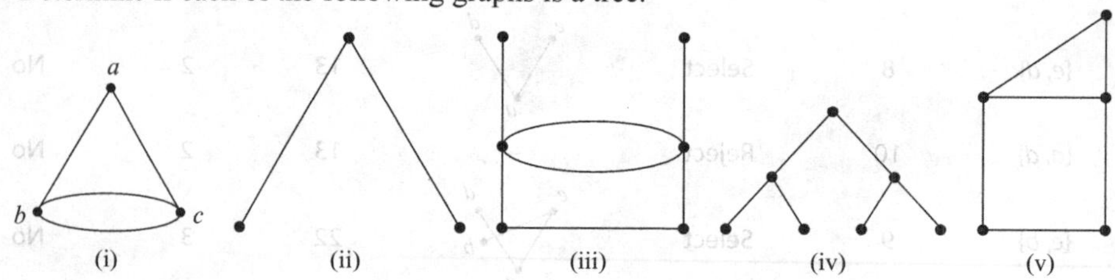

2. Is the following graph a tree?

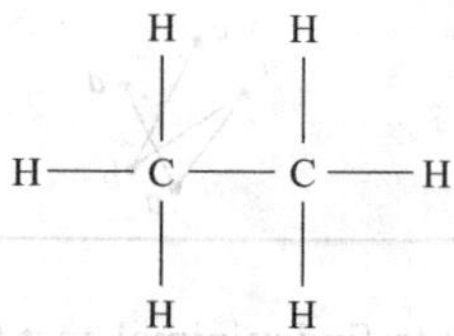

3. Draw the tree defined by $T = (V, E)$ where $V = \{a, b, c, d, e, f, g\}$ and $E = \{(a, b), (a, c), (b, d), (b, e), (c, f), (c, g)\}$. List vertices at level 2. What is its root?

4. Draw the tree defined by $T = (V, E)$ where $V = \{a, b, c, d, e, f, g, h\}$ and $E = \{(a, b), (a, c), (a, d), (b, e), (e, g), (d, f), (d, h)\}$. Find its root and siblings of d. Also find the tree $T(d)$.

5. Let $V = \{1, 2, 3, 4, 5, 6\}$. Determine if T is a tree. If so, find the root.
   (i) $T = \{(1, 4), (1, 5), (2, 3), (2, 6)\}$
   (ii) $T = \{(1, 2), (1, 3), (2, 4), (2, 5), (3, 6)\}$
   (iii) $T = \{(4, 1), (4, 3), (6, 2), (5, 6), (5, 4)\}$
   Where T is the set of edges (u, v).

6. Consider the rooted tree (T, v_0) shown below:

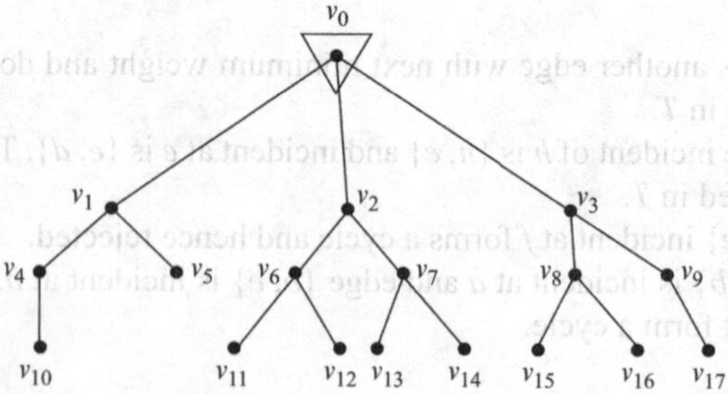

(i) List all 2nd-level vertices.
(ii) List all leaves.
(iii) List the siblings of v_6 and v_7.
(iv) Who are the descendants of v_3 and ancestors of v_{15}.
(v) Find $T(v_1)$ and $T(v_2)$.

7. What is the maximum height for a tree on $V = \{a, b, c, d\}$?

8. Find the maximum height for a complete binary tree on $V = \{a, b, c, d, e, f\}$.

9. Prove that the maximum number of vertices in a binary tree of height k is $2^{k+1} - 1$.

10. Can there be a tree with degree sequence $\{1, 1, 2, 2, 3, 3\}$.

11. Show that there is no tree with degree sequence.
(i) $\{1, 1, 2, 2, 2, 2, 3, 3, 3, 3, 3\}$
(ii) $\{1, 1, 1, 1, 1, 2, 3, 4, 4\}$

12. Construct the tree for the following algebraic expressions:
(i) $((7 + x) - 5x) - (x + 3)$
(ii) $(5 + 3)(5 - 3)/(2 + x)$
(iii) $[10 + (15 - 6)] + [x - (y + 1)]$

13. Find the result of performing preorder, inorder and postorder path traversals of the tree.

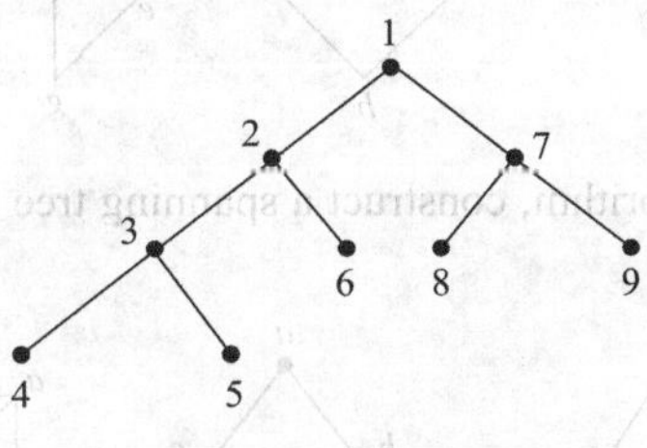

14. Find the result of performing preorder and inorder path traversals of the tree.

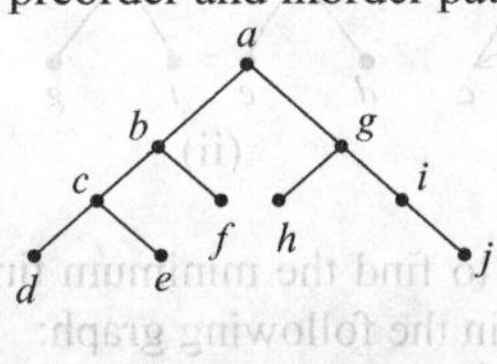

15. Find the result of performing postorder path traversal (search) for the tree.

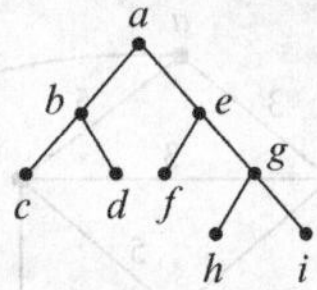

16. Write the algebraic expression

$$E = \{[(a + b) \div c] * (d + e)\} - [f - (g * h)]$$

as a tree and then express the result in operator prefix notation.

17. Represent the following expression by a tree:

$$2 * (5 + 7)/4 - 1$$

18. Evaluate the following expression represented by the binary tree:

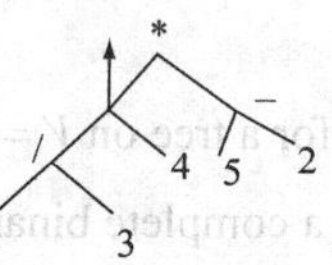

19. Evaluate the prefix expression – ↑*/3 + 57238.

20. Find all the spanning trees of the graph.

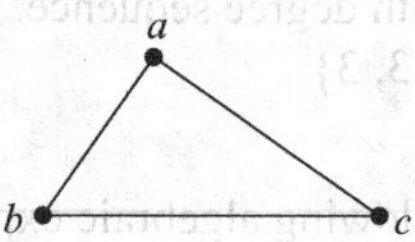

21. Find two spanning trees of the graph, which involves the edge $\{a, b\}$.

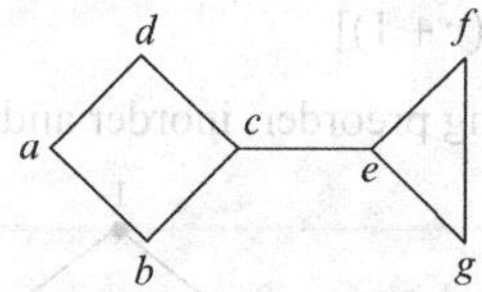

22. Using Krushkal's algorithm, construct a spanning tree for the following graphs starting at vertex a:

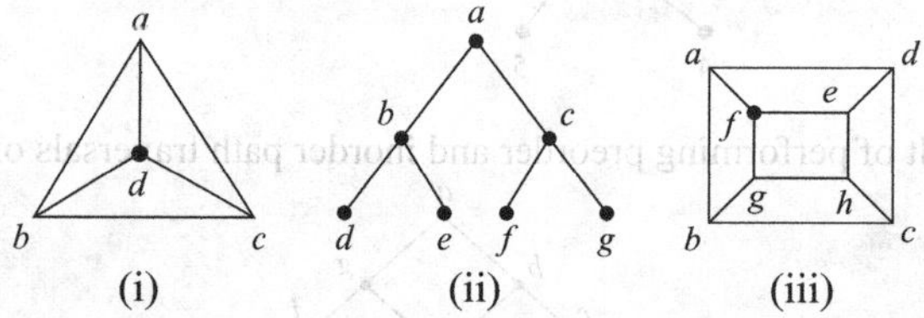

23. Use Krushkal's algorithm to find the minimum time (in hour) for travelling five cities whose road map is shown in the following graph:

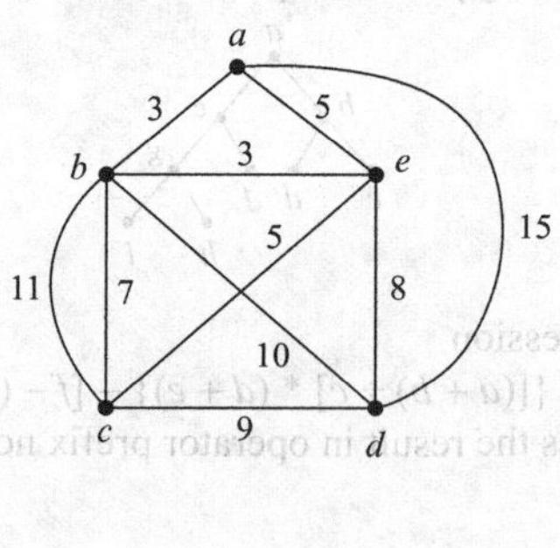

24. Find a minimal spanning tree for the graph using Prim's algorithm.

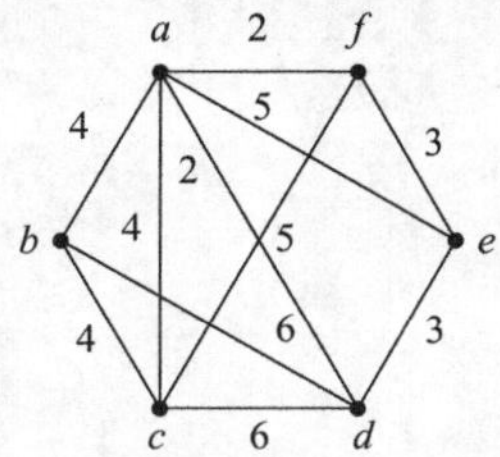

25. Using Prim's algorithm find a minimal spanning tree for the graph.

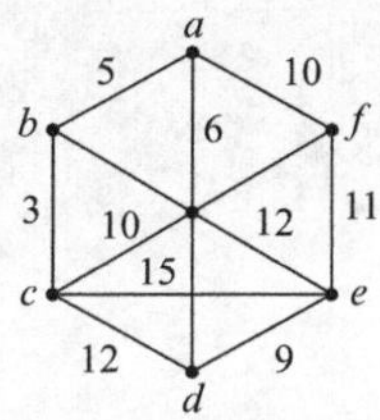

12

Vector Spaces

LEARNING OBJECTIVES

After reading this chapter, you will be conversant with:

- Introduction
- Definition
- Vector Subspace
- Calculus of Subspaces
- Linear Dependence and Indepence of Vectors
- Dimension and Basis
- Linear Transformation
- Normed and Inner Product Spaces

12.1 INTRODUCTION

Algebraic structures such as semigroups, monoids, rings, and fields provide a powerful tool in the theory as well as applications to computer science. Another algebraic structure essentially required for thorough understanding of graph theory-related problems and their solutions is the concept of Vector spaces.

12.2 DEFINITION

A vector space consists of the following:

1. A field $(F, +, \cdot)$ whose elements are called scalars.
2. A set V of objects called vectors on which is defined a binary operation (+) called vector addition and an operation of scalar multiplication, that is, to every pair of elements a and $\underset{\sim}{\alpha}, a \in F$ *and* $\underset{\sim}{\alpha} \in V$, there is a unique element $a\,\underset{\sim}{\alpha} \in V$.
3. $(V, +)$ is an abelian group. That is
   (i) $\underset{\sim}{\alpha} + \underset{\sim}{\beta} = \underset{\sim}{\beta} + \underset{\sim}{\alpha}$ (addition is commutative)

(ii) $(\underset{\sim}{\alpha}+\underset{\sim}{\beta})+\underset{\sim}{\gamma}=\underset{\sim}{\alpha}+(\underset{\sim}{\beta}+\underset{\sim}{\gamma})$ (addition is associative)

(iii) There exists a unique vector $\underset{\sim}{0}$ in V called zero vector such that

$$\underset{\sim}{0}+\underset{\sim}{\alpha}=\underset{\sim}{\alpha}+\underset{\sim}{0}=\underset{\sim}{\alpha} \qquad \forall \alpha \in V$$

(iv) For each vector $\underset{\sim}{\alpha}$ in V, there is a unique vector $-\underset{\sim}{\alpha}$ in V such that

$$-\underset{\sim}{\alpha}+\underset{\sim}{\alpha}=\underset{\sim}{\alpha}+(-\underset{\sim}{\alpha})=\underset{\sim}{0}$$

4. The following four laws of scalar multiplication are satisfied:

(i) $a\,(\underset{\sim}{\alpha}+\underset{\sim}{\beta})=a\,\underset{\sim}{\alpha}+a\,\underset{\sim}{\beta},\ \forall a\in F, \alpha, \beta \in V$ (multiplication by a scalar is distributive with respect to vector addition)

(ii) $(a+b)\,\underset{\sim}{\alpha}=a\,\underset{\sim}{\alpha}+b\,\underset{\sim}{\alpha},\ \forall a, b\in F, \alpha\in V,$ (multiplication by a vector is distributive with respect to scalar addition)

(iii) $a\,(b\underset{\sim}{\alpha})=(ab)\underset{\sim}{\alpha}$ (multiplication by scalars is associative)

(iv) $1\underset{\sim}{\alpha}=\underset{\sim}{\alpha}$, where 1 is the unity element of F and $\alpha \in V$.

A vector space V over a field F is denoted by $V(F)$. A vector space is also called a linear space.

Note 12.1

1. A vector space is said to be **real vector space** when the scalar field F is the field of real numbers. If the scalar field F is the field of complex numbers, it is called a **complex vector space**.
2. The name 'vector' applied to the elements of the set V is a simply a matter of convenience. The elements of V are not vectors in the real sense of the term as these can be matrices, polynomials and real-valued functions, etc. The name is motivated as the axioms are abstractions of the properties of three-dimensional vectors.

Example 1 Let V be the set of all ordered n-tuples over the field F of real numbers, that is,

$$V=\{(a_1, a_2, a_3, \ldots, a_i, \ldots, a_n), a_i \in R\}$$

Define addition and scalar multiplication on V by

$$\underset{\sim}{\alpha}+\underset{\sim}{\beta} = (a_1, a_2, a_3, \ldots, a_i, \ldots, a_n)+(b_1, b_2, b_3, \ldots, b_i, \ldots, b_n)$$

or

$$\underset{\sim}{\alpha}+\underset{\sim}{\beta} = (a_1+b_1, a_2+b_2, \ldots, a_i+b_i, \ldots, a_n+b_n) = (a_i+b_i)$$

and

$$c\ \underset{\sim}{\alpha} = (ca_1, ca_2, \ldots, ca_i, \ldots, ca_n) = (ca_i)\ c\in F,\ \alpha \in V$$

Now we shall see that V is a vector space over F.

Let the vectors $\underset{\sim}{\alpha}+\underset{\sim}{\beta}$ be denoted by (a_i) and (b_i), respectively.

1. $(V, +)$ is an abelian group.

(i) Addition is commutative.

$$\underset{\sim}{\alpha}+\underset{\sim}{\beta}=(a_i)+(b_i)=(a_i+b_i)=(b_i+a_i)=(b_i)+(a_i)=\underset{\sim}{\beta}+\underset{\sim}{\alpha}$$

(ii) Addition is associative

$$(\underset{\sim}{\alpha}+\underset{\sim}{\beta})+\underset{\sim}{\gamma}=(a_i+b_i)+(c_i)=(a_i+b_i+c_i)=(a_i)+(b_i+c_i)=\underset{\sim}{\alpha}+(\underset{\sim}{\beta}+\underset{\sim}{\gamma})$$

(iii) Existence of identity in V

There exists a zero vector, that is, (0, 0,..., 0) = (0) = $\underset{\sim}{0}$, such that

$$\underset{\sim}{\alpha}+\underset{\sim}{0}=(a_i)+(0)=(a_i+0)=(0+a_i)=(0)+(a_i)=\underset{\sim}{0}+\underset{\sim}{\alpha}=\underset{\sim}{\alpha}$$

(iv) Existence of additive inverse.

To each vector $\underset{\sim}{\alpha}=(a_i)$, there exists a vector $-\underset{\sim}{\alpha}=(-a_i)\in V$

such that $\underset{\sim}{\alpha}+(-\underset{\sim}{\alpha})=(a_i)+(-a_i)=(a_i-a_i)=(0)=\underset{\sim}{0}$ and

$$-\underset{\sim}{\alpha}+\underset{\sim}{\alpha}=(-a_i)+(a_i)=(-a_i+a_i)=(0)=\underset{\sim}{0}$$

2. To each real number $c\in R$ and a vector $\underset{\sim}{\alpha}=(a_i)$, there exists a unique vector $c\underset{\sim}{\alpha}=(ca_i)\in V$, which satisfies the following laws of scalar multiplication.

(i) $c_1(\underset{\sim}{\alpha}+\underset{\sim}{\beta})=c_1(a_i+b_i)=(c_1a_i+c_1b_i)=(c_1a_i)+(c_1b_i)=c_1\underset{\sim}{\alpha}+c_1\underset{\sim}{\beta},\ c_1\in F$

(ii) $(c_1+c_2)\underset{\sim}{\alpha}=(c_1+c_2)(a_i)=((c_1+c_2)a_i)=(c_1a_i+c_2a_i)=(c_1a_i)+(c_2a_i),=c_1\underset{\sim}{\alpha}+c_2\underset{\sim}{\alpha},\ c_1,c_2\in F$

(iii) $(c_1c_2)\underset{\sim}{\alpha}=(c_1c_2)(a_i)=(c_1c_2a_i)=c_1(c_2a_i)=c_1(c_2\underset{\sim}{\alpha})$

(iv) 1 is the unity element of the field of real numbers and $\underset{\sim}{\alpha}=(a_i)\in V$, and hence, $1\underset{\sim}{\alpha}=1(a_i)=(1a_i)=(a_i)=\underset{\sim}{\alpha}$.

Hence, V is a vector space over F, that is, field of real numbers R. It is often denoted by R^n.

Note 12.2 We can show in a similar manner that n-tuples of complex numbers form a vector space over the field of complex number, which is denoted by C^n.

Example 2 Show that the set V of all $m\times n$ matrices with elements as real numbers is a vector space over the field of real numbers $(R,+,\cdot)$ with respect to operations of matrix addition and multiplication of a matrix by a real number.

Let $A=[a_{ij}]$, $B=[b_{ij}]$ and $C=[c_{ij}]$ be matrices of order $m\times n$ with their elements as real numbers, that is, $a_{ij}, b_{ij}, c_{ij}\in R$.

(1) $(V,+)$ is an abelian group.

(i) Addition is commutative

$$A+B=(a_{ij}+b_{ij})=[a_{ij}+b_{ij}]=[b_{ij}+a_{ij}]=[b_{ij}]+[a_{ij}]=B+A$$

(ii) Addition is associative

$$(A+B)+C=(a_{ij}+b_{ij})+(c_{ij})=(a_{ij}+b_{ij}+c_{ij})=(a_{ij})+(b_{ij}+c_{ij})=A+(B+C)$$

(iii) Existence of identity in V

There exists a zero matrix of order $m\times n$ denoted by 0, such that

$$A+\mathbf{0}=[a_{ij}]+[0]=[a_{ij}+0]=[a_{ij}]=A$$
$$\mathbf{0}+A=[0]+[a_{ij}]=[0+a_{ij}]=A$$

Thus, $A+\mathbf{0}=\mathbf{0}+A=A$

(iv) Existence of additive inverse

For each matrix $A=[a_{ij}]$, there exists a matrix $-A=[-a_{ij}]$ such that

$$A+(-A)=[a_{ij}]+[-a_{ij}]=[a_{ij}-a_{ij}]=[0]=\mathbf{0}$$
$$-A+A=[-a_{ij}]+[a_{ij}]=[-a_{ij}+a_{ij}]=[0]=\mathbf{0}$$

Thus, $A+(-A)=(-A)+A=\mathbf{0}$

(2) To each real number $c \in R$ and a matrix $A = [a_{ij}]$, there exists a unique matrix $cA = [ca_{ij}] \in V$, which satisfies the following laws of scalar multiplication of a matrix.

(i) $c_1(A+B) = c_1[a_{ij} + b_{ij}] = [c_1(a_{ij} + b_{ij})] = [c_1a_{ij} + c_1b_{ij}] = [c_1a_{ij}] + [c_1b_{ij}] = c_1A + c_1B,\ c_1 \in R$

(ii) $(c_1 + c_2)\,A = (c_1 + c_2)[a_{ij}] = [(c_1 + c_2)a_{ij}] = [c_1a_{ij} + c_2a_{ij}] = [c_1a_{ij}] + [c_2a_{ij}] = c_1A + c_2A,$ $c_1, c_2 \in R$

(iii) $(c_1c_2)A = (c_1c_2)[a_{ij}] = [(c_1c_2)a_{ij}] = [c_1(c_2a_{ij})] = c_1[c_2a_{ij}] = c_1(c_2A)$

(iv) 1 is the unity element of the field of real numbers R, and $A = [a_{ij}]$ is any element in V, then

$$1A = 1[a_{ij}] = [1a_{ij}] = [a_{ij}] = A$$

As all the postulates of a vector space are satisfied, V is a vector space over the field or real numbers.

Example 3 Let V be the set of all polynomials in x with real coefficients of degree n or less, then V is a vector space over the field of real numbers under polynomial addition and multiplication by a real number.

If $p(x)$ and $q(x)$ are polynomials in x of degree n or less $\in V$, then $p(x) + q(x)$ is also a polynomial in x of degree n or less $\in V$. Further, if c is a real number, then $cp(x)$ is also a polynomial of degree n or less and hence belongs to V.

(1) $(V, +)$ is an abelian group

$$\text{Let } p(x) = \sum_i a_ix^i, q(x) = \sum_i b_ix^i, r(x) = \sum_i c_ix^i \in V$$

(i) Addition is commutative

$$\begin{aligned} p(x) + q(x) &= \sum_i a_ix^i + \sum_i b_ix^i = \sum_i (a_i + b_i)x^i \\ &= \sum_i (b_i + a_i)x^i = \sum_i b_ix^i + \sum_i a_ix^i = q(x) + p(x) \end{aligned}$$

(ii) Addition is associative

$$\begin{aligned} \{p(x) + q(x)\} + r(x) &= \sum_i (a_i + b_i)x^i + \sum_i c_ix^i \\ &= \sum_i \{(a_i + b_i) + c_i\}x^i, \qquad a_i, b_i, c_i \in R \\ &= \sum_i (a_i + b_i + c_i)x^i \\ &= \sum_i \{a_i + (b_i + c_i)\}x^i \\ &= \sum_i a_ix^i + \sum_i (b_i + x_i)x^i = p(x) + \{q(x) + r(x)\} \end{aligned}$$

(iii) There exists in V a zero polynomial denoted by $\mathbf{0}$

i.e., $$\mathbf{0} = 0 + 0x + 0x^2 + \ldots + 0x^n = \sum_i 0.x^i$$

such that $p(x) + \mathbf{0} = \sum_i a_i x^i + \sum_i 0x^i = \sum_i (a_i + 0)x^i = \sum_i a_i x^i = p(x)$

and $\mathbf{0} + p(x) = \sum_i 0x^i + \sum_i a_i x^i = \sum_i (0 + ai)x^i = \sum_i a_i x^i = p(x)$

Thus, we have $p(x) + \mathbf{0} = \mathbf{0} + p(x) = p(x)$

(iv) For each polynomial $p(x) = \sum_i a_i x^i$, there exists a polynomial $-p(x) = \sum_i (-a_i)x^i$ such that

$$p(x) + \{-p(x)\} = \{-p(x)\} + p(x) = \sum_i (a_i - a_i)x^i = \sum_i 0x^i = \mathbf{0}$$

(2) The multiplication of matrix by a real number satisfies the following postulates:

(i) $c_1[p(x) + q(x)] = c_1 \sum_i (a_i + b_i)x^i = \sum_i c_1(a_i + b_i)x^i = \sum_i (c_1 a_i + c_1 b_i)x^i$

$$= \sum_i c_1 a_i x^i + \sum c_1 b_i x^i = c_1 p(x) + c_1 q(x)$$

(ii) $(c_1 + c_2)p(x)$

$$= (c_1 + c_2) \sum_i a_i x^i = \sum_i (c_1 + c_2) a_i x^i = \sum_i (c_1 a_i + c_2 a_i)x^i$$

$$= \sum_i c_1 a_i x^i + \sum_i c_2 a_i x^i = c_1 p(x) + c_2 p(x)$$

(iii) $(c_1 c_2)\, p(x) = (c_1 c_2) \sum_i a_i x^i = \sum_i c_1 c_2 a_i x^i = c_1 \sum_i c_2 a_i x^i = c_1 \left(c_2 p(x^i)\right)$

(iv) 1 is the unity element of the field of real numbers and $p(x) = \sum_i a_i x^i \in V$, then $1p(x) = 1 \sum_i a_i x^i = \sum_i 1 a_i x^i = a_i x^i = p(x)$

As all the postulates of a vector space are satisfied, V is a vector space over R.

Example 4 Let R be the field of real numbers and V the set of all real functions f over the domain R. Show that V is a vector space over R.

Let f and g be two real functions $\in V$, then $f + g$ is also a real function over $R \in V$.

(1) $(V, +)$ is an abelian group

(i) Addition is commutative

$$(f + g)\, x = f(x) + g(x) \qquad f, g \in V, x \in R$$
$$= g(x) + f(x) = (g + f)x$$
$$\therefore\ f + g = g + f$$

(ii) Addition is associative.
Let f, g, and h be real functions in V, then

$$
\begin{aligned}
[(f+g)+h]\,x &= (f+g)x + h(x) \\
&= [f(x) + g(x)] + h(x) \\
&= f(x) + [g(x) + h(x)] \\
&= [f + (g+h)]x
\end{aligned}
$$

$$\therefore (f+g)+h = f+(g+h)$$

(iii) There exists a zero function V (denoted by $\mathbf{0}$), which assigns to each real number the number 0 in R such that

$$f + \mathbf{0} = \mathbf{0} + f = f$$

as $$(f+\mathbf{0})x = f(x) + 0x = f(x) + \mathbf{0} = f(x)$$

$$\therefore f + \mathbf{0} = f$$

and $$(\mathbf{0} + f)x = 0x + f(x) = \mathbf{0} + f(x) = f(x)$$

$$\therefore \mathbf{0} + f = f + \mathbf{0} = f$$

(iv) For each function $f \in V$, there exists a function $f \in V$ such that

$$[f + (-f)]\,x = f(x) + (-f)x = f(x) - f(x) = 0 = 0x \text{ and}$$

$$(-f) + f]x = (-f)x + f(x) = -f(x) + f(0) = 0 = 0x.$$

Thus,

$$f + (-f) = (-f) + f = \mathbf{0}$$

(2) Since multiplication by real numbers is distributive and associative, we have

(i) $c_1(f+g)x = c_1[f(x) + g(x)] = c_1f(x) + c_1g(x) = (c_1f + c_1g)x$

$$\therefore c_1(f+g) = c_1f + c_1g$$

(ii) $(c_1 + c_2)f(x) = c_1f(x) + c_2f(x) = (c_1f + c_2f)x$

$$\therefore (c_1 + c_2)f = c_1f + c_2f$$

(iii) $(c_1c_2f)x = c_1c_2f(x) = c_1[c_2f(x)] = c_1(c_2f)x$

$$\therefore (c_1c_2)f = c_1(c_2f)$$

(iv) 1 is the unity element of the field of real numbers and $(1f)x = 1f(x) = f(x)$

$$\therefore 1f = f$$

Hence, V is a vector space over the field of real numbers R.

Note 12.3 Let V be the set of real-valued functions over the domain [0, 1], then V is a vector space over R with respect to addition and scalar multiplication of functions.

12.2.1 Some Simple Properties of a Vector Space

THEOREM 12.1 Let $V(F)$ be a vector space and $\underset{\sim}{0}$ be the zero vector of V. We then have

(1) $a\underset{\sim}{0}=\underset{\sim}{0}, \quad a\in F$

(2) $0\underset{\sim}{\alpha}=\underset{\sim}{0}, \quad \underset{\sim}{\alpha}\in V$

(3) $a(-\underset{\sim}{\alpha})=-(a\underset{\sim}{\alpha}), \quad a\in F,\underset{\sim}{\alpha}\in V$

(4) $a(\underset{\sim}{\alpha}-\underset{\sim}{\beta})=a\underset{\sim}{\alpha}-a\underset{\sim}{\beta}, \quad a\in F,\underset{\sim}{\alpha},\underset{\sim}{\beta}\in V$

(5) $a\,a=\underset{\sim}{0}\Rightarrow a=0$ or $\underset{\sim}{\alpha}=\underset{\sim}{0}$

Proof:

(1) $\underset{\sim}{0}+\underset{\sim}{0}=\underset{\sim}{0}, \quad \underset{\sim}{0}\in V$

Therefore,

$$a(\underset{\sim}{0}+\underset{\sim}{0})=a\underset{\sim}{0}$$

$$a\underset{\sim}{0}+a\underset{\sim}{0}=a\underset{\sim}{0}+\underset{\sim}{0}\Rightarrow a\underset{\sim}{0}=\underset{\sim}{0}$$

By left cancellation law as V is an abelian group w. r. t. addition.

(2) $0+0=0, \quad 0\in F$

Therefore,

$$(0+0)\underset{\sim}{\alpha}=0\underset{\sim}{\alpha}$$

$$0\underset{\sim}{\alpha}+0\underset{\sim}{\alpha}=0\underset{\sim}{\alpha}+\underset{\sim}{0}\Rightarrow 0\underset{\sim}{\alpha}=\underset{\sim}{0}$$

By left cancellation law as V is an abelian graph w. r. t. addition.

(3) Now

$$a[\underset{\sim}{\alpha}+(-\underset{\sim}{\alpha})]=a\underset{\sim}{\alpha}+a(-\underset{\sim}{\alpha})$$

$$a\underset{\sim}{0}=a\underset{\sim}{\alpha}+a(-\underset{\sim}{\alpha})$$

$$\underset{\sim}{0}=a\underset{\sim}{\alpha}+a(-\underset{\sim}{\alpha})$$

which shows that $a(-\underset{\sim}{\alpha})$ is the additive inverse of $a\underset{\sim}{\alpha}$.

i.e.,

$$a(-\underset{\sim}{\alpha})=-(a\underset{\sim}{\alpha})$$

(4) $a(\underset{\sim}{\alpha}-\underset{\sim}{\beta})+a\underset{\sim}{\beta}=a[\underset{\sim}{\alpha}-\underset{\sim}{\beta}+\underset{\sim}{\beta}]=a(\underset{\sim}{\alpha}+\underset{\sim}{0})$

$$=a\underset{\sim}{\alpha}+a\underset{\sim}{0}$$

$$=a\underset{\sim}{\alpha}+\underset{\sim}{0}$$

$$=a\underset{\sim}{\alpha}$$

Hence,

$$a(\underset{\sim}{\alpha}-\underset{\sim}{\beta})=a\underset{\sim}{\alpha}-a\underset{\sim}{\beta}$$

(5) Let $a\underset{\sim}{\alpha} = \underset{\sim}{0}$ and $a \neq 0$.

Then a^{-1} will exist as a is a non-zero element in F.
Hence,

$$a^{-1}(a\underset{\sim}{\alpha}) = a^{-1}\underset{\sim}{0}$$

$$(a^{-1}a)\alpha = \underset{\sim}{0}$$

$$1\alpha = \underset{\sim}{0} \Rightarrow \alpha = \underset{\sim}{0}$$

Again, let $a\underset{\sim}{\alpha} = \underset{\sim}{0}$ and $\underset{\sim}{\alpha} \neq \underset{\sim}{0}$
We are required to prove that $a = 0$.
Let us assume that $a \neq 0$, then a^{-1} will exist.

$$\therefore a\underset{\sim}{\alpha} = \underset{\sim}{0} \Rightarrow a^{-1}(a\underset{\sim}{\alpha}) = a^{-1}\underset{\sim}{0} = \underset{\sim}{0}$$

$$\Rightarrow (a^{-1}a)\underset{\sim}{\alpha} = \underset{\sim}{0}$$

$$\Rightarrow 1\underset{\sim}{\alpha} = \underset{\sim}{0}$$

$$\Rightarrow \underset{\sim}{\alpha} = \underset{\sim}{0}$$

which contradicts the assumption $\underset{\sim}{\alpha} \neq \underset{\sim}{0}$ Hence, $a = 0$.

THEOREM 12.2 In a vector space $V(F)$

(1) $\underset{\sim}{x} + \underset{\sim}{y} = \underset{\sim}{x} + \underset{\sim}{z} \Rightarrow \underset{\sim}{y} = \underset{\sim}{z}, \quad \underset{\sim}{x}, \underset{\sim}{y}, \underset{\sim}{z} \in V$

(2) $\underset{\sim}{y} + \underset{\sim}{x} = \underset{\sim}{z} + \underset{\sim}{x} \Rightarrow \underset{\sim}{y} = \underset{\sim}{z}$

(3) $a\underset{\sim}{x} = b\underset{\sim}{x} \Rightarrow a = b$, $a, b \in F$, and $\underset{\sim}{x}$ is a non-zero element in V.

Proof:

(1) Given $\underset{\sim}{x} + \underset{\sim}{y} = \underset{\sim}{x} + \underset{\sim}{z} \qquad \underset{\sim}{x}, \underset{\sim}{y}, \underset{\sim}{z} \in V$
Adding $-\underset{\sim}{x}$ to both sides, we have

$$-\underset{\sim}{x} + (\underset{\sim}{x} + \underset{\sim}{y}) = -\underset{\sim}{x} + (\underset{\sim}{x} + \underset{\sim}{z})$$

$$(-\underset{\sim}{x} + \underset{\sim}{x}) + \underset{\sim}{y} = (-\underset{\sim}{x} + \underset{\sim}{x}) + \underset{\sim}{z}$$

$$\underset{\sim}{0} + \underset{\sim}{y} = \underset{\sim}{0} + \underset{\sim}{z} \Rightarrow \underset{\sim}{y} = \underset{\sim}{z}$$

(2) Given $\underset{\sim}{y} + \underset{\sim}{x} = \underset{\sim}{z} + \underset{\sim}{x} \qquad \underset{\sim}{x}, \underset{\sim}{y}, \underset{\sim}{z} \in V$
Adding $-\underset{\sim}{x}$ to both sides, we have

$$(\underset{\sim}{y} + \underset{\sim}{x}) + (-\underset{\sim}{x}) = \underset{\sim}{z} + \underset{\sim}{x} + (-\underset{\sim}{x})$$

$$\underset{\sim}{y} + [\underset{\sim}{x} + (-\underset{\sim}{x}] = \underset{\sim}{z} + [\underset{\sim}{x} + (-\underset{\sim}{x})]$$

$$\underset{\sim}{y} + \underset{\sim}{0} = \underset{\sim}{z} + \underset{\sim}{0} \Rightarrow \underset{\sim}{y} = \underset{\sim}{z}$$

(3) Given $a\underset{\sim}{x} = b\underset{\sim}{x} \Rightarrow a\underset{\sim}{x} - b\underset{\sim}{x} = \underset{\sim}{0}$

$$\Rightarrow (a-b)\underset{\sim}{x} = \underset{\sim}{0}$$

$$\Rightarrow a - b = 0 \text{ as } \underset{\sim}{x} \neq \underset{\sim}{0}$$

$\therefore a = b$

12.3 VECTOR SUBSPACE

If V is a vector space over a field F, it is often possible to find a subset S of V, which itself is a vector space under the operations of V and F.

Definition 12.1 A non-empty subset W of a vector space $V(F)$ is called a subspace of V if W is itself a vector space over F with respect to the addition and scalar multiplication of V.

By definition, every vector space $V(F)$ has at least two vector subspaces: The vector space V itself and the null space $\{0\}$.

THEOREM 12.3 The necessary and sufficient condition for a non-empty subset W of $V(F)$ to be a subspace is that W is closed under vector addition and scalar multiplication in V.

Proof: Let W be a subspace of $V(F)$, then W itself is a vector space and hence W is closed under vector addition and scalar multiplication in V.

Conversely, let W be a non-empty subset of $V(F)$ closed under vector addition and scalar multiplication, that is,

(1) $\underset{\sim}{x}, \underset{\sim}{y} \in W \Rightarrow \underset{\sim}{x} + \underset{\sim}{y} \in W$

(2) $a \in F, \underset{\sim}{x} \in W \Rightarrow a\underset{\sim}{x} \in W$

Since $W \subset V$, its elements obey associative law of addition.

$$(\underset{\sim}{x} + \underset{\sim}{y}) + \underset{\sim}{z} = \underset{\sim}{x} + (\underset{\sim}{y} + \underset{\sim}{z}), \ \forall \underset{\sim}{x}, \underset{\sim}{y}, \underset{\sim}{z} \in W$$

(3) Now $\underset{\sim}{x} \in W$ and 1 is the unity element of field F, $-1 \in F$

$$-1 \in F, \underset{\sim}{x} \in W \Rightarrow (-1)\underset{\sim}{x} \in W$$

$$\Rightarrow -\underset{\sim}{x} \in W$$

Thus, every element $\underset{\sim}{x} \in W$ has its additive inverse $-\underset{\sim}{x} \in W$.

(4) $\underset{\sim}{x} \in W, -\underset{\sim}{x} \in \Rightarrow x + (-x) \in W$

$$\Rightarrow \underset{\sim}{0} \in W$$

Hence, $\underset{\sim}{0}$ identity of V is also the identity of W. Further, all the elements of W are also the elements of V and hence obey commutativity of addition and so $\underset{\sim}{x} + \underset{\sim}{y} = \underset{\sim}{y} + \underset{\sim}{x} \ \forall \underset{\sim}{x}, \underset{\sim}{y} \in W$

Thus, $(W, +)$ is an abelian group. Further, $a \in F$, $\underset{\sim}{x} \in W \Rightarrow a\underset{\sim}{x} \in W$, scalar multiplication is defined. All the postulates for scalar multiplication hold for W, which hold for V.

Thus, W is a vector subspace of $V(f)$. Consequently, the given conditions are sufficient.

THEOREM 12.4 The necessary and sufficient condition for a non-empty subset W of a vector space $V(F)$ to be a subspace is $a, b \in F,\ \underset{\sim}{\alpha}, \underset{\sim}{\beta} \in W \Rightarrow a\underset{\sim}{\alpha} + b\underset{\sim}{\beta} \in W$

Proof: Let W be a subspace of $V(F)$, then

$$a \in F,\ \underset{\sim}{\alpha} \in W \Rightarrow a\underset{\sim}{\alpha} \in W$$

$$b \in F,\ \underset{\sim}{\beta} \in W \Rightarrow b\underset{\sim}{\beta} \in W$$

Further, $a\underset{\sim}{\alpha} \in W, b\underset{\sim}{\beta} \in W \Rightarrow a\underset{\sim}{\alpha} + b\underset{\sim}{\beta} \in W$ Conversely, let the given condition hold, that is

$$a, b \in F,\ \underset{\sim}{\alpha}, \underset{\sim}{\beta} \in W \Rightarrow a\underset{\sim}{\alpha} + b\underset{\sim}{\beta} \in W$$

Take $a = 1$, $b = 1$, then $\underset{\sim}{\alpha}, \underset{\sim}{\beta} \in W \Rightarrow 1\alpha + 1\beta \in W$

$$\Rightarrow \underset{\sim}{\alpha} + \underset{\sim}{\beta} \in W$$

Hence, W is closed under addition.

Again, take $b = 0$, then $\underset{\sim}{\alpha}, \underset{\sim}{\beta} \in W \Rightarrow a\underset{\sim}{\alpha} + 0\underset{\sim}{\beta} \in W$

$$\Rightarrow a\underset{\sim}{\alpha} \in W$$

Therefore, $a \in F,\ \underset{\sim}{\alpha} \in W \Rightarrow a\underset{\sim}{\alpha} \in W$ and hence W is closed under scalar multiplication. Hence, by Theorem 12.3, W is a subspace of V. Consequently, the given condition is sufficient.

Example 1 The subset W of n-tuples $(a_1, a_2, \ldots, a_{n-1}, 0)$, $a_i \in R$ is a subspace of the vector space of R^n over the field of real numbers.

Solution: Let $\underset{\sim}{\alpha} = (a_1, a_2, \ldots, a_{n-1}, 0)$

and $\qquad \underset{\sim}{\beta} = (b_1, b_2, \ldots, b_{n-1}, 0)$

Then $\qquad a\underset{\sim}{\alpha} + b\underset{\sim}{\beta} = (aa_1 + bb_1, aa_2 + bb_2 + \ldots + aa_{n-1} + bb_{n-1}, 0) \in W$

Hence, W is a subspace.

Example 2 The subset W of n-tuples $(a_1, a_2, \ldots, a_n)$, $a_i \in R$ with a_1 an integer, is *not* a subspace of R^n over the field of real numbers.

Solution: Let $\underset{\sim}{\alpha}$ and $\underset{\sim}{\beta}$ be two vectors with first component equal to an integer.

If $a_1 = p$ and $b_1 = q$, p, and q are integers, then the first component of $\underset{\sim}{\alpha} + \underset{\sim}{\beta}$ is $a_1 + b_1 = p + q$ is an integer and hence $\underset{\sim}{\alpha} + \underset{\sim}{\beta} \in W$. But $a\,\underset{\sim}{\alpha} = (ap, aa_2, \ldots, aa_n)$, $a \in R$.

The first component $= aa_1 = ap$ is not an integer for every real number a, and hence $a\,\underset{\sim}{\alpha} \in W$. Hence, W is not a subspace.

Example 3 Let W be the subset of n-tuples $(x_1, x_2, \ldots, x_n)$, $x_i \in R$ of R^n satisfying

$$a_1x_1 + a_2x_2 + \ldots + a_nx_n = 0, \qquad a_i \in R$$

Prove that W is a subspace of R^n.

Solution: Let $\underset{\sim}{x} = (x_1, x_2, \ldots, x_n)$ and $\underset{\sim}{y} = (y_1, y_2, \ldots, y_n)$ be two such vectors, then

$$a_1x_1 + a_2x_2 + \ldots + a_nx_n = 0$$

and

$$a_1y_1 + a_2y_2 + \ldots + a_ny_n = 0$$

Therefore,

$$a_1(x_1 + y_1) + a_2(x_2 + y_2) + \ldots + a_n(x_n + y_n) = 0$$

which shows that $\underset{\sim}{x} + \underset{\sim}{y} \in W$

Also

$$a_1(cx_1) + a_2(cx_2) + \ldots + a_n(cx_n) = 0$$

by the given condition. This shows that $c\underset{\sim}{x} \in W$.

Therefore, W is a subspace.

Example 4 Show that $W = \{(x_1, x_2, x_3) | x_2 = x_3\}$ is a subspace of R^3.

Solution: Let $\underset{\sim}{x} = (a, b, b)$ and $\underset{\sim}{y} = (c, d, d) \in W$

then $\underset{\sim}{x} + \underset{\sim}{y} = (a + c, b + d, b + d) \in W$, that is, W is closed under addition.

Also, $p\underset{\sim}{x} = (pa, pb, pb)$ where $p \in R$ belongs to W. This shows that W is closed under scalar multiplication. Therefore, W is subspace of R^3.

Example 5 Let W be the set of all polynomials of degree less than or equal to n with the property $p(0) = 0$, where $p(x) \in W$. Show that W is a subspace of $P_n(x)$, that is, all polynomials in x of degree less than or equal to n.

Solution:

(1) If $p(x)$ and $q(x)$ are elements of W, then $(p + q)x = p(0) + q(0) = 0 + 0 = 0$ and hence $p + q \in W$.

(2) If $p(x) \in W$ and c is a scalar, then $cp(0) = c.0 = 0$ and hence $cp(x) \in W$. Therefore, W is a subspace.

Example 6 Let $V = \{(x, y, z): x, y, z \in R\}$, where R is the field of real numbers. Show that if $W = \{(x, 2x, x + 1)\}, x \in R$, then it is not a vector subspace of $V(R)$.

Solution: Let $\underset{\sim}{\alpha}$ and $\underset{\sim}{\beta}$ be any two elements in W, then

$$\underset{\sim}{\alpha} = (x,\ 2x, x + 1), x \in R$$

$$\underset{\sim}{\beta} = (y,\ 2y, y + 1), y \in R$$

$$\therefore\ \underset{\sim}{\alpha} + \underset{\sim}{\beta} = (x + y,\ 2(x + y), x + y + 2)$$

Clearly, $\underset{\sim}{\alpha} + \underset{\sim}{\beta}$ does not belong to W as its third component is not $x + y + 1$. Hence, W is not closed with respect to vector addition. Therefore, W is not a subspace of $V(R)$.

12.4 CALCULUS OF SUBSPACES

THEOREM 12.5 The intersection of any two subspaces of a vector space is a subspace.

Proof: Let W_1 and W_2 be any two subspaces of a vector space $V(F)$. Let $\underset{\sim}{\alpha}, \underset{\sim}{\beta} \in W_1 \cap W_2$ and a, $b \in F$.
Then

$$\underset{\sim}{\alpha} \in W_1 \cap W_2 \Rightarrow \underset{\sim}{\alpha} \in W_1 \text{ and } \underset{\sim}{\alpha} \in W_2$$

and

$$\underset{\sim}{\beta} \in W_1 \cap W_2 \Rightarrow \underset{\sim}{\beta} \in W_1 \text{ and } \underset{\sim}{\beta} \in W_2$$

Now

$$\underset{\sim}{\alpha} \in W_1, \underset{\sim}{\beta} \in W_1 \Rightarrow a\underset{\sim}{\alpha} + b\underset{\sim}{\beta} \in W_1$$

and

$$\underset{\sim}{\alpha} \in W_2, \underset{\sim}{\beta} \in W_2 \Rightarrow a\underset{\sim}{\alpha} + b\underset{\sim}{\beta} \in W_2$$

Hence, $a\underset{\sim}{\alpha} + b\underset{\sim}{\beta} \in W_1 \cap W_2$, which shows that $W_1 \cap W_2$ is a subspace by Theorem 12.4.

Note 12.4

1. It can be observed that $W_1 \cap W_2$ is the largest subspace contained in W_1 and W_2 as it contains every subspace contained in W_1 and W_2. The theorem can be generalized and we can say that the intersection of any family of subspaces of a vector space is a subspace.
2. The union of two spaces W_1 and W_2, that is, $W_1 \cup W_2$ of a vector space may not be a subspace. For example, let R be the field of real numbers and $W_1 = \{(x, 0, 0): x \in R\}$ and $W_2 = \{(0, y, 0): y \in R\}$ be the subspace of $V_3(R)$.

By definition

$$\underset{\sim}{\alpha} = (x, 0, 0) \in W_1 \Rightarrow \alpha \in W_1 \cup W_2$$

$$\underset{\sim}{\beta} = (0, y, 0) \in W_2 \Rightarrow \beta \in W_1 \cup W_2$$

Now, if a, $b \in R$, then $a\underset{\sim}{\alpha} + \underset{\sim}{\beta} = (ax, by, 0)$

which belongs to neither W_1 nor W_2. Hence, $a\underset{\sim}{\alpha} + b\underset{\sim}{\beta} \notin W_1 \cup W_2$.
Therefore, $W_1 \cup W_2$ is not a subspace of $V_3(R)$.

12.4.1 Linear Sum of Two Subspaces

Definition 12.2 Let W_1 and W_2 be subspaces of vector space $V(F)$, then linear sum denoted by $W_1 + W_2$ is defined by,

$$W_1 + W_2 = \{\underset{\sim}{\alpha}_1 + \alpha_2, \alpha_1 \in W_1 \text{ and } \alpha_2 \in W_2\}$$

Since W_2 contains zero vector $\underset{\sim}{0}$, therefore if $\underset{\sim}{\alpha} \in W_1$, then

$$\alpha_1 + \underset{\sim}{0} = \underset{\sim}{\alpha}_1 \in W_1 + W_2$$

Thus,

$$\underset{\sim}{\alpha}_1 \in W_1 \Rightarrow \underset{\sim}{\alpha}_1 \in W_1 + W_2$$

Hence,

$$W_1 \subseteq W_1 + W_2$$

Similarly,

$$W_2 \subseteq W_1 + W_2$$

THEOREM 12.6 If W_1 and W_2 are subspaces of a vector space $V(F)$, then $W_1 + W_2$ is a subspace of $V(F)$.

Proof:

Let $\underset{\sim}{\alpha}_1, \underset{\sim}{\beta}_1 \in W_1$ and $\underset{\sim}{\alpha}_2, \underset{\sim}{\beta}_2 \in W_2$

Then $\underset{\sim}{\alpha}_1 + \underset{\sim}{\alpha}_2 \in W_1 + W_2$

and $\underset{\sim}{\beta}_1 + \underset{\sim}{\beta}_2 \in W_1 + W_2$

Let $a, b \in F$, then

$$a(\underset{\sim}{\alpha}_1 + \underset{\sim}{\alpha}_2) + b(\underset{\sim}{\beta}_1 + \underset{\sim}{\beta}_2) = (a\underset{\sim}{\alpha}_1 + b\underset{\sim}{\beta}_1) + (a\underset{\sim}{\alpha}_2 + b\underset{\sim}{\beta}_2)$$

since W_1 and W_2 are subspaces of V,

$$a\underset{\sim}{\alpha}_1 + b\underset{\sim}{\beta}_1 \in W_1 \text{ and } a\underset{\sim}{\alpha}_2 + b\underset{\sim}{\beta}_2 \in W_2$$

Hence, $(a\underset{\sim}{\alpha}_1 + b\underset{\sim}{\beta}_1) + (a\underset{\sim}{\alpha}_2 + b\underset{\sim}{\beta}_2) \in W_1 + W_2$

Therefore, $W_1 + W_2$ is a subspace of $V(F)$.

12.5 LINEAR DEPENDENCE AND INDEPENCE OF VECTORS

Definition Linear Combination: Let V be a vector space over the field F. If, $\underset{\sim}{x}_1, \underset{\sim}{x}_2, \ldots, \underset{\sim}{x}_m$ be vectors of V and $c_1, c_2, \ldots, c_m$ be scalars, then the vector sum of the form

$$c_1\underset{\sim}{x}_1 + c_2\underset{\sim}{x}_2 + \cdots + c_m\underset{\sim}{x}_m$$

is called a linear combination of the vector $\underset{\sim}{x}_1, \underset{\sim}{x}_2, \ldots, \underset{\sim}{x}_m$.

Then $\underset{\sim}{x}_1 + 2\underset{\sim}{x}_2 - 3\underset{\sim}{x}_3$ is a linear combination of vectors $\underset{\sim}{x}_1, \underset{\sim}{x}_2$, and $\underset{\sim}{x}_3$.

Similarly, $2\underset{\sim}{x}_1 - 5\underset{\sim}{x}_2 + 3\underset{\sim}{x}_3 - \underset{\sim}{x}_4$ is a linear combination of vectors $\underset{\sim}{x}_1, \underset{\sim}{x}_2, \underset{\sim}{x}_3$, and $\underset{\sim}{x}_4$.

We observe that the sum of any two linear combinations of $\underset{\sim}{x}_1, \underset{\sim}{x}_2, ..., \underset{\sim}{x}_m$ is another linear combination of these vectors and so also is a $a(c_1\underset{\sim}{x}_1 + c_2\underset{\sim}{x}_2 + ... + c_m\underset{\sim}{x}_m),\ a \in F$.

Hence, the set U of all linear combination of vectors $\underset{\sim}{x}_1, \underset{\sim}{x}_2, ..., \underset{\sim}{x}_m$ in a vector space V, is a subspace of V.

We also say that the subspace U is spanned or generated by vectors $\underset{\sim}{x}_1, \underset{\sim}{x}_2, ..., \underset{\sim}{x}_m$.

If S denotes the set of vectors $\underset{\sim}{x}_1, \underset{\sim}{x}_2, ..., \underset{\sim}{x}_m$. Then U is also called Linear span of S and is denoted $L(s)$. Thus,

$$L(s) = \{c_1\underset{\sim}{x}_1 + c_2\underset{\sim}{x}_2 + \cdots + c_m\underset{\sim}{x}_m) : c_i \in F,\ x_i \in V, \text{m a positive integer}\}$$

Example 1 Determine the subspace spanned by

$$S = \{(2, 3, 1), (1, 0, 2)\} \text{ in } R^3$$

The required vector space U is given by,

$$\text{U} = \{a\,(2, 3, 1) + b(1, 0, 2),\ a,\ b \in R\}$$

$$\text{or} \quad \text{U} = \{(2a + \text{b},\ 3a,\ a + 2b),\ a,\ b \in R\}$$

Example 2 Determine the subspace spanned by,

$$S = \{(1, -1, 2), (2, -1, 3)\} \text{ in } R^3$$

The required vector space U is given by,

$$\text{U} = L(s) = \{a\,(1, -1, 2) + b\,(2, -1, 3),\ a,\ b \in R\}$$
$$= \{(a + 2b,\ -a - b,\ 2a + 3b),\ a,\ b \in R\}$$

12.5.1 Definition of Linear Dependence

The vectors $\underset{\sim}{x}_1, \underset{\sim}{x}_2, ..., \underset{\sim}{x}_n$ of a vector space $V(F)$ are said to be linearly dependent if there exist scalars $c_1, c_2, \ldots, c_n$ not all zero such that

$$c_1\underset{\sim}{x}_1 + c_2\underset{\sim}{x}_2 + \cdots + c_n\underset{\sim}{x}_n = \underset{\sim}{0}$$

$$\text{If } c_1\underset{\sim}{x}_1 + c_2\underset{\sim}{x}_2 + \cdots + c_n\underset{\sim}{x}_n = \underset{\sim}{0} \Rightarrow c_1 = c_2 = \cdots = c_n = 0$$

then vectors $\underset{\sim}{x}_1, \underset{\sim}{x}_2, \ldots, \underset{\sim}{x}_n$ are said to be linearly independent.

Example 1 Show that the set of vectors

$\underset{\sim}{x}_1 = (1, 2, 3), \underset{\sim}{x}_2 = (2, -1, 2), \underset{\sim}{x}_3 = (1, 7, 7)$ in the vector space R^3 is linearly dependent.

Solution:

$$\underset{\sim}{x}_2 - 2\underset{\sim}{x}_1 = (0, -5, -4)$$

$$\underset{\sim}{x}_3 - \underset{\sim}{x}_1 = (0, 5, 4)$$

$$\therefore \quad (\underset{\sim}{x}_3 - \underset{\sim}{x}_1) + (\underset{\sim}{x}_2 - 2\underset{\sim}{x}_1) = (0, 0, 0) = \underset{\sim}{0}$$

$$\text{or} \quad \underset{\sim}{x}_3 + \underset{\sim}{x}_2 - 3\underset{\sim}{x}_1 = \underset{\sim}{0}$$

which shows that the vectors are linearly dependent.

Alt: Let us assume that scalars c_1, c_2, and c_3 exist such that

$$c_1\underset{\sim}{x}_1 + c_2\underset{\sim}{x}_2 + c_3\underset{\sim}{x}_3 = \underset{\sim}{0}$$

$$\text{or} \quad c_1(1, 2, 3) + c_2(2, -1, 2) + c_3(1, 7, 7) = (0, 0, 0) \quad \text{(i)}$$

This gives
$$c_1 + 2c_2 + c_3 = 0 \quad \text{(ii)}$$
$$2c_1 - c_2 + 7c_3 = 0 \quad \text{(iii)}$$
$$3c_1 + 2c_2 + 7c_3 = 0 \quad \text{(iv)}$$

The last two equations give $c_1 + 3c_2 = 0$; $c_1 = -3c_2$. Substituting for c_1 in the first equation, we get $c_3 = c_2$. Substitution of values c_1 and c_3 in (i) gives

$$-3c_2(1, 2, 3) + c_2(2, -1, 2) + c_2(1, 7, 7) = \underset{\sim}{0}$$

$$\text{or} \quad -3\underset{\sim}{x}_1 + \underset{\sim}{x}_2 + \underset{\sim}{x}_3 = \underset{\sim}{0}$$

This shows that the given set of vectors is linearly dependent.

Example 2 Show that the vectors

$$\underset{\sim}{x}_1 = (1, 2, 1), \underset{\sim}{x}_2 = (2, 1, 4), \underset{\sim}{x}_3 = (-1, 4, -5)$$

in the vector space R^3 are linearly dependent.

$$\underset{\sim}{x}_2 - 2\underset{\sim}{x}_1 = (0, -3, 2)$$

$$\underset{\sim}{x}_3 + \underset{\sim}{x}_1 = (0, 6, -4)$$

$$\therefore \left(\underset{\sim}{x}_3 + \underset{\sim}{x}_1\right) + 2\left(\underset{\sim}{x}_2 - 2\underset{\sim}{x}_1\right) = (0, 0, 0) = \underset{\sim}{0}$$

$$\text{or} \quad \underset{\sim}{x}_3 + \underset{\sim}{x}_1 + 2\underset{\sim}{x}_2 - 4\underset{\sim}{x}_1 = \underset{\sim}{0}$$

$$\text{or} \quad \underset{\sim}{x}_3 + 2\underset{\sim}{x}_2 - 3\underset{\sim}{x}_1 = \underset{\sim}{0}$$

which shows that the given vectors are linearly dependent.
Alt: Let us assume that scalars c_1, c_2, and c_3 exist such that

$$c_1\underset{\sim}{x}_1 + c_2\underset{\sim}{x}_2 + c_3\underset{\sim}{x}_3 = \underset{\sim}{0}$$

$$\text{or} \quad c_1(1, 2, 1) + c_2(2, 1, 4) + c_3(-1, 4, -5) = (0, 0, 0)$$

this gives
$$c_1 + 2c_2 - c_3 = 0 \quad \text{(i)}$$
$$2c_1 + c_2 + 4c_3 = 0 \quad \text{(ii)}$$
$$c_1 + 4c_2 - 5c_3 = 0 \quad \text{(iii)}$$

From relations (i) and (iii) we have $2c_2 - 4c_3 = 0 \Rightarrow c_2 = 2c_3$
Substituting for c_2 in (ii) we have

$$2c_1 + 2c_3 + 4c_3 = 0 \Rightarrow c_1 = -3c_3$$

Hence we have

$$-3c_3\underset{\sim}{x}_1 + 2c_3\underset{\sim}{x}_2 + c_3\underset{\sim}{x}_3 = \underset{\sim}{0}$$

$$\text{or} \quad \underset{\sim}{x}_3 + 2\underset{\sim}{x}_2 - 3\underset{\sim}{x}_1 = \underset{\sim}{0}$$

which shows that the given vectors are linearly dependent.

Example 3
Show that the vectors (1, 0, 0), (0, 1, 0), and (0, 0, 1) are linearly independent in vector space R^3.

Solution: Let

$$c_1(1, 0, 0) + c_2(0, 1, 0) + c_3(0, 0, 1) = (0, 0, 0)$$

this implies
$$(c_1, 0, 0) + (0, c_2, 0) + (0, 0, c_3) = (0, 0, 0)$$
or
$$(c_1, c_2, c_3) = (0, 0, 0)$$
Hence
$$c_1 = c_2 = c_3 = 0$$
which shows that the given set of vectors is linearly independent.

Example 4 Show that the vectors (1, 1, 1), (1, 1, 0), and (1, 0, 0) are linearly independent in the vector space R^3.

Solution: Let $c_1(1, 1, 1) + c_2(1, 1, 0) + c_3(1, 0, 0) = (0, 0, 0)$. This gives $c_1 + c_2 + c_3 = 0$, $c_1 + c_2 = 0$, $c_3 = 0$. Its solution is $c_1 = c_2 = c_3 = 0$, which shows that the given set of vectors is linearly independent

Example 5 Are the vectors
(2, –2, –4, 0), (2, 2, 2, 4), (4, –2, –5, 2), and (4, 2, 1, 6) linearly dependent in R^4?

Solution: Let

$$\underset{\sim}{x}_1 = (2, -2, -4, 0)$$
$$\underset{\sim}{x}_2 = (2, 2, 2, 4)$$
$$\underset{\sim}{x}_3 = (4, -2, -5, 2)$$
$$\underset{\sim}{x}_4 = (4, 2, 1, 6)$$

Now

$$\underset{\sim}{x}_2 - \underset{\sim}{x}_1 = (0, 4, 6, 4)$$
$$\underset{\sim}{x}_3 - 2\underset{\sim}{x}_1 = (0, 2, 3, 2)$$
$$\underset{\sim}{x}_4 - 2\underset{\sim}{x}_1 = (0, 6, 9, 6)$$

Further

$$(\underset{\sim}{x}_2 - \underset{\sim}{x}_1) - 2(\underset{\sim}{x}_3 - 2\underset{\sim}{x}_1) = (0, 0, 0, 0)$$
$$\underset{\sim}{x}_2 - \underset{\sim}{x}_1 - 2\underset{\sim}{x}_3 + 4\underset{\sim}{x}_1 = \underset{\sim}{0}$$
or
$$3\underset{\sim}{x}_1 + \underset{\sim}{x}_2 - 2\underset{\sim}{x}_3 = \underset{\sim}{0} \qquad \text{(i)}$$
Also
$$(\underset{\sim}{x}_4 - 2\underset{\sim}{x}_1) - 3(\underset{\sim}{x}_3 - 2\underset{\sim}{x}_1) = \underset{\sim}{0}$$
or
$$\underset{\sim}{x}_4 - 2\underset{\sim}{x}_1 - 3\underset{\sim}{x}_3 + 6\underset{\sim}{x}_1 = \underset{\sim}{0}$$
or
$$\underset{\sim}{x}_4 - 3\underset{\sim}{x}_3 + 4\underset{\sim}{x}_1 = \underset{\sim}{0} \qquad \text{(ii)}$$
Multiplying (ii) by 2 and expressing $\underset{\sim}{x}_3$ from (i) in terms of $\underset{\sim}{x}_1$ and $\underset{\sim}{x}_2$, we have

$$2x_4 - 3(3x_1 + x_2) + 4x_1 = 0$$
$$2x_4 - 9x_1 - 3x_2 + 4x_1 = 0$$
$$2x_4 - 3x_2 - 5x_1 = 0 \qquad \text{(iii)}$$

From relation (i), x_1, x_2, x_3 are linearly dependent and from relation (iii) x_1, x_2, x_4 are linearly dependent. The vectors x_1 and x_2 are linearly independent as $c_1x_1 + c_2x_2 = 0 \Rightarrow c_1 = c_2 = 0$. Therefore, the given vectors are linearly dependent in R^4.

12.5.2 Method for Testing Linear Dependence of Vectors

To test the linear dependence of n vectors $x_1, x_2, \ldots, x_n$, we write them as row vectors and add such multiples of one vector to the remaining vectors so that the resulting vectors have their first component zero. This gives $(n-1)$ new vectors all having their first component zero. Choosing one of these new vectors, add its suitable multiples to others so that the resulting vectors have their first two components zero. Reducing step by step the successive components to zero, if we obtain a sum of vectors whose all components are zero, then the given original vectors are linearly dependent. However, if the final reduction does not lead to a sum whose components are *not* all zero, then the original vectors are linearly independent.

Consider the set of vectors

$$x_1 = (-1, 2, 3),\ x_2 = (2, -1, -4),\ x_3 = (1, 1, 3),\ x_4 = (8, 2, -4) \qquad (12.1)$$

Adding suitable multiples of the first vector to reduce the first component zero, we have

$$x_2 + 2x_1 = (0, 3, 2),\quad x_3 + x_1 = (0, 3, 6),\quad x_4 + 8x_1 = (0, 18, 20) \qquad (12.2)$$

The next step is to reduce the second components to zero by adding suitable multiples of $x_2 + 2x_1$. This gives

$$(x_3 + x_1) - (x_2 + 2x_1) = (0, 0, 4) \qquad (12.3)$$
$$(x_4 + 8x_1) - 6(x_2 + 2x_1) = (0, 0, 8)$$

or $\qquad x_3 - x_2 - x_1 = (0, 0, 4),\ x_4 - 6x_2 - 4x_1 = (0, 0, 8)$

Reducing the third component to zero, we have

$$(x_4 - 6x_2 - 4x_1) - 2(x_3 - x_2 - x_1) = (0, 0, 0)$$

or $\qquad x_4 - 2x_3 - 4x_2 - 2x_1 = (0, 0, 0) = 0 \qquad (12.4)$

Which shows that the given set of vectors is linearly dependent. The above procedure can be applied by using the concept of *matrix* and employing elementary row operations as is explained below writing each vector as a row vector and reducing the matrix to echelon form. The processes (12.1), (12.2), (12.3) and (12.4) can be presented in the matrix form as

$$\begin{bmatrix} x_1 \\ x_2 \\ x_3 \\ x_4 \end{bmatrix} \sim \begin{bmatrix} x_1 \\ x_2 + 2x_1 \\ x_3 + x_1 \\ x_4 + 8x_1 \end{bmatrix} \sim \begin{bmatrix} x_1 \\ x_2 + 2x_1 \\ x_3 - x_2 - x_1 \\ x_4 - 6x_2 - 4x_1 \end{bmatrix} \sim \begin{bmatrix} x_1 \\ x_2 + 2x_1 \\ x_3 - x_2 - x_1 \\ x_4 - 2x_3 - 4x_2 - 2x_1 \end{bmatrix}$$

We have been able to obtain a zero vector $\underset{\sim}{x}_4-2\underset{\sim}{x}_3-4\underset{\sim}{x}_2-2\underset{\sim}{x}_1$ and hence the vectors $\underset{\sim}{x}_1$, $\underset{\sim}{x}_2$, $\underset{\sim}{x}_3$, and $\underset{\sim}{x}_4$ are linearly dependent.

Thus, we have

$$\begin{bmatrix}-1 & 2 & 3\\ 2 & -1 & -4\\ 1 & 1 & 3\\ 8 & 2 & -4\end{bmatrix}\underset{R_4+8R_1}{\overset{R_2+2R_1}{\underset{\sim}{R_3+R_1}}}\begin{bmatrix}-1 & 2 & 3\\ 0 & 3 & 2\\ 0 & 3 & 6\\ 0 & 18 & 20\end{bmatrix}\underset{R_4-6R_2}{\overset{R_3-R_2}{\sim}}\begin{bmatrix}-1 & 2 & 3\\ 0 & 3 & 2\\ 0 & 0 & 4\\ 0 & 0 & 8\end{bmatrix}\overset{R_4-2R_3}{\sim}\begin{bmatrix}-1 & 2 & 3\\ 0 & 3 & 2\\ 0 & 0 & 4\\ 0 & 0 & 0\end{bmatrix}$$

12.5.3 Echelon form of a Matrix

A matrix is said to be in echelon form if

1. all the non-zero rows, if any, precede zero rows;
2. number of zeros preceding the first non-zero element in a row is greater than the number of such zeros in the preceding rows.

Rank of a Matrix: The number of non-zeros in the echelon form is defined as the rank of a matrix. This is the number of linearly independent rows.

Example 1 Determine whether the following set of vectors

$$\underset{\sim}{x}_1=(-1,\ 2,\ 7,\ 6), \underset{\sim}{x}_2=(1,\ -2,\ 3,\ 4), \text{and } \underset{\sim}{x}_3=(1,\ -2,\ 3,\ 5)$$

are linearly dependent or independent.

Solution: Writing the given vectors as rows in the matrix, and applying elementary row operation, we have

$$\begin{bmatrix}-1 & 2 & 7 & 6\\ 1 & -2 & 3 & 4\\ 1 & -2 & 2 & 5\end{bmatrix}\underset{R_3+R_1}{\overset{R_2+R_1}{\sim}}\begin{bmatrix}-1 & 2 & 7 & 6\\ 0 & 0 & 10 & 10\\ 0 & 0 & 9 & 11\end{bmatrix}\overset{R_2(1/10)}{\sim}$$

$$\sim\begin{bmatrix}-1 & 2 & 7 & 6\\ 0 & 0 & 1 & 1\\ 0 & 0 & 9 & 11\end{bmatrix}\overset{R_3-9R_2}{\sim}\begin{bmatrix}-1 & 2 & 7 & 6\\ 0 & 0 & 1 & 1\\ 0 & 0 & 0 & 2\end{bmatrix}$$

Since the final reduction does not lead to a zero vector, the given vectors $\underset{\sim}{x}_1$, $\underset{\sim}{x}_2$, $\underset{\sim}{x}_3$ are linearly independent.

Example 2 Are the vectors (4, –1, 1), (1, 2, 4), and (1, –1, –1) linearly dependent or independent.

Solution: Writing the given vectors as row vectors of a matrix and applying elementary row operations, we have

$$\begin{bmatrix}1 & -1 & -1\\ 1 & 2 & 4\\ 4 & -1 & 1\end{bmatrix}\underset{R_3-4R_1}{\overset{R_2-R_1}{\sim}}\begin{bmatrix}1 & -1 & -1\\ 0 & 3 & 5\\ 0 & 3 & 5\end{bmatrix}\overset{R_3-R_2}{\sim}\begin{bmatrix}1 & -1 & -1\\ 0 & 3 & 5\\ 0 & 0 & 0\end{bmatrix}$$

Since the last row are all zeros, the given vectors are linearly dependent.

Note 12.5 The results given below follow from the definitions.

1. If the vectors $\underset{\sim}{x}_1, \underset{\sim}{x}_2, \ldots \underset{\sim}{x}_m$ are linearly dependent, then one of them can be expressed as a linear combination of the remaining vectors.
2. Conversely if $\underset{\sim}{x}_m$ can be expressed as a linear combination of $\underset{\sim}{x}_1, \underset{\sim}{x}_2, \cdots, \underset{\sim}{x}_{m-1}$ then the vectors $\underset{\sim}{x}_1, \underset{\sim}{x}_2, \cdots, \underset{\sim}{x}_m$ are linearly dependent.
3. A set of vectors $\underset{\sim}{0}, \underset{\sim}{x}_2, \underset{\sim}{x}_3, \ldots, \underset{\sim}{x}_m$ that contains zero vector is linearly dependent. As in this case, there exist scalars $c_1, c_2, \ldots, c_m (c_1 \neq 0)$ such that

$$c_1\underset{\sim}{x}_1 + c_2\underset{\sim}{x}_2 + c_3\underset{\sim}{x}_3 + \cdots + c_m\underset{\sim}{x}_m = \underset{\sim}{0}$$

THEOREM 12.7 Any set that contains a linearly dependent set is linearly dependent.

Proof: Let $\underset{\sim}{x}_1, \underset{\sim}{x}_2, \ldots, \underset{\sim}{x}_m$ be a set of linearly dependent vectors and let its superset be $\underset{\sim}{x}_1, \underset{\sim}{x}_2, \ldots, \underset{\sim}{x}_m, \underset{\sim}{x}_{m+1}, \underset{\sim}{x}_{m+2}, \ldots, \underset{\sim}{x}_n$

Since $\underset{\sim}{x}_1, \underset{\sim}{x}_2, \ldots, \underset{\sim}{x}_m$ are linearly dependent, there will exist scalars $c_1, c_2, \ldots, c_m$ not all zero such that

$$c_1\underset{\sim}{x}_1 + c_2\underset{\sim}{x}_2 + \ldots + c_m\underset{\sim}{x}_m = \underset{\sim}{0}$$

This can also be written as $c_1\underset{\sim}{x}_1 + c_2\underset{\sim}{x}_2 + \cdots + c_m\underset{\sim}{x}_m + c_{m+1}\underset{\sim}{x}_{m+1} + c_{m+2}\underset{\sim}{x}_{m+2} + \cdots + c_n\underset{\sim}{x}_n = \underset{\sim}{0}$ where $c_{m+1} = c_{m+2} = \ldots = c_n = 0$. Hence, the superset of vector is linearly dependent.

THEOREM 12.8 Any subset of a linearly independent set is linearly independent.

Proof: Let the set of vectors $\underset{\sim}{x}_1, \underset{\sim}{x}_2, \ldots, \underset{\sim}{x}_m, \underset{\sim}{x}_{m+1}, \underset{\sim}{x}_{m+2}, \ldots, \underset{\sim}{x}_n$ be linearly independent.

Consider its subset $\{\underset{\sim}{x}_1, \underset{\sim}{x}_2, \ldots, \underset{\sim}{x}_m\}$ $(m < n)$.

Let us assume that these are linearly dependent: Hence, there will exist scalars $c_1, c_2, \ldots, c_m$ not all zero such that $c_1\underset{\sim}{x}_1 + c_2\underset{\sim}{x}_2 + \cdots + c_m\underset{\sim}{x}_m = \underset{\sim}{0}$.

This can also be written as $c_1\underset{\sim}{x}_1 + c_2\underset{\sim}{x}_2 + \cdots + c_m\underset{\sim}{x}_m + 0\underset{\sim}{x}_{m+1} + \cdots + c_n\underset{\sim}{x}_n = \underset{\sim}{0}$, which shows that the given set is linearly dependent.

This contradicts the assumption that the set of vectors $\underset{\sim}{x}_1, \underset{\sim}{x}_2, \ldots, \underset{\sim}{x}_n$ are linearly independent. Hence, the subset of vectors $\underset{\sim}{x}_1, \underset{\sim}{x}_2, \ldots, \underset{\sim}{x}_m$ is linearly independent.

THEOREM 12.9 Let $X_1, X_2, \ldots, X_i, \ldots, X_n$ be n vectors in R^n with $X_i = (x_{1i}, x_{2i}, x_{ni})^T$ for $i = 1, 2, \ldots, n$. Let $X = (x_{ij})_{n \times n}$, that is, X_i forms ith column of X.
The vectors $X_1, X_2, \ldots, X_i, \ldots, X_n$ will be linearly dependent if X is singular.

Proof: The vectors $X_1, X_2, \ldots, X_n$ are linearly dependent if there exist scalars $c_1, c_2, \ldots, c_n$ not all zero such that

$$c_1X_1 + c_2X_2 + \ldots + c_nX_n = \tilde{O}$$

or

$$c_1\begin{pmatrix} x_{11} \\ x_{21} \\ \vdots \\ x_{n1} \end{pmatrix} + c_2\begin{pmatrix} x_{12} \\ x_{22} \\ \vdots \\ x_{n2} \end{pmatrix} + \cdots + c_n\begin{pmatrix} x_{1n} \\ x_{2n} \\ \vdots \\ x_{nn} \end{pmatrix} = \begin{pmatrix} 0 \\ 0 \\ \vdots \\ 0 \end{pmatrix}$$

This is equivalent to the system of homogeneous equations

$$c_1x_{11} + c_2x_{12} + \ldots + c_nx_{1n} = 0$$
$$c_1x_{21} + c_2x_{22} + \ldots + c_nx_{2n} = 0$$
$$\vdots$$
$$c_1x_{n1} + c_2x_{n2} + \ldots + c_nx_{nn} = 0$$

Which can be written in the matrix form as $XC = 0$
where

$$C = (c_1, c_2, \ldots, c_n)T \text{ and } X = (X_1, X_2, \ldots, X_n).$$

These equations will have a non-zero solution if and only if X is singular. Thus, the vectors X_1, $X_2, \ldots, X_n$ are linearly dependent if and only if X is singular.
i. e., $|X| = 0$

$$\begin{vmatrix} x_{11} & x_{12} & \cdots & x_{1n} \\ x_{21} & x_{22} & \cdots & x_{2n} \\ \vdots & \vdots & & \vdots \\ x_{n1} & x_{n2} & \cdots & x_{nn} \end{vmatrix} = 0$$

That is, the determinant whose columns are the given vectors is zero.
If $|X| \neq 0$, the vectors are linearly independent.

Example 1 Determine whether the following vectors in R^3 are linearly dependent

$$(1, -2, 8)^T, (1, 1, -1)^T, (3, 1, 3)^T$$

Solution: Consider the determinant whose columns are the given vectors, that is,

$$\begin{vmatrix} 1 & 1 & 3 \\ 1 & -2 & 1 \\ -1 & 8 & 3 \end{vmatrix} \begin{matrix} R_2 - R_1 \\ R_3 + R_1 \\ = \end{matrix} \begin{vmatrix} 1 & 1 & 3 \\ 0 & -3 & -2 \\ 0 & 9 & 6 \end{vmatrix} = 0$$

And hence the vectors are linearly dependent.

Example 2 Determine whether the following vectors in R^3 are linearly dependent.

$$(1, -2, 3)^T, (2, 3, -4)^T, (4, -1, 5)^T$$

Solution: Consider the determinant whose columns are the given vectors, that is,

$$\begin{vmatrix} 1 & 2 & 4 \\ -2 & 3 & -1 \\ 3 & -4 & 5 \end{vmatrix} \begin{matrix} R_2 + 2R_1 \\ R_3 - 3R_1 \\ = \end{matrix} \begin{vmatrix} 1 & 2 & 4 \\ 0 & 7 & 7 \\ 0 & -10 & -7 \end{vmatrix} = -49 + 70 = 21 \neq 0$$

Hence, the given vectors are linearly independent.

Example 3 Find whether the following vectors of $p_3(x)$ $1 - x^2$, $x + 2$, x^2 are linearly dependent.

Solution: Let $c_1(1 - x^2) + c_2(x + 2) + c_3x^2 = 0$

$$\Rightarrow -c_1 + c_3 = 0,\ c_2 = 0,\ c_1 + 2c_2 = 0$$

$$\Rightarrow c_1 = c_2 = c_3 = 0$$

Hence, the given vectors are linearly independent.

Note 12.6 Determinants can also be used to determine whether a set of vectors (i.e., functions) is linearly independent in $C^{n-1}[a, b]$.

Let $f_1, f_2, \ldots, f_n$ be elements of $C^{n-1}[a, b]$, then these vectors are linearly dependent if there exist scalars $c_1, c_2, \ldots, c_n$ not all zero such that

$$c_1 f_1(x) + c_2 f_2(x) + \ldots + c_n f_n(x) = 0$$

Taking derivatives w. r. t. x of both sides, we end up with the system

$$c_1 f_1'(x) + c_2 f_2'(x) + \ldots + c_n f_n'(x) = 0$$

$$c_1 f_1''(x) + c_2 f_2''(x) + \ldots + c_n f_n''(x) = 0$$

$$\vdots$$

$$c_1 f_1^{n-1}(x) + c_2 f_2^{n-1}(x) + \cdots + c_n f_n^{n-1}(x) = 0$$

Thus, for each fixed x in $[a, b]$, the system of equations can be written in the matrix form as

$$\begin{bmatrix} f_1(x) & f_2(x) & \cdots & f_n(x) \\ f_1'(x) & f_2'(x) & \cdots & f_n'(x) \\ f_1''(x) & f_2''(x) & \cdots & f_n^{n-1}(x) \\ \vdots & & & \\ f_1^{n-1}(x) & f_2^{n-1}(x) & & f_n^{n-1}(x) \end{bmatrix} \begin{bmatrix} c_1 \\ c_1 \\ c_3 \\ \vdots \\ c_n \end{bmatrix} = \begin{bmatrix} 0 \\ 0 \\ 0 \\ \vdots \\ 0 \end{bmatrix}$$

Hence, if $f_1, f_2, \ldots, f_n$ are linearly dependent in $C^{n-1}[a, b]$, then for each fixed x in $[a, b]$, the coefficient matrix is singular, that is, the determinant of the coefficient matrix is zero.

12.5.4 Definition of Wronskian

Let $f_1, f_2, \ldots, f_n$ be functions in $C^{n-1}[a, b]$. Define the function $W[f_1, f_2, \ldots, f_n]$ on $[a, b]$ by,

$$W[f_1, f_2, \ldots, f_n] = \begin{bmatrix} f_1(x) & f_2(x) & \cdots & f_n(x) \\ f_1'(x) & f_2'(x) & \cdots & f_n'(x) \\ \vdots & & & \\ f_1^{n-1}(x) & f_2^{n-1}(x) & \cdots & f_n^{n-1}(x) \end{bmatrix}$$

The function $W[f_1, f_2, \ldots, f_n]$ is called Wronskian of $f_1, f_2, \ldots, f_n$.

THEOREM 12.10 If $f_1, f_2, \ldots, f_n$ are elements in $C^{n-1}[a, b]$ and if there exists a point x_0 in $[a, b]$ such that $W[f_1, f_2, \ldots, f_n]x_0 \neq 0$, then $f_1, f_2, \ldots, f_n$ are linearly independent.

However, if $W[f_1, f_2, \ldots, f_n]x = 0$ for each x in $[a, b]$, then $f_1, f_2, \ldots, f_n$ are linearly dependent.

Example 1 Show that the vectors $1, x, x^2, x^3$ of the vector space of all polynomials in x of degree ≤ 3 are linearly independent.

Solution:

$$W[1, x, x^2, x^3] = \begin{vmatrix} 1 & x & x^2 & x^3 \\ 0 & 1 & 2x & 3x^2 \\ 0 & 0 & 2 & 6x \\ 0 & 0 & 0 & 6 \end{vmatrix} = 12 \neq 0$$

Hence, the vectors $1, x, x^2, x^3$ are linearly independent.

Example 2 Show that e^x and e^{-x} are linearly independent in $(-\infty, \infty)$.

Solution:

$$W[e^x, e^{-x}] = \begin{vmatrix} e^x & e^{-x} \\ e^x & -e^{-x} \end{vmatrix} = -2$$

is not identically zero and hence e^x and e^{-x} are linearly independent.

Example 3 Show that the vectors e^x, e^{-x}, e^{2x} are linearly independent in $C[-1, 1]$.

Solution:

$$W[e^x, e^{-x}, e^{2x}] = \begin{vmatrix} e^x & e^{-x} & e^{2x} \\ e^x & -e^{-x} & 2e^{2x} \\ e^x & e^{-x} & 4e^{2x} \end{vmatrix} \begin{matrix} \\ R_2 - R_1 \\ R_3 - R_1 \end{matrix}$$

$$= \begin{vmatrix} e^{x} & e^{-x} & e^{2x} \\ 0 & -2e^{-x} & e^{2x} \\ 0 & 0 & 3e^{2x} \end{vmatrix} = (3e^{2x})(-2)e^{x}.e^{-x} = -6e^{2x} \neq 0$$

is not identically zero. Hence, the given vectors e^x, e^{-x}, x^{2x} are linearly independent.

12.6 DIMENSION AND BASIS

If in a vector space V there exists a set of n linearly independent vectors, while every set of $n+1$ vectors is linearly dependent, then the number n is called the Dimension of V.

A vector space in which an arbitrarily large number of linearly independent vectors exist is called Infinite Dimensional.

BASIS: Let V be a vector space over the field F. A non-empty subset S of V is called a BASIS of V if and only if

1. S is linearly independent.
2. S spans V, that is, every vector in V can be expressed as a linear combination of the vectors in S.

If $S = \{\underset{\sim}{x}_1, \underset{\sim}{x}_2, \ldots, \underset{\sim}{x}_n\}$, then n, the number of vectors in the basis, is called the dimension of V. Such a vector space is called finite dimensional vector space.

Example 1 Show that the vectors $\underset{\sim}{e}_1 = (1, 0, 0)$, $\underset{\sim}{e}_2 = (0, 1, 0)$, $\underset{\sim}{e}_3 = (0, 0, 1)$ in R^3 form a basis for the vector space R^3.

Solution:

(1) The set of vectors $\{\underset{\sim}{e}_1, \underset{\sim}{e}_2, \underset{\sim}{e}_3\}$ is linearly independent as $c_1\underset{\sim}{e}_1 + c_2\underset{\sim}{e}_2 + c_3\underset{\sim}{e}_3 = \underset{\sim}{0}$

$$\Rightarrow c_1(1, 0, 0) + c_2(0, 1, 0) + c_3(0, 0, 1) = (0, 0, 0)$$

$$\Rightarrow (c_1, c_2, c_3) = (0, 0, 0)$$

$$\Rightarrow c_1 = c_2 = c_3 = 0$$

(2) Any vector $\underset{\sim}{x} = (x_1, x_2, x_3) \in R^3$ can be expressed as a linear combination of basis vector $\underset{\sim}{e}_1, \underset{\sim}{e}_2, \underset{\sim}{e}_3$.

$$\underset{\sim}{x} = (x_1, x_2, x_3) = x_1\underset{\sim}{e}_1 + x_2\underset{\sim}{e}_2 + x_3\underset{\sim}{e}_3$$

Note 12.7 The set of vectors $\{\underset{\sim}{e}_1, \underset{\sim}{e}_2, \underset{\sim}{e}_3, \ldots, \underset{\sim}{e}_n\}$ forms a basis for n-dimensional vector space R^n. This basis is called standard basis for R^n where $e_1 = (1, 0, 0, \ldots, 0)$, $e_2 = (0, 1, 0, \ldots, 0), \ldots,$ $e_n = (0, 0, \ldots, 1)$.

The basis for a vector space is not unique. There can be several bases for a vector space. For example, $\{(1, 0, 1), (1, 1, 0), (1, 1, 1)\}$ is a basis for R^3. Similarly, $\{(1, 2, 2), (-1, 1, -1), (1, 5, 2)\}$ is another basis. Any basis in R^3 must have exactly three vectors.

Example 2 Let $V(R)$ be a vector space of all square matrices of order 2 with real elements with respect to matrix addition and multiplication of a matrix by a real number.

Show that the set of matrices $\{E_{11}, E_{12}, E_{21}, E_{22}\}$

where $E_{11} = \begin{bmatrix} 1 & 0 \\ 0 & 0 \end{bmatrix}, E_{12} = \begin{bmatrix} 0 & 1 \\ 0 & 0 \end{bmatrix}, E_{21} = \begin{bmatrix} 0 & 0 \\ 1 & 0 \end{bmatrix}, E_{22} = \begin{bmatrix} 0 & 0 \\ 0 & 1 \end{bmatrix}$

forms a basis for $V(R)$.

If $$c_1E_{11} + c_2E_{12} + c_3E_{21} + c_4E_{22} = \mathbf{0}$$

then $$\begin{bmatrix} c_1 & c_2 \\ c_3 & c_4 \end{bmatrix} = \begin{bmatrix} 0 & 0 \\ 0 & 0 \end{bmatrix}$$

Hence, $c_1 = c_2 = c_3 = c_4 = 0$. Therefore, $E_{11}, E_{12}, E_{21}, E_{22}$ are linearly independent.

Further, if $A = \begin{bmatrix} a_{11} & a_{12} \\ a_{21} & a_{22} \end{bmatrix}$ is any element in $V(R)$, then $A = a_{11}E_{11} + a_{12}E_{12} + a_{13}E_{13} + a_{14}E_{14}$.

Thus, $E_{11}, E_{12}, E_{21}, E_{22}$ span $V(R)$ and hence form a basis of vector space V.

Example 3 Let P be the vector space of all polynomials. We claim that P is infinite dimensional. Let us assume that P is finite dimensional, say of dimension n. Then any set of $n + 1$ vector will be linearly dependent. But $1, x, x^2, \ldots, x^n$ are linearly independent as $W[1, x, x^2, \ldots, x^n] > 0$. Hence, P cannot be of dimension n. Since n is arbitrary, P must be infinite dimensional.

Applying the above argument, it can be shown that the vector space of all real-valued continuous functions $C[a, b]$ is infinite dimensional.

THEOREM 12.11 If V is a vector space of dimension $n > 0$, then

(1) Any set of n linearly independent vectors spans V.
(2) Any n vectors that span V are linearly independent.

Proof: (1) Suppose that vectors $\underset{\sim}{x}_1, \underset{\sim}{x}_2, \ldots, \underset{\sim}{x}_n$ are linearly independent and $\underset{\sim}{x}$ is any other vector in V. Then vectors $\underset{\sim}{x}_1, \underset{\sim}{x}_2, \ldots, \underset{\sim}{x}_n, \underset{\sim}{x}$ are linearly dependent and hence there will exist scalars $c_1, c_2, \ldots, c_n, c_{n+1}$ not all zero such that $c_1\underset{\sim}{x}_1 + c_2\underset{\sim}{x}_2 + \cdots + c_n\underset{\sim}{x}_n + c_{n+1}\underset{\sim}{x} = \underset{\sim}{0}$.

The scalar c_{n+1} cannot be zero, for then the vectors $\underset{\sim}{x}_1, \underset{\sim}{x}_2, \ldots, \underset{\sim}{x}_n$ will be linearly dependent. The above relation can be written as

$$\underset{\sim}{x} = k_1\underset{\sim}{x}_1 + k_2\underset{\sim}{x}_2 + \cdots + k_n\underset{\sim}{x}_n$$

where $k_i = -\dfrac{c_i}{c_{n+1}}$ for $i = 1, 2, \ldots, n$.

Since $\underset{\sim}{x}$ is an arbitrary vector in V, it follows that $\underset{\sim}{x}_1, \underset{\sim}{x}_2, \ldots, \underset{\sim}{x}_n$ span V.

To prove (2) suppose that n vectors $x_1, x_2, \ldots, x_n$ span V.

If $\underset{\sim}{x}_1, \underset{\sim}{x}_2, \ldots, \underset{\sim}{x}_n$ are linearly dependent, then one of these vectors, say x_n, can be expressed as a linear combination of others. It follows that $\underset{\sim}{x}_1, \underset{\sim}{x}_2, \ldots, \underset{\sim}{x}_{n-1}$ will span V. If $\underset{\sim}{x}_1, \underset{\sim}{x}_2, \ldots, \underset{\sim}{x}_{n-1}$ are linearly dependent, we can eliminate another vector and still have the spanning set. Continuing this process, we can have a linearly independent set of $k < n$ vectors, which contradicts dim $V = n$. Therefore, $\underset{\sim}{x}_1, \underset{\sim}{x}_2, \ldots, \underset{\sim}{x}_n$ must be linearly independent.

THEOREM 12.12 If vectors $\underset{\sim}{x}_1, \underset{\sim}{x}_2, \ldots, \underset{\sim}{x}_n$ form a basis for a vector space V, then any collection of m vectors of V, where $m > n$, is linearly dependent.

Proof: Let $\underset{\sim}{v}_1, \underset{\sim}{v}_2, \ldots, \underset{\sim}{v}_m$ be m vectors in V where $m > n$. Since $\underset{\sim}{x}_1, \underset{\sim}{x}_2, \ldots, \underset{\sim}{x}_n$ is a basis, it spans V and so we have

$$\underset{\sim}{v}_i = a_{i1}\,\underset{\sim}{x}_1 + a_{i2}\underset{\sim}{x}_2 + \cdots + a_{ij}\,\underset{\sim}{x}_j + \cdots + a_{in}\,\underset{\sim}{x}_n$$

for $i = 1, 2, \ldots, m$

Hence, the linear combination $c_1\,\underset{\sim}{v}_1 + c_2\,\underset{\sim}{v}_2 + \cdots + c_m\,\underset{\sim}{v}_m$ can be written as

$$\sum_{i=1}^{m} c_i\,\underset{\sim}{v}_i = \sum_{i=1}^{m} c_i\left(\sum_{j=1}^{n} a_{ij}\,\underset{\sim}{x}_j\right)$$

$$= \sum_{j=1}^{n}\left(\sum_{i=1}^{m} a_{ij}\,c_i\right)\underset{\sim}{x}_j \quad \text{(rearranging terms)}$$

Now the system of equation

$$\sum_{i=1}^{m} a_{ij}\,c_i = 0,\ j = 1, 2, \ldots, n$$

when written in full, is

$$a_{11}c_1 + a_{21}c_2 + \ldots + a_{m1}c_m = 0$$

$$a_{12}c_1 + a_{22}c_2 + \ldots + a_{m2}c_m = 0$$

$$\vdots$$

$$a_{1n}c_1 + a_{2n}c_2 + \ldots + a_{mn}c_m = 0$$

This is a homogeneous system of equations with more unknowns than the number of equations and so there must be some free variables. Hence, the vectors $\underset{\sim}{v}_1, \underset{\sim}{v}_2, \ldots, \underset{\sim}{v}_m$ are linearly dependent.

THEOREM 12.13 If $\{\underset{\sim}{v}_1, \underset{\sim}{v}_2, \ldots, \underset{\sim}{v}_m\}$ and $\{\underset{\sim}{x}_1, \underset{\sim}{x}_2, \ldots, \underset{\sim}{x}_n\}$ are both bases for a vector space V, then $n = m$.

Proof: Let $\{\underset{\sim}{x}_1, \underset{\sim}{x}_2, \ldots, \underset{\sim}{x}_n\}$ and $\{\underset{\sim}{v}_1, \underset{\sim}{v}_2, \ldots, \underset{\sim}{v}_m\}$ be bases for a vector space V. Since $\underset{\sim}{x}_1, \underset{\sim}{x}_2, \ldots, \underset{\sim}{x}_n$ are linearly independent, $n \le m$ by Theorem 12.12. Similarly, if $\underset{\sim}{v}_1, \underset{\sim}{v}_2, \ldots, \underset{\sim}{v}_m$ are linearly independent, $m \le n$. Thus, we have $n = m$.

Example 1 Show that the vectors $\{(1, 1, 2), (2, 1, 1), (3, 1, 5)\}$ form a basis of R^3.

Solution: If the given vectors form a basis, they must be linearly independent.

Let $\quad a(1, 1, 2) + b(2, 1, 1) + c(3, 1, 5) = (0, 0, 0)$

i.e., $\quad (a + 2b + 3c, a + b + c, 2a + b + 5c) = (0, 0, 0)$

$$\therefore a + 2b + 3c = 0 \quad \text{(i)}$$

$$a + b + c = 0 \quad \text{(ii)}$$

$$2a + b + 5c = 0 \quad \text{(iii)}$$

On solving we get $a = b = c = 0$

Hence, the vectors linearly independent.

If $(x_1, x_2, x_3) \in R^3$, then let

$$(x_1, x_2, x_3) = c_1(1, 1, 2) + c_2(2, 1, 1) + c_3(3, 1, 5),\ c_1, c_2, c_3 \in R$$

This gives

$$c_1 + 2c_2 + 3c_3 = x_1$$

$$c_1 + c_2 + c_3 = x_2$$

$$2c_1 + c_2 + 5c_3 = x_3$$

On solving, we get

$$c_1 = \frac{(-4x_1 + 7x_2 + x_3)}{5}, c_2 = \frac{3x_1 + x_2 - 2x_3}{5}, c_3 = \frac{x_1 - 3x_2 + x_3}{5}$$

Thus, any vector $(x_1, x_2, x_3) \in R^3$ can be expressed as a linear combination of the given vectors. Therefore, the given vectors form a basis of R^3.

Example 2 Show that two vectors $(1, 0, 0)$ and $(0, 0, 1)$ are linearly independent but do not form a basis of R^3.

Solution: Consider the linear combination

$$c_1(1, 0, 0) + c_2(0, 0, 1) = \underset{\sim}{0} = (0, 0, 0)$$

This implies $(c_1, 0, c_2) = (0, 0, 0)$

$$\Rightarrow c_1 = 0, c_2 = 0$$

Hence, the given vectors are linearly independent. The linear combination of the given vectors has it second component zero. Hence, a vector like $(2, 3, 5)$ of R^3 cannot be expressed as a linear combination of $(1, 0, 0)$ and $(0, 0, 1)$. Thus, these two vectors do not generate R^3 and so cannot form a basis for R^3.

Example 3 Show that the vectors $(1, 0)$ and $(0, 1)$ form a basis for the vector space R^2.

Solution: Consider the linear combination

$$c_1(1, 0) + c_2(0, 1) = \underset{\sim}{0} = (0, 0)$$

$$\Rightarrow (c_1, c_2) = (0, 0)$$

$$\Rightarrow c_1 = c_2 = 0$$

which shows that the given vectors are linearly independent.

Also, let $(x_1, x_2) = a\,(1, 0) + b\,(0, 1)$, $\quad (x_1, x_2) \in R^2$

$$= (a, b)$$

$$\therefore a = x_1,\ b = x_2$$

Which shows that any vector $(x_1, x_2) \in R^2$ can be expressed as a linear combination of the given vectors. That is, given vectors span R^2.
Therefore, the given vectors form a basis for R^2.

Example 4 Given vectors $\underset{\sim}{x}_1 = \begin{pmatrix} -2 \\ 3 \\ 4 \end{pmatrix}, \underset{\sim}{x}_2 = \begin{pmatrix} 2 \\ -3 \\ -4 \end{pmatrix}, \underset{\sim}{x}_3 = \begin{pmatrix} -6 \\ 9 \\ 12 \end{pmatrix}$.

Find the dimension of span $(\underset{\sim}{x}_1, \underset{\sim}{x}_2, \underset{\sim}{x}_3)$.

Solution: To find the dimension of the vector space spanned by the given vectors, we find the number of linearly independent vectors.

Writing the given vectors as row vectors of a matrix and reducing to echelon form by means of elementary row operations, we have

$$\begin{bmatrix} -2 & 3 & 4 \\ 2 & -3 & -4 \\ -6 & 9 & 12 \end{bmatrix} \begin{matrix} R_2+R_1 \\ R_3-3R_1 \\ = \end{matrix} \begin{bmatrix} -2 & 3 & 4 \\ 0 & 0 & 0 \\ 0 & 0 & 0 \end{bmatrix}$$

There is only one non-zero row. This shows that the given vectors are linearly dependent. Only one vector is linearly independent. Hence, span $(\underset{\sim}{x}_1, \underset{\sim}{x}_2, \underset{\sim}{x}_3)$ has dimension 1.

Example 5

(1) Find whether the following vectors in R^4 are linearly independent.

$$\underset{\sim}{x}_1 = (-1, 2, 5, 2), \underset{\sim}{x}_2 = (3, 0, 4, -1), \underset{\sim}{x}_3 = (1, 1, -2, 1)$$

(2) Does the vector $(4, -5, 9, -7)$ lie in the subspace of R^4 spanned by the given vectors $\underset{\sim}{x}_i$ $(i = 1, 2, 3)$?

Solution:

(1) Consider the matrix whose rows are the given vectors.

$$\begin{bmatrix} -1 & 2 & 5 & 2 \\ 3 & 0 & 4 & -1 \\ 1 & 1 & -2 & 1 \end{bmatrix} \begin{matrix} R_2+3R_1 \\ R_3+R_1 \\ \sim \end{matrix} \begin{bmatrix} -1 & 2 & 5 & 2 \\ 0 & 6 & 19 & 5 \\ 0 & 3 & 3 & 3 \end{bmatrix} \sim$$

$$\sim R_{23} \begin{bmatrix} 1 & 2 & 5 & 2 \\ 0 & 3 & 3 & 3 \\ 0 & 6 & 19 & 5 \end{bmatrix} \begin{matrix} R_3-2R_2 \\ \sim \end{matrix} \begin{bmatrix} 1 & -2 & -5 & -2 \\ 0 & 3 & 3 & 3 \\ 0 & 0 & 13 & -1 \end{bmatrix}$$

As no zero row occurs in the echelon form of the matrix, the set of vectors $\underset{\sim}{x}_1, \underset{\sim}{x}_2, \underset{\sim}{x}_3$ is linearly independent.

(2) If the given vector lies in the subspace spanned by the vectors $\underset{\sim}{x}_1, \underset{\sim}{x}_2, \underset{\sim}{x}_3$ then there will exist scalars c_1, c_2, c_3 not all zero such that

$$(4, -5, 9, -7) = c_1(-1, 2, 5, 2) + c_2(3, 0, 4, -1) + c_3(1, 1, -2, 1)$$

This gives the following system of equations:

$$-c_1 + 3c_2 + c_3 = 4 \qquad \text{(i)}$$

$$2c_1 + c_3 = -5 \qquad \text{(ii)}$$

$$5c_1 + 4c_2 - 2c_3 = 9 \qquad \text{(iii)}$$

$$2c_1 - c_2 + c_3 = -7 \qquad \text{(iv)}$$

Its solution is $c_1 = -1$, $c_2 = 2$, $c_3 = -3$, which shows that the vector (4, –5, 9, –7) lies in the subspace of R^4 spanned by the given vectors.

Example 6 Let V be the vector space of real functions. Determine whether $f, g, h \in V$ are linearly dependent or independent.

(1) $f(t) = \sin t$, $g(t) = \cos t$, $h(t) = t$
(2) $f(t) = \sin^2 t$, $g(t) = \cos 2t$, $h(t) = 1$

Solution:

(1) The Wronskian $W[f, g, h]$ is

$$W[f, g, h] = \begin{vmatrix} \sin t & \cos t & t \\ \cos t & -\sin t & 1 \\ -\sin t & -\cos t & 0 \end{vmatrix} \overset{R_3 + R_1}{=} \begin{vmatrix} \sin t & \cos t & t \\ \cos t & -\sin t & 1 \\ 0 & 0 & t \end{vmatrix}$$

$= t(-\sin^2 t - \cos^2 T) = -t \neq 0$ for all values of t. Hence, the functions are linearly independent.

(2) $$W[f, g, h] = W[\sin^2 t, \cos 2t, 1] = \begin{vmatrix} \sin^2 t & \cos 2t & 1 \\ \sin 2t & -2\sin 2t & 0 \\ 2\cos 2t & -4\cos 2t & 0 \end{vmatrix} = 0$$

which shows that the functions are linearly dependent.

Example 7 Find the dimension and basis spanned by the vectors in R^4.

$$\underset{\sim}{x}_1 = (-1, 2, 3, 4), \underset{\sim}{x}_2 = (2, -1, -4, 3)$$

$$\underset{\sim}{x}_3 = (3, 0, -5, 10), \underset{\sim}{x}_4 = (8, -7, -18, 1)$$

Solution: Writing the vectors as rows of a matrix and applying elementary row operations to reduce it to echelon form, we have

$$\begin{bmatrix} -1 & 2 & 3 & 4 \\ 2 & -1 & -4 & 3 \\ 3 & 0 & -5 & 10 \\ 8 & -7 & -18 & 1 \end{bmatrix} \begin{matrix} R_2+2R_1 \\ R_3+3R_1 \\ \sim \\ R_4+8R_1 \end{matrix} \begin{bmatrix} -1 & 2 & 3 & 4 \\ 0 & 3 & 2 & 11 \\ 0 & 6 & 4 & 22 \\ 0 & 9 & 6 & 33 \end{bmatrix} \begin{matrix} R_3-2R_2 \\ R_4-3R_2 \\ \sim \end{matrix} \begin{bmatrix} -1 & 2 & 3 & 4 \\ 0 & 3 & 2 & 11 \\ 0 & 0 & 0 & 0 \\ 0 & 0 & 0 & 0 \end{bmatrix}$$

There are only two non-zero rows and so its rank is 2. Hence, its dimension is 2. The basis vectors are (–1, 2, 3, 4) and (0, 3, 2, 11).

Definition 12.3 The rank of the matrix with rows as the vectors of a vector space is defined as the dimension of the vector space spanned by the given vectors. The non-zero rows of the echelon form will form the basis.

THEOREM 12.14 (Extension Theorem) Any set of linearly independent vectors in a vector space can be extended to a basis.

Proof: Let $V(F)$ be a vector space of dimension n. Let $S = \{x_1, x_2, \ldots, x_m\}$, $m < n$, be a set of linearly independent vectors of V. Then $L(S) \subset V$.

Let $\underset{\sim}{y} \in V - L(S)$, then the set of vectors

$S_1 = \{\underset{\sim}{x}_1, \underset{\sim}{x}_2, \ldots, \underset{\sim}{x}_m, \underset{\sim}{y}\}$ is linearly independent. In order to prove it, consider the linear combination

$$c_1\underset{\sim}{x}_1 + c_2\underset{\sim}{x}_2 + \ldots + c_m\underset{\sim}{x}_m + c_{m+1}\underset{\sim}{y} = \underset{\sim}{0} \qquad \text{(i)}$$

Here $c_{m+1} = 0$ otherwise $\underset{\sim}{y}$ can be expressed as the linear combination of $\underset{\sim}{x}_1, \underset{\sim}{x}_2, \ldots, \underset{\sim}{x}_m$ and will be in $L(S)$ which will contradict our assumption.

For $c_{m+1} = 0$, we have the linear combination (i) equal to zero for $c_1 = c_2 = \ldots = c_m = c_{m+1} = 0$. Thus, the vectors $\underset{\sim}{x}_1, \underset{\sim}{x}_2, \ldots, \underset{\sim}{x}_m, \underset{\sim}{y}$ are linear independent.

If the set $S_1 = \{\underset{\sim}{x}_1, \underset{\sim}{x}_2, \ldots, \underset{\sim}{x}_m, \underset{\sim}{y}\}$ is a basis, the theorem is proved. If not, we choose another vector $\underset{\sim}{z}$ from $V - L(S_1)$ and let $S_2 = \{\underset{\sim}{x}_1, \underset{\sim}{x}_2, \ldots, \underset{\sim}{x}_m, \underset{\sim}{y}, \underset{\sim}{z}\}$.

If S_2 is a basis, the theorem is proved. If not, we follow the above steps till a basis is obtained.

Example: Extend the set $S = \{(2, 0, 1), (1, 1, 2)\}$ to be a basis of R^3.

Solution: $L(S) = \{a(2, 0, 1) + b(1, 1, 2);\ a, b \in R\}$

$= \{(2a + b, b, a + 2b;\ a, b \in R\}$

Now any vector $(2a + b, b + 1, a + 2b) \in R^3 - L(S)$
Taking any set of values for a, b, say $a = 1$, $b = 1$, we get the vector (3, 2, 3).
Now consider the set $S_1 = \{(2, 0, 1), (1, 1, 2), (3, 2, 3)\}$
If S_1 is a basis, then the set of vectors S_1 is linear independent.

Consider the matrix whose rows are these vectors and reduce it to echelon form. This gives

$$\begin{bmatrix} 1 & 1 & 2 \\ 2 & 0 & 1 \\ 3 & 2 & 3 \end{bmatrix} \underset{R_3-3R_1}{\overset{R_2-2R_1}{\sim}} \begin{bmatrix} 1 & 1 & 2 \\ 0 & -2 & -3 \\ 0 & -1 & -3 \end{bmatrix} \underset{R_{23}}{\overset{\substack{R_2(-1)\\R_3(-1)}}{\sim}} \begin{bmatrix} 1 & 1 & 2 \\ 0 & 1 & 3 \\ 0 & 2 & 3 \end{bmatrix} \overset{R_3-2R_2}{\sim} \begin{bmatrix} 1 & 1 & 2 \\ 0 & 2 & 3 \\ 0 & 0 & -3 \end{bmatrix}$$

As there are 3 non-zero rows, its rank is 3. Hence, the vectors are linearly independent.
Let (x_1, x_2, x_3) be any vector in R^3 and let

$$(x_1, x_2, x_3) = a\,(1, 1, 2) + b\,(2, 0, 1) + c\,(3, 2, 3)$$

This gives

$$\therefore a + 2b + 3c = x_1$$

$$a + 2c = x_2$$

$$2a + b + 3c = x_3$$

Solving these equations, we get

$$a = \frac{1}{3}[-\,2x_1 - 3x_2 + 4x_3]$$

$$b = \frac{1}{3}[x_1 - 3x_2 + x_3]$$

$$c = \frac{1}{3}[x_1 + 3x_2 - 2x_3]$$

Showing that every vector in R^3 can be expressed as a linear combination of the extended set of vectors. Hence, the extended set

$$\{(1, 1, 2), (3, 0, 1), (4, 2, 3)\}$$

is a basis of R^3.

THEOREM 12.15 If W_1 and W_2 are finite dimensional subspaces of a vector space V, then $W_1 + W_2$ is finite dimensional and $\dim (W_1 + W_2) = \dim W_1 + \dim W_2 - \dim (W_1 \cap W_2)$.

Proof: Let $S = \{\underset{\sim}{x}_1, \underset{\sim}{x}_2, \ldots, \underset{\sim}{x}_k\}$ be a basis of $W_1 \cap W_2$.
Since $S \subseteq W_1$, S can be extended to form basis of W_1.
Let $S_1 = \{\underset{\sim}{x}_1, \underset{\sim}{x}_2, \ldots, \underset{\sim}{x}_k, \underset{\sim}{y}_1, \underset{\sim}{y}_2, \ldots, \underset{\sim}{y}_m\}$ be the basis of W_1 and hence $\dim W_1 = k + m$
Similarly $S \subseteq W_2$ and the basis S can be extended to form the basis of W_2.
Let $S_2 = \{\underset{\sim}{x}_1, \underset{\sim}{x}_2, \ldots, \underset{\sim}{x}_k, \underset{\sim}{z}_1, \underset{\sim}{z}_2, \ldots, \underset{\sim}{z}_n\}$ be the basis of W_2 and hence $\dim W_2 = k + n$.
The space $W_1 + W_2$ is spanned by the vectors $\underset{\sim}{x}_1, \underset{\sim}{x}_2, \ldots, \underset{\sim}{x}_k, \underset{\sim}{y}_1, \underset{\sim}{y}_2, \ldots, \underset{\sim}{y}_m, \underset{\sim}{z}_1, \underset{\sim}{z}_2, \ldots, \underset{\sim}{z}_n$ and these vectors form an independent set.

Therefore

$$\sum_i \alpha_i \underset{\sim}{x}_i + \sum_j \beta_j \underset{\sim}{y}_j + \sum_r \gamma_r \underset{\sim}{z}_r = \underset{\sim}{0}$$

Then

$$-\sum_r \gamma_r \underset{\sim}{z}_r = \sum_i \alpha_i \underset{\sim}{x}_i + \sum_j \beta_j \underset{\sim}{y}_j$$

Which shows that $\sum_r \gamma_r \underset{\sim}{z}_r$ belongs to W_1. Also, $\sum_r \gamma_r \underset{\sim}{z}_r$ belongs to W_2. Hence, it follows that $\sum_r \gamma_r \underset{\sim}{z}_r = \sum_i c_i \underset{\sim}{x}_i$ for some scalars $c_1, c_2, \ldots, c_k$.

As the set $\{\underset{\sim}{x}_1, \underset{\sim}{x}_2, \ldots, \underset{\sim}{x}_k, \underset{\sim}{z}_1, \underset{\sim}{z}_2, \ldots, \underset{\sim}{z}_n\}$ is independent each of scalars $\gamma_r = 0$. Thus,

$$\sum_i \alpha_i \underset{\sim}{x}_i + \sum_j \beta_j \underset{\sim}{y}_j = 0$$

Further, since $\{\underset{\sim}{x}_1, \underset{\sim}{x}_2, \ldots, \underset{\sim}{x}_k, \underset{\sim}{y}_1, \underset{\sim}{y}_2, \ldots, \underset{\sim}{y}_m\}$ is also an independent set, each $\alpha_i = 0$ and each $\beta_j = 0$. Thus,

$$\{\underset{\sim}{x}_1, \underset{\sim}{x}_2, \ldots, \underset{\sim}{x}_k, \underset{\sim}{y}_1, \underset{\sim}{y}_2, \ldots, \underset{\sim}{y}_m, \underset{\sim}{z}_1, \underset{\sim}{z}_2, \ldots, \underset{\sim}{z}_n\}$$

is a basis for $W_1 + W_2$. Finally

$$\begin{aligned} \dim W_1 + \dim W_2 &= (k+m) + (k+n) \\ &= k + (m+n+k) \\ &= \dim (W_1 \cap W_2) + \dim (W_1 + W_2) \end{aligned}$$

Hence,

$$\dim (W_1 + W_2) = \dim W_1 + \dim V_2 - \dim (W_1 \cap W_2)$$

Example 1 Find a basis and dimension of $S \cap T$ where S and T are subspaces of R^3 given by,

$$S = \{(x_1, x_2, x_3) \in R^3 : x_1 + x_2 + x_3 = 0\}$$
$$\text{and } T = \{(x_1, x_2, x_3) \in R^3 : 2x_1 + 5x_2 + 4x_3 = 0\}$$

Solution: Let (a, b, c) be any vector in $S \cap T$, then

$$a + b + c = 0$$
$$2a + 5b + 4c = 0$$
$$\therefore \frac{a}{-1} = \frac{-b}{+2} = \frac{c}{3}$$
$$\text{or } \frac{a}{1} = \frac{b}{2} = \frac{c}{-3}$$
$$\therefore S \cap T = L\,(1, 2, -3)$$

Hence, dimension of $S \cap T$ is 1 and basis is $(1, 2, -3)$.

Example 2 Let V be the vector space of 2×2 matrices over the field of real numbers and let

$$S = \left\{ \begin{bmatrix} a & b \\ c & d \end{bmatrix} : a + c = 0 \right\}$$

and
$$T = \left\{ \begin{bmatrix} a & b \\ c & d \end{bmatrix} : b + d = 0 \right\}$$

be its subspaces. Find the dimension of S, T and $S \cap T$. Also, find the dimension of $S + T$.

Solution: Dimension of S

Let $A = \begin{bmatrix} a & b \\ -a & d \end{bmatrix} \in S$, $a, b, d \in R$.

Then
$$A = a\begin{bmatrix} 1 & 0 \\ 0 & 0 \end{bmatrix} + a\begin{bmatrix} 0 & 0 \\ -1 & 0 \end{bmatrix} + b\begin{bmatrix} 0 & 1 \\ 0 & 0 \end{bmatrix} + d\begin{bmatrix} 0 & 0 \\ 0 & 1 \end{bmatrix}$$

$$= a\begin{bmatrix} 1 & 0 \\ -1 & 0 \end{bmatrix} + b\begin{bmatrix} 0 & 1 \\ 0 & 0 \end{bmatrix} + d\begin{bmatrix} 0 & 0 \\ 0 & 1 \end{bmatrix}$$

$$= aE_1 + bE_2 + dE_3$$

where
$$E_1 = \begin{bmatrix} 1 & 0 \\ -1 & 0 \end{bmatrix}, E_2 = \begin{bmatrix} 0 & 1 \\ 0 & 0 \end{bmatrix}, E_3 = \begin{bmatrix} 0 & 0 \\ 0 & 1 \end{bmatrix}$$

which shows that any matrix $A \in S$ can be expressed as the linear combination of matrices E_1, E_2, and E_3. Further, the matrices E_1, E_2, E_3 are linearly independent as

$$c_1E_1 + c_2E_2 + c_3E_3 = 0$$

$$\Rightarrow c_1\begin{bmatrix} 1 & 0 \\ -1 & 0 \end{bmatrix} + c_2\begin{bmatrix} 0 & 1 \\ 0 & 0 \end{bmatrix} + c_3\begin{bmatrix} 0 & 0 \\ 0 & 1 \end{bmatrix} = \begin{bmatrix} 0 & 0 \\ 0 & 0 \end{bmatrix}$$

$$\Rightarrow \begin{bmatrix} c_1 & c_2 \\ -c_1 & c_3 \end{bmatrix} = \begin{bmatrix} 0 & 0 \\ 0 & 0 \end{bmatrix} \Rightarrow c_1 = c_2 = c_3 = 0$$

Hence, $\dim(S) = 3$.

Similarly, it can be shown that $\dim(T) = 3$, a basis being matrices $\left\{ \begin{bmatrix} 0 & 1 \\ 0 & -1 \end{bmatrix}, \begin{bmatrix} 1 & 0 \\ 0 & 0 \end{bmatrix}, \begin{bmatrix} 0 & 0 \\ 1 & 0 \end{bmatrix} \right\}$.
Let B be any matrix in $S \cap T$, then

$$B = \begin{bmatrix} a & b \\ -a & -b \end{bmatrix} = a\begin{bmatrix} 1 & 0 \\ -1 & 0 \end{bmatrix} + b\begin{bmatrix} 0 & 1 \\ 0 & -1 \end{bmatrix} = aE_1' + bE_2'$$

where $E_1' = \begin{bmatrix} 1 & 0 \\ -1 & 0 \end{bmatrix}$ and $E_2' = \begin{bmatrix} 0 & 1 \\ 0 & -1 \end{bmatrix}$, which shows that the matrix $B \in S \cap T$ is a linear combination of matrices E_1' and E_2'.

Further, E_1' and E_2' are linearly independent as $\lambda_1 E_1' + \lambda_2 E_2' = 0$

$$\Rightarrow \lambda_1 \begin{bmatrix} 1 & 0 \\ -1 & 0 \end{bmatrix} + \lambda_2 \begin{bmatrix} 0 & 1 \\ 0 & -1 \end{bmatrix} = \begin{bmatrix} 0 & 0 \\ 0 & 0 \end{bmatrix}$$

$$\Rightarrow \begin{bmatrix} \lambda_1 & 0 \\ -\lambda_1 & 0 \end{bmatrix} + \begin{bmatrix} 0 & \lambda_2 \\ 0 & \lambda_2 \end{bmatrix} = \begin{bmatrix} 0 & 0 \\ 0 & 0 \end{bmatrix}$$

$$\Rightarrow \begin{bmatrix} \lambda_1 & \lambda_2 \\ -\lambda_1 & -\lambda_2 \end{bmatrix} = \begin{bmatrix} 0 & 0 \\ 0 & 0 \end{bmatrix}$$

$$\Rightarrow \lambda_1 = \lambda_2 = 0$$

Hence, dim $(S \cap T) = 2$, a basis being matrices

$$\left\{ \begin{bmatrix} 1 & 0 \\ -1 & 0 \end{bmatrix}, \begin{bmatrix} 0 & 1 \\ 0 & -1 \end{bmatrix} \right\}$$

Therefore, dim $(S + T) =$ dim $S +$ dim $T -$ dim $S \cap T$

$$= 3 + 3 - 2$$
$$= 4$$

12.6.1 Coordinates

A useful feature of a basis in a vector space V is to be able to introduce coordinates similar to natural coordinates.

$$\text{If} \quad \underset{\sim}{x} = (x_1, x_2, \ldots, x_n) = \sum_{i=1}^{n} x_i \underset{\sim}{e}_i$$

Then $\{e_i\}$ is a standard basis. A given vector $\underset{\sim}{x}$ has a unique representation as a linear combination of the basis vectors. $x_1, x_2, \ldots, x_n$ are called *coordinates* of the vector $\underset{\sim}{x}$ relative to the basis. There is a *natural ordering* of vectors in the standard basis as there is a rule to know which is the first vector, or second vector and so on in the basis.

Definition 12.4 *An ordered basis* for a finite dimensional vector space V is a finite sequence of vectors, which are linearly independent and span V.

Let the sequence $\underset{\sim}{u}_1, \underset{\sim}{u}_2, \ldots, \underset{\sim}{u}_n$ be an ordered basis for V.

Then ordered basis is the set $\{\underset{\sim}{u}_1, \underset{\sim}{u}_2, \ldots, \underset{\sim}{u}_n\}$. Given a vector $\underset{\sim}{u} \in V$, there will exist unique scalars $(c_1, c_2, \ldots, c_n)$ such that

$$\underset{\sim}{u} = \sum_{i=1}^{n} c_i \underset{\sim}{u}_i$$

The scalars are unique because if we also have $\underset{\sim}{u} = \sum_{i=1}^{n} c_i' \underset{\sim}{u}_i$

then
$$\sum_{i=1}^{n} (c_i - c_i')\underset{\sim}{u}_i = \underset{\sim}{0}$$

As $\underset{\sim}{u}_i$ $(i = 1, 2, \ldots, n)$ are linearly independent, $c_i - c_i' = 0$ for all i, that is $c_i' = c_i$. We shall call c_i, the i th coordinate of $\underset{\sim}{u}$ relative to ordered basis $\{\underset{\sim}{u}_1, \underset{\sim}{u}_2, \ldots, \underset{\sim}{u}_n\}$.

If $\underset{\sim}{v} = \sum_{i=1}^{n} d_i \underset{\sim}{u}_i$, then $\underset{\sim}{u} + \underset{\sim}{v} = \sum_{i=1}^{n} (c_i + d_i)\underset{\sim}{u}_i$ and $c_i + d_i$ is the i th coordinate of $\underset{\sim}{u} + \underset{\sim}{v}$ relative to ordered basis $\{\underset{\sim}{u}_1, \underset{\sim}{u}_2, \ldots, \underset{\sim}{u}_n\}$.

12.7 LINEAR TRANSFORMATION

Linear mappings from one vector space to another have acquired an important role in mathematics. Each linear transformation can be represented by a matrix, which enables us to work with the matrix in place of the transformation.

Definition 12.5 Let U and V be vector spaces over the same field F. A mapping $T: U \rightarrow V$ is said to be a linear transformation if

1. $T(\underset{\sim}{x} + \underset{\sim}{y}) = T\underset{\sim}{x} + T\underset{\sim}{y}, \quad \forall\, \underset{\sim}{x}, \underset{\sim}{y} \in U$
2. $T(\alpha \underset{\sim}{x}) = \alpha T\underset{\sim}{x}, \quad \forall\, \alpha \in F, \underset{\sim}{x} \in U$

The conditions (1) and (2) can be combined into a single condition

$$T(\alpha \underset{\sim}{x} + \beta \underset{\sim}{y}) = \alpha T\underset{\sim}{x} + \beta T\underset{\sim}{y}, \qquad \forall\, \alpha, \beta \in F, \underset{\sim}{x}, \underset{\sim}{y} \in U$$

Identity and Zero Transformation: Let U be a vector space over the field F. The identity Transformation $I: U \rightarrow U$ is defined by,

$$I\underset{\sim}{x} = \underset{\sim}{x} \quad \forall\, \underset{\sim}{x} \in U$$

Identity transformation is a linear transformation as

$$I(\alpha \underset{\sim}{x} + \beta \underset{\sim}{y}) = \alpha \underset{\sim}{x} + \beta \underset{\sim}{y} = \alpha I \underset{\sim}{x} + \beta I \underset{\sim}{y}, \qquad \alpha, \beta \in F \text{ and } \underset{\sim}{x}, \underset{\sim}{y} \in U$$

The zero transformation 0 from a vector space U into U is defined by $0\,\underset{\sim}{x} = \underset{\sim}{0}$ is a linear transformation as

$$\begin{aligned} 0(\alpha \underset{\sim}{x} + \beta \underset{\sim}{y}) &= \alpha\, 0 \underset{\sim}{x} + \beta\, 0 \underset{\sim}{y} \\ &= \alpha \underset{\sim}{0} + \beta \underset{\sim}{0} \\ &= \underset{\sim}{0} + \underset{\sim}{0} \\ &= \underset{\sim}{0} \end{aligned}$$

Examples 1

(1) The mapping $T: R^2 \to R$ defined by $T(x_1, x_2) = 3x_1 + 4x_2$ is linear.
(2) The mapping $T: R^2 \to R^3$ defined by $T(x_1, x_2) = (x_2, x_1, x_1 + x_2)$ is linear.
(3) The mapping $T: R^3 \to R^3$ defined by $T(x_1, x_2, x_3) = (x_1 + x_2, x_2 + x_3, x_3 + x_1)$ is linear.

Solution:

(1) $T(\alpha \underset{\sim}{x} + \beta \underset{\sim}{y}) = T(\alpha x_1 + \beta y_1, \alpha x_2 + \beta y_2) = 3(\alpha x_1 + \beta y_1) + 4(\alpha x_2 + \beta y_2)$

$$= \alpha(3x_1 + 4x_2) + \beta(3y_1 + 4y_2) = \alpha Tx + \beta Ty$$

which shows that T is linear.

(2) Now $T\left(\alpha \underset{\sim}{x} + \beta \underset{\sim}{y}\right) = (\alpha x_1 + \beta y_1, \alpha x_2 + \beta y_2)$

$$= (\alpha x_2 + \beta y_2, \alpha x_1 + \beta y_1, \alpha x_1 + \beta y_1 + \alpha x_2 + \beta y_2)$$

$$= (\alpha x_2, \alpha x_1, \alpha x_1 + \alpha x_2) + (\beta y_2, \beta y_1, \beta y_1 + \beta y_2)$$

$$= \alpha(x_2, x_1, x_1 + x_2) + \beta(y_2, y_1, y_1 + y_2)$$

$$= \alpha T\underset{\sim}{x} + \beta T\underset{\sim}{y}$$

which shows that T is linear.

(3) The mapping $T: R^3 \to R^3$ is linear since

(i) $T\left(\underset{\sim}{x} + \underset{\sim}{y}\right) = T(x_1 + y_1, x_2 + y_2, x_3 + y_3)$

$$= (x_1 + y_1 + x_2 + y_2, x_2 + y_2 + x_3 + y_3, x_3 + y_3 + x_1 + y_1)$$

$$= (x_1 + x_2, x_2 + x_3, x_3 + x_1) + (y_1 + y_2, y_2 + y_3, y_3 + y_1)$$

$$= T\underset{\sim}{x} + T\underset{\sim}{y}$$

(ii) $T(\alpha \underset{\sim}{x}) = T(\alpha x_1, \alpha x_2, \alpha x_3)$

$$= (\alpha x_1 + \alpha x_2, \alpha x_2 + \alpha x_3, \alpha x_3 + \alpha x_1)$$

$$= \alpha(x_1 + x_2, x_2 + x_3, x_3 + x_1) = \alpha \underset{\sim}{x}$$

which shows that T is linear.

Example 2 Show that the mapping $T: R^3 \to R^3$ defined by $T(x_1, x_2, x_3) = (x_1 + 1, x_2, x_3 - 1)$ is not linear.

Solution: $T(\alpha \underset{\sim}{x}) = T(\alpha x_1, \alpha x_2, \alpha x_3)$

$$= (\alpha x_1 + 1, \alpha x_2, \alpha x_3 - 1)$$

$\neq \alpha T\underset{\sim}{x}$, which shows that T is not linear.

Example 3 Find whether the transformation $T: R \to R^3$ defined by $Tx = (x, x^2, x^2 + 1)$ is linear.

Solution: $T(\alpha x) = (\alpha x, \alpha^2 x^2, \alpha^2 x^2 + 1) \neq \alpha Tx$, which shows that T is not linear.

Example 4 Show that the transform $T: R^3 \rightarrow R^2$ defined by $T(x_1, x_2, x_3) = (x_2, x_1 + x_3)$ is linear.

Solution: Let $\underset{\sim}{x} = (x_1, x_2, x_3)$ and $\underset{\sim}{y} = (y_1, y_2, y_3) \in R^3$

Then
$$\underset{\sim}{x} + \underset{\sim}{y} = (x_1 + y_1, x_2 + y_2, x_3 + y_3)$$

$$\begin{aligned}\therefore T(\underset{\sim}{x} + \underset{\sim}{y}) &= (x_2 + y_2, x_1 + y_1 + x_3 + y_3)\\ &= (x_2, x_1 + x_3) + (y_2, y_1 + y_3)\\ &= T\underset{\sim}{x} + T\underset{\sim}{y}\end{aligned}$$

Also,
$$\begin{aligned}T(\alpha \underset{\sim}{x}) &= T(\alpha x_1, \alpha x_2, \alpha x_3)\\ &= (\alpha x_2, \alpha x_1 + \alpha x_3)\\ &= \alpha (x_1, x_2 + x_3)\\ &= \alpha T\underset{\sim}{x}\end{aligned}$$

Hence, T is a linear transformation.

Example 5 Find the linear transformation $T: R^2 \rightarrow R^3$ such that $T(1, -1) = (1, 3, 0)$ and $T(2, 3) = (7, 1, 5)$.

Solution: The vectors $(1, -1)$ and $(2, 3)$ in R^2 are linearly independent and hence form a basis. Let $(x_1, x_2) \in R^2$, then

$$\begin{aligned}(x_1, x_2) &= a\,(2, 3) + b\,(1, -1)\\ &= (2a + b, 3a - b)\end{aligned}$$

This gives
$$\begin{aligned}2a + b &= x_1\\ 3a - b &= x_2\end{aligned}$$

Its solution is
$$a = \frac{x_1 + x_2}{5}, b = \frac{3x_1 - 2x_2}{5}$$

Hence,
$$(x_1, x_2) = \frac{x_1 + x_2}{5}(2,3) + \frac{3x_1 - 2x_2}{5}(1,-1)$$

Since T is a linear transformation, we have

$$\begin{aligned}T(x_1, x_2) &= \left(\frac{x_1 + x_2}{5}\right)T(2, 3) + \left(\frac{3x_1 - 2x_2}{5}\right)T(1, -1)\\ &= \left(\frac{x_1 + x_2}{5}\right)(7, 1, 5) + \frac{3x_1 - 2x_2}{5}(1,3,0)\\ &= \left(\frac{7(x_1 + x_2)}{5} + \frac{3x_1 - 2x_2}{5}, \frac{x_1 + x_2}{5} + \frac{3(3x_1 - 2x_2)}{5}, x_1 + x_2\right)\end{aligned}$$

$$\therefore \quad T(x_1, x_2) = (2x_1 + x_2, 2x_1 - x_2, x_1 + x_2)$$

12.7.1 Properties of Linear Transformation

THEOREM 12.16 Let $T: V \rightarrow W$ be a linear transformation, then

(1) $T(\underset{\sim}{0}_V) = \underset{\sim}{0}_W$ (where $\underset{\sim}{0}_V$ and $\underset{\sim}{0}_W$ are zero vectors in vectors spaces V and W)

(2) $T(-\underset{\sim}{v}) = -T(\underset{\sim}{v})$ for all $\underset{\sim}{v} \in V$

(3) $T(\underset{\sim}{v}_1 - v_2) = T\underset{\sim}{v}_1 - T\underset{\sim}{v}_2$ for all $\underset{\sim}{v}_1, \underset{\sim}{v}_2 \in V$

Proof: (1) For any vector $\underset{\sim}{v} \in V$

$$T(\underset{\sim}{v} + \underset{\sim}{0}_V) = T(\underset{\sim}{v}) + T(\underset{\sim}{0}_V) \quad \text{(since } T \text{ is linear)}$$

or $$T(\underset{\sim}{v}) = T(\underset{\sim}{v}) + T(\underset{\sim}{0}_V)$$

or $$T(\underset{\sim}{v}) + \underset{\sim}{0}_w = T(\underset{\sim}{v}) + T(\underset{\sim}{0}_V)$$

$$\Rightarrow T(\underset{\sim}{0}_V) = \underset{\sim}{0}_W \quad \text{(by cancellation law)}$$

Alternatively, since T is linear

$$T(c\underset{\sim}{v}) = c(T\underset{\sim}{v}) \quad ..., \text{c} \in \text{F}$$

Putting scalar c = 0, we have $T(0\underset{\sim}{v}) = 0T(\underset{\sim}{v})$

$$\therefore \quad T(\underset{\sim}{0}_V) = \underset{\sim}{0}_W$$ Proved

(2) $T[\underset{\sim}{v} + (-\underset{\sim}{v})] = T(\underset{\sim}{v}) + T(-\underset{\sim}{v})$ (since T is linear)

$$T(\underset{\sim}{0}_V) = T(\underset{\sim}{v}) + T(-\underset{\sim}{v})$$

or $$\underset{\sim}{0}_W = T(\underset{\sim}{v}) + T(-\underset{\sim}{v})$$

Therefore, $T(-\underset{\sim}{v})$ is the additive inverse of $T(-\underset{\sim}{v})$

That is, $$T(-\underset{\sim}{v}) = -T(v)$$ Proved

(3) $T(\underset{\sim}{v}_1 - \underset{\sim}{v}_2) = \left[T\underset{\sim}{v}_1 + (-\underset{\sim}{v}_2)\right]$ (Since T is linear)

$$= T(\underset{\sim}{v}_1) - T(-\underset{\sim}{v}_2)$$

$$= T(\underset{\sim}{v}_1) - T(-\underset{\sim}{v}_2) \text{ by (2)}$$

12.7.2 Matrix Representation of a Linear Transformation

Let T: $R^n \to R^m$ be a linear transformation. Let $\underset{\sim}{e}_1, \underset{\sim}{e}_2, \ldots, \underset{\sim}{e}_n$ be the standard basis in R^n . Let us define $L\underset{\sim}{e}_j = \underset{\sim}{a}_j$ and $A = (\underset{\sim}{a}_1, \underset{\sim}{a}_2, \ldots, \underset{\sim}{a}_n)$

Now if $\underset{\sim}{x} \in R^n$, then we have

$$T\underset{\sim}{x} = T(x_1\underset{\sim}{e}_1 + x_2\underset{\sim}{e}_2 + \cdots + x_n\underset{\sim}{e}_n)$$

$$= x_1\underset{\sim}{a}_1 + x_2\underset{\sim}{a}_2 + \cdots + x_n\underset{\sim}{a}_n = (\underset{\sim}{a}_1, \underset{\sim}{a}_2, \ldots, \underset{\sim}{a}_n)\underset{\sim}{x} = A\underset{\sim}{x}$$

Thus, linear transformation T can be represented by $m \times n$ matrix A.

Matrix of Linear Transformation: Let V and W be finite dimensional vector spaces over the same field F with dimensions n and m, respectively. Also, let $E_1 = \{\underset{\sim}{v}_1, \underset{\sim}{v}_2, \ldots, \underset{\sim}{v}_n\}$ and $E_2 = \{\underset{\sim}{w}_1, \underset{\sim}{w}_2, \ldots, \underset{\sim}{w}_n\}$ be two ordered bases for V and W, respectively.

An element $\underset{\sim}{v} \in V$ can be uniquely expressed as $\underset{\sim}{v} = x_1\underset{\sim}{v}_1 + x_2\underset{\sim}{v}_2 + \cdots + x_n\underset{\sim}{v}_n$ where $x_1, x_2, \ldots, x_n$ are called coordinates of $\underset{\sim}{v}$ relative to basis sectors of E_1. Thus, each vector $\underset{\sim}{v} \in V$ can be associated with n-tuple $(x_1, x_2, \ldots, x_n)^T \in R^n$. It is called coordinate vector of $\underset{\sim}{v}$ relative to basis E_1 and we write $\underset{\sim}{x} = (x_1, x_2, \ldots, x_n)^T = (\underset{\sim}{v})_{E_1}$. The same can be defined in W relative to E_2.

Now we introduce the elements of the matrix A associated with the linear transformation T: $V \to W$ as follows:

Let

$$T\underset{\sim}{v}_j = a_{1j}\underset{\sim}{w}_1 + a_{1j}\underset{\sim}{w}_2 + \cdots + a_{mj}\underset{\sim}{w}_m = \sum_{\gamma=1}^{m} a_{ij}\underset{\sim}{w}_i$$

Then

$$T\underset{\sim}{v} = T(x_1\underset{\sim}{v}_1 + x_2\underset{\sim}{v}_2 + \cdots + x_n\underset{\sim}{v}_n)$$

$$= T\sum_{j=1}^{n} x_j\underset{\sim}{v}_j$$

$$= \sum_{j=1}^{n} x_j(T\underset{\sim}{v}_j) \quad \text{(since } T \text{ is linear)}$$

$$= \sum_{j=1}^{n} x_j(\sum_{a=1}^{m} a_{ij}\underset{\sim}{w}_i)$$

$$= \sum_{i=1}^{m}(\sum_{j=1}^{n} a_{ij}x_j)\underset{\sim}{w}_i$$

$$= \sum_{i=1}^{m} y_i\underset{\sim}{w}_i$$

where $y_i = \sum_{j=1}^{n} a_{ij} x_j, i = 1, 2, \ldots, m$

Writing in full

$$y_1 = a_{11}x_1 + a_{12}x_2 + \ldots + a_{1n}x_n$$
$$y_2 = a_{21}x_2 + a_{22}x_2 + \ldots + a_{2n}x_n.$$
$$\vdots$$
$$y_m = a_{m1}x_1 + a_{m2}x_2 + \ldots + a_{1mn}x_n$$

Using matrices, we can write the system of equations as

$$\begin{bmatrix} y_1 \\ y_2 \\ \vdots \\ y_m \end{bmatrix} = \begin{bmatrix} a_{11} & a_{12} & \cdots & a_{1n} \\ a_{21} & a_{22} & \cdots & a_{2n} \\ \vdots & & & \\ y_{m1} & a_{m2} & \cdots & a_{mn} \end{bmatrix} \begin{bmatrix} x_1 \\ x_2 \\ \vdots \\ x_n \end{bmatrix}$$

$$\text{or} \quad \underset{\sim}{y} = A\underset{\sim}{x}, \qquad \underset{\sim}{a}_j = \left(T\underset{\sim}{v}_j\right)E_2$$

$$\text{or} \quad \left(T\underset{\sim}{v}\right)_{E_2} = A\left(\underset{\sim}{v}_j\right)_{E_1} \quad \text{for each } \underset{\sim}{v} \in V.$$

A is the matrix representing the transformation T relative ordered bases E_1 and E_2 where

$$\underset{\sim}{a}_j = \left(T\underset{\sim}{v}_j\right)_{E_2}, \qquad j = 1, 2, \ldots, n$$

This can be understood easily from the following diagrams.

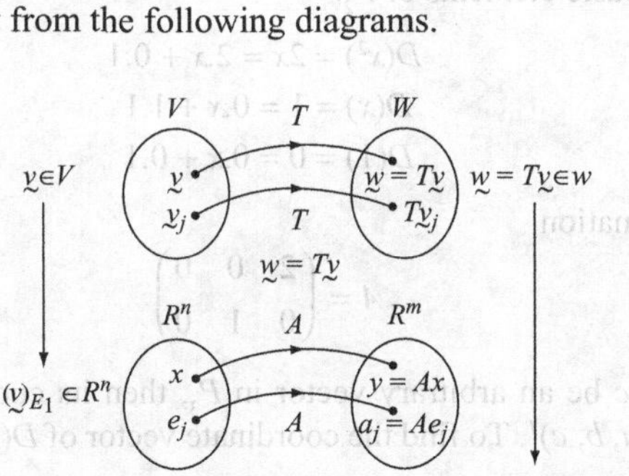

Thus in order to find matrix A for a linear transformation T: $R^n \rightarrow R^m$ with respect to ordered bases $E_1 = \{\underset{\sim}{v}_1, \underset{\sim}{v}_2, \ldots, \underset{\sim}{v}_n\}$ and $E_2 = \left(\underset{\sim}{w}_1, \underset{\sim}{w}_2, \underset{\sim}{w}_3, \ldots, \underset{\sim}{w}_m\right)$, we must express each vector $T\underset{\sim}{v}_j$ in terms of basis sectors $(\underset{\sim}{w}_1, \underset{\sim}{w}_2, \underset{\sim}{w}_3, \ldots, \underset{\sim}{w}_m)$. The coordinate vectors of $T\underset{\sim}{v}_j$ will give jth column of matrix A. The coordinates of $T\underset{\sim}{v}$ and $\underset{\sim}{v}$ will then be related by $y_i = \sum_j a_{ij}\, x_j$ or $\underset{\sim}{y} = A\underset{\sim}{x}$

Example 1 Find the matrix of the linear transformation T: $R^3 \rightarrow R^2$ defined by,

$$T(x_1, x_2, x_3) = (x_1 + x_2, x_2 + x_3)$$

Relative to the bases $\{\underset{\sim}{e}_1, \underset{\sim}{e}_2, \underset{\sim}{e}_3\}$ and $\{b_1, b_2\}$

where $\quad \underset{\sim}{e}_1 = (1, 0, 0),\ \underset{\sim}{e}_2 = (0, 1, 0),\ \underset{\sim}{e}_3 = (0, 0, 1)$ and $\underset{\sim}{b}_1 = (1, 1),\ b_2 = (1, -1)$.

Solution:

$$T(\underset{\sim}{e}_1) = T(1, 0, 0) = (1, 0) = \frac{1}{2}\underset{\sim}{b}_1 + \frac{1}{2}\underset{\sim}{b}_2$$

$$T(\underset{\sim}{e}_2) = T(0, 1, 0) = (1, 1) = 1\underset{\sim}{b}_1 + 0\underset{\sim}{b}_2$$

$$T(\underset{\sim}{e}_3) = T(0, 0, 1) = (0, 1) = \frac{1}{2}\underset{\sim}{b}_1 - \frac{1}{2}\underset{\sim}{b}_2$$

The matrix of transformation is $A = \begin{pmatrix} \frac{1}{2} & 1 & \frac{1}{2} \\ \frac{1}{2} & 0 & -\frac{1}{2} \end{pmatrix}$

Example 2 The linear transformation D defined by $Dp(x) = p'(x)$ maps $P_3(x)$ into $P_2(x)$. Find the matrix representation for D relative to bases $\{x^2, x, 1\}$ and $\{x, 1\}$ for P_3 and P_2, respectively.

Solution: We apply transformation D to each of the basic elements of P_3 and express them as linear combination of basic elements of P_2.

$$D(x^2) = 2x = 2.x + 0.1$$

$$D(x) = 1 = 0.x + 1.1$$

$$D(1) = 0 = 0.x + 0.1$$

The matrix of transformation

$$A = \begin{pmatrix} 2 & 0 & 0 \\ 0 & 1 & 0 \end{pmatrix}$$

Let $p(x) = ax^2 + bx + c$ be an arbitrary vector in P_3, then its coordinate vector relative to the ordered basis of P_3 is $(a, b, c)^T$. To find the coordinate vector of $D(p)$, with respect to the ordered basis of P_2, we simply multiply it by matrix A. This gives $\begin{pmatrix} 2 & 0 & 0 \\ 0 & 1 & 0 \end{pmatrix}\begin{pmatrix} a \\ b \\ c \end{pmatrix} = \begin{pmatrix} 2a \\ b \end{pmatrix}$.

Thus, $$D(ax^2 + bx + c) = 2ax + b$$

Example 3 Find the matrix of the linear transformation $T: R^2 \to R^3$ defined by, $T(x_1, x_2) = (x_1 + 2x_2, 2x_1, 2x_1 - x_2)$ with respect to ordered bases $\{(1, 2), (2, 1)\}$ and $\{(1, 1, 1), (1, 2, 3), (2, 0, -1)\}$ of R^2 and R^3, respectively.

Solution: We apply transformation to each basis of R^2 and express it as a linear combination of basic vectors of R^3.

$$T(1, 2) = a_1(1, 1, 1) + b_1(1, 2, 3) + c_1(2, 0, -1)$$

or
$$T(1, 2) = (5, 2, 0) = 4(1, 1, 1) - 1(1, 2, 3) + 1(2, 0, -1)$$

and
$$T(2, 1) = a_2(1, 1, 1) + b_2(1, 2, 3) + c_2(2, 0, -1)$$

or
$$T(2, 1) = (4, 4, 3) = 8(1, 1, 1) - 2(1, 2, 3) - 1(2, 0, -1)$$

The matrix of transformation

$$A = \begin{pmatrix} 4 & 8 \\ -1 & -2 \\ 1 & -1 \end{pmatrix}$$

Example 4 Given matrix $\begin{pmatrix} 1 & 2 \\ -1 & 3 \\ 2 & -1 \end{pmatrix}$, find the linear transformation $T: R^2 \to R^3$ relative to the ordered bases $E_1 = \{(1, 0), (0, 1)\}$ and $E_2 = \{(1, 0, 0), (0, 1, 0), (0, 0, 1)\}$ of R^2 and R^3, respectively.

Solution:

$$T(1, 0) = 1(1, 0, 0) + (-1)(0, 1, 0) + 2(0, 0, 1)$$
$$= (1, 0, 0) + (0, -1, 0) + (0, 0, 2) = (1, -1, 2)$$

And
$$T(0, 1) = 2(1, 0, 0) + 3(0, 1, 0) - 1(0, 0, 1)$$
$$= (2, 0, 0) + (0, 3, 0) + (0, 0, -1) = (2, 3, -1)$$

Let
$$(x_1, x_2) \in R^2, \text{ then } (x_1, x_2) = a(1, 0) + b(0, 1) = (a, b)$$

$$\therefore\ a = x_1, b = x_2$$

Now

$$T(x_1, x_2) = aT(1, 0) + bT(0, 1)$$
$$= x_1(1, -1, 2) + x_2(2, 3, -1)$$

Or $T(x_1, x_2) = (x_1 + 2x_2, -x_1 + 3x_2, 2x_1 - x_2)$

12.7.3 Change of Basis-transition Matrix

Many engineering problems are simplified by changing from one coordinate system to another. For example, the motion of a particle moving in a plane can be described by employing a unit tangent vector $\underset{\sim}{\hat{t}}$ and a unit normal vector $\underset{\sim}{\hat{n}}$ or by using unit vectors along the radius vector and transverse direction (when moving on a circular path) instead of standard basis $\underset{\sim}{e}_1, \underset{\sim}{e}_2$. We show that this is obtained by multiplying the vector $\underset{\sim}{x}$ by a non-singular matrix S. The product $\underset{\sim}{y} = S\underset{\sim}{x}$ is the coordinate vector for the new coordinate system.

Let V be a vector space of dimension n and $E_1 = [\underset{\sim}{v}_1, \underset{\sim}{v}_2, \ldots, \underset{\sim}{v}_n]$ and $E_2 = [\underset{\sim}{w}_1, \underset{\sim}{w}_2, \ldots, \underset{\sim}{w}_n]$ be two ordered bases in it. Let a vector $\underset{\sim}{v} \in V$ be expressed as a linear combination of vectors of E_1 and E_2 as follows:

$$\underset{\sim}{v} = \sum_{j=1}^{n} x_j \underset{\sim}{v}_j \text{ and } \underset{\sim}{v} = \sum_{j=1}^{n} y_i \underset{\sim}{w}_i$$

Then $x_j'^s$ are called coordinates of $\underset{\sim}{v}$ relative to E_1 and $y_i'^s$ are the coordinates of $\underset{\sim}{v}$ relative to to E_2. It is required to find a relation between them.

Let $\underset{\sim}{v}_j = s_{1j}\underset{\sim}{w}_1 + s_{2j}\underset{\sim}{w}_2 + \ldots + s_{nj}\underset{\sim}{w}_n = \sum_{i=1}^{n} s_{ij}\underset{\sim}{w}_i$. Thus, the jth column of the matrix $S = [s_{ij}]_{n\times n}$ gives the coordinates of $\underset{\sim}{v}_j$ relative to basis vectors E_2.

Now

$$\underset{\sim}{v} = \sum_{j=1}^{n} x_j \underset{\sim}{v}_j = \sum_{j=1}^{n} x_j \sum_{i=1}^{n} s_{ij}\underset{\sim}{w}_i = \sum_{i=1}^{n}\left(\sum_{j=1}^{n} s_{ij}x_j\right)\underset{\sim}{w}_i = \sum_{i=1}^{n} y_i \underset{\sim}{w}_i$$

Hence,

$$y_i = \sum_{j=1}^{n} s_{ij}x_j, \quad i = 1, 2, \ldots, n$$

$$\text{or} \quad \underset{\sim}{y} = S\underset{\sim}{x}, \qquad \underset{\sim}{x}, \underset{\sim}{y} \in R^n$$

$$\text{or} \quad (\underset{\sim}{v})_{E_2} = S(\underset{\sim}{v})_{E_1}$$

It can be shown that S is non-singular.

Example 1 Find the transition matrix from $[1, x, x^2]$ to $[1, 4x, 4x^2 + 1]$ in the vector spaces of polynomials $P_3(x)$.

Solution: Since $[1, x, x^2]$ is the standard basis for P_3, it is easy to find transition matrix from $[1, 4x, 4x^2 + 1]$ to $[1, x, x^2]$. Its inverse will be the desired matrix. Since

$$1 = 1.1 + 0.x + 0.x^2$$
$$4x = 0.1 + 4.x + 0.x^2$$
$$4x^2 + 1 = 1.1 + 0.x + 4.x^2$$

The transition matrix from $[1, 4x, 4x^2 + 1]$ to $[1, x, x^2]$ is

$$S = \begin{bmatrix} 1 & 0 & 1 \\ 0 & 4 & 0 \\ 0 & 0 & 4 \end{bmatrix}$$

$$\therefore S^{-1} = \begin{bmatrix} 1 & 0 & -\frac{1}{4} \\ 0 & \frac{1}{4} & 0 \\ 0 & 0 & \frac{1}{4} \end{bmatrix}$$

Note 12.8 The coordinates of polynomial $p(x) = a + bx + cx^2$ relative to $[1, 4x, 4x^2 + 1]$ will be

$$\begin{bmatrix} 1 & 0 & -\frac{1}{4} \\ 0 & \frac{1}{4} & 0 \\ 0 & 0 & \frac{1}{4} \end{bmatrix}\begin{bmatrix} a \\ b \\ c \end{bmatrix} = \begin{bmatrix} a-\frac{1}{4}c \\ \frac{1}{4}b \\ \frac{1}{4}c \end{bmatrix}$$

Hence,

$$p(x) = \left(a - \frac{1}{4}c\right).1 + \left(\frac{1}{4}b\right)4x + \left(\frac{1}{4}c\right).\left(4x^2+1\right).$$

Example 2 Find the transition matrix from $[1, x, x^2]$ to $[1, 2x, 4x^2 - 2]$ in the vector spaces of polynomials P_3.

Solution: Since $[1, x, x^2]$ is the standard basis for P_3, it is easy to find transition matrix from $[1, 2x, 4x^2 - 2]$ to $[1, x, x^2]$. Its inverse will be the desired matrix. Since

$$1 = 1.1 + 0.x + 0.x^2$$
$$2x = 0.1 + 2.x + 0.x^2$$
$$4x^2 - 2 = -2.1 + 0.x + 4.x^2$$

The transition matrix from $[1, 2x, 4x^2 - 2]$ to $[1, x, x^2]$ is

$$S = \begin{bmatrix} 1 & 0 & -2 \\ 0 & 2 & 0 \\ 0 & 0 & 4 \end{bmatrix}$$

$$\therefore S^{-1} = \begin{bmatrix} 1 & 0 & \frac{1}{2} \\ 0 & \frac{1}{2} & 0 \\ 0 & 0 & \frac{1}{4} \end{bmatrix}$$

Note 12.9 The coordinates of polynomial $p(x) = a + bx + cx^2$ relative to $[1, 2x, 4x^2 - 2]$ will be

$$\begin{bmatrix} 1 & 0 & \frac{1}{2} \\ 0 & \frac{1}{2} & 0 \\ 0 & 0 & \frac{1}{4} \end{bmatrix} \begin{bmatrix} a \\ b \\ c \end{bmatrix} = \begin{bmatrix} a - \frac{1}{2}c \\ \frac{1}{2}b \\ \frac{1}{4}c \end{bmatrix}$$

Hence,

$$p(x) = \left(a + \frac{1}{2}c\right)\cdot 1 + \left(\frac{1}{2}b\right)2x + \left(\frac{1}{4}c\right)\cdot\left(4x^2 - 2\right).$$

12.7.4 Range and Kernel of Linear Transformation

Associated with every linear transformation T, there are two concepts of particular importance: one is its range denoted by $R(T)$ and another is the kernel (or null space) denoted by $N(T)$.

Definition 12.6 Let $T: V \to W$ be a linear transformation. The range of T is defined by,

$$R(T) = \{\underset{\sim}{w} \in W : \underset{\sim}{w} = T(v), v \in V\}$$

Thus, range of T is all T-images of the elements of V.

Definition 12.7 The kernel or null space of T is defined by,

$$\text{Ker}(T) = N(T) = \{\underset{\sim}{v} \in V : T(\underset{\sim}{v}) = \underset{\sim}{o}_W\}$$

THEOREM 12.17 Let $T: V \to W$ be a linear transformation and U be a subspace of V, then

(1) $T(U)$ is a subspace of W
(2) $N(T)$ is a subspace of V

Proof:

(1) Let $\underset{\sim}{w} \in T(U)$, then $\underset{\sim}{w} = T(\underset{\sim}{v})$ for some $\underset{\sim}{v} \in U$.
For any scalar c,

$$c\underset{\sim}{w} = cT(\underset{\sim}{v}) = T(c\underset{\sim}{v})$$

Now since U is a subspace, $c\,\underset{\sim}{v} \in T(U)$ and so $T(U)$ is closed under scalar multiplication. Also, if $\underset{\sim}{w}_1, \underset{\sim}{w}_2 \in T(U)$, then there will exist $\underset{\sim}{v}_1, \underset{\sim}{v}_2 \in U$ such that $T(\underset{\sim}{v}_1) = \underset{\sim}{w}_1$ and $T(\underset{\sim}{v}_2) = \underset{\sim}{w}_2$
Hence,

$$\underset{\sim}{w}_1 + \underset{\sim}{w}_2 = T(\underset{\sim}{v}_1) + T(\underset{\sim}{v}_2) = T(\underset{\sim}{v}_1 + \underset{\sim}{v}_2)$$

and hence $T(U)$ is closed under addition. Therefore, $T(U)$ is a subspace of W.

(2) To prove that $N(T)$ is a subspace, we must prove that $N(T)$ is closed under addition and scalar multiplication.

Let $\underset{\sim}{v}_1, \underset{\sim}{v}_2 \in N(T)$, then $T(\underset{\sim}{v}_1) = \underset{\sim}{0}_W$ and $T(\underset{\sim}{v}_2) = \underset{\sim}{0}_W$

Now

$$T(\underset{\sim}{v}_1 + \underset{\sim}{v}_2) = T\underset{\sim}{v}_1 + T\underset{\sim}{v}_2 = \underset{\sim}{0}_W + \underset{\sim}{0}_W = \underset{\sim}{0}_W$$

$\therefore \underset{\sim}{v}_1 + \underset{\sim}{v}_2 \in N(T)$, and hence, $N(T)$ is closed under addition.

Also, $$T(c\underset{\sim}{v}) = cT(\underset{\sim}{v}) = c.\underset{\sim}{0}_W = \underset{\sim}{0}_W$$

where c is any scalar.

Hence, $N(T)$ is closed under scalar multiplication.

12.7.5 Rank-nullity Theorem

Definition 12.8 If V is a finite dimensional vector space, then rank of T is the dimension of the range of T and the nullity of T is the dimension of the null space of T.

THEOREM 12.18 If V and W are finite dimensional vector spaces over the field F and T is a linear transformation from V into W, then rank (T) + nullity (T) = dim V

Proof: Let dim $V = n$ and nullity $= N(T) = k$. Since $N(T)$ is a subspace of V, $k \le n$.
Let $\{\underset{\sim}{v}_1, \underset{\sim}{v}_2, \ldots, \underset{\sim}{v}_K\}$ be a basis of $N(T)$. This basis can be extended to form the basis of V.

We shall now prove that the set $B = \{T\underset{\sim}{v}_{k+1}, T\underset{\sim}{v}_{k+2}, \ldots, T\underset{\sim}{v}_n\}$ of $(n-k)$ elements is a basis for the range of T.

(1) B spans range of T, i.e., $L(B) = R(T)$

The vectors $T\underset{\sim}{v}_1, T\underset{\sim}{v}_2, \ldots, T\underset{\sim}{v}_n$ span the range of T. But $T\underset{\sim}{v}_i = \underset{\sim}{0}$ for $i \le k$. Hence, $T\underset{\sim}{v}_{k+1}, T\underset{\sim}{v}_{k+2}, \ldots, T\underset{\sim}{v}_n$ span the range, that is, $L(B) = R(T)$.

(2) The set B is linearly independent. Suppose that scalars c_i exist such that

$$\sum_{i=k+1}^{n} c_i T(\underset{\sim}{v}_i) = \underset{\sim}{0}$$

$$\Rightarrow \left(\mathrm{T} \sum_{i=k+1}^{n} c_i \underset{\sim}{v}_i \right) = \underset{\sim}{0}$$

which shows that the vector $\underset{\sim}{v} = \sum_{i=k+1}^{n} c_i \underset{\sim}{v}_i \in N(T)$

since $\underset{\sim}{v}_1, \underset{\sim}{v}_2, \ldots, \underset{\sim}{v}_K$ from basis for $N(T)$, there will exist scalars $a_1, a_2, \ldots, a_k$ such that

$$\underset{\sim}{v} = \sum_{j=1}^{k} a_j \underset{\sim}{v}_j$$

Thus

$$\sum_{j=1}^{k} a_j \underset{\sim}{v}_j = \sum_{i=k+1}^{n} c_i \underset{\sim}{v}_i$$

$$\text{or} \quad \sum_{j=1}^{k} a_j \underset{\sim}{v}_j - \sum_{i=k+1}^{n} c_i \underset{\sim}{v}_j = \underset{\sim}{0}$$

since $\underset{\sim}{v}_1, \underset{\sim}{v}_2, \ldots, \underset{\sim}{v}_n$ are linearly independent, we have

$$a_1 = a_2 = \ldots = a_k = c_{k+1} = c_{k+2} = \ldots = c_n = 0$$

which shows that B is linearly independent.
Hence, dim $R(T) = n - k =$ dim $V -$ dim $N(T)$ or dim $V =$ dim $R(T) +$ dim $N(T)$.

12.8 NORMED AND INNER PRODUCT SPACES

In this section, we consider only vector spaces over the field of real and complex numbers. The object is to study vector spaces in which the concepts of 'length' or magnitude of a vector and the angle between two vectors can be defined leading to practical applications.

12.8.1 Norm

The concept of norm provides a way of measuring the 'length' of vector. It also defines the distance between the vectors of a vector space.

Definition 12.9 Let V be a vector space over the field of real or complex numbers.

A norm on V is a function that assigns to each vector $\underset{\sim}{x}$ in V a non-negative real number satisfying the following axioms:

1. $\|\underset{\sim}{x}\| > 0$ if $\underset{\sim}{x} \neq \underset{\sim}{0}$, $\|\underset{\sim}{x}\| = 0$ if $\underset{\sim}{x} = \underset{\sim}{0}$
2. $\|\underset{\sim}{x} + \underset{\sim}{y}\| \leq \|\underset{\sim}{x}\| + \|\underset{\sim}{y}\|, \forall \underset{\sim}{x}, \underset{\sim}{y} \in V$ (triangle inequality)
3. $\|\alpha \underset{\sim}{x}\| = |\alpha| . \|\underset{\sim}{x}\| \;\; \forall \underset{\sim}{x} \in V$ and $\alpha \in F$

A vector space on which a norm is defined for every vector is called a normed vector space.

Example 1 For the vector space R^3, the norm is defined by $\|\underset{\sim}{x}\| = |x_1| + |x_2| + |x_3|$ where $\underset{\sim}{x} = (x_1, x_2, x_3)$. Show that norm axioms are satisfied.

Solution:

(1) if $\underset{\sim}{x} = (0,0,0) = \underset{\sim}{0}$, then

$$\|\underset{\sim}{x}\| = 0 + 0 + 0 = 0$$

When

$$\underset{\sim}{x} \neq \underset{\sim}{0}, \|\underset{\sim}{x}\| = |x_1| + |x_2| + |x_3| > 0$$

Since x_1, x_2, x_3 are not all zero.

(2) Let $\underset{\sim}{x} = (x_1, x_2, x_3)$ and $\underset{\sim}{y} = (y_1, y_2, y_3)$ be any two vectors in R^3, then

$$\underset{\sim}{x} + \underset{\sim}{y} = (x_1 + y_1, x_2 + y_2, x_3 + y_3)$$

$$\|\underset{\sim}{x} + \underset{\sim}{y}\| = |x_1 + y_1| + |x_2 + y_2| + |x_3 + y_3|$$

$$\leq |x_1| + |y_1| + |x_2| + |y_2| + |x_3| + |y_3|$$
$$\leq (|x_1| + |x_2| + |x_3|) + (|y_1| + |y_2| + |y_3|)$$
$$\leq (\|\underset{\sim}{x}\| + \|\underset{\sim}{y}\|)$$

(3) $\|\alpha\underset{\sim}{x}\| = |\alpha x_1| + |\alpha x_2| + |\alpha x_3|$

$$= |\alpha|.|x_1| + |\alpha|.|x_2| + |\alpha|.|x_3|$$
$$= |\alpha|(|x_1| + |x_2| + |x_3|)$$
$$= |\alpha| \, \|\underset{\sim}{x}\|$$

Thus, all the axioms of norm are satisfied.

Examples of different norms defined on $V = R^n$ or C^n

1. $$\|\underset{\sim}{x}\|_p = \left(\sum_{i=1}^{n} |x_i|^p\right)^{1/p}$$ (for any real number $p \geq 1$)

2. $$\|\underset{\sim}{x}\|_1 = \sum_{i=1}^{n} |x_i|$$

3. $$\|\underset{\sim}{x}\|_2 = \left(\sum_{i=1}^{n} |x_i|^2\right)^{1/2}$$ (Euclidean norm)

4. $$\|\underset{\sim}{x}\|_\infty = \max_{1\leq i\leq n} |x_i|$$ (Uniform norm)

Example 2 Let $V = C\,[a, b]$ and $p \geq 1$, then $\|\underset{\sim}{f}\|_p \left(\int_a^b |f(x)|^p \, dx\right)^{1/p}$
In particular

$$\|\underset{\sim}{f}\|_1 \left(\int_a^b |f(x)| \, dx\right)$$

$$\|\underset{\sim}{f}\|_2 \left(\int_a^b |f(x)|^2 \, dx\right)^{1/2}$$

$$\|\underset{\sim}{f}\|_\infty = \max_{a\leq x\leq b} |f(x)|$$ (Uniform norm)

Example 3 For the vector $\underset{\sim}{x} = (2, -3, 6)$ in the vector space R^3, compute $\|\underset{\sim}{x}\|_1$, $\|\underset{\sim}{x}\|_2$, and $\|\underset{\sim}{x}\|_\infty$.

Solution: $\|\underset{\sim}{x}\|_1 = |x_1| + |x_2| + |x_3| = |2| + |-3| + |6| = 2 + 3 + 6 = 11$

$$\|\underset{\sim}{x}\|_2 = \left(\sum_{i=1}^{2} |x_i|^2\right)^{1/2} = (|2|^2 + |-3|^2 + |6|^2)^{1/2} = \sqrt{4+9+36} = 7$$

$$\|\underset{\sim}{x}\|_\infty = \max_{1\leq i\leq 3} |x_i| = \max\{|2|, |-3|, |6|\} = 6$$

Example 4 On the space $C[a, b]$, norm is defined by $\|\underset{\sim}{f}\| = \int_a^b | f(x) | \, dx$. Verify norm axioms.

Solution:

(1) If $f(x) = 0$ for all $x \in C[a, b]$, then $\|\underset{\sim}{f}\| = \int_a^b | f(x) | \, dx = \int_a^b 0.dx = 0$

When $f \neq \underset{\sim}{0}$, then $|f(x)|$ is positive for all x in $[a, b]$ or in some subinterval of $[a, b]$ and therefore

$$\|\underset{\sim}{f}\| = \int_a^b | f(x) | \, dx > 0$$

(2) Since $|f(x) + g(x)| \leq |f(x)| + |g(x)|$ for every x, $\int_a^b | f(x)+g(x) | \, dx \leq \int_a^b | f(x) | \, dx + \int_a^b | g(x) | \, dx$

i.e., $\|\underset{\sim}{f} + \underset{\sim}{g}\| \leq \|\underset{\sim}{f}\| + \|\underset{\sim}{g}\|$

(3) $\|\alpha \underset{\sim}{f}\| = \int_a^b | \alpha f(x) | \, dx = |\alpha| \int_a^b | f(x) | \, dx, \alpha \in F$

$= |\alpha| \, \| \underset{\sim}{f} \|$

Hence, the norm axioms are verified.

12.8.2 Inner Product

The angle between two non-zero vectors in either R^2 or R^3 can be defined by using the concept of scalar product or dot product. The angle θ between vectors $\underset{\sim}{x}$ and $\underset{\sim}{y}$ is determined by,

$$\underset{\sim}{x} . \underset{\sim}{y} = \|\underset{\sim}{x}\| \|\underset{\sim}{y}\| \cos\theta$$

Scalar products are of importance not only in R^n but also in general vector spaces.

Definition 12.10 Let V be a vector space over the field F of real or complex numbers. An inner product on V is a function that assigns to each ordered pair of vectors $\underset{\sim}{x}$ and $\underset{\sim}{y}$ in V, a scalar $< \underset{\sim}{x}, \underset{\sim}{y} >$ in F such that

(1) $< \underset{\sim}{x}, \underset{\sim}{y} > = \overline{< \underset{\sim}{y}, \underset{\sim}{x} >}$ for $\forall \underset{\sim}{x}, \underset{\sim}{y} \in V$

(2) $< \alpha \underset{\sim}{x} + \beta \underset{\sim}{y}, \underset{\sim}{z} > = \overline{\alpha} < \underset{\sim}{x}, \underset{\sim}{z} > + \overline{\beta} < \underset{\sim}{y}, \underset{\sim}{z} >$, $\alpha, \beta \in F, \underset{\sim}{x}, \underset{\sim}{y}, \underset{\sim}{z} \in V$

(3) $< \underset{\sim}{x}, \underset{\sim}{x} >> \underset{\sim}{0}$ if $\underset{\sim}{x} \neq \underset{\sim}{0}$ and $< \underset{\sim}{x}, \underset{\sim}{x} >= 0$ if $\underset{\sim}{x} = \underset{\sim}{0}$.

From above it follows that

$$< \underset{\sim}{x}, \alpha \underset{\sim}{y} > = \overline{< \alpha \underset{\sim}{y} \underset{\sim}{x} >} = \overline{(\overline{\alpha})} < \underset{\sim}{y}, \underset{\sim}{x} > = \alpha < \underset{\sim}{y}, \underset{\sim}{x} > .$$

A vector space with inner product is called Inner product space. For a vector space over the field of real numbers, the above conditions are

(1) $<\underset{\sim}{x},\underset{\sim}{y}>=<\underset{\sim}{y},\underset{\sim}{x}>$ (symmetry)

(2) $<\alpha\underset{\sim}{x}+\beta\underset{\sim}{y},\underset{\sim}{z}>=\alpha<\underset{\sim}{x},\underset{\sim}{z}>+\beta<\underset{\sim}{y},\underset{\sim}{z}>$ (Homogeneity)

(3) $<\underset{\sim}{x},\underset{\sim}{x}>> 0$ if $\underset{\sim}{x}\neq\underset{\sim}{0}$, and $=0$ if $\underset{\sim}{x}=\underset{\sim}{0}$. (non-negativity)

Note 12.10 The axiom $<\alpha\underset{\sim}{x}+\beta\underset{\sim}{y},\underset{\sim}{z}>$ can be represented by

(1) $<\underset{\sim}{x}+\underset{\sim}{y},\underset{\sim}{z}>=<\underset{\sim}{x},\underset{\sim}{z}>+<\underset{\sim}{y},\underset{\sim}{z}>$

(2) $<\alpha\underset{\sim}{x},\underset{\sim}{z}>=\bar{\alpha}<\underset{\sim}{x},\underset{\sim}{z}>$ for convenience in working out problems.

Example 1 For the vector space R^n, define inner product $<\underset{\sim}{x},\underset{\sim}{y}>=x_1y_1+x_2y_2+\ldots+x_ny_n$. Show that inner product axioms are satisfied.

Solution:

(1) $$\begin{aligned}<\underset{\sim}{x},\underset{\sim}{y}> &= x_1y_1+x_2y_2+\ldots+x_ny_n\\ &= y_1x_1+y_2x_2+\ldots+y_nx_n\\ &= <\underset{\sim}{y},\underset{\sim}{x}>\end{aligned}$$

(2) $$\begin{aligned}<\underset{\sim}{x}+\underset{\sim}{y},\underset{\sim}{z}> &= (x_1+y_1)z_1+(x_2+y_2)z_2+\ldots+(x_n+y_n)z_n\\ &= (x_1z_1+x_2z_2+\ldots+x_nz_n)+(y_1z_1+y_2z_2+\ldots+y_nz_n)\\ &= <\underset{\sim}{x},\underset{\sim}{z}>+<\underset{\sim}{y},\underset{\sim}{z}>\end{aligned}$$

(3) $$\begin{aligned}<\alpha\underset{\sim}{x},\underset{\sim}{y}> &= \alpha x_1y_1+\alpha x_2y_2+\ldots+\alpha x_ny_n\\ &= \alpha(x_1y_1+x_2y_2+\ldots+x_ny_n)\\ &= \alpha<\underset{\sim}{x},\underset{\sim}{y}> \qquad \alpha\in R\end{aligned}$$

(4) $$\begin{aligned}<\underset{\sim}{x},\underset{\sim}{x}> &= x_1x_2+x_2x_2+\ldots+x_nx_n\\ &= x_1^2+x_2^2+\ldots+x_n^2>0\end{aligned}$$

If $\underset{\sim}{x}\neq\underset{\sim}{0}$ and equal to zero if $\underset{\sim}{x}=\underset{\sim}{0}$. Thus, all the axioms of inner product are satisfied.

Example 2 Let $C[a, b]$ be the vector space of all real finite functions $f(x)$ integrable on internal a to b. Show that

$$<\underset{\sim}{f},\underset{\sim}{g}>=\int_a^b f(x)g(x)\,dx$$

defines an inner product.

Solution: Since the functions are real, the inner product is a real number.

(1) $<\underset{\sim}{f},\underset{\sim}{g}>=\int_a^b f(x)g(x)\,dx=\int_a^b g(x)f(x)\,dx=<\underset{\sim}{g},\underset{\sim}{f}>$

(2) $< \underset{\sim}{f}+\underset{\sim}{g}, \underset{\sim}{h}> = \int_a^b [f(x)+g(x)]h(x)\,dx$

$$= \int_a^b [f(x)h(x)+g(x)h(x)]\,dx$$

$$= \int_a^b f(x)h(x)\,dx + \int_a^b g(x)h(x)\,dx$$

$$=< \underset{\sim}{f},\underset{\sim}{h}> + < \underset{\sim}{g},\underset{\sim}{h}>$$

(3) $< \alpha \underset{\sim}{f}, \underset{\sim}{g}> = \int_a^b \alpha f(x)g(x)\,dx$

$$= \alpha \int_a^b f(x)g(x)\,dx$$

(4) $< \underset{\sim}{f}, \underset{\sim}{f}> = \int_a^b \{f(x)\}^2\,dx > 0$ if $f(x) \neq 0$ for all x

$= 0$ if $f(x) = 0$ for all x, i.e., if $\underset{\sim}{f} = \underset{\sim}{0}$

In $C[a,b]$, we may also define inner product by,

$$< \underset{\sim}{f}, \underset{\sim}{g}> = \int_a^b w(x)f(x)g(x)\,dx$$

Where $w(x)$ is positive continuous function on $[a, b]$ and is called weight function.

Example 3 For the vector space C^n, that is, ordered n-tuples of complex numbers the inner product is defined by,

$$< \underset{\sim}{x}, \underset{\sim}{y}> = \bar{x}_1 y_1 + \bar{x}_2 y_2 + \cdots + \bar{x}_n y_n$$

Show that inner product axioms are satisfied.

Solution:

(1) $< \underset{\sim}{x}, \underset{\sim}{y}> = \bar{x}_1 y_1 + \bar{x}_2 y_2 + \cdots + \bar{x}_n y_n$

$$= (\overline{\bar{y}_1 x_1}) + (\overline{\bar{y}_2 x_2}) + \cdots + (\overline{\bar{y}_n x_n})$$

$$= \overline{(\bar{y}_1 x_1 + \bar{y}_2 x_2 + \cdots + \bar{y}_n x_n)}$$

$$=\overline{< \underset{\sim}{y}, \underset{\sim}{x}>} \qquad , \forall \underset{\sim}{x}, \underset{\sim}{y} \in C^n$$

(2) $< \underset{\sim}{x}+\underset{\sim}{y}, z> = (x_1 + y_1)z_1 + (x_2 + y_2)z_2 + \cdots + (x_n + y_n)z_n$

$$= (\bar{x}_1 z_1 + \bar{x}_2 z_2 + \cdots + \bar{x}_n z_n) + (\bar{y}_1 z_1 + \bar{y}_2 z_2 + \cdots + \bar{y}_n z_n)$$

$$= < \underset{\sim}{x}, \underset{\sim}{z}> + < \underset{\sim}{y}, \underset{\sim}{z}>$$

(3) $<\alpha \underset{\sim}{x}, \underset{\sim}{y}> = (\overline{\alpha x_1})y_1 + (\overline{\alpha x_2})y_2 + \cdots + (\overline{\alpha x_n})y_n$

$= \bar{\alpha}(\bar{x}_1 y_1 + \bar{x}_2 y_2 + \cdots + \bar{x}_n y_n)$

$= \bar{\alpha} < \underset{\sim}{x}, \underset{\sim}{y}> \qquad \alpha \in c$

(4) $< \underset{\sim}{x}, \underset{\sim}{x} > = \bar{x}_1 x + \bar{x}_2 x_2 + \cdots + \bar{x}_n x_n$

$= |\bar{x}_1|^2 + |\bar{x}_2|^2 + \cdots + |\bar{x}_n|^2$

> 0 if $\underset{\sim}{x} \neq \underset{\sim}{0}$, and $= 0$ if $\underset{\sim}{x} = \underset{\sim}{0}$

Thus, all the axioms are satisfied.

12.8.3 Inner Product Norm

Let V be a vector space and let $<\underset{\sim}{x}, \underset{\sim}{y}>$ be an inner product defined on V. Then the relation

$$\| \underset{\sim}{x} \| = \sqrt{<\underset{\sim}{x}, \underset{\sim}{x}>}$$

defines norm on V. We say that $\|.\|$ is a norm on V induced by the inner product. The above definition satisfies all the axioms of norm.

1. $\| \underset{\sim}{x} \| = \sqrt{<\underset{\sim}{x}, \underset{\sim}{x}>} > 0$ if $\underset{\sim}{x} \neq \underset{\sim}{0}$ and $= 0$ if $\underset{\sim}{x} = \underset{\sim}{0}$

2. $\|\alpha \underset{\sim}{x}\| = \sqrt{<\alpha \underset{\sim}{x}, \alpha \underset{\sim}{x}>} = \sqrt{\bar{\alpha}\alpha <\underset{\sim}{x}, \underset{\sim}{x}>} = \sqrt{|\alpha|^2 <\underset{\sim}{x}, \underset{\sim}{x}>}$

$= \|\alpha\| \sqrt{<\underset{\sim}{x}, \underset{\sim}{x}>}$

$= |\alpha| . \| \underset{\sim}{x} \|$

3. $\|\underset{\sim}{x} + \underset{\sim}{y}\|^2 = <\underset{\sim}{x} + \underset{\sim}{y}, \underset{\sim}{x} + \underset{\sim}{y}>$

$= <\underset{\sim}{x}, \underset{\sim}{x}> + <\underset{\sim}{x}, \underset{\sim}{y}> + <\underset{\sim}{y}, \underset{\sim}{x}> + <\underset{\sim}{y}, \underset{\sim}{y}>$

$= \|\underset{\sim}{x}\|^2 + <\underset{\sim}{x}, \underset{\sim}{y}> + \overline{<\underset{\sim}{x}, \underset{\sim}{y}>} + \|\underset{\sim}{y}\|^2$

$= \|\underset{\sim}{x}\|^2 + 2\,\mathrm{Re} <\underset{\sim}{x}, \underset{\sim}{y}> + \|\underset{\sim}{y}\|^2$

$\leq \|\underset{\sim}{x}\|^2 + 2|<\underset{\sim}{x}, \underset{\sim}{y}>| + \|\underset{\sim}{y}\|^2$

$\leq \|\underset{\sim}{x}\|^2 + 2\|\underset{\sim}{x}\| . \|\underset{\sim}{y}\| + \|\underset{\sim}{y}\|^2$

$\leq (\|\underset{\sim}{x}\| + \|\underset{\sim}{y}\|)^2$ (Cauchy–Schwarz inequality)

Thus,

$$\|\underset{\sim}{x} + \underset{\sim}{y}\| \leq \|\underset{\sim}{x}\| + \|\underset{\sim}{y}\|$$

12.8.4 Cauchy–Schwarz Inequality

In an inner product space V

$$|<\underset{\sim}{x},\underset{\sim}{y}>|^2 \le <\underset{\sim}{x},\underset{\sim}{x}><\underset{\sim}{y},\underset{\sim}{y}> \qquad \underset{\sim}{x},\underset{\sim}{y}\in V$$

The equality holds only if $\underset{\sim}{x}$ and $\underset{\sim}{y}$ are linearly dependent.

Proof: Let $\underset{\sim}{x},\underset{\sim}{y}\in V$ and λ be any scalar. If $\underset{\sim}{x}$ is not a scalar multiple of $\underset{\sim}{y}, \underset{\sim}{x}+\lambda \underset{\sim}{y}$ is a non-zero vector and hence

$$<\underset{\sim}{x}+\lambda\underset{\sim}{y},\underset{\sim}{x}+\lambda\underset{\sim}{y}> > 0$$

$$\text{or} \quad <\underset{\sim}{x},\underset{\sim}{x}> + <\underset{\sim}{x},\lambda\underset{\sim}{y}> + <\lambda\underset{\sim}{y},\underset{\sim}{x}> + <\lambda\underset{\sim}{y},\lambda\underset{\sim}{y}> > 0$$

$$\text{or} \quad <\underset{\sim}{x},\underset{\sim}{x}> + \lambda<\underset{\sim}{x},\underset{\sim}{y}> + \bar{\lambda}<\underset{\sim}{y},\underset{\sim}{x}> + \lambda\bar{\lambda}<\underset{\sim}{y},\underset{\sim}{y}> > 0$$

Let us choose $\lambda = -\dfrac{<\underset{\sim}{y},\underset{\sim}{x}>}{<\underset{\sim}{y},\underset{\sim}{y}>}$, then we have

$$<\underset{\sim}{x},\underset{\sim}{x}> - \frac{<\underset{\sim}{x},\underset{\sim}{y}><\underset{\sim}{y},\underset{\sim}{x}>}{<\underset{\sim}{y},\underset{\sim}{y}>} > 0$$

$$\text{or} \quad <\underset{\sim}{x},\underset{\sim}{x}> - \frac{<\underset{\sim}{x},\underset{\sim}{y}>\overline{<\underset{\sim}{x},\underset{\sim}{y}>}}{<\underset{\sim}{y},\underset{\sim}{y}>} > 0$$

$$\text{or} \quad <\underset{\sim}{x},\underset{\sim}{x}><\underset{\sim}{y},\underset{\sim}{y}> > \left|<\underset{\sim}{x},\underset{\sim}{y}>\right|^2$$

When $\underset{\sim}{x} = \lambda\,\underset{\sim}{y}$, substitution shows that

$$\text{LHS} = |<\underset{\sim}{x},\underset{\sim}{y}>|^2 = <\underset{\sim}{x},\underset{\sim}{y}>\overline{<\underset{\sim}{x},\underset{\sim}{y}>} = <\underset{\sim}{x},\underset{\sim}{y}><\underset{\sim}{y},\underset{\sim}{x}>$$

$$= <\lambda\,\underset{\sim}{y},\underset{\sim}{y}><\underset{\sim}{y},\lambda\,\underset{\sim}{y}>$$

$$= \lambda\bar{\lambda}<\underset{\sim}{y},\underset{\sim}{y}><\underset{\sim}{y},\underset{\sim}{y}> \qquad (12.5)$$

$$\text{RHS} = <\underset{\sim}{x},\underset{\sim}{x}><\underset{\sim}{y},\underset{\sim}{y}> = <\lambda y,\lambda y><y,y>$$

$$= \lambda\bar{\lambda}<\underset{\sim}{y},\underset{\sim}{y}><\underset{\sim}{y},\underset{\sim}{y}> \qquad (12.6)$$

From (12.5) and (12.6), we have

$$\left|<\underset{\sim}{x},\underset{\sim}{y}>\right|^2 = <\underset{\sim}{x},\underset{\sim}{x}><\underset{\sim}{y},\underset{\sim}{y}>$$

Cauchy Schwarz inequality can also be written as

$$<\underset{\sim}{x},\underset{\sim}{y}> \le \|\underset{\sim}{x}\|.\|\underset{\sim}{y}\|$$

Examples of Induced Norms:

1. In the vector space, R^n, an inner product is defined by,

$$< \underset{\sim}{x}, \underset{\sim}{y} >= x_1 y_1 + x_2 y_2 + + x_n y_n = \sum_{i=1}^{n} x_i y_i$$

Hence, the norm induced by this will be

$$\|\underset{\sim}{x}\| = \left(\sum_{i=1}^{n} x_i^2 \right)^{1/2}$$

2. In the vector space C^n, an inner product is defined by,

$$< \underset{\sim}{x}, \underset{\sim}{y} >= \overline{x_1}\, y_1 + \overline{x_2}\, y_2 + \quad + \overline{x_n}\, y_n = \sum_{i=1}^{n} \overline{x_i}\, y_i$$

Hence, the norm induced by this will be

$$\|x\| = \left(\sum_{i=1}^{n} | x_i |^2 \right)^{1/2}$$

3. In the vector space $C\,[a, b]$, an inner product defined by $< \underset{\sim}{f}, g >= \int_a^b w(x) f(x) g(x)\, dx$, will induce norm

$$\|\underset{\sim}{f}\| = \left(\int_a^b w(x) f^2(x)\, dx \right)^{1/2}$$

Parallelogram Law: The norm associated with an inner product satisfies parallelogram law

$$\|\underset{\sim}{x} + \underset{\sim}{y}\|^2 + \|\underset{\sim}{x} - \underset{\sim}{y}\|^2 = 2(\|\underset{\sim}{x}\|^2 + \|\underset{\sim}{y}\|^2)$$

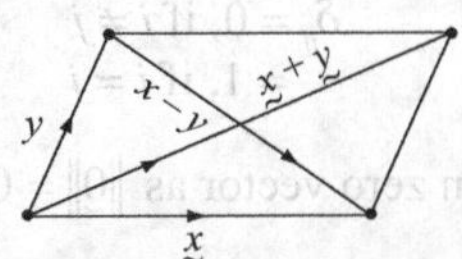

We are now in a position to visualize that norm defines distance between vectors of a sector space.
Let us consider the vector space $v_3(R)$, that is,

$$R^3 = \{ \underset{\sim}{x} = (x_1, x_2, x_3), x_i \in R \}$$

Then the length (or magnitude) of any vector $\underset{\sim}{x}$ is

$$\|\underset{\sim}{x}\| = (< \underset{\sim}{x}, \underset{\sim}{x} >)^{\frac{1}{2}} = \sqrt{x_1^2 + x_2^2 + x_3^2}$$

Further, the distance between two vector $\underset{\sim}{x}$ and y is

$$d(\underset{\sim}{x}, \underset{\sim}{y}) = \|\underset{\sim}{x} - \underset{\sim}{y}\| = \sqrt{< \underset{\sim}{x} - \underset{\sim}{y}, \underset{\sim}{x} - \underset{\sim}{y} >}$$

$$= \sqrt{(x_1 - y_1)^2 + (x_2 - y_2)^2 + (x_3 - y_3)^2}$$

Example 1 Compute $< \underset{\sim}{x}, \underset{\sim}{y} >$, $\|\underset{\sim}{x}\|$ and $d(\underset{\sim}{x}, \underset{\sim}{y})$ where $\underset{\sim}{x} = (1, -2, 4), \underset{\sim}{y} = (3, 1, -2)$

Solution:

$$< \underset{\sim}{x}, \underset{\sim}{y} > = 1.3 + (-2).1 + 4.(-2) = 3 - 2 - 8 = -7$$

$$\|\underset{\sim}{x}\| = \sqrt{(1)^2 + (-2)^2 + 16} = \sqrt{1 + 14 + 16} = \sqrt{21}$$

$$d(\underset{\sim}{x}, \underset{\sim}{y}) = \|\underset{\sim}{x} - \underset{\sim}{y}\| = \|(-2, -3, 6)\|$$

$$= \sqrt{4 + 9 + 36} = 7$$

12.8.5 Orthogonal Vectors

Two vectors $\underset{\sim}{x}$ and $\underset{\sim}{y} \in R^3$ are said to be orthogonal if the angle θ between them is $\frac{\pi}{2}$. This gives

$$< \underset{\sim}{x}, \underset{\sim}{y} > = 0 \qquad \text{(by 12.8.2)}$$

A set of vectors $\underset{\sim}{x}_1, \underset{\sim}{x}_2, \ldots, \underset{\sim}{x}_n$ is said to be an orthogonal set if

$$< \underset{\sim}{x}_i, \underset{\sim}{x}_j > = 0 \text{ for every } i \neq j \qquad (12.7)$$

If in addition $\|\underset{\sim}{x}_i\| = 1$ for every i, then the set is called an orthonormal set. (12.8)

The above conditions (12.7) and (12.8) can be combined into $< \underset{\sim}{x}_i, \underset{\sim}{x}_j > = \delta_{ij}$.
Where δ_{ij} known as Kronecker delta is defined by,

$$\delta_{ij} = 0, \text{ if } i \neq j$$
$$= 1, \text{ if } i = j$$

An orthonormal set does not contain zero vector as $\|\underset{\sim}{0}\| = 0$.

The set of vectors $\underset{\sim}{e}_1 = (1, 0, 0)$, $\underset{\sim}{e}_2 = (0, 1, 0)$, and $\underset{\sim}{e}_3 = (0, 0, 1)$ forms an orthonormal basis in R^3.

Example 1 Show that the set of vectors $\{(1, -2, 1), (3, 4, 5), (7, 1, -5)\}$ is orthogonal in R^3. Also find the orthonormal set.

Solution: Let $\underset{\sim}{x}_1 = (1, -2, 1)$, $\underset{\sim}{x}_2 = (3, 4, 5)$, $\underset{\sim}{x}_3 = (7, 1, -5)$
Then

$$< \underset{\sim}{x}_1, \underset{\sim}{x}_2 > = 1.3 + (-2).4 + 1.5$$

$$= 3 - 8 + 5 = 0$$

$$<\underset{\sim}{x}_1, \underset{\sim}{x}_3> = 1.7 + (-2).1 + 1.(-5)$$
$$= 7 - 2 - 5 = 0$$
$$<\underset{\sim}{x}_2, \underset{\sim}{x}_3> = 3.7 + 4.1 + 5(-5)$$
$$= 21 + 4 - 25 = 0$$

Hence, the vectors in the set are orthogonal. To find an orthonormal set, we compute $\|\underset{\sim}{x}_1\|, \|\underset{\sim}{x}_2\|$, and $\|\underset{\sim}{x}_3\|$.

Now

$$\|\underset{\sim}{x}_1\|^2 = <\underset{\sim}{x}_1, \underset{\sim}{x}_1> = 1 + 4 + 1 = 6 \therefore \|\underset{\sim}{x}_1\| = \sqrt{6}$$
$$\|\underset{\sim}{x}_2\|^2 = <\underset{\sim}{x}_2, \underset{\sim}{x}_2> = 9 + 16 + 25 = 50 \therefore \|\underset{\sim}{x}_2\| = \sqrt{50} = 5\sqrt{2}$$
$$\|\underset{\sim}{x}_3\|^2 = <\underset{\sim}{x}_3, \underset{\sim}{x}_3> = 49 + 1 + 25 = 75 \therefore \|\underset{\sim}{x}_3\| = \sqrt{75} = 5\sqrt{3}$$

The orthonormal set is $\left\{ \frac{\underset{\sim}{x}_1}{\|\underset{\sim}{x}_1\|}, \frac{\underset{\sim}{x}_2}{\|\underset{\sim}{x}_2\|}, \frac{\underset{\sim}{x}_3}{\|\underset{\sim}{x}_3\|} \right\}$

i.e., $\left\{ \frac{1}{\sqrt{6}}(1, -2, 1), \frac{1}{5\sqrt{2}}(3, 4, 5), \frac{1}{5\sqrt{3}}(7, 1, -5) \right\}$

Example 2 Show that in $C[-\pi, \pi]$ with inner product $<f, g> = \int_{-\pi}^{\pi} f(x)g(x)dx$, the set $\{1, \cos x, \sin x\}$ is an orthogonal set. Also find the orthonormal set.

Solution:

$$<1, \cos x> = \int_{-\pi}^{\pi} \cos x dx = 0, <1, \sin x> = \int_{-\pi}^{\pi} \sin x dx = 0, <\cos x, \sin x> = \int_{-\pi}^{\pi} \sin x \cos x \, dx = 0$$

Hence, the given set of vector is an orthogonal set. To find orthonormal set, we compute normal of each vector.

$$\|1\|^2 = <1, 1> = \int_{-\pi}^{\pi} 1 dx = 2\pi; \ \|\cos x\|^2 = <\cos x, \cos x> = \int_{-\pi}^{\pi} \cos^2 x dx = \pi, \ \|\sin x\|^2 = \int_{-\pi}^{\pi} \sin^2 x dx = \pi$$

Hence $\left\{ \frac{1}{\sqrt{2\pi}}, \frac{1}{\sqrt{\pi}} \cos x, \frac{1}{\sqrt{\pi}} \sin x \right\}$ is an orthonormal set.

THEOREM 12.19 A set of orthogonal vectors is linearly independent.

Proof: Let $\{\underset{\sim}{x}_1, \underset{\sim}{x}_2, \ldots, \underset{\sim}{x}_n\}$ be a set of orthogonal vectors. Suppose further that scalars $c_1, c_2, \ldots, c_n$ exist such that

$$c_1\underset{\sim}{x}_1 + c_2\underset{\sim}{x}_2 + \cdots + c_k\underset{\sim}{x}_k + \cdots + c_n\underset{\sim}{x}_n = \underset{\sim}{0}$$

Forming inner product with $\underset{\sim}{x}_k$, we get

$$c_1 < \underset{\sim}{x}_k, \underset{\sim}{x}_1 > + c_2 < \underset{\sim}{x}_k, \underset{\sim}{x}_2 > + \cdots + c_k < \underset{\sim}{x}_k, \underset{\sim}{x}_k > + \cdots + c_n < \underset{\sim}{x}_k, \underset{\sim}{x}_n > = \underset{\sim}{0}$$

Since the given vectors are orthogonal, each of these inner products is zero except $< \underset{\sim}{x}_k, \underset{\sim}{x}_k >$. Hence, we have

$$c_k < \underset{\sim}{x}_k, \underset{\sim}{x}_k > = 0 \Rightarrow c_k = 0$$

Giving values $k = 1, 2, \ldots, n$ in turn, we obtain $c_1 = c_2 = \ldots = c_n = 0$.
Therefore, the given set of vectors is linearly independent.

12.8.6 Gram–Schmidt Orthogonalization Process

Let V be a vector space whose basis is given, that is, a set of linearly independent vectors is given. From the given set, we can construct an orthogonal set. The process is known as Gram–Schmidt orthogonalization, which is discussed below.

Let $\{\underset{\sim}{x}_1, \underset{\sim}{x}_2, \ldots, \underset{\sim}{x}_n\}$ be the given basis.

Take $\underset{\sim}{y}_1 = \underset{\sim}{x}_1$ and $\underset{\sim}{y}_2 = \underset{\sim}{x}_2 - \lambda_1 \underset{\sim}{y}_1$

Then $\underset{\sim}{y}_1, \underset{\sim}{y}_2$ are orthogonal if $< \underset{\sim}{y}_1, \underset{\sim}{y}_2 > = 0$

$$\text{i.e., } < \underset{\sim}{y}_1, \underset{\sim}{x}_2 - \lambda \underset{\sim}{y}_1 > = 0$$

$$\text{or } < \underset{\sim}{y}_1, \underset{\sim}{x}_2 > - \lambda < \underset{\sim}{y}_1, \underset{\sim}{y}_1 > = 0 \Rightarrow \lambda_1 = \frac{< \underset{\sim}{y}_1, \underset{\sim}{x}_2 >}{< \underset{\sim}{y}_1, \underset{\sim}{y}_1 >}$$

Next let $\underset{\sim}{y}_3 = \underset{\sim}{x}_3 - \mu_1 \underset{\sim}{y}_1 - \mu_2 \underset{\sim}{y}_2$

The constants μ_1 and μ_2 are determined from the conditions that $\underset{\sim}{y}_3$ is orthogonal to both $\underset{\sim}{y}_1$ and $\underset{\sim}{y}_2$. Hence

$$< \underset{\sim}{y}_1, \underset{\sim}{y}_3 > = 0 \Rightarrow < \underset{\sim}{y}_1, \underset{\sim}{x}_3 - \mu_1 \underset{\sim}{y}_1 - \mu_2 \underset{\sim}{y}_2 > = 0$$

$$\Rightarrow < \underset{\sim}{y}_1, \underset{\sim}{x}_3 > - \mu_1 < \underset{\sim}{y}_1, \underset{\sim}{y}_1 > - \mu_2 < \underset{\sim}{y}_1, \underset{\sim}{y}_2 > = 0$$

$$\Rightarrow \mu_1 = \frac{< \underset{\sim}{y}_1, \underset{\sim}{x}_3 >}{< \underset{\sim}{y}_1, \underset{\sim}{y}_1 >} \qquad (\text{since } < \underset{\sim}{y}_1, \underset{\sim}{y}_2 > = 0)$$

and

$$< \underset{\sim}{y}_2, \underset{\sim}{y}_3 > = 0, \Rightarrow < \underset{\sim}{y}_2, \underset{\sim}{x}_3 - \mu_1 \underset{\sim}{y}_1 - \mu_2 \underset{\sim}{y}_2 > = 0$$

$$\Rightarrow < \underset{\sim}{y}_2, \underset{\sim}{x}_3 > - \mu_1 < \underset{\sim}{y}_2, \underset{\sim}{y}_1 > - \mu_2 < \underset{\sim}{y}_2, \underset{\sim}{y}_2 > = 0$$

$$\Rightarrow \mu_2 = \frac{< \underset{\sim}{y}_2, \underset{\sim}{x}_3 >}{< \underset{\sim}{y}_2, \underset{\sim}{y}_2 >}$$

Therefore

$$y_3 = x_3 - \frac{<y_1, x_3>}{<y_1, y_1>} y_1 - \frac{<y_2, x_3>}{<y_2, y_2>} y_2.$$

Processing in this manner, we get

$$y_j = x_j - \frac{<y_1, x_j>}{<y_1, y_1>} y_1 - \frac{<y_2, x_j>}{<y_2, y_2>} y_2 - \cdots\cdots \frac{<y_{j-1}, x_j>}{<y_{j-1}, y_{j-1}>} y_{j-1}$$

$$j = 1, 2, \ldots, n$$

An orthonormal set is obtained by dividing each of the above sectors y_j by its norm $\| y_j \|$. The above process is known as Gram–Schmidt process.

Example 1 Construct an orthogonal set of vectors from $x_1 = (1, -2, 1)$ $x_2 = (4, 1, 2)$, and $x_3 = (3, 1, 2)$.
Also compute orthonormal set.

Solution: Let $y_1 = x_1 = (1, -2, 1)$

Take $y_2 = x_2 - \lambda y_1$. Since y_2 must be orthogonal to $y_1, < y_1, y_2 > = 0$

$$\Rightarrow < y_2,\ x_2 - \lambda y_1 > = 0$$

$$\Rightarrow \lambda = \frac{< y_1,\ x_2 >}{< y_1,\ y_1 >}$$

$$= \frac{<1, -2, 1> <4, 1, 2>}{<1, -2, 1> <1, -2, 1>}$$

$$= \frac{4-2+2}{1+4+1} = \frac{4}{6} = \frac{2}{3}$$

$$= y_2 = (4, 1, 2) - \frac{2}{3}(1, -2, 1)$$

$$= \left(\frac{10}{3}, \frac{7}{3}, \frac{4}{3}\right)$$

To find the third orthogonal vector, we can put

$y_3 = x_3 - \mu_1 y_1 - \mu_2 y_2$ and determine μ_1 and μ_2 by applying condition

$$< y_1, y_3 > = 0 \text{ and } < y_2, y_3 > = 0$$

These condition given $\mu_1 = \dfrac{<y_1,\ x_3 >}{<y_1, y_1 >}$ and $\mu_2 = \dfrac{<y_2,\ x_3 >}{<y_2, y_2 >}$

We can also obtain $\underset{\sim}{y}_3$ in a simpler way by assuming $\underset{\sim}{y}_3 = (a, b, c)$ and applying the condition of orthogonality.
This gives $a - 2b + c = 0$, $10a + 7b + 4c = 0$

Its solution gives
$$\frac{a}{-8-7} = \frac{-b}{4-10} = \frac{7}{7+20}$$

$$\text{or} \quad \frac{a}{-15} = \frac{b}{6} = \frac{c}{27} \quad \text{or} \quad \frac{a}{-5} = \frac{b}{2} = \frac{c}{9}$$

$$\therefore \underset{\sim}{y}_3 = (-5, 2, 9)$$

Now

$$\|\underset{\sim}{y}_1\| = \sqrt{1+4+1} = \sqrt{6}, \|\underset{\sim}{y}_2\| = \frac{1}{3}\sqrt{100+49+16} = \frac{1}{3}\sqrt{165}$$

$$\|\underset{\sim}{y}_3\| = \sqrt{25+4+81} = \sqrt{110}$$

Therefore, the orthonormal set is

$$\left\{\frac{1}{\sqrt{6}}(1, -2, 1), \frac{3}{\sqrt{165}}\left(\frac{10}{3}, \frac{7}{3}, \frac{4}{3}\right), \frac{1}{\sqrt{110}}(-5, 2, 9)\right\}$$

$$\text{or} \quad \left\{\frac{1}{\sqrt{6}}(1, -2, 1), \frac{1}{\sqrt{165}}(10, 7, 4), \frac{1}{\sqrt{110}}(-5, 2, 9)\right\}$$

Example 2 Find an orthogonal basis for the subspace generated by vectors (1, 1, 1, 1), (–1, 4, 4, 1), and (0, 2, –2, 1) using Gram–Schmidt process.

Solution: Let $\underset{\sim}{x}_1 = (1, 1, 1, 1)$, $\underset{\sim}{x}_2 = (-1, 4, 4, 1)$, and $\underset{\sim}{x}_3 = (0, 2, -2, 1)$
Then

$$\underset{\sim}{y}_1 = \underset{\sim}{x}_1 = (1, 1, 1, 1)$$

$$\underset{\sim}{y}_2 = \underset{\sim}{x}_2 - \frac{<\underset{\sim}{x}_2, \underset{\sim}{y}_1>}{<\underset{\sim}{y}_1, \underset{\sim}{y}_1>}\underset{\sim}{y}_1$$

$$= (-1, 4, 4, 1) - \frac{8}{4}(1, 1, 1, 1)$$
$$= (-1, 4, 4, 1) - 2(1, 1, 1, 1)$$
$$= (-3, 2, 2, -1)$$

And
$$\underset{\sim}{y}_3 = \underset{\sim}{x}_3 - \frac{<\underset{\sim}{x}_3, \underset{\sim}{y}_1>}{<\underset{\sim}{y}_1, \underset{\sim}{y}_1>}\underset{\sim}{y} - \frac{<\underset{\sim}{x}_3, \underset{\sim}{y}_2>}{<\underset{\sim}{y}_2, \underset{\sim}{y}_2>}\underset{\sim}{y}_2$$

$$= (0, 2, -2, 1) - \frac{1}{4}\ (1, 1, 1, 1) - \frac{-1}{18}\ (-3, 2, 2, -1)$$

$$= (0,2,-2,1) - \left(\frac{1}{4},\frac{1}{4},\frac{1}{4},\frac{1}{4}\right) + \left(-\frac{1}{6},\frac{1}{9},\frac{1}{9},-\frac{1}{18}\right)$$

$$= \left(0-\frac{1}{4}-\frac{1}{6}, 2-\frac{1}{4}+\frac{1}{9}, -2-\frac{1}{4}+\frac{1}{9}, 1-\frac{1}{4}-\frac{1}{18}\right)$$

$$\underset{\sim}{y}_3 = \left(-\frac{15}{36},\frac{67}{36},-\frac{77}{36},\frac{25}{36}\right)$$

$$= \frac{1}{36}\ (-15, 67, -77, 25)$$

Hence, the orthogonal set of vector is

$$= (1, 1, 1, 1), (-3, 2, 2, -1), \frac{1}{36}\ (-15, 67, -77, 25)$$

Example 3 On the space $P_3(x)$ of all polynomials of degree ≤ 3, define inner product by,

$$< \underset{\sim}{f}, \underset{\sim}{g} > = \int_{-1}^{1} f(x)g(x)dx$$

Construct an orthonormal basis with the help of linearly independent set $\{1, x, x^2, x^3\}$.

Solution: Let the orthogonal basis be $\underset{\sim}{y}_1, \underset{\sim}{y}_2, \underset{\sim}{y}_3, \underset{\sim}{y}_4$ Then

$$\underset{\sim}{y}_1 = 1$$

$$\underset{\sim}{y}_2 = x - \frac{<x, \underset{\sim}{y}_1>}{<\underset{\sim}{y}_1, \underset{\sim}{y}_1>}\underset{\sim}{y}_1 = x - \frac{<x, 1>}{<1, 1>} = x - \frac{\int_{-1}^{1} xdx}{\int_{-1}^{1} 1dx} = x - 0 = x$$

$$\underset{\sim}{y}_3 = x^2 - \frac{<x^2, \underset{\sim}{y}_1>}{<\underset{\sim}{y}_1, \underset{\sim}{y}_1>}\underset{\sim}{y}_1 - \frac{<x^2, \underset{\sim}{y}_2>}{<\underset{\sim}{y}_2, \underset{\sim}{y}_2>}\underset{\sim}{y}_2$$

$$= x^2 \frac{\int_{-1}^{1} x^2 dx}{\int_{-1}^{1} 1dx} - \frac{\int_{-1}^{1} x^2 .xdx}{\int_{-1}^{1} x.xdx}$$

$$= x^2 - \frac{2}{3.2} - 0 = x^2 - \frac{1}{3}$$

And

$$\underset{\sim}{y_4} = x^3 - \frac{<x^3, \underset{\sim}{y_1}>}{<\underset{\sim}{y_1}, \underset{\sim}{y_1}>}\underset{\sim}{y_1} - \frac{<x^3, \underset{\sim}{y_2}>}{<\underset{\sim}{y_2}, \underset{\sim}{y_2}>}\underset{\sim}{y_2} - \frac{<x^3, \underset{\sim}{y_3}>}{<\underset{\sim}{y_3}, \underset{\sim}{y_3}>}\underset{\sim}{y_3}$$

$$= x^3 - \frac{\int_{-1}^{1} x^3 \cdot 1dx}{\int_{-1}^{1} 1dx} - \frac{\int_{-1}^{1} x^3 \cdot xdx}{\int_{-1}^{1} x^2 dx}x - \frac{\int_{-1}^{1} x^3 (x^2 - \frac{1}{3})dx}{\int_{-1}^{1} (x^2 - \frac{1}{3})^2 dx}(x^2 - \frac{1}{3})$$

$$= x^3 - 0 - \frac{1/5}{1/3}x - 0 = x^3 - \frac{3}{5}x$$

To find orthonormal set, we compute

$$\|1\|^2 = \int_{-1}^{1} 1dx = 2 \quad \therefore \|1\|^2 = \sqrt{2}$$

$$\|\underset{\sim}{x}\|^2 = \int_{-1}^{1} x^2 dx = \frac{2}{3} \quad \therefore \|\underset{\sim}{x}\| = \frac{\sqrt{2}}{\sqrt{3}}$$

$$\left\|x^2 - \frac{1}{3}\right\|^2 = \int_{-1}^{1} (x^4 - \frac{2}{3}x^2 + \frac{1}{9})dx = 2\left[\frac{1}{5} - \frac{2}{9} + \frac{1}{9}\right] = 2\left[\frac{1}{5} - \frac{1}{9}\right] = \frac{8}{45}$$

$$\therefore \left\|x^2 - \frac{1}{3}\right\| = \frac{2\sqrt{2}}{3\sqrt{5}}$$

And $$\left\|x^3 - \frac{3}{5}x\right\|^2 = \int_{-1}^{1} (x^3 - \frac{3}{5}x)^2 dx = 2\int_{0}^{1}\left(x^6 - \frac{6}{5}x^4 + \frac{9}{25}x^2\right)dx = 2\int_{0}^{1}\left(x^6 - \frac{6}{5}x^4 + \frac{9}{25}x^2\right)dx$$

$$= 2\left[\frac{1}{7} - \frac{6}{5.5} + \frac{3}{25}\right]$$

$$= 2\left[\frac{1}{7} - \frac{3}{25}\right]$$

$$= \frac{8}{175}$$

$$\therefore \left\|x^3 - \frac{3}{5}x\right\| = \frac{2\sqrt{2}}{5\sqrt{7}}$$

Hence, the required orthonormal basis is

$$\frac{1}{\sqrt{2}}, \frac{\sqrt{3}}{\sqrt{2}}x, \frac{\sqrt{5}}{2\sqrt{2}}\left(3x^2 - 1\right), \frac{\sqrt{7}}{2\sqrt{2}}\left(5x^2 - 3x\right)$$

EXERCISE – A

1. Show that the set of complex numbers $C = \{x + iy: x, y \in R\}$ forms a vector space over the field of real numbers with respect to addition and scalar multiplication.
2. Show that the set $R^{m \times n}$ of matrices with real elements forms a vector space over the field of real numbers with respect to matrix addition and scalar multiplication.
3. Let V be the set of all ordered pairs of real numbers. Define addition and scalar multiplication on V by $(x_1, x_2) + (y_1, y_2) = (0, x_2 + y_2)$. Show that V is not a vector space. Specify the axiom that is not satisfied. Hint: Identity for addition not satisfied, that is, $\underset{\sim}{x} + \underset{\sim}{o} \neq \underset{\sim}{x}$.
4. Show that the set V of all sequences of real numbers with addition and scalar multiplication defined by $\{a_n\} + \{b_n\} = \{a_n + b_n\}$ and $\alpha\{a_n\} = \{\alpha a_n\}$, $a_n, b_n, \alpha \in R$ is a vector space.
5. If V is a vector space over the field F. Show that $\alpha \underset{\sim}{x} = \underset{\sim}{o}$, if either $\alpha = 0$ or $\underset{\sim}{x} = \underset{\sim}{o}$ $\alpha \in F$, $\underset{\sim}{x} \in v$.
6. Let $S = \{(x_1, x_2, 0): x_1, x_2 \in R\}$ be a subset of R^3. Prove that S is a subspace with respect to addition and scalar multiplication.
7. Show that the set $S = \{(x_1, x_2, x_3): x_1 + x_2 + x_3 = 0, x_i \in R\}$ is a subspace of R^3 with respect to addition and scalar multiplication defined by
$$\underset{\sim}{x} + \underset{\sim}{y} = (x_1 + y_1,\ x_2 + y_2,\ x_3 + y_3)$$
$$\alpha \underset{\sim}{x} = (\alpha x_1, \alpha x_2, \alpha x_3),\ \alpha \in R$$
8. Show that the set $S = \{(x_1, x_2, x_3): x_1 = x_2, x_i \in R\}$ is a subspace of R^3 with usual addition and scalar multiplication.
9. Show that $S = \{(x_1, x_2, 1): x_i \in R\}$ is not a subspace of R^3 under usual addition and scalar multiplication.
10. Show that the set of functions in $C[-1, 1]$ such that $f(-1) = f(1)$ is a subspace of the vector space $C[-1, 1]$.
11. Show that the vectors (2, 0, 1), (1, –1, 2), (3, –1, 1) are linearly independent in R^3.
12. Show that the vectors (1, 1, 1, 1), (1, 1, 1, 0), (1, 1, 0, 0), (1, 0, 0, 0) are linearly independent in R^4.
13. Show that the following set of vectors (1, –1, 2, 1), (–1, 2, 3, 1), and (1, 0, 5, 3) is linearly independent in R^4.
14. Show that the set of vectors (2, 1, 3), (1, –1, 2), and (–1, 4, –3) is linearly dependent.
15. Show that the vectors (1, 2, –3, 1), (2, –1, 0, 4), and (4, 3, –6, 6) are linearly dependent.
16. Express the vector (2, –1, 6) as a linear combination of the vectors (1, 2, 3), (2, –1, 2), and (3, 1, 1).
17. Express the vector (2, 2, 3) as a linear combination of vectors (1, –1, 1), (2, 0, 1), (1, 1, 2) in $V_3(R)$.

18. Show that the vectors 1, $e^x + e^{-x}$, $e^x - e^{-x}$ are linearly independent in $C[-1, 1]$.

19. Determine whether the following vectors are linearly independent in $P_3(x)$:
    (i) $2, x^2, x^2 - 2$
    (ii) $x + 2, x + 5, x^2 + 1$

20. Show that the vectors $2, 2x, x^2, x^3$ are linearly independent in $P_4(x)$.

21. Show that the functions x^2 and $x|x|$ are linearly independent in $C[-1, 1]$.

EXERCISE – B

1. Show that the subset $S = \{(1, 2), (2, 3)\}$ of R^2 is its basis.
2. Show that the set of vectors (1, 1, 1), (1, 1, 0), (1, 0, 0) is a basis of R^3.
3. Show that the set $\{E_{11}, E_{12}, E_{21}, E_{22}\}$ of square matrices of order 2 is a basis for the vector space $R^{2 \times 2}$ of matrices with real elements with respect to matrix addition and scalar multiplication.
4. Show that the vectors (2, 1, 1), (3, 2, 5), (−1, 1, 2) form a basis of R^3.
5. Determine whether or not the following matrices

$$\begin{bmatrix} 1 & 0 \\ 0 & 1 \end{bmatrix}, \begin{bmatrix} 0 & 1 \\ 0 & 0 \end{bmatrix}, \begin{bmatrix} 0 & 0 \\ 1 & 0 \end{bmatrix} \text{ are linearly independent in } R^{2\times 2}.$$

6. Find a basis for the subspace of R^4 consisting of all vectors of the form $(2a - b, a + b - c, 3b, c)^T$ where a, b, and c are real numbers. Also, find the dimension of the subspace.
7. Let W be the surface of P_3 consisting of all polynomials of the form $ax^2 + 2bx + 3a + 5b$. Find a basis for W.
8. Show that T: $R^3 \to R^3$ defined by $T(x_1, x_2, x_3) = (x_1 + 2x_2, x_2 + 2x_3, x_3 - x_1)$ is a linear transformation.
9. Show that T: $R^3 \to R^2$ defined by $T(x_1, x_2, x_3) = (x_1 + x_3, x_2 - x_3)$ is a linear transformation.
10. Determine which of the following are linear transformations from $R^3 \to R^3$.
    (i) $T(x_1, x_2, x_3) = (0, x_2 + 1, x_3 + 2)$
    (ii) $T(x_1, x_2, x_3) = (x_1, x_2, 0)$
    (iii) $T(x_1, x_2, x_3) = (\sin x_1, \cos x_2, 1)$
11. Let T: $R^2 \to R$ be a linear transformation defined by $T(1,1)^T = 1$ and $T(1,0)^T = 4$. Find $T(x_1, x_2)^T$.
12. Find the matrix of linear transformation T: $R^3 \to R^2$ defined by $T(x_1, x_2, x_3)^T = (x_1, x_3)^T$ with respect to standard bases.
13. Find the matrix A of the linear transformation T such that $Tx = Ax$ for every x in R^3 where T: $R^3 \to R^3$ is defined by $T(x_1, x_2, x_3)^T = (x_1 + 2x_3, x_2 - x_3, 2x_1 + 3x_2)^T$ with respect to standard basis.

14. Let T: $R^2 \to R^3$ be the linear transformation defined by $T(x_1, x_2)^T = (x_2, x_1 + x_2, x_1 - x_2)^T$. Find the matrix representation of T with respect to ordered bases $[\underset{\sim}{u}_1, \underset{\sim}{u}_2]$ and $[\underset{\sim}{v}_1, \underset{\sim}{v}_2, \underset{\sim}{v}_3]$ where $\underset{\sim}{u}_1 = (1, 2)^T$, $\underset{\sim}{u}_2 = (3, 1)^T$, and $\underset{\sim}{v}_1 = (1, 0, 0)^T$, $\underset{\sim}{v}_2 = (1, 1, 0)^T$, $\underset{\sim}{v}_3 = (1, 1, 1)^T$.

15. Find the transition matrix corresponding to the change of basis from $[\underset{\sim}{v}_1, \underset{\sim}{v}_2]$ to $[\underset{\sim}{e}_1, \underset{\sim}{e}_2]$ for the following:

    (i) $\underset{\sim}{v}_1 = (1, 1)^T$, $\underset{\sim}{v}_2 = (1, -1)^T$

    (ii) $\underset{\sim}{v}_1 = (1, 1)^T$, $\underset{\sim}{v}_2 = (-1, 1)^T$

    (iii) $\underset{\sim}{v}_1 = (2, 3)^T$, $\underset{\sim}{v}_2 = (3, 5)^T$

16. Find the transition matrix representing the change of ordered basis $[1, x, x^2]$ of P_3 to the ordered basis $[1, 1 + x, 1 + x + x^2]$.

17. For the vector $\underset{\sim}{x} = (2, 6, -3)$ in R^3, compute $\| \underset{\sim}{x} \|_1, \| \underset{\sim}{x} \|_2$, and, $\| \underset{\sim}{x} \|_\infty$.

18. Let $\underset{\sim}{x} = (4, 1, 3)$ and $\underset{\sim}{y} = (2, 3, 5)$ be vectors in R^3. Compute $\| \underset{\sim}{x} - \underset{\sim}{y} \|_1$ and $\| \underset{\sim}{x} - \underset{\sim}{y} \|_\infty$.

19. Prove that for any vectors $\underset{\sim}{u}$ and $\underset{\sim}{v}$ in an inner product space $\| \underset{\sim}{u} + \underset{\sim}{v} \|^2 + \| \underset{\sim}{u} - \underset{\sim}{v} \|^2 = 2 \| \underset{\sim}{u} \|^2 + 2 \| \underset{\sim}{v} \|^2$

20. For the vector space of real n-tuples $\underset{\sim}{x} = (x_1, x_2, \ldots, x_n) \in R^n$, $x_i \in R$, the norm is defined by $\| \underset{\sim}{x} \| = | x_1 | + | x_2 | + \ldots + | x_n |$ Show that norm axioms are satisfied.

21. For the vector space of real n-tuples $(x_1, x_2, \ldots, x_n)$ the norm is defined by $\| \underset{\sim}{x} \| = \max\limits_{1 \le i \le n} | x_i |$. Show that norm axioms are satisfied.

22. Show that the functions $v_1(x) = \dfrac{1}{\sqrt{\pi}} \cos x$ and $\underset{\sim}{v}_2(x) = \dfrac{1}{\sqrt{\pi}} \sin x$ form an orthonormal set in $-C[-\pi, \pi]$ if inner product is defined by $<f, g> = \int_{-\pi}^{\pi} f(x) g(x) dx$.

23. Show that the set of functions $\{\sin t, \sin 2t, \ldots, \sin nt\}$ is orthonormal on the space $C[-\pi, \pi]$ with inner product $<f, g> = \int_{-\pi}^{\pi} f(t) g(t) dt$.

24. Obtain an orthonormal basis by applying Gram–Schmidt process to the set of vectors (1, 1, 0), (–1, 0, 1), and (2, 3, 4) in R^3 with standard inner product $< \underset{\sim}{x}, \underset{\sim}{y} > = x_1 y_1 + x_2 y_2 + x_3 y_3, \underset{\sim}{x}, \underset{\sim}{y} \in R^3$.

25. Using Gram–Schmidt process, find an orthonormal basis for the subspace spanned by $\{1, x, x^2\}$ in the vector space $C[-1, 1]$ with inner product $<f, g> = \int_{-1}^{1} f(x) g(x) dx$.

Appendix

ANSWER TO EXERCISES

Chapter 1

1. (a) $\{1, 2, 3\}$ (b) $\{-1, 1\}$ (c) $\{3, 4, 5, 6\}$ (d) $\{-\sqrt{2}, \sqrt{2}\}$
   (e) $\{-2, -1, 0, 1, 2\}$
2. $\{6, 12, 30\}$
3. (a) yes (b) yes (c) yes (d) yes
4. The relation is true
5. (i) equal (ii) equal
6. $A = \{(1, 2, 3), (1, 2, 4), (1, 3, 4), (2, 3, 4)\}$
7. (a) True (b) True
8. Not true as 5 is not even
9. (a) $A \cup B = \{1, 2, 3, 5, 6, 7, 8\}, A \cap B = \{5\}$
   (b) $A \cup C = \{1, 2, 3, 4, 5, 6, 8\}, A \cap C = \{2\}$
   (c) $B \cup C = \{2, 4, 5, 6, 7, 8\}, B \cap C = \{6, 8\}$
10. $A \cap B = \{5\} = A \cap C$ but $B \neq C$
12. (a) $\{5, 6\}$ (b) $\{1, 2, 3\}$, (c) ϕ (d) $\{1, 2, 3\}$
    (e) $\{1, 3, 5\}$ (f) $\{1, 2, 3, 5, 6\}$
15. $P(A) = \{\phi, \{1\}, \{2\}, \{3\}, \{4\}, \{1, 2\}, \{1, 3\}, \{1, 4\}, \{2, 3\}, \{2, 4\}, \{3, 4\}, \{1, 2, 3\}, \{2, 3, 4\}, \{1, 3, 4\}, \{1, 2, 4\}, \{1, 2, 3, 4\}\}$, 16 elements
16. (a), (b), (e), (f) and (g) are true and (c) and (d) are not true

17. (a) $A \times B = \{(2, 4), (2, 5), (2, 6), (3, 4), (3, 5), (3, 6)\}$
    (b) $B \times A = \{(4, 2), (4, 3), (5, 2), (5, 3), (6, 2), (6, 3)\}$
    (c) No
    (d) Six elements
18. $A_1 \times A_2 \times A_3$ has n_1, n_2, n_3 elements.
19. $P \times Q \times R = \{(1, a, p), (1, b, p), (1, c, p), (2, a, p), (2, b, p), (2, c, p), (1, a, q), (1, b, q), (1, c, q), (2, a, q), (2, b, q), (2, c, q)\}$

Chapter 2

1. (a) $a = 2$ (b) $b = 4$ (c) $b = 2$
   (d) $x = 3$ (e) $x =$ JAVA, $y =$ C ++ (f) $x = 2, y = 2$.
2. Dom $(R) = \{a, b, c)$ Range $(R) = \{1, 2, 3, 4\}$
3. Dom $(R) = \{1, 2, 3\}$, Range $(R) = \{2, 3, 4\}$
4. (a) $RoS = \{(1, 3), (1, 4), (3, 1), (4, 1)\}$
   (b) $SoR = \{(2, 1), (2, 3), (3, 2), (3, 4), (4, 1)\}$
   (c) $RoR = \{(1, 1), (1, 2), (1, 3), (1, 4), (3, 2)\}$
   (d) $SoS = \{(3, 4), (3, 1)\}$
5. (a) $R = \{(1, 9), (2, 6), (3, 3)\}$, Dom $(R) = \{1, 2, 3\}$, Range $(R) = \{3, 6, 9\}$
   (b) $R^{-1} = \{(9, 1), (6, 2), (3, 3)\}$
   (c) $RoR = \{(3, 3)\}$
6. $R = \{(1, 1), (2, 4), (4, 16)\}$, Dom $(R) = \{1, 2, 4\}$, Range $(R) = \{1, 4, 16\}$

$$M_R = \begin{array}{c} \\ 1 \\ 2 \\ 3 \\ 4 \end{array}\!\!\begin{array}{c} \begin{array}{ccccc} 1 & 4 & 5 & 6 & 16 \end{array} \\ \begin{bmatrix} 1 & 0 & 0 & 0 & 0 \\ 0 & 1 & 0 & 0 & 0 \\ 0 & 0 & 0 & 0 & 0 \\ 0 & 0 & 0 & 0 & 1 \end{bmatrix} \end{array}$$

7. $a\,R\,b$ if a/b

   $R = \{(1, 1), (1, 4), (1, 9), (1, 12), (2, 4), (2, 12), (3, 9), (3, 12), (4, 4), (4, 12)\}$
   (i) Dom $(R) = \{1, 2, 3, 4\}$
   (ii) Range $(R) = \{1, 4, 9, 12\}$

$$M_R = \begin{array}{c} \\ 1 \\ 2 \\ 3 \\ 4 \end{array}\!\!\begin{array}{c} \begin{array}{cccc} 1 & 4 & 9 & 12 \end{array} \\ \begin{bmatrix} 1 & 1 & 1 & 1 \\ 0 & 1 & 0 & 1 \\ 0 & 0 & 1 & 1 \\ 0 & 1 & 0 & 1 \end{bmatrix} \end{array}$$

8. (a) Reflexive and symmetric
   (b) Irreflexive, not symmetric, and transitive.

9. Reflexive, symmetric, and transitive.

10. (a) $R = \{(1, 1), (2, 2), (3, 3), (2, 1), (2, 3), (3, 1)\}$

    (b) $R = \{(1, 2), (1, 4), (3, 2), (3, 4), (2, 5), (4, 5)\}$

11.

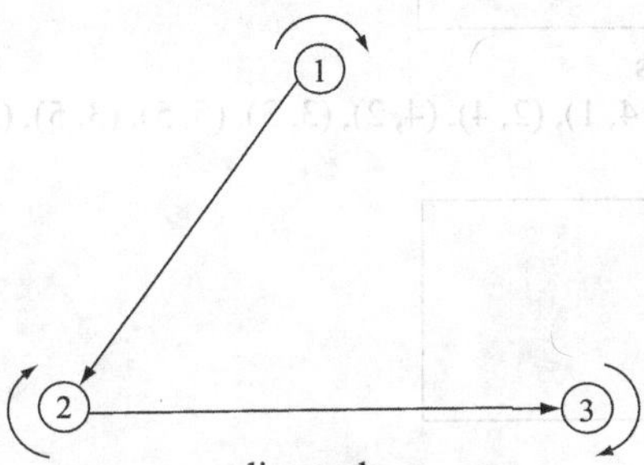

diagraph

$$M_R = \begin{array}{c} \\ 1 \\ 2 \\ 3 \end{array}\begin{array}{c} \begin{matrix} 1 & 2 & 3 \end{matrix} \\ \begin{bmatrix} 1 & 1 & 0 \\ 0 & 1 & 1 \\ 0 & 0 & 1 \end{bmatrix} \end{array}, \; R^2 = ROR = \{(1,1),(1,2),(1,3),(2,2),(2,3),(3,3)\}$$

(a) reflexive (b) not symmetric
(c) not transitive (d) R is not an equivalence relation.

12. **Hint:** Let R be irreflexive and transitive. Assume $a\ R\ b$ and $b\ R\ a$. This implies that $a\ R\ a$, that is, R is reflexive which contradicts the assumption that R is irreflexive and $a\ R\ b$ and bRa is not true. Hence if aRb, then $b \not R\ a$ and so R is asymmetric

13. **Hint:** Let aR^2b and hence an element $c \in A$ exists such that $a\ R\ c$ and $c\ R\ b$. As R is symmetric $a\ R\ c \Rightarrow c\ R\ a$ and $c\ R\ b \Rightarrow b\ R\ c$. Now, $b\ R\ c$ and $c\ R\ a \Rightarrow b\ R^2\ a$, that is, R^2 is symmetric.

14. The matrix M_R represents relation $R = \{(a, a), (a, b), (b, a), (b, b), (c, c)\}$. Clearly the relation R is reflexive, symmetric, and transitive, and hence, R is an equivalence relation.

15. The relation $R = \{(a, a), (b, b), (c, c), (a, b), (b, c), (c, a)\}$.
    The relation is reflexive but neither symmetric nor transitive.

    $(a, b) \in R$ but $(b, a) \notin R$ and $(a, b) \in R$, $(b, c) \in R \not\Rightarrow (a, c) \in R$, R is not an equivalence relation.

16. R is an equivalence relation. Further

$$A = \{(1, 1), (1, 2), (1, 3), (2, 1), (2, 2), (2, 3), (3, 1), (3, 2), (3, 3)\}$$

The sum of elements can be 2, 3, 4, 5, and 6; hence, the corresponding equivalence classes are $\{1, 1\}$, $\{(1, 2), (2, 1)\}$, $\{(1, 3), (2, 2), (3, 1)\}$, $\{(2, 3), (3, 2)\}$ and the quotient set A/R is the set whose elements are these classes, that is, sets.

17. $A = \{a, b, c, d\}$.
    The equivalence relation is $R = \{(a, a), (b, b), (c, c), (d, d), (a, b), (b, a), (c, d), (d, c)\}$.

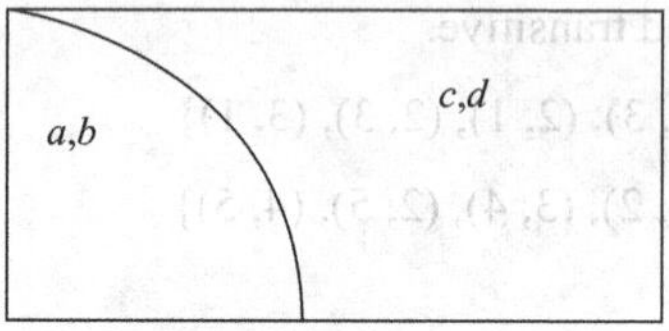

18. $A = \{1, 2, 3, 4, 5\}$, the equivalence relation is
$R = \{(1, 1), (2, 2), (4, 4), (1, 2), (2, 1), (1, 4), (4, 1), (2, 4), (4, 2), (3, 3), (5, 5), (3, 5), (5, 3)\}$

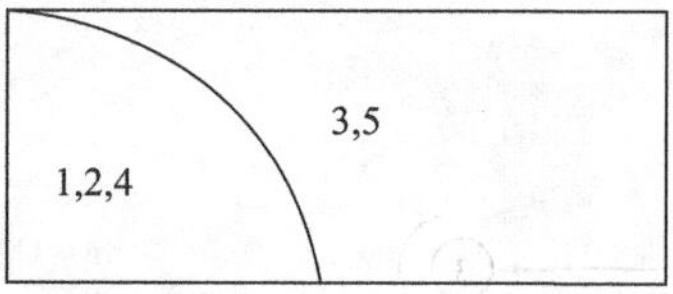

19. (a) $R \cap S = \{(1, 2), (4, 4)\}$
(b) $R \cup S = \{(1, 1), (1, 2), (1, 4), (2, 3), (2, 4), (4, 4), (1, 3), (2, 1)\}$
(c) $\overline{R} = \{(1, 3), (2, 1), (2, 2), (3, 1), (3, 2), (3, 3), (3, 4), (4, 1), (4, 2), (4, 3)\}$
(d) $RoS = \{(1, 2), (1, 3), (1, 1), (1, 4), (2, 4), (4, 4)\}$
(e) $S^2 = SoS = \{(1, 1), (2, 2), (2, 3), (4, 4)\}$

Chapter 3

ANSWERS – A

2. (i) function (ii) function (iii) function
4. (i) not a function (ii) not a function (iii) function
5. (i) function (ii) not a function (iii) not a function (iv) function
6. (i) relation is not a function (ii) range $(f) = \{-1, 1, 3, 5, 7\}$
7. (i) g is not a function
(ii) g is not a function
(iii) g is a function whose domain is A and range is B
(iv) g is a function whose domain is A and range is $\{1, 3\}$
9. (i) one-one onto (ii) many-one into
(iii) one-one onto (iv) one-one onto
(v) one-one onto (vi) one-one into
10. (i) one-one onto (ii) one-one into
(iii) many-one into (as negative integers have no pre-image)
11. (i) one-one onto (ii) many-one into as $4 \in$ B does not have pre image
12. (i) $f = \{(1, 1), (2, 2), (3, 3), (4, 4)\}$
(ii) $f = \{(1, 2), (2, 4), (3, 4), (4, 2)\}$
13. Domain $(f) = \{x: x \in R \text{ and } -\frac{\pi}{2} \le x \le \frac{\pi}{2}\}$, Co-domain $(f) = \{y: y \in R \text{ and } -1 \le y \le 1\}$
Domain $(g) = (x: x \in R \text{ and } 0 \le x \le \pi\}$, Co-domain $(g) = \{y: y \in R \text{ and } -1 \le y \le 1\}$

14. $f^{-1}(y) = y/2$, where $y = f(x) = 2x$

15. $g(x) = ax + b = y \therefore g^{-1}(y) = x = (y - b)/a$ and hence g^{-1} exists.

16. $f^{-1} = \{(x, a), (u, v), (z, c), (y, d)\}$
    $g^{-1} = \{(u, a), (x, b), (y, c), (z, d)\}$

17. $(gof)x = x^2 + 2$, $(fog)x = x^2 + 4x + 4$; $gof(1) = 3$, $fog(1) = 9$, $gof \neq fog$.

18. (i) $(gof)x = \sin^2 x$, $(fog)x = \sin x^2$, $gof \neq fog$.
    (ii) $(gof)x = 4x^2 + 8x + 7$, $(fog)x = 2x^2 + 8$, $gof \neq fog$ in both cases (i) & (ii).

19. (i) $(gof)5 = 17$ (ii) $(fog)5 = 14$ (iii) $gof(-5) = -3$
    (iv) $fog(-5) = -6$ (v) $(gof)x = 2x + 7$, (vi) $(fog)x = 2x + 4$.

20. $(gof)^{-1}z = x = \dfrac{4z-1}{3}$, $f^{-1}(y) = x = (y - 1/3)$, $g^{-1}(z) = y = 4z$,

    $(f^{-1}og^{-1})z = f^{-1}(g^{-1}(z)) = f^{-1}(4z) = \dfrac{4z-1}{3}$, $(gof)^{-1} = f^{-1}og^{-1}$

21. (i) 8 (ii) −9 (iii) −5
22. (i) 9 (ii) −8 (iii) −5
23. (i) 4 (ii) 0 (iii) 9
24. (i) 3 (ii) 4 (iii) 6
25. (i) 2 (ii) 3 (iii) −3

ANSWERS – B

1. (i) The number of steps remains 5001.
   (ii) The number of steps doubles.
   (iii) The number of steps quadruples.
   (iv) The number of steps increases eightfold.

2. $|f(n)| = |n!| = |1.\,2.3.........(n-1).\,n|$
   $\leq n.n.n.................n.n,$ for $n \geq 1$
   $\leq n^n$
   $\therefore f(n) = O(n^n)$

3. $|g(n)| = |1 + 2 + 3 + \ldots + n| \leq n + n + n + \ldots + n$
   $\leq n.n \leq n^2$
   $\therefore g(n) = O(n^2)$

4. $|f(n)| = |5n + \lg n| \leq |5n + n| = 6n,$ $n \geq 1$
   $\therefore f(n) = O(n)$

5. $|g(n)| = |n^2 (7n^2 - 2n)| = |7n^4 - 2n^3| \le |7n^4 + 2n^3|$

$$\le |7n^4 + 2n^4| = 9n^4, \qquad n \ge 1$$

$\therefore\ g(n) = O(n^4)$

6. $f(n) = 2n^2 + 5n,\ g(n) = n^2 + 10n$

$|f(n)| = |2n^2 + 5n| \le |2n^2 + 20n| \le 2|(n^2 + 10n)| \le 2|g(n)|$

$\therefore\ f(n) = O(g(n))$

Also, $|g(n)| = |n^2 + 10n| \le |2n^2 + 20n|, \qquad n \ge 1$

$$\le |2n^2 + 5n + 15n| \le |2n^2 + 5n| + |15n|$$
$$\le |2n^2 + 5n| + |2n^2 + 15n|$$
$$\le |2n^2 + 5n| \cdot 2$$
$$\le 2\,|f(n)|$$

$\therefore\ g(n) = O(f(n))$

Hence, $f(n)$ and $g(n)$ are of the same order.

7. $f(n) = 5n^4 - 2n^3,\ g(n) = n^4$

$|f(n)| = |5n^4 - 2n^3| \le |5n^4 + 2n^3| \le |5n^4 + 2n^4| \le 7n^4, \qquad n \ge 1$

$$\le 7\,|g(n)|$$

$\therefore\ f(n) = O[g(n)]$

Also, $g(n) = n^4 = 5n^4 - 4n^4 \le 5n^4 - 2n^3, \qquad n \ge 1$

Or $|g(n)| \le |5n^4 - 2n^3|, \qquad n \ge 1$

$$\le |f(n)|$$

$\therefore\ g(n) = O(f(n))$

Hence, $f(n)$ and $g(n)$ are of the same order.

8. (i) Let $f(n) = 3n + 5$, then $|f(n)| = |3n + 5| \ge |n|, \qquad n \ge 1$

$\therefore\ f(n) = \Omega(n)$

(ii) Let $f(n) = 5n^2 + 3n - 2$

Then $|f(n)| = |5n^2 + 3n - 2| \ge |5n^2|, \qquad n \ge 1$

$$\ge 5n^2$$
$$\ge cg(n) \text{ where } g(n) = n^2$$

Hence, $f(n) = \Omega[g(n)] = \Omega(n^2)$

9. $f(n) = 25n^3 + 5n + 15 \ge 25n^3, \qquad n \ge 1$

Hence, $f(n) = \Omega(n^3)$

10. Let $f(n) = 25n^5 - 2n^4 + 5n^3 + 2n + 1$

then $|f(n)| = |25n^5 - 2n^4 + 5n^3 + 2n + 1|$

$$\le |25n^5| + |2n^4| + |5n^3| + |2n| + |1| \qquad n \ge 1$$
$$\le 25n^5 + 2n^4 + 5n^3 + 2n + 1$$

$$\leq 25n^5 + 2n^5 + 5n^5 + 2n^5 + n^5$$
$$\leq 35n^5$$
$$\therefore f(n) = O(n^5)$$

11. $f(n) = \dfrac{n(n+1)}{2} = \dfrac{1}{2}(n^2 + n) \leq \dfrac{1}{2}(n^2 + n^2), \qquad n \geq 1$

$$\leq n^2$$
$$\leq g(n) \text{ where } g(n) = n^2$$

$\therefore \qquad f(n) = O[g(n)]$

Also since $n + 1 \geq n$ for every $n \geq 1$,

$$\frac{n\,(n+1)}{2} \geq \frac{n^2}{2}$$
$$\therefore f(n) = \Omega\,[g(n)]$$

Thus, $\qquad f(n) = \Theta\,[g(n)] = \Theta\,(n^2)$

12. $f(n) = 2^{n+1} = 2.2^n \leq c.2^n$, for some constant c, for $n \geq 1$. Hence, $2^{n+1} = O(2^n)$.

13. $f(n) = 2^{2n} = 2^n.2^n$
$\leq c.\,2^n$, for some constant c and $n \geq n_0$.
Here the inequality gives the value of $c = 2^n$ that depends upon n which is not possible for a constant. Hence, 2^{2n} is not of order 2^n.

14. $f(n) = 25 + 2 \lg n$
Now $|f(n)| = |25 + 2 \lg n| \quad \geq 2 \lg n, \quad$ for all $n > 1$
$\geq \lg n, \quad$ for all $n > 1$
$\therefore f(n) = \Omega(\lg n)$

15. Let $f(n) = 1^2 + 2^2 + 3^2 + \ldots + n^2$
$\leq n^2 + n^2 + n^2 + \ldots + n^2$
$\leq n.n^2$
$\leq n^3$
$\therefore f(n) = O(n^3)$

16. Let $f(n) = 1^k + 2^k + 3^k + \ldots + n^k$
$\leq n^k + n^k + n^k + \ldots + n^k, \qquad n \geq 1$
$\leq n.n^k$
$\leq n^{k+1}$
$= O(n^{k+1})$

17. $R^\infty = \{(1, 1), (1, 2), (1, 3), (2, 3), (3, 3)\}$

18. $R^\infty = \{(1, 1), (1, 2), (1, 3), (2, 1), (2, 2), (2, 3), (3, 3)\}$

19. $R^\infty = \{(1, 1), (1, 2), (1, 3), (2, 1), (2, 2), (2, 3), (3, 1), (3, 2), (3, 3)\}$

20. $R^{\infty} = \{(1, 1), (1, 2), (1, 3), (1, 4), (2, 1), (2, 2), (2, 3), (2, 4), (3, 1), (3, 2), (3, 3), (3, 4), (4, 1), (4, 2), (4, 3), (4, 4)\}$

21. $R^{\infty} = \{(1, 1), (1, 2), (1, 3), (1, 4), (2, 1), (2, 3), (2, 4), (3, 3), (3, 4), (4, 3), (4, 4)\}$

22. $W_1 = \begin{bmatrix} 1 & 0 & 0 & 1 & 0 \\ 0 & 1 & 0 & 0 & 0 \\ 0 & 0 & 0 & 1 & 1 \\ 1 & 0 & 0 & 1 & 0 \\ 0 & 1 & 0 & 0 & 1 \end{bmatrix} W_2 = W_1 = W_3$

Chapter 4

ANSWERS – A

3. (a), (b), (c), (f) are not propositions, (d), (e) are propositions as the statements are true. (g) is neither true nor false and hence is not a proposition.

4. (a), (b), (c), (d) are propositions, (e) is not.

5. (a) (i) $p \wedge q$ (ii) $p \wedge \sim q$ (iii) $\sim p \vee \sim q$
   (b) (i) Politician and scholar
   (ii) Politician or scholar
   (iii) Neither politician nor scholar
   (iv) Neither politician nor scholar

6. (a) (i) F $\wedge$ T is false
   (ii) T $\wedge$ T is true
   (iii) T $\wedge$ F is false
   (iv) T $\wedge$ T is true
   (b) (i) $\sim p$: Gopal is not an engineer
   (ii) $p \wedge q$: Gopal is an engineer and Ashok is a reputed doctor
   (iii) $p \vee q$: Gopal is an engineer or Ashok is a reputed doctor
   (iv) $p \vee \sim q$: Gopal is an engineer or Ashok is not a reputed doctor.

8. (a) Tautology (b) Contradiction

9. (a) $\sim p \vee t$ is true (b) $p \wedge \sim t$ is false

10. (a) Tautology (b) Tautology (c) Contradiction

11. Truth tables of (a) $\sim (p \wedge q)$ and (b) $(p \to q) \wedge (q \to p)$

(a)

p	q	p ∧ q	~p ∧ q
T	T	T	F
T	F	F	T
F	T	F	T
F	F	F	T

(b)

$p \to q$	$q \to p$	$(p \to q) \wedge (q \to p)$
T	T	T
F	T	F
T	F	F
T	T	T

16. (a) $p \vee q$

$\sim p$

$\therefore q$

		Premises		Conclusion
p	q	$p \vee q$	$\sim p$	q
T	T	T	F	T
T	F	T	F	F
F	T	T	T	T
F	F	F	T	F

The critical row 3 contains true conclusion and hence the argument is valid.

(b) $p \vee q$, we construct table of premises and conclusion.

$\sim q$

$\therefore \sim p$

(c) $p \vee q$

$\sim q$

p

		Premises		Conclusion	
p	q	$p \vee q$	$\sim q$	$\sim p$	p
T	T	T	F	F	T
T	F	T	T	F	T
F	T	T	F	T	F
F	F	F	T	T	F

For (b) the critical row 2 does not contain true conclusion, and hence, the argument is invalid. For (c) the critical row 2 contains true conclusion, and hence, the argument is valid.

(d) $(p \to q, p \to r) \therefore p \to q \wedge r$

We construct the truth table.

			Premises		Conclusion	
p	q	r	$p \to q$	$p \to r$	$q \wedge r$	$p \to q \wedge r$
T	T	T	T	T	T	T
T	T	F	T	F	F	F
T	F	T	F	T	F	F
T	F	F	F	F	F	F
F	T	T	T	T	T	T
F	T	F	T	T	F	T
F	F	T	T	T	F	T
F	F	F	T	T	F	T

Critical rows 1, 5, 6, 7 and 8 show that conclusion is true and hence the argument is valid.

17. Argument is valid by *Modus Ponens*, that is, $p \to q$ and p is true and hence q is true.
18. The truth table for $(p \to q \text{ and } q) \to p$ is as follows:

	Premises		Conclusion
p	q	$p \to q$	$(p \to q \text{ and } q) \to p$
T	T	T	T
T	F	F	F
F	T	T	T
F	F	T	F

The critical rows 1 and 3 contain true conclusion and hence the argument is valid.

19. Let p: man is bachelor, q: man is unhappy, r: man dies young.
Now premises are: $p \to q$, $q \to r$
Conclusion: $p \to r$ is true by the law of syllogism.
20. (a) Converse of p: If the insurance company pays me, then a cyclone destroys my house or a tsunami destroys my house.
(b) Contrapositive of p: If the insurance company does not pay me, then a cyclone does not destroy my house and a tsunami does not destroy my house.
21. (a) All students do not live in hostels.
(b) In a university, every faculty member is not a PhD.
(c) It is not raining.
(d) It is not an unusual summer time.

22. Let p: Anand is rich, and q: Anand is humble. Negation of $p \wedge q$ is $\sim(p \wedge q)$
Truth table for $\sim(p \wedge q)$

p	q	p ∧ q	~(p ∧ q)
T	T	T	F
T	F	F	T
F	T	F	T
F	F	F	T

23. (a) For every x, $P(x)$ is odd.
(b) For every x, $Q(x)$ such that x is a perfect square.
(c) For every x and y, $R(x, y)$, that is, $x + y$ is even.

24. (a) $\exists x\,[p(x) \wedge q(x)]$
(b) $\forall x\,[t(x) \rightarrow \sim s(x)]$
(c) $\forall x\,[r(x) \rightarrow \sim s(x)]$
(d) $\forall x\,[t(x) \rightarrow \sim s(x)]$
(e) $\forall x\,[q(x) \wedge s(x)]$
(f) $\exists x\,[\sim p(x)]$

ANSWERS – B

1. (a) proposition (b) No (c) No (d) proposition
(e) proposition (f) proposition (g) No

2. (a) T (b) F (c) F (d) F

3. (a) It is not very cold today.
(b) Ganga is not a sacred river of India.
(c) Cow milk is not regarded as the best milk for health.
(d) Someone wants to buy my car.
(e) All people in the village have a house.

4. (a) $p \wedge q$ (b) $\sim p \wedge q$ (c) $p \wedge \sim q$
(d) $\sim(p \wedge q)$ (e) $\sim p \vee q$ (f) $p \vee \sim q$

5. (a) $p \wedge q$ (b) $\sim p \wedge q$ (c) $p \wedge \sim q$ (d) $\sim p \wedge \sim q$

6. (a) It is not very cold.
(b) Either it is very cold or temp 1°C.
(c) It is very cold and the temp is 1°C.
(d) It is not true that it is very cold and the temp is 1°C.

7. (a) T (b) T (c) F

8. (a) Truth table for $\sim p \wedge q$

p	q	~p	~p ∧ q
T	T	F	F
T	F	F	F
F	T	T	T
F	F	T	F

(b) Truth table for $p \vee \sim q$

p	q	~q	p ∨ ~q
T	T	F	T
T	F	T	T
F	T	F	F
F	F	T	T

(c) Truth table for $\sim p \wedge \sim q$

p	q	~q	~p ∧ ~q
T	T	F	F
T	F	T	F
F	T	F	F
F	F	T	T

(d) Truth table for $\sim(p \vee q) \vee (\sim p \wedge \sim q)$

p	q	~p	~q	p ∨ q	~(p ∧ q)	~p ∧ ~q	~(p ∧ q) ∨ (~p ∧ ~q)
T	T	F	F	T	F	F	F
T	F	F	T	T	F	F	F
F	T	T	F	T	F	F	F
F	F	T	T	F	T	T	T

9. (a) x is not divisible by 3.
   (b) x is even or divisible by 3.
   (c) x is neither even nor divisible by 3.

10. (a) $\exists x\left[P(x) \wedge Q(x)\right]$
    (b) $\exists x\left[P(x) \wedge \sim Q(x)\right]$
    (c) $\forall x\left[P(x) \vee Q(x)\right]$

Chapter 5

ANSWERS – A

1. 585×10^4 2. 24 3. 720 4. 30,240
5. 4536 6. 420 7. 35 8. (i) 2^8 (ii) 16
9. 60 10. 48 11. 42×10^2 12. 60
13. 3^{20} 14. 6^3 or 216 15. 210 16. 700
17. 5×10^4 18. 330 19. 3003, 2002
20. (i) 1820 (ii) 375 (iii) 25
21. (i) 286 (ii) 165 (iii) 80 (iv) 276
22. 14!, 13!, 2 × 13! 23. 2880 24. 2520
25. (i) 1, 26, 126 (ii) $15!/(5!)^3$

ANSWERS – B

1. (i) 0.4 (ii) 0.5 (iii) 0.7 (iv) 0.3
2. 8/15
3. (i) 1/13 (ii) 1/4 (iii) 3/13 (iv) 3/52 (v) 4/13
4. (i) 1/221 (ii) 13/102
5. (i) 1/18 (ii) 1/6 (iii) 7/12
6. 7/8 7. 101/168 8. 1/270725
9. (i) 42/143 (ii) 27/143
10. (i) 45/91 (ii) 24/91 (iii) 67/91
11. 7/10 12. 1/7
13. (i) $p(A/B) = 1/4$ and $p(B/A) = 1/5$ (ii) A and B are independent.
14. (i) 2/5 (ii) 1/15
15. (i) 1/2 (ii) 1/3 (iii) 2/5 (iv) 3/5
16. (i) 0.02169 (ii) 0.2410 (iii) 0.5422

ANSWERS – C

1. (i) 0.0729 (ii) 0.5905 (iii) 0.9186
2. (i) 0.4096 (ii) 0.0016 (iii) 0.1808
3. (i) 0.0778 (ii) 0.2592 (iii) 0.9222

4. (i) 0.0778 (ii) 0.6826 (iii) 0.2304
5. 0.0170
6. 0.5551
7. (i) 0.1240 (ii) 0.9743
8. 0.2461
9. 0.3151
10. 0.6786

ANSWERS – D

1. Mean of disribution $m = 0.5$, Poisson distribution is $200 \times \dfrac{e^{-0.5} \times (0.5)^r}{r!}$, $r = 0, 1, 2, 3, 4.$

   The theoretical frequencies for 0, 1, 2, 3, and 4 deaths are 122, 61, 15, 2, and 0, respectively.
2. $m = 1.2$, (i) 0.3012 (ii) 0.2169
3. 0.6283
4. 0.9998
5. $m = 2$, (i) 0.2706 (ii) 0.2706 (iii) 0.8647 (iv) 0.4059
6. 0.0190
7. (i) The probability that neither car is in demand is 0.2231 and its number of days in a year is 81.

   (ii) The probability that car demand is refused is 0.1913 and its number of days in a year is 70.
8. $m = 0.5625$, Poisson distribution is $\dfrac{Ne^{-m}m^r}{r!} = 400 \times \dfrac{e^{-0.5625} \times (0.5625)^r}{r!}$, $r = 0,1,2,3.$

   The theoretical frequencies (number of pages) having 0,1,2, and 3 errors are 228, 128, 36 and 7.
9. $m = 4$ (i) 2 days (ii) 91 days (iii) 43 days (iv) 39 days
10. $m = 0.02$,

    The number of packets containing 0, 1, and 2 defective blades are, respectively, Ne^{-m}, Nme^{-m}, and $N\dfrac{m^2e^{-m}}{2!}$ (where $N = 10{,}000$, $e^{-m} = e^{-0.02} = 0.9802$), i.e. 9802, 196, 2.

Chapter 6

ANSWERS

1. (i) $a_n = (-1)^{n+1}$

   (ii) $a_n = \dfrac{1}{n}$

   (iii) $a_n = n$

(iv) $a_n = 3^n, n = 0.1,...$

(v) $a_n = 1$ if n is odd and 0 if n is even

2. (i) $(1 + z)^{-1}$

(ii) $(1 - z^2)^{-1}$

(iii) $(1 - z)^{-1}$

(iv) e^z

3. (i) Linear homogeneous of degree 1

(ii) Linear homogeneous of degree 2

(iii) No

(iv) No

(v) No

4. (i) $a_n = 10(3.5)^{n-1}$

(ii) $\frac{1}{4}(9.5^{n-1}-1)$

5. (i) $a_n = A.10^n$

(ii) $a_n = A_1 3^n + A_2(-1)^n$

(iii) $b_n = 3.5^n + 2(-2)^n$

(iv) $a_n = \frac{1}{\sqrt{5}}\left(\frac{1+\sqrt{5}}{2}\right)^n - \frac{1}{\sqrt{5}}\left(\frac{1-\sqrt{5}}{2}\right)^n$

(v) $a_n = (1 + 3n)\, 2^{-n}$

(vi) $a_n = 2(1)^n + 3(2)^n$

6. (i) $a_n = c_1 2^n + c_2 3n + \frac{1}{6}5^n + n + \frac{3}{2}$

(ii) $a_n = (C_1 + C_2 n)5^n + n^2 + 6n + \frac{75}{8}$

(iii) $a_n = (C_1 + C_2 n)1^n + n^2 - 8n + 20$

(iv) $a_n = (C_1 + C_2 n)2^n + \frac{1}{8}n^2 2^n$

7. (i) $a_n = \left(C_1\left(\frac{2}{3}\right)^n + 15\right)$

(ii) $a_n = C_1 2^n + C_2 3^n + \frac{1}{6}5^n$

(iii) $a_n = (C_1 + C_2 n)2^{-n} + n^2 - 3n$

(iv) $a_n = (C_1 + C_2 n)3^n + (n^2 + 8n + 28)2^n$

8. (i) $a_n = 6.3^n - 2^n$

(ii) $a_n = \frac{1}{2}3^{n+1} - 3.2^n + \frac{5}{2}$

9. (i) $a_n = \frac{11}{9}(1)^n + \frac{2}{3}(n+1) + \frac{14}{9}(-1)^n 2^n$

   (ii) $a_n = \frac{1}{2}5^n + \frac{1}{2}7^n$

   (iii) $a_n = 2^n + \frac{1}{2}3^n - \frac{1}{2}1^n$

10. 6_{c_2} i.e., 15

11. 7_{c_2} i.e., 21

12. 13_{c_3} i.e., 286

Chapter 7

ANSWERS

1. Standard result
2. (i) Semigroup
   (ii) Semigroup
   (iii) Not a semigroup
   (iv) Not a semigroup
   (v) Semigroup
4. (i) Not a monoid
   (ii) Monoid
   (iii) Monoid
9. (i) Group
   (ii) Group
   (iii) Group
   (iv) Group
12. The set of positive rational numbers do not form a group as binary operation is not associative.
15. $(-i)$ is also its generator.
19. $G = \{1, -1, i, -i\}$ is a multiplicative group and $G' = \{\{0\}, \{1\}, \{2\}, \{3\}\}$ is additive group of residue classes modulo 4.
    If f: $G \to G'$ is an isomorphism, then identities and inverses correspand.
    Let $f\{1\} = \{0\}, f\{-1\} = \{2\}$ since inverses correspand, -1 is its own inverse and so also is $\{2\}$.
    $f(i) = \{1\}, f(-i) = \{3\}$
    The mapping is clearly one to one and ONTO.
26. (i) The set N of natural numbers is not a ring.
    (ii) The set of numbers of the form a $\sqrt{3}$ where $a \in Q$ is not a ring as binary operation is not defined, for example, $2\sqrt{3} \times 4\sqrt{3} = 24 \notin a\sqrt{3}$ with a rational.
    (iii) The set of numbers of the form $a + b\sqrt{3}$, $a,b \in Q$ is a commutative ring with unity element, is an integral domain and is also a field.

Chapter 8

ANSWERS

1. The relation R is a partial order relation.
2. (i) Not a partial order relation
   (ii) R is a partial order relation
3. The relation R is an equality relation.
4. The relation '$\geq$' is a partial order relation on the set of real numbers. The relation $\geq$ is linear order.
5. The relation of divisibility is linear order.
6. The poset B is linear.
7.

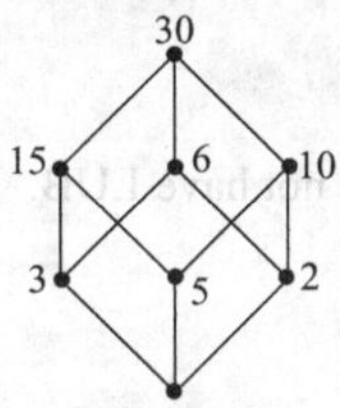

(a)

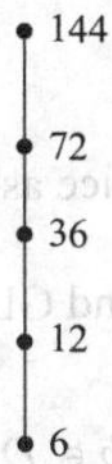

(b)

8.

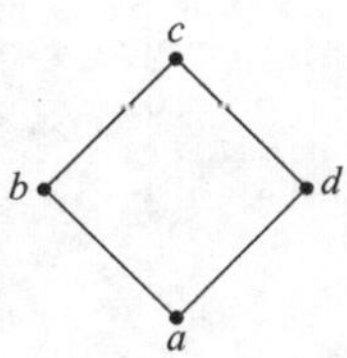

Minimal: a
Maximal: c

(a)

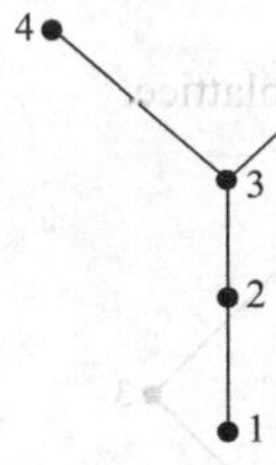

Minimal: 1
Maximal: 4,5

(b)

9.

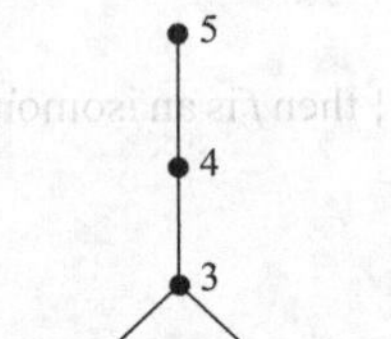

Greatest element: 5
Least element: none

(a)

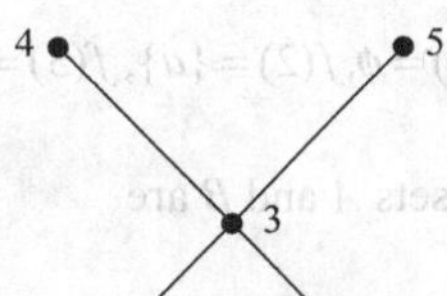

No greatest and no least element

(b)

10. $A = \{x: x \in R \text{ and } 0 < x < 2\}$, No greatest and no least element
    $B = \{x: x \in R \text{ and } 0 \leq x \leq 2\}$, 2 is greatest element and 0 is least element
11. 72 is greatest element and 2 is least element
12. A is lattice
13. lattice
14. 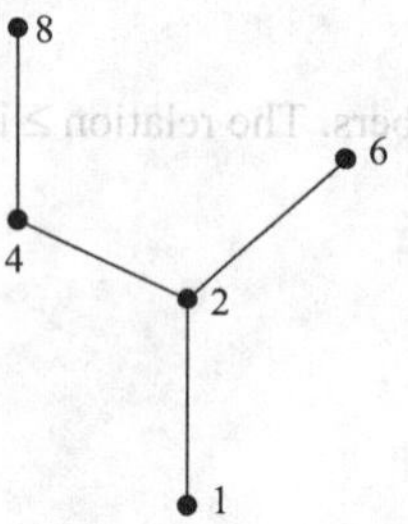

    Hasse diagram of poset A. The poset A is not a lattice as (6, 8) does not have LUB.
15. (i) is not a lattice as LUB of (c, d) does not exit
    (ii) is a lattice as every pair of element has LUB and GLB
16. D_{36} is a lattice
17. $D_n \subseteq Z^+$ is a sublattice as every pair of element $a, b \in D_n$ has its GLB and LUB in D_n.
18. (i) The set B is a lattice
    (ii) The set C is a lattice
19. The subset S_1 is a sublattice but S_2 is not a sublattice.
20. 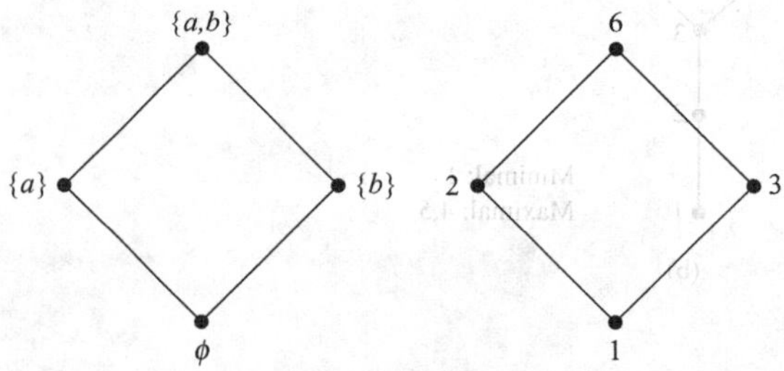

    Define $f: L \to P(A)$ by $f(1) = \phi, f(2) = \{a\}, f(3) = \{b\}, f(6) = \{a, b\}$ then f is an isomorphism and $L \simeq P(A)$.
21. The Hasse diagram of posets A and B are
    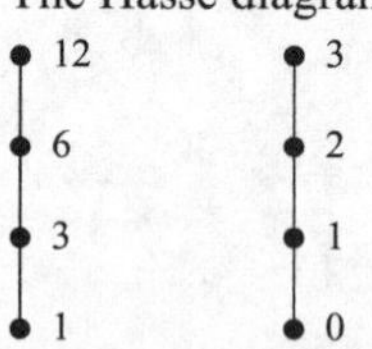

    Define $f: A \to B$ by $f(1) = 0, f(3) = 1, f(6) = 2, f(12) = 3$. Hasse diagram shows that if $a \leq b$ then $f(a) \leq' f(b)$

Chapter 9

ANSWERS

2. (i) $(a+b)' = a'.b'$ (ii) $(a.b)' = a' + b'$
3. (i) $A'.(BC)'$ (ii) $A.(BC)'$
4. (i) xy (ii) xy (iii) x (iv) x
5. 1, 0
6. 1, 1
7.

x	y	x′	y′	x + y′	x′ + y	f(x, y)
0	0	1	1	1	1	1
0	1	1	0	0	1	0
1	0	0	1	1	0	0
1	1	0	0	1	1	1

8. (iii) (a) 70 (b) 1 (c) 7 (d) 10 (e) 5 (f) 2
10. (i) $xy + xy' + x'y$ (ii) xy'
11. (i) $xyz + xyz' + xy'z$
    (ii) $xyz + x'yz' + x'yz + xyz' + xy'z$
    (iii) $xyz + xy'z + xyz' + xy'z' + x'yz' + x'y'z'$
    (iv) $xy'z + xy'z' + x'yz + x'yz'$
    (v) $xyz + x'yz + x'yz' + x'y'z' + x'y'z$
12. $x'yz + xy'z + xyz' + xyz$
13. (i) $(x+y+z)(x+y+z')(x+y'+z)(x+y'+z')(x'+y+z)(x'+y'+z)$
    (ii) $(x+y+z)(x'+y+z)(x+y+z')(x+y'+z)(x+y'+z')(x'+y+z')$
14. (i) 0 (ii) 1
15. (i) (a) 1 (b) 1
    (ii) (a) 1 (b) 0
16. (i) $E(x, y) = xy' + x'y$
    (ii) $E(x, y) = x' + y'$
    (iii) $E = ABC + ABC' + A'B'C$
    (iv) $E = A'B'CD + A'B'CD' + AB'C'D' + AB'CD'$
    (v) $f(x, y, z) = xyz + xyz' + x'yz + x'yz'$
17. (i) $x = 0$ (ii) $x = 1 = y$ (iii) $x = 1$ or $y = 1$
18. (a) $w = xy' + y'z$ (b) $w = xy' + xz'$

19.

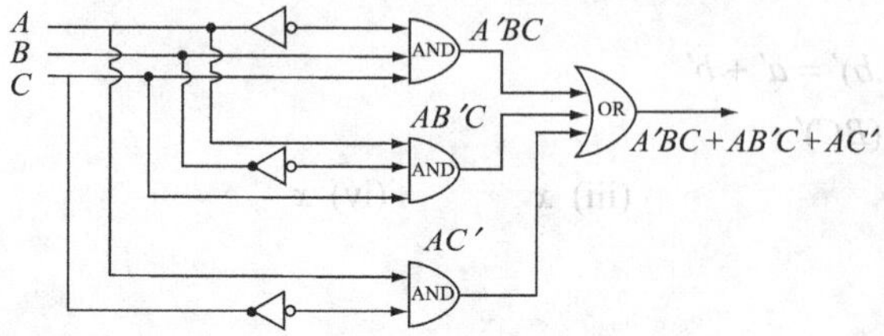

20.

(i)

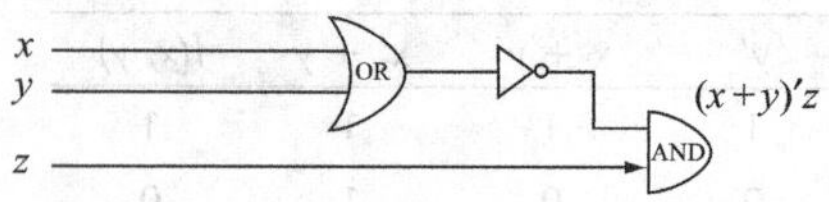

(ii)

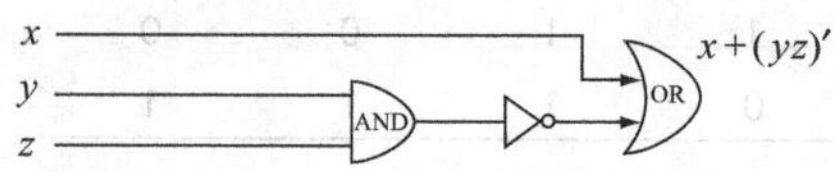

(iii)

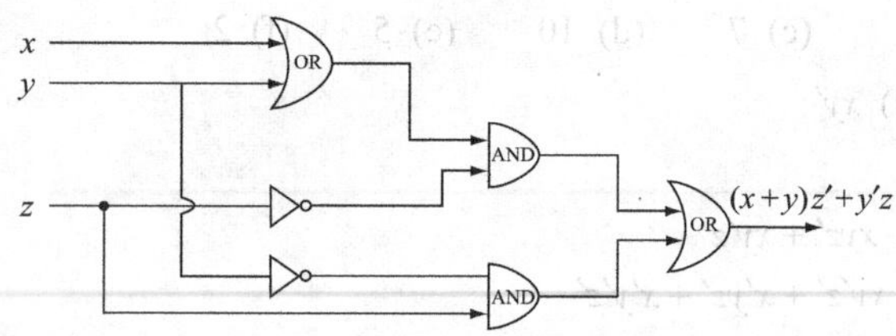

Chapter 10

ANSWERS – A

1. (i) $V = \{A, B, C, D\}$; $E = \{(A, A), (A, B), (A, C), (A, D), (B, C), (C, D)\}$
   (ii) $V = \{A, B, C, D\}$; $E = \{(A, B), (B, C), (C, D), (D, D)\}$
   (iii) $V = \{A, B, D, F, G\}$; $E = \{(A, B), (B, D), (A, D), (F, D), (G, D)\}$
   (iv) $V = \{A, B, C, D\}$; $E = \{(A, B), (B, C), (C, D), (D, A)\}$; All edges are double edges.
   (v) $V = \{A, B, C, D, E\}$; $E = \{(A, B), (A, D), (A, E), (B, C), (B, D), (C, D), (D, E)\}$
   (vi) $V = \{1, 2, 3, 4, 5\}$; $E = \{(1, 2), (1, 4), (1, 5), (2, 4), (2, 3), (3, 4), (4, 5)\}$
2. Degree of vertices of graph 1(a) are deg(A) = 5, deg(B) = 2, deg(C) = 3, and deg(D) = 2.
   Degree of vertices of graph 1(e) are deg(A) = 3, deg(B) = 3, deg(C) = 2, deg(D) = 4, and deg(E) = 2.
   Degree of vertices of graph 1(f) are deg(1) = 3, deg(2) = 3, deg(3) = 2, deg(4) = 4, and deg(5) = 2.
3. The null graph N_3 of three vertices is a graph consisting of three vertices and zero edges, i.e.

• • •

4.

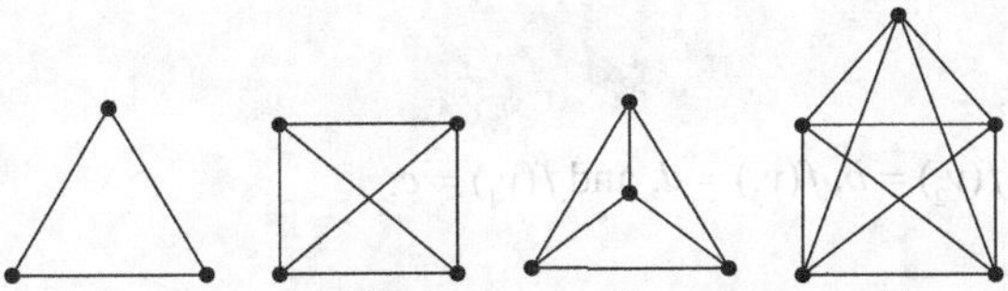

K_3 K_4 K_4 K_5

5. (i) In a regular graph all vertices are of equal degree.
Regular graphs of degree 2 (with three and four vertices) are

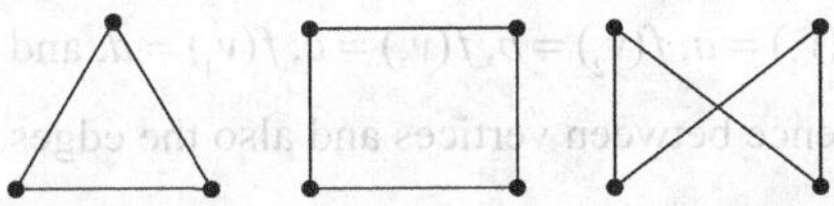

2-regular 2-regular 2-regular

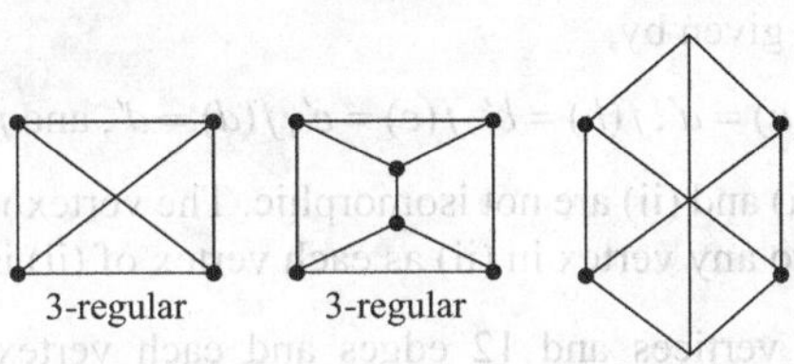

3-regular 3-regular 3-regular

(ii) Yes, K_n is a regular graph of degree $n - 1$.

6. (i) The isolated vertices are v_5 and v_6 and pendant vertices are v_3 and v_9.
(ii) (a) There are ${}^5C_2 = 10$ ways of choosing two vertices from the set V of graph G and hence the number of edges are 10.
(b) In a multigraph, multiple edges are permitted, and hence the graph G can have any number of edges. No maximum number of edges exists.
(iii) No. In a simple graph of n vertices, the maximum number of edges is $\dfrac{n(n-1)}{2}$
$= \dfrac{7 \times 6}{2} = 21$ edges and not 25 edges.

7. (i) Isomorphic

$$v_1 \to a, v_2 \to b, v_3 \to c, v_4 \to d, v_5 \to e$$

The vertices in correspondence have same degrees and also match the corresponding edges. Both have 5 vertices and 5 edges: 2 vertices are of degree 3, 1 vertex is of degree, and 2 are pendant vertices.
(ii) Isomorphic

$$v_1 \to e, v_2 \to b, v_3 \to d, v_4 \to c, v_5 \to a$$

The vertices as well as edges are in correspondence.

8. Not isomorphic as vertices a and c are pendant vertices and they do not correspond to v_1 or v_2 or v_3.

9. Isomorphic
   The isomorphism f is given by,

$$f(v_4) = a, f(v_2) = b, f(v_3) = d, \text{ and } f(v_1) = c$$

10. Not isomorphic
    If $f(v_4) = e$, then $f(v_3) = d$, which is *not* possible as v_3 is of degree 3 and d is of degree 2.

11. Isomorphic
    The isomorphism f is given by,

$$f(v_1) = a, f(v_2) = b, f(v_3) = c, f(v_4) = d, \text{ and } f(v_5) = e$$

    showing correspondence between vertices and also the edges.

12. Isomorphic
    The isomorphism f is given by,

$$f(a) = a', f(b) = b', f(c) = c', f(d) = d', \text{ and } f(e) = e'$$

13. The graphs given in (i) and (ii) are not isomorphic. The vertex v_4 in (i) is of degree 3 which does not correspond to any vertex in (ii) as each vertex of (ii) is of degree 2.

14. Both graphs have 8 vertices and 12 edges and each vertex is of degree 3 and so a correspondence of vertices and edges is possible. Hence the graphs are isomorphic.

15. Adjacency matrix $A = [a_{ij}]$ is given by,

    (a)

$$A = \begin{array}{c} \\ v_1 \\ v_2 \\ v_3 \\ v_4 \\ v_5 \end{array}\begin{array}{c} \begin{array}{ccccc} v_1 & v_2 & v_3 & v_4 & v_5 \end{array} \\ \begin{bmatrix} 0 & 1 & 1 & 1 & 0 \\ 1 & 0 & 1 & 0 & 0 \\ 1 & 1 & 0 & 1 & 1 \\ 1 & 0 & 1 & 0 & 1 \\ 0 & 0 & 1 & 1 & 0 \end{bmatrix} \end{array}$$

    (b)

$$A = \begin{array}{c} \\ v_1 \\ v_2 \\ v_3 \\ v_4 \end{array}\begin{array}{c} \begin{array}{cccc} v_1 & v_2 & v_3 & v_4 \end{array} \\ \begin{bmatrix} 0 & 1 & 1 & 2 \\ 1 & 1 & 1 & 1 \\ 1 & 1 & 1 & 1 \\ 2 & 1 & 1 & 0 \end{bmatrix} \end{array}$$

16. The adjacency matrix $A = [a_{ij}]$ is given by,

$$A = \begin{array}{c} \\ 1 \\ 2 \\ 3 \\ 4 \end{array}\begin{array}{c} \begin{array}{cccc} 1 & 2 & 3 & 4 \end{array} \\ \begin{bmatrix} 0 & 1 & 2 & 1 \\ 1 & 0 & 1 & 0 \\ 2 & 1 & 0 & 1 \\ 1 & 0 & 1 & 0 \end{bmatrix} \end{array}$$

The incidence matrix $M = [m_{ij}]$ is given by,

$$A = \begin{array}{c} \\ 1 \\ 2 \\ 3 \\ 4 \end{array}\begin{array}{c} \begin{matrix} e_1 & e_2 & e_3 & e_4 & e_5 & e_6 \end{matrix} \\ \begin{bmatrix} 1 & 0 & 0 & 1 & 1 & 1 \\ 1 & 1 & 0 & 0 & 0 & 0 \\ 0 & 1 & 1 & 0 & 1 & 1 \\ 0 & 0 & 1 & 1 & 0 & 0 \end{bmatrix} \end{array}$$

17. The graph corresponding to adjacency matrix consists of 4 vertices connected as follows:

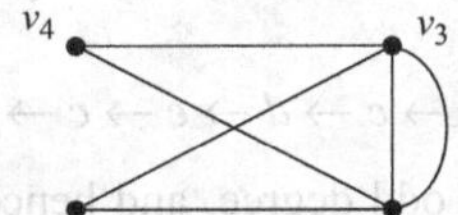

18. (a) Simple path (A, B, C, F) since all the vertices are distinct
    (b) (A, B, C, F, E, D, A) Closed Path, i.e., circuit
    (c) (A, B, F, E, D, B, A) Closed Walk
    (d) (A, B, F, E, D) Open Walk (simple path)
    (e) (D, E, F, C, B, D) Circuit
19. The possible circuits are 1, 2, 3, 1; 1, 3, 4, 1; 1, 4, 3, 1; 1, 4, 5, 1;
    1, 2, 3, 4, 1; 1, 6, 5, 1; 1, 6, 5, 4, 1; 1, 5, 4, 1.
20. (a) The sum of degrees of all vertices is $2 + 3 + 4 = 9$, which is odd and hence no graph is possible.
    (b) The sum of all vertices is $2 + 2 + 2 + 2 = 8$, which is even and hence there might be a graph with $8/2 = 4$ edges.
    A possible graph is 
    (c) The sum of degrees of four vertices is $4 + 5 + 2 + 5 = 16$ which is even and hence there might be a graph with $\frac{16}{2} = 8$ edges.
    A possible graph is

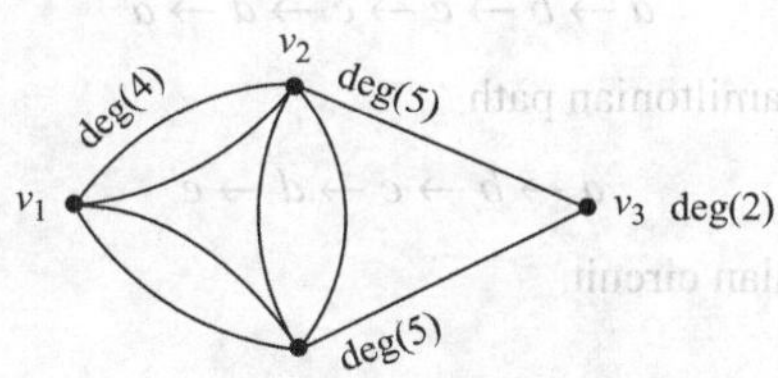

21. Yes, the given graph is bipartite. The two sets of vertices are $V_1 = \{v_1, v_2\}$ and $V_2 = \{v_3, v_4, v_5\}$.
22. The graphs in Figs (a) and (b) are both bipartite
    (a) $V_1 = \{v_1, v_3, v_5\}$, $V_2 = \{v_2, v_4, v_6\}$
    (b) $V_1 = \{A, B, C\}$, $V_2 = \{D, E, F, G\}$

23. (i) Each vertex is of even degree, and hence, the graph is an Eulerian graph. The Eulerian circuit is $a \to b \to c \to a$.
    (ii) Each vertex is of even degree, and hence, the graph is an Eulerian graph. The Eulerian circuit is $a \to b \to c \to d \to a$.
    (iii) There are three vertices of odd degree, and hence, there is neither Eulerian path nor Eulerian circuit.
    (iv) There are four vertices of odd degree, and hence, there is neither Eulerian path nor Eulerian circuit.
    (v) Each vertex is of even degree, and hence, the graph is an Eulerian graph. The Euler circuit exists, that is,

$$a \to b \to c \to d \to e \to c \to f \to a$$

    (vi) There are four vertices of odd degree, and hence, there is neither Eulerian path nor Eulerian circuit.
    (vii) There are exactly two vertices of odd degree, and hence, an Eulerian path is possible

$$v_1 \xrightarrow{e_1} v_2 \xrightarrow{e_2} v_3 \xrightarrow{e_3} v_4 \xrightarrow{e_4} v_1 \xrightarrow{e_5} v_5 \xrightarrow{e_6} v_4 \xrightarrow{e_7} v_3 \xrightarrow{e_8} v_5 \xrightarrow{e_9} v_2$$

24. (i) Each vertex is of even degree, and hence, the graph is Eulerian graph.
    The Euler circuit is

$$a \to b \to c \to d \to b \to e \to d \to f \to e \to a$$

    (ii) Traversable, since there are exactly two vertices of odd degree. The Eulerian path must begin at one of the odd vertices and will end at the other.
    The Eulerian path is

$$a \to b \to c \to d \to e \to f \to g \to h \to f \to c \to a$$

    The graph is not an Eulerian graph as it has no Eulerian circuit.

25. (i) No Hamiltonian path because each vertex is *not* of degree $\geq \frac{5}{2}$ where 5 is the number of vertices.
    (ii) The graph is Hamiltonian as its every vertex is of degree $\geq \frac{5}{2}$ (where 5 is the number of vertices). It has Hamiltonian circuit.

$$a \to b \to e \to c \to d \to a$$

    (iii) The graph has a Hamiltonian path

$$a \to b \to c \to d \to e$$

    but not a Hamiltonian circuit.

26. The graph 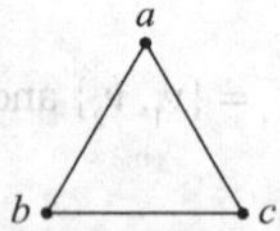

is both Eulerian and Hamiltonian.

27. Each vertex of the graph is of even degree and hence the graph is Eulerian.
The number of vertices $n = 6$

$$\therefore \frac{n}{2} = \frac{6}{2} = 3$$

As each vertex is not of degree $\geq \frac{n}{2}$, i.e., 3, the graph is *not* Hamiltonian.

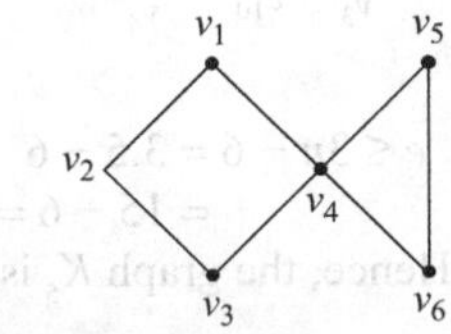

28. The number of vertices $n = 4$

$$\therefore \frac{n}{2} = \frac{4}{2} = 2$$

The graph is Hamiltonian as each vertex is of degree $\geq \frac{n}{2}$, i.e., 2.

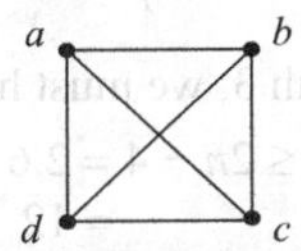

The graph is not Eulerian as all vertices are of odd degree.

29. The graph is not Eulerian as its vertices are not of even degree.
The number of vertices $n = 6$

$$\therefore \frac{n}{2} = \frac{6}{2} = 3$$

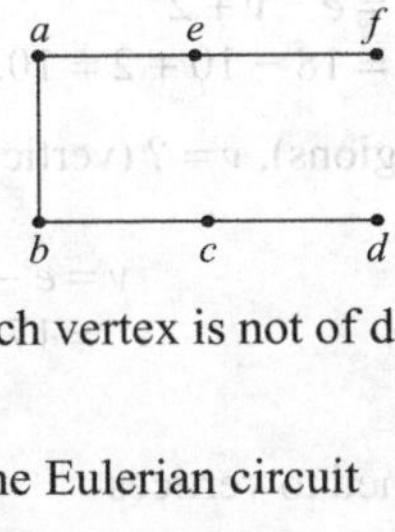

The graph is not Hamiltonian as each vertex is not of degree $\geq \frac{n}{2}$, i.e., 3 (the vertices f and d are of degree 1).

30. Fleury's Algorithm will result in the Eulerian circuit

$$A \to B \to C \to D \to E \to F \to G \to D \to C \to H \to A$$

31. The complete graph of 5 vertices is K_5.
The no. of vertices $n = 5$
The no. of edges $e = 10$

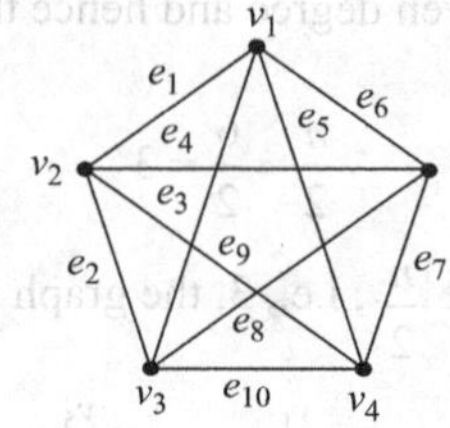

For a planar graph

$$\begin{aligned} e &\leq 3n - 6 = 3.5 - 6 \\ &= 15 - 6 = 9 \end{aligned}$$

which is not satisfied as $e = 10$. Hence, the graph K_5 is non-planar.

32. The complete graph $K_{3,3}$ has $(3 + 3 = 6)$ vertices and $3.3 = 9$ edges.

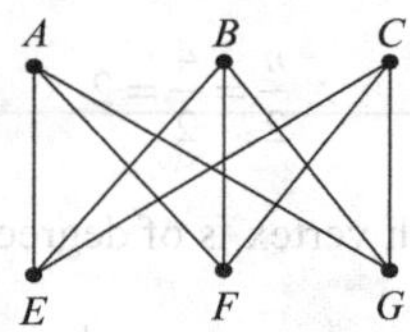

As the graph has no circuit of length 3, we must have

$$\begin{aligned} e &\leq 2n - 4 = 2.6 - 4 \\ &= 12 - 4 = 8 \end{aligned}$$

But the number of edges $e = 9$.
Hence, the condition is not satisfied and so the graph is non-planar.

33. The graph has $r = 10$ (regions), $v = 10$ (vertices), and $e = 18$ (edges).
According to Euler's law

$$\begin{aligned} r &= e - v + 2 \\ &= 18 - 10 + 2 = 10, \text{ which is satisfied.} \end{aligned}$$

34. Here $e = 17$ (edges), $r = 9$ (regions), $v = ?$ (vertices)
According to Euler's law

$$\begin{aligned} v &= e - r + 2 \\ &= 17 - 9 + 2 = 10 \end{aligned}$$

i.e., the graph has 10 vertices.

35. The colours that can be assigned to vertices

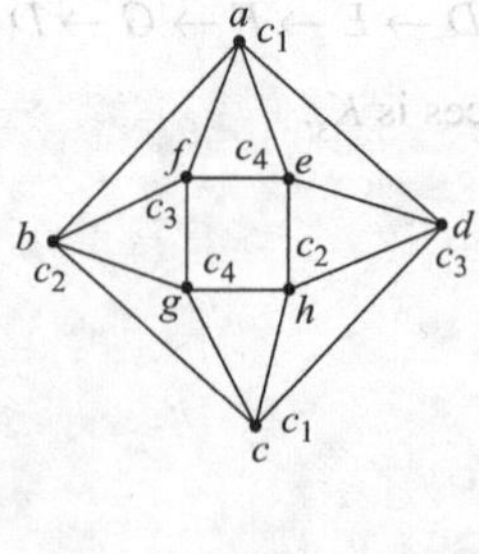

$$\begin{array}{l} a\ b\ c\ d\ e\ f\ g\ h \\ \text{are } c_1\ c_2\ c_1\ c_3\ c_4\ c_3\ c_4\ c_2 \end{array}$$

Four colours are required for proper colouring.

36. Every vertex in K_n is adjacent to each of the remaining $(n - 1)$ vertices and hence n distinct colours are needed for proper colouring of K_n. Hence, for K_4, four distinct colours are needed. Therefore, chromatic number of K_4 is 4.

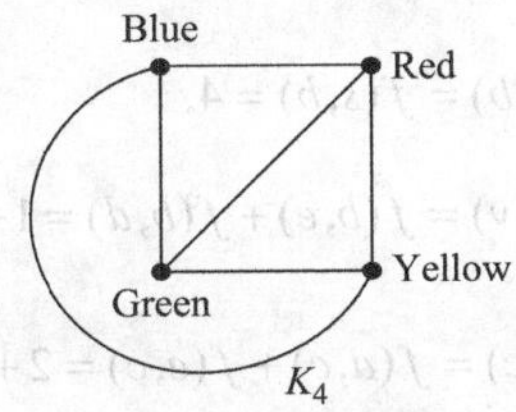

37. The vertex set in the graph $K_{m,n}$ can be partitioned into two disjoint subsets V_1 and V_2 such that every vertex in V_1 is adjacent to every vertex in V_2 and vice versa. The vertices in V_1 and V_2 are not adjacent. So, every vertex in V_1 can be assigned one colour c_1 and every vertex in V_2 can be assigned second colour c_2. Hence, chromatic colour of $K_{m,\,n}$ is two. In particular, the chromatic number of $K_{2,\,3}$ is 2 as is clear from the following colouring:

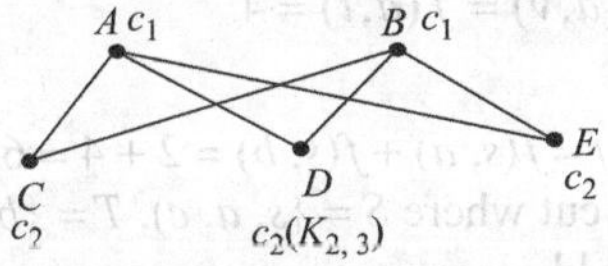

38. (i) Chromatic number of the graph is 2 as only two colours c_1 and c_2 are needed for proper colouring.

(ii) Chromatic number is 4 as four different colours are required for its proper colouring.

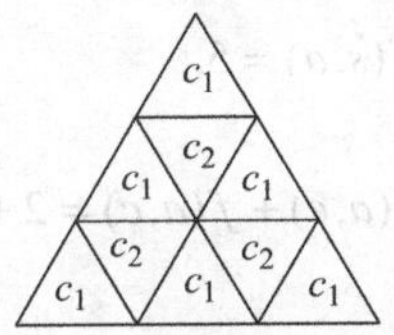

(i)

ANSWERS – B

1. (i) a, c: 7, 4; s, b: 3, 2; a, c: 7, 4; c, t: 5, 4

   (ii) s, b: 4, 2; a, b: 3, 2; c, d: 2, 1; d, t: 2, 1

2. (i) At vertex a, $\sum_v f(a,v) = f(a,d) + f(a,c) = 4 + 2 = 6$, $\sum_v f(v,a) = f(s,a) = 6$

   At vertex c, $\sum_v f(c,v) = f(c,d) + f(c,e) = 1 + 6 = 7$

   $$\sum f(v,c) = f(s,c) + f(a,c) = 5 + 2 = 7$$

   (ii) The value of flow $f = f(s, a) + f(s, b) + f(s, c) = 6 + 5 + 2 = 13$

(iii) There is no flow augmenting path as the edges (s, a), (s, c), and (b, e) are saturated.
(iv) Capacity for cuts = $\{s, a\}$ is

$$C(s, b) + C(s, c) + C(a, c) + C(a, d) = 3 + 5 + 3 + 7 = 18$$

and capacity for cut $S = \{s, b\}$ is

$$C(s, a) + C(s, c) + C(b, e) = 6 + 5 + 2 = 13$$

3. (i) At the vertex b,

$$\sum_v f(v,b) = f(s,b) = 4,$$

$$\sum_v f(b,v) = f(b,e) + f(b,d) = 1 + 3 = 4$$

At the vertex c, $\sum_v f(v,c) = f(a,c) + f(e,c) = 2 + 0 = 2$

$$\sum_v f(c,v) = f(c,t) = 2$$

At the vertex d, $\sum_v f(v,d) = f(e,d) + f(b,d) = 1 + 3 = 4$

$$\sum_v f(d,v) = f(d,t) = 4$$

(ii) The value of the flow $f = f(s, a) + f(s, b) = 2 + 4 = 6$
(iii) The capacity of (S, T) cut where $S = \{s, a, c)$, $T = \{b, d, e, t\}$ is $C(s, b) + C(a, e) + C(c, t) = 5 + 1 + 5 = 11$.
(iv) No, as the path $d - t$ is saturated.
(v) The given flow is not maximum as there is an unsaturated path $s - a - c - t$.
(vi) The flow along the path $s - a - c - t$ can be increased by 1.

4. (i) At vertex a, $\sum_v f(v,a) = f(s,a) = 5,$

$$\sum_v f(a,v) = f(a,e) + f(a,c) = 2 + 3 = 5$$

At vertex b, $\sum_v f(v,b) = f(s,b) = 5,$

$$\sum_v f(b,v) = f(b,e) + f(b,d) = 2 + 3 = 5$$

At vertex c, $\sum_v f(v,c) = f(a,c) + f(e,c) = 3 + 2 = 5,$

$$\sum_v f(c,v) = f(c,t) = 5$$

(ii) The value of flow $f = f(s, a) + f(s, b) + f(s, e) = 5 + 5 + 3 = 13$.
(iii) The capacity of (S, T) cut = $C(c, t) + C(e, t) + C(d, t) = 6 + 2 + 6 = 14$.

(iv) The saturated arcs are (s, e), (e, t), (d, t).

(v) The given flow is not saturated as there is a path $s - a - c - t$ along which flow can be increased by 1.

5. (i) At vertex a, $\sum_v f(v,a) = f(s,a) = 4, \sum_v f(a,v) = f(a,c) + f(a,b) = 3+1 = 4$

At vertex b, $\sum_v f(v,b) = f(s,b) + f(a,b) = 4+1 = 5$

$\sum_v f(b,v) = f(b,c) + f(b,d) = 2+3 = 5$

At vertex d, $\sum_v f(v,d) = f(b,d) + f(c,d) = 3+1 = 4$

$\sum_v f(d,v) = f(d,t) = 4$

(ii) The value of the flow $f = f(s, a) + f(s, b) = 4 + 4 = 8$.

(iii) The flow is not maximum as the path $s - a - c - t$ is unsaturated and so the flow can be increased.

(iv) The flow along the path $s - b - c - t$ cannot be increased as the arc (b, c) is saturated.

(v) The capacity of (S, T) cut where $S = \{s, a, b\}$, $T = \{c, d, t\}$ is $C(a, c) + C(b, c) + C(b, d) = 4 + 2 + 4 = 10$.

6.

	S	T	Capacity of (S, T)
1.	{s}	{a, b, c, d, t}	7
2.	{s, a}	{b, c, d, t}	12
3.	{s, b}	{a, c, d, t}	11
4.	{s, c}	{a, b, d, t}	18
5.	{s, d}	{a, b, c, t}	21
6.	{s, a, b}	{c, d, t}	12
7.	{s, a, c}	{b, d, t}	23
8.	{s, a, d}	{b, c, t}	21
9.	{s, b, c}	{a, d, t}	16
10.	{s, b, d}	{a, c, t}	18
11.	{s, c, d}	{a, b, t}	23
12.	{s, a, b, c}	{d, t}	17
13.	{s, a, b, d}	{c, t}	14
14.	{s, a, c, d}	{b, t}	23
15.	{s, b, c, d}	{a, t}	14
16.	{s, a, b, c, d}	{t}	10

The maximum flow corresponds to minimum cut value and hence maximum flow = 7.

7. (i) The value of flow $f = 10$.
   (ii) The flow along path s *bct* can be augmented by excess minimum capacity [3, 1, 2] = 1. No further augmentation is possible due to saturation of edges (a, c), (b, c), and (d, t). Hence, maximum flow $= f(s, a) + f(s, b) = 6 + 5 = 11$.
   (iii) The maximum flow corresponds to (S, T) cut where $S = \{s, a, b, d\}$ and $T = \{c, t\}$. Capacity of (S, T) cut is

$$C(a, c) + C(b, c) + C(d, t) = 6 + 1 + 4 = 11$$

8. The shortest path is $a - b - c - h$. Its length is $3 + 2 + 2 = 7$.

9. Using nearest neighbourhood method, the path is $A \to B \to C \to H \to I \to D \to E$. Its length is $5 + 4 + 3 + 4 + 3 + 2 = 21$

10. The path of minimum distance is a Hamiltonian circuit

$$a \to b \to c \to j \to d \to e \to g \to f \to h \to i \to a$$

11. The minimum distances from the vertex v_1 to other vertices are

Vertex v	v_1	v_2	v_3	v_4	v_5	v_6
$L(v)$	0	4	8	8	4	12

12. The minimum distances from the vertex a to other vertices are

Vertex v	a	b	c	d	e	f	g	h
$L(v)$	0	3	6	6	7	1	5	8

13. (i) The bipartite graph for the given relationship is

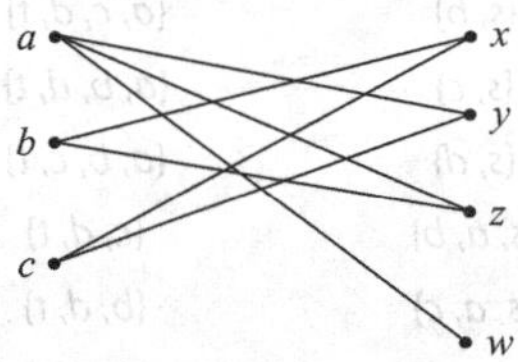

   (ii) The five different solutions of the problem are
   1. $\{(a, w), (b, x), (c, y)\}$ 2. $\{(a, w), (b, z), (c, x)\}$
   3. $\{(a, w), (b, z), (c, y)\}$ 4. $\{(a, z), (b, x), (c, y)\}$
   5. $\{(a, y), (b, z), (c, x)\}$
   (iii) The marriage condition $|X| \le |A(X)|$ for every $X \le \{a, b, c\}$

X:	a	b	c	ab	bc	ac	abc
|X|:	1	1	1	2	2	2	3
|A(X)|:	3	2	2	4	3	4	4

Thus, $|X| \le |A(X)|$ for each subset of $\{a, b, c\}$ is satisfied.

14. The bipartite graph for the assignment problem is

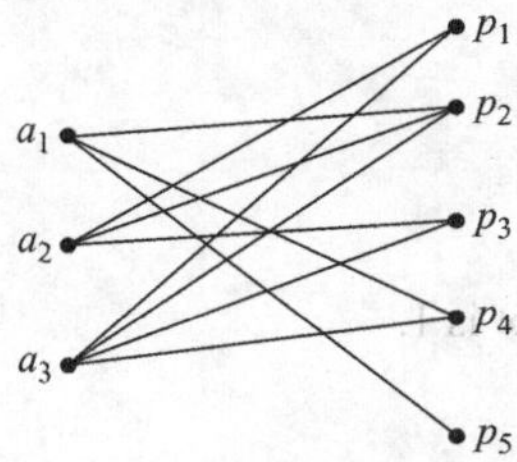

The required two solutions are
(i) $\{(a_1, p_2), (a_2, p_1), (a_3, p_3)\}$
(ii) $\{(a_1, p_5), (a_2, p_3), (a_3, p_1)\}$

Chapter 11

1. (i) not a tree as there is a cycle.
   (ii) is a tree.
   (iii) not a tree as there is a cycle.
   (iv) not a tree as there is a cycle.
2. is a tree
3. 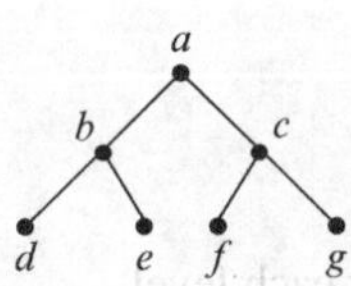

   Root is vertex a. The vertices at level 2 are d, e, f, and g.

4. 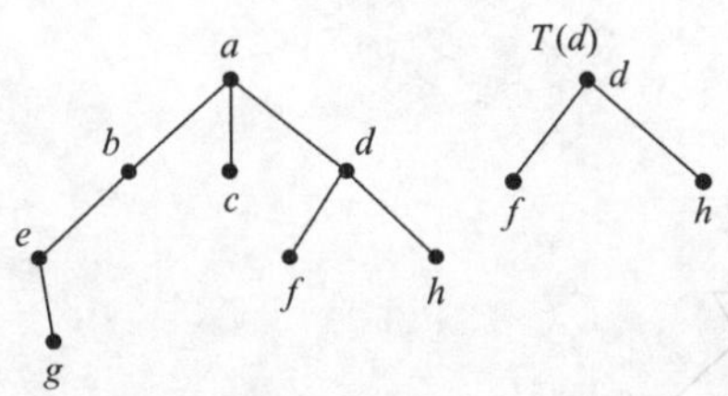

   Root vertex is a, siblings of d are f and h.

5. (i) 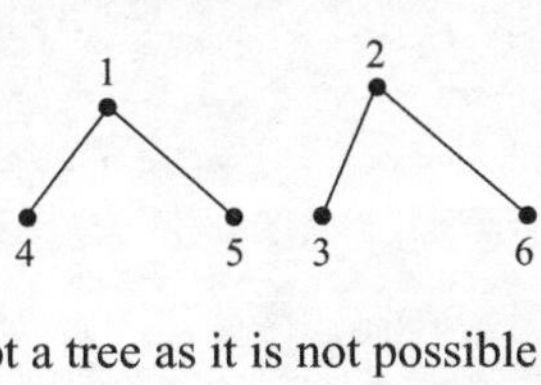

   Not a tree as it is not possible to go from vertex 1 to 2.

(ii)

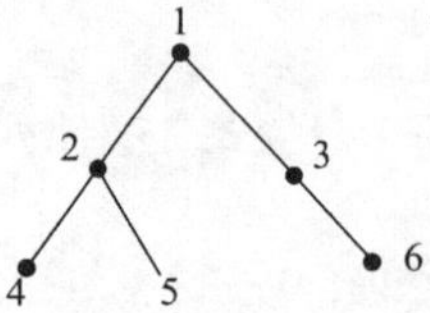

The graph is a tree and its root is 1.

(iii)

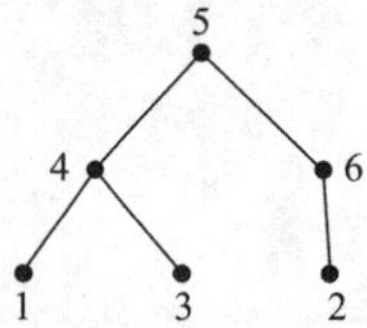

The graph is a tree and its root is 5.

6. (i) The second-level vertices are v_4, v_5, v_6, v_7, v_8, and v_9.
   (ii) The leaves are v_5, v_{10}, v_{11}, v_{12}, v_{13}, v_{14}, v_{15}, v_{16}, and v_{17}.
   (iii) The siblings of v_6 are v_{11} and v_{12}.
   (iv) The descendants of v_3 are v_8, v_{15}, v_{16} and ancestors of v_{15} are v_8, v_3, and v_0.

   (v)

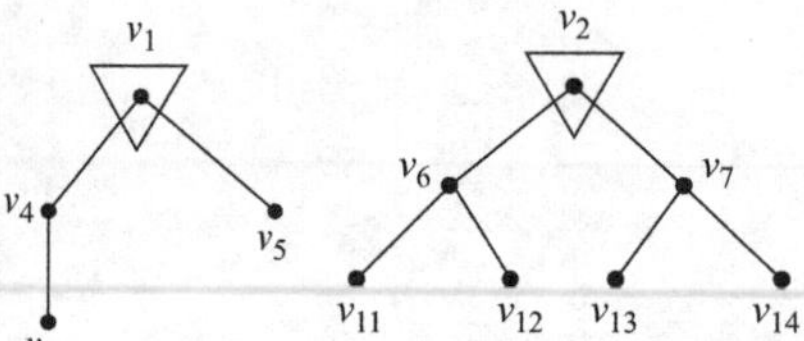

$T(v_1)$ $T(v_2)$

7. The maximum height of the tree is three as it requires one vertex at each level

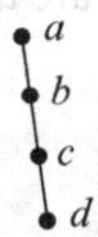

8. The complete binary tree is

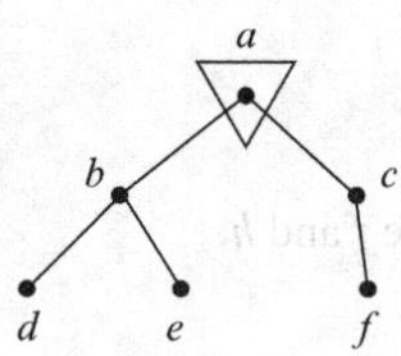

The maximum height is 2.

10. No such tree is possible

12. (i)

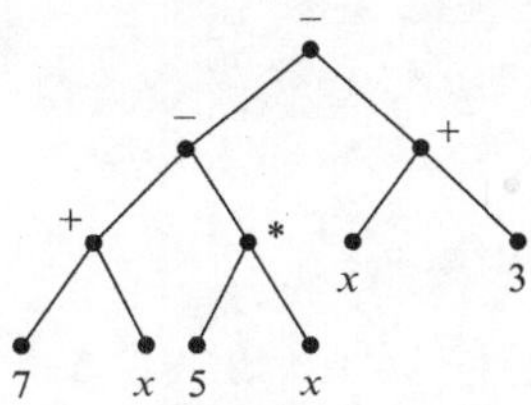

(ii)

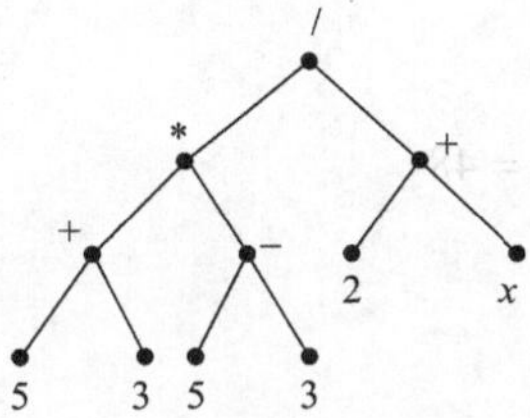

(iii)

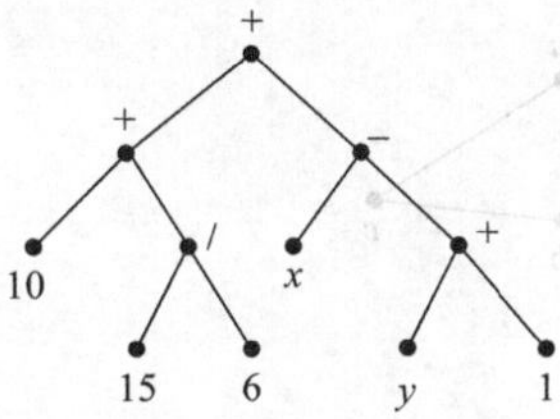

13. Preorder: 1–2–3–4–5–6–7–8–9
    Inorder: 4–3–5–2–6–1–8–7–9
    Postorder: 4–5–3–6–2–8–7–9–1
14. Preorder: *a–b–c–d–e–f–g–h–i–j*
    Inorder: *d–c–e–b–f–a–h–g–i–j*
15. Postorder: *c–d–b–f–h–i–g–e–a*
16.

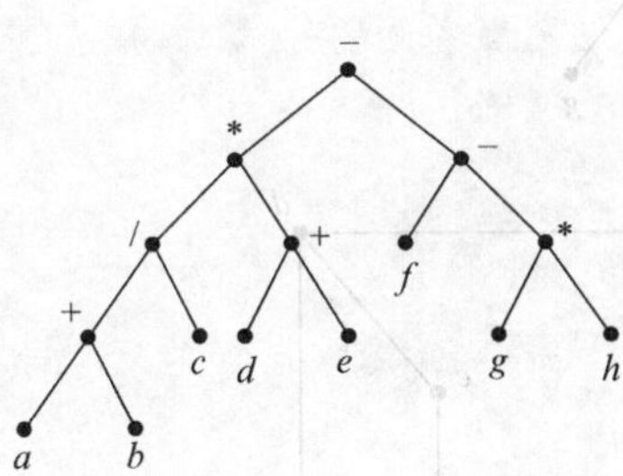

Prefix order notation gives

$$- ** \div + abc + de - f * gh$$

17.

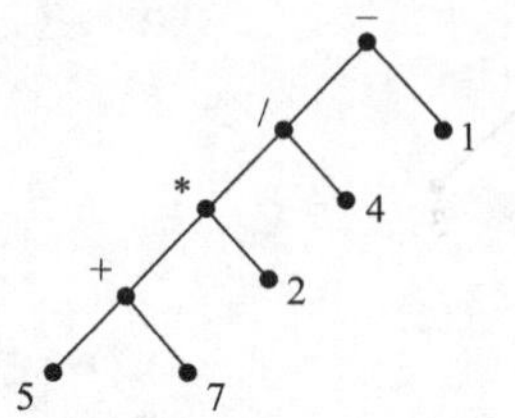

18. The simplification is

$$(6/3)^4 \times (5-2) = 16 \times 3 = 48$$

19. $\{((5+7)/3)\times 2\}^3 - 8 = 512 - 8 = 504$

20. The spanning trees are

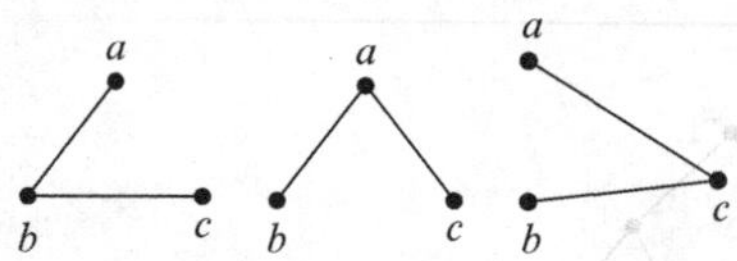

21. The spanning trees are

(i)

(ii)

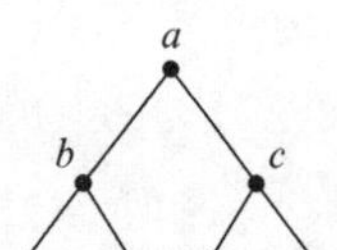

(iii)

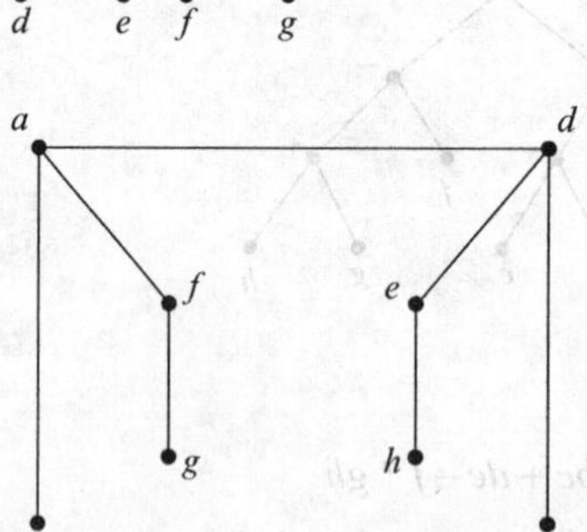

22. Kruskal's algorithm gives the spanning tree as

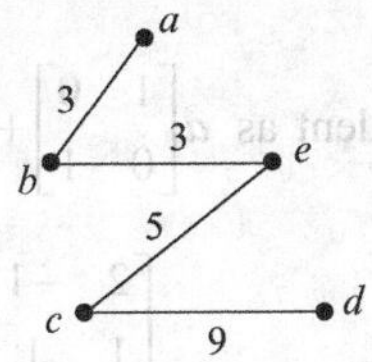

Minimum time = 3 + 3 + 5 + 9 = 20 hours

23. The Prim's algorithm gives

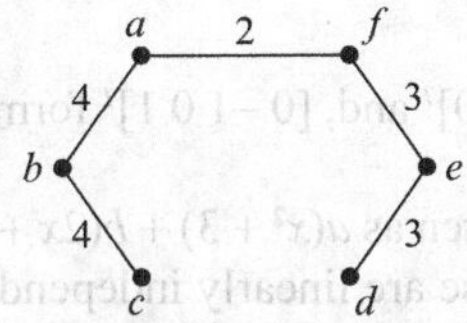

24. The minimal spanning tree using Prim's algorithm is

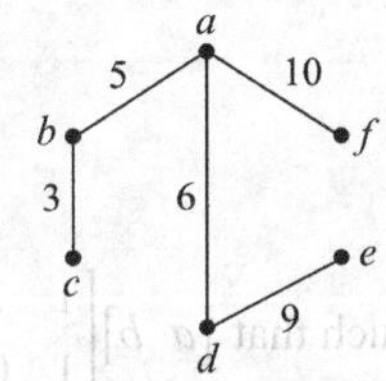

Chapter 12

Answers – A

16. $(2, -1, 6) = a(1, 2 ,3) + b(2, -1, 2) + c(3, 1, 1)$ where $a = 1, b = 2, c = -1$

17. $(2, 2, 3) = a(1, -1, 1) + b(2, 0, 1) + c(1, 1, 2)$ where $a = -\frac{1}{2}, b = \frac{1}{2}, c = \frac{3}{2}$

18. $W[1, e^x + e^{-x}, e^x - e^{-x}] = -4 \neq 0$ and hence the vectors are linearly independent.

19. (i) $W[2, x^2, x^2 - 2] = \begin{vmatrix} 2 & x^2 & x^2-2 \\ 0 & 2x & 2x \\ 0 & 2 & 2 \end{vmatrix} = 0$, and hence linearly dependent.

(ii) $W[x+2, x+5, x^2+1] = \begin{vmatrix} x+2 & x+5 & x^2+1 \\ 1 & 1 & 2x \\ 0 & 0 & 2 \end{vmatrix} = -6 \neq 0$, and hence are linearly independent.

ANSWERS – B

5. The given matrices are independent as $a\begin{bmatrix}1 & 0\\0 & 1\end{bmatrix}+b\begin{bmatrix}0 & 1\\0 & 0\end{bmatrix}+c\begin{bmatrix}0 & 0\\1 & 0\end{bmatrix}=\begin{bmatrix}0 & 0\\0 & 0\end{bmatrix}$ implies $a=b=c=0$

6. The matrix associated with the subspace is $\begin{bmatrix}2 & -1 & 0\\1 & 1 & -1\\0 & 3 & 0\\0 & 0 & 1\end{bmatrix}\begin{bmatrix}a\\b\\c\end{bmatrix}=\begin{bmatrix}2a-b\\a+b-c\\3b\\c\end{bmatrix}$

   The vectors $[2\ 1\ 0\ 0]^T$, $[-1\ 1\ 3\ 0]^T$ and, $[0\ -1\ 0\ 1]^T$ form a basis. The dimension of subspace is three.

7. Each element in W can be written as $a(x^2+3)+b(2x+5)$, which is a linear combination of vectors x^2+3 and $2x+5$. These are linearly independent as $W[x^2+3, 2x+5] \neq 0$. Hence, $\{x^2+3, 2x+5\}$ is a basis for W.

10. (i) Not linear as $T(a\underset{\sim}{x}) \neq aT\underset{\sim}{x}$

    (ii) Linear

    (iii) Not linear as $T(a\underset{\sim}{x}) \neq aT\underset{\sim}{x}$

11. There exists a matrix $A=[a\ \ b]$ such that $[a\ \ b]\begin{bmatrix}1 & 1\\1 & 0\end{bmatrix}=\begin{bmatrix}1\\4\end{bmatrix}$, which gives $a+b=1$, $a=4$ $\therefore\ b=-3$.

    Hence, the required matrix $A=[4\ \ -3]$.

$$\therefore T\begin{bmatrix}x_1\\x_2\end{bmatrix}=[4\ \ -3]\begin{bmatrix}x_1\\x_2\end{bmatrix}=4x_1-3x_2$$

12. The matrix of transformation is $\begin{bmatrix}1 & 0 & 0\\0 & 0 & 1\end{bmatrix}$

13. $A=\begin{bmatrix}1 & 0 & 2\\1 & 0 & -1\\2 & 3 & 0\end{bmatrix}$

14. $\begin{bmatrix}-1 & -3\\4 & 2\\-1 & 2\end{bmatrix}$

15. (i) $\begin{bmatrix}1 & 1\\1 & -1\end{bmatrix}$, (ii) $\begin{bmatrix}1 & 1\\1 & -1\end{bmatrix}$, (iii) $\begin{bmatrix}2 & 3\\3 & 5\end{bmatrix}$

16. $\begin{pmatrix} 1 & -1 & 0 \\ 0 & 1 & -1 \\ 0 & 0 & 1 \end{pmatrix}$

17. $\| \underset{\sim}{x} \|_1 = 11, \| \underset{\sim}{x} \|_2 = 7, \| \underset{\sim}{x} \|_\infty = 6$

18. $\| \underset{\sim}{x} - \underset{\sim}{y} \| = 6, \| \underset{\sim}{x} - \underset{\sim}{y} \|_\infty = 2$

24. The orthonormal basis is

$$\frac{1}{\sqrt{2}}(1, 1, 0), \frac{1}{\sqrt{6}}(-1, 1, 1), \frac{1}{\sqrt{3}}(1, -1, 1)$$

25. The orthonormal basis is

$$\frac{1}{\sqrt{2}}, \frac{\sqrt{3}}{\sqrt{2}}x, \frac{\sqrt{5}}{2\sqrt{2}}\left(3x^2 - 1\right)$$

References

1. Lipschutz, S. *Set Theory and Topics Schaum Series*, New York, 1964.
2. Herstein, I.N. *Topics in Algebra*, Blaisdell Publishing Company, New York, 1964.
3. Mishra, R.S. and Bhattacharya, N.N.A *Course in Modern Algebra*, Pothishala Pvt. Ltd., Allahabad, 1967.
4. Leon, S.J. *Linear Algebra with Applications* (3rd Edition), Macmillan Publishing Company, New York, 1990.
5. Hoffman, K. and Kunze, R. *Linear Algebra* (2nd edition), Pearson Education, New Delhi, 2003.
6. Vasistha, A.R. and Vasistha, A.K. *Modern Algebra*, Krishna Prakashan Media Private Ltd, Meerut, 2008.
7. Krishnamurthy, V. *Combinatorics: Theory and Application*, East – West Press, New Delhi, 1986.
8. Walpole, R.E. and Myers, R.H. *Probability and Statistics for Engineers and Scientists* (4th Edition), Macmillan Publishing Company, New York, 1985.
9. Miller, I. and Freund, J.E. *Probability and Statistics for Engineers* (3rd Edition), Prentice Hall, Englewood Cliffs, NJ, Prentice Hall, 1985.
10. Prasad, C. *Advanced Mathematics for Engineers* (4th Edition), Prasad Mudranalaya, Allahabad, 1974.
11. Gupta, Radhey.S. *Elements of Numerical Analysis*, Macmillan India Ltd, New Delhi, 2009.
12. Klir, G.J. and Folger, T.A. *Fuzzy Sets, Uncertainty and Information*, Prentice Hall of India, Pvt. Ltd., New Delhi, 2000.
13. Klir, G.J. and Yuan, B.O. *Fuzzy Sets and Fuzzy Logic, Theory and Applications*, Prentice Hall of India, Pvt. Ltd., New Delhi, 2002.
14. Kolman, B., Busby, R.C. and Ross, S. *Discrete Mathematical Structures* (3rd Edition), Prentice Hall of India, Pvt. Ltd., New Delhi, 1999.
15. Anderson, Ian *First Course in Discrete Mathematics*, Springer – Verlag, London, 2001.
16. Mott, J.L., Kandal, A. and Baker, T.P. *Discrete Mathematics for Computer Scientists and Mathematicians*, Prentice Hall of India, Pvt. Ltd., New Delhi, 2001.
17. Lipschutz, S. and Lipson, M.L. *Discrete Mathematics* (2nd Edition), Tata McGraw Hill, New Delhi, 2002.
18. Liu, C.L. *Elements of Discrete Mathematics* (2nd Edition), Tata McGraw Hill, New Delhi, 2002.

19. Goodaire, E.G. and Paramenter, M.M. *Discrete Mathematics with Graph Theory* (2nd Edition), Prentice Hall of India, Pvt. Ltd., New Delhi, 2002.
20. Koshy, T. *Applied Discrete Mathematics with Applications*, Academic Press, an imprint of Elsevier, 2004.
21. Chakraborty, S.K. and Sarkar, B.K. *Discrete Mathematics*, Oxford University Press, 2011.
22. Rao Bhism, G.S.S. *Discrete Structures and Graph Theory* (3rd Edition), Scitech, New Delhi, 2009.
23. Sarkar, S.K. *A Text Book of Discrete Mathematics*, S. Chand and Company Ltd., New Delhi, 2012.
24. Kenneth, H.R. *Discrete Mathematics and its Applications* (Special Indian Edition), Tata McGraw Hill, New Delhi, 2007.
25. Sharma, J.K. *Discrete Mathematics* (2nd Edition), Macmillan Publishing Company, 2008.
26. Wilson, R.J. *Introduction to Graph Theory* (4th Edition), Pearson Education, 2007.
27. Balakrishnan, R. and Ranganathan, K. *A Text Book of Graph Theory*, Springer (India) Pvt. Ltd., New Delhi, 2008.
28. Deo, N. *Graph Theory with Application to Engineers and Computer Science*, PHI Learning Pvt. Ltd., New Delhi, 2010.
29. Kumar, S. and Chaudhary, S. *Applied Discrete Mathematics*, Ram Prasad and Sons (India) Educational Publisher, Agra, 2007.
30. Rathor, S.K. *Discrete Mathematics and Graph Theory*, EPH, New Delhi, 2004.
31. Shankar Rao, G. *Discrete Mathematical Structure*, New Age International Publication, New Delhi, 2009.

Index

Q

R

S

T

U

V

W

X